Texts and Monographs in Physics

Springer-Verlag Berlin Heidelberg GmbH

Texts and Monographs in Physics

Series Editors: R. Balian W. Beiglböck H. Grosse E. H. Lieb
N. Reshetikhin H. Spohn W. Thirring

Maciej Błaszak

Multi-Hamiltonian Theory of Dynamical Systems

With 9 Figures

 Springer

Professor Maciej Błaszak

Physics Department
A. Mickiewicz University
Umultowska 85
61-614 Poznań, Poland

ISBN 978-3-642-63780-3   ISBN 978-3-642-58893-8 (eBook)
DOI 10.1007/ 978-3-642-58893-8

Library of Congress Cataloging-in-Publication Data. Multi-Hamiltonian theory of dynamical systems /
M. Błaszak. p.cm. – (Texts and monographs in physics, ISSN 0172-5998) Includes bibliographical
references and index. ISBN 3-540-64251-X (Berlin: acid free paper) 1. Hamiltonian systems.
2. Nonlinear theories. 3. Differentiable dynamical systems. 4. Mathematical physics.
I. Błaszak, Maciej. II. Series. QC20.7.H35M85 1998 514'.74–dc21 98-22162

© Springer-Verlag Berlin Heidelberg 1998
Originally published by Springer-Verlag Berlin Heidelberg New York in 1998

Typesetting: Data conversion by Satztechnik Katharina Steingraeber, Heidelberg
Cover design: *design & production* GmbH, Heidelberg
SPIN: 10651992   55/3144-5 4 3 2 1 0 - Printed on acid-free paper

To my wife Elizabeth

Preface

The book aims to provide a comprehensive and easy presentation of modern algebraic theory of integrable nonlinear dynamical systems. This relatively new field of mathematical physics, originating from the theory of solitons, has been intensely developed in the last two decades. Particular attention in this book has been paid to modern multi-Hamiltonian formalism on manifolds and associative Lie algebras, which seem very powerful tools in the investigation of integrable nonlinear systems of different kinds, e.g. field, lattice and mechanical ones.

The book is addressed to graduates of physics and applied mathematics who have elementary background in classical mechanics and differential geometry. My intention was to make the book as easy to read as possible, so many considerations are explained from scratch, all calculation formalisms are explained in detail, almost all theorems and lemmas are proved, and each chapter is supplemented with numerous examples.

I am much indebted to all my co-workers but in particular to two of them. The first person is Professor Benno Fuchssteiner from Paderborn University, who inspired my interest in algebraic methods of soliton theory and who implanted in me his passion for the work in this field. The other person whom I must mention is Professor Stefan Rauch-Wojciechowski from the Linköping University, with whom I have been working for many years and who shared with me many original ideas on algebraic theory of integrable finite-dimensional systems.

Finally, I wish to thank Professor Wolf Beiglböck from Springer-Verlag for his enthusiastic encouragement to write this book, which otherwise might not have been written.

Poznań, April 1998   Maciej Błaszak

Contents

1. Preliminary Considerations

The main subject of our interest in this book is equations of the form

$$u_t = K(u), \tag{1.1}$$

where $K(u)$ denotes a *vector field* on a certain manifold M and u is a point of this manifold, which we shall refer to as a *dynamical system*. We do not impose any restrictions on the dimensionality of M. When M is finite dimensional, (1.1) is represented by a set of first order ordinary differential equations called a *finite dimensional dynamical system*. When M is infinite dimensional but with a countable number of degrees of freedom, (1.1) turns into a set of differential-difference equations known as a *lattice dynamical system*. Finally, when M is infinite dimensional and such that each point $u = u(x) \in M$ is represented by a function of the x-variable, $K = K(u, u_x, ...)$ takes the form of a differential function of u, hence, (1.1) turns into a system of partial differential equations and we refer to it as a *field dynamical system*. We are going to consider all three types of dynamical systems (1.1) with K depending on u in a nonlinear way. We assume that the reader is familiar with the concept of differential manifolds. Fortunately, the majority of the interesting field dynamical systems can be considered in a *topological linear space* V which is reflexive, i.e. $V^{**} = V$, like in the finite dimensional case, instead of an arbitrary manifold. Hence, we can avoid the problems connected with differential geometry on manifolds of infinite dimension. Actually, for most of our further considerations it is sufficient to assume that V consists of, in general complex-valued, $\mathbb{C}^\infty$–functions f of a real variable $x \in \mathbb{R}$ such that f and all its derivatives vanish 'rapidly' at $\pm\infty$. 'Rapidly' means for example faster than any rational function. Typical examples of such a V are the Schwartz space $S(\mathbb{R})$ or $L_1(\mathbb{R})$ space, respectively. In the case of lattice functions, the continuous space variable $x \in \mathbb{R}$ is replaced by a discrete integer variable $n \in \mathbb{Z}$. In such a case, this means that to be an element of $L_1(\mathbb{Z})$ the series $\sum_{n=-\infty}^{n=+\infty} f(u)$ must be absolutely convergent. Moreover, all differential formulas are introduced in a way which is required for our further considerations. For a more detailed discussion of differential calculus in a topological vector space we refer the reader to [191].

Roughly speaking, to solve the dynamical system (1.1) means to find *integral curves* of a vector field K, i.e. smooth parametrized curves such that

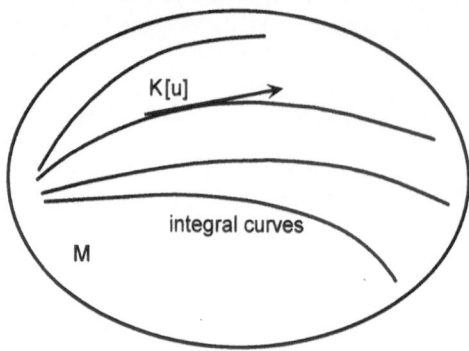

Fig. 1.1. Flow on M

for an arbitrary curve a tangent vector at any point coincides with the value of K at the same point, for all values of the parameter. Integral curves constitute a *flow* on the underlying manifold M (Fig 1.1). Unfortunately, to solve the Cauchy problem for arbitrary initial data, i.e. to find an appropriate integral curve, especially when (1.1) is a nonlinear one, is in general a hopeless task. Nevertheless, there exists an important class of dynamical systems for which the Cauchy problem can be solved, at least in principle. Suppose that we have been lucky enough to find a pair of matrices (operators in the infinite dimensional case) L and A, called a *Lax pair*, whose entries depend on the dynamical variables u_i in such a way that (1.1) is equivalent to the equation

$$L_t = [A, L], \tag{1.2}$$

called the *isospectral Lax equation*, where $[.,.]$ means the commutator of L and A. With the Lax pair we can relate two linear equations (the *isospectral problem*)

$$L\psi = \lambda\psi$$

$$\psi_t = A\psi \tag{1.3}$$

in some Hilbert space, where the first equation represents the *spectral equation* for L and the second one gives the time evolution for eigenfunctions ψ. Condition (1.2) guarantees that the eigenvalues λ of $L(t)$ are therefore time independent or, equivalently, $L(t)$ undergoes an *isospectral deformation*. Actually, differentiating the spectral equation with respect to t

$$L_t\psi + L\psi_t = \lambda_t\psi + \lambda\psi_t \tag{1.4}$$

and applying the second relation of (1.3)

$$L_t\psi + LA\psi = \lambda_t\psi + AL\psi, \tag{1.5}$$

we get the desired result

$$(L_t - [A, L])\,\psi = \lambda_t \psi. \tag{1.6}$$

As we will see later, the eigenvalues λ are simultaneously the constants of motion of the underlying dynamical system (1.1).

Remark 1.1 Sometimes the isospectral problem takes the form

$$\phi_x = U(u; \lambda)\phi, \qquad \lambda_t = 0$$

$$\phi_t = V(u; \lambda)\phi, \tag{1.7}$$

and then the isospectral Lax equation (1.2) takes the form of the so-called *zero-curvature equation*

$$U_t - V_x + [U, V] = 0. \tag{1.8}$$

Very often both Lax representations are admissible for a given system.

Example 1.1 The celebrated Korteweg-de Vries (KdV) equation

$$u_t = u_{xxx} + auu_x, \qquad a = \text{const.} \tag{1.9}$$

is an example of a nonlinear field system which we will consider many times for the sake of illustration of the theory presented in this book. The Lax pair (L, A) takes the form

$$L = 4\partial_x^2 + \frac{2}{3}au, \qquad A = 4\partial_x^3 + au\partial_x + \frac{1}{2}au_x, \qquad \partial_x \equiv \frac{\partial}{\partial x}, \tag{1.10}$$

where ∂_x is a differential operator whose action on an arbitrary function $v(x)$ is given by $\partial_x v = v_x + v\partial_x$ and, hence, we find

$$L_t + [L, A] = \frac{2}{3}a\left(u_t - u_{xxx} - auu_x\right) = 0. \tag{1.11}$$

Notice that in this particular case the spectral equation (1.3) is exactly the Schrödinger equation with the potential $u(x, t)$. This means that there is an equivalence between the potentials which undergo an isospectral deformation and solutions of the KdV equation.

Example 1.2 The famous Toda lattice, whose equation of motion reads

$$q(n)_{tt} = \exp[q(n) - q(n+1)] - \exp[q(n-1) - q(n)], \qquad n \in \mathbb{Z}, \tag{1.12}$$

where $q(n)$ is the displacement of the n-th particle from its equilibrium position, is an excellent example of a nonlinear lattice dynamical system. Introducing new variables $v(n) = \exp[q(n-1) - q(n)]$ and $p(n) = q(n)_t$, the system (1.12) takes the form

$$v(n)_t = v(n)[p(n) - p(n-1)]$$
$$p(n)_t = v(n+1) - v(n), \tag{1.13}$$

i.e. becomes a two component lattice dynamical system (1.1). The Lax pair (L, A) reads

$$L = v(n)\mathcal{E}^{-1} + p(n) + \mathcal{E}, \qquad A = p(n) + \mathcal{E}, \tag{1.14}$$

where $\mathcal{E}$ stands for a shift operator satisfying a simple commutation rule $\mathcal{E}^k u(n) = u(n+k)\mathcal{E}^k$. It is a simple task to verify the equivalence between (1.2) and the dynamics (1.14).

Example 1.3 The generalized Henon–Heiles system [14]

$$q_{1_{tt}} = -3q_1^2 - \frac{1}{2}q_2^2,$$

$$q_{2_{tt}} = -q_1 q_2 + \frac{\alpha}{q_2^3}, \qquad \alpha = \text{const.} \tag{1.15}$$

is an example of a finite dimensional dynamical system with the Lax pair

$$L = \begin{pmatrix} -4\lambda q_{1_t} + q_2 q_{2_t} & 16\lambda^2 + 8\lambda q_1 - q_2^2 \\ 16\lambda^3 - 8\lambda^2 q_1 + 4\lambda q_1^2 + \lambda q_2^2 + q_{2_t}^2 + \alpha/q_2^2 & 4\lambda q_{1_t} - q_2 q_{2_t} \end{pmatrix},$$
$$\tag{1.16}$$

$$A = \begin{pmatrix} 0 & 1 \\ \lambda - q_1 & 0 \end{pmatrix}.$$

Of key importance for solving the Cauchy problem for the dynamical system (1.1) with the Lax representation (1.2), (1.8), are the linear equations (1.3), (1.7) and the so-called *spectral transform method*. The details of this analytic approach are sufficient for a separate book so we refer the reader to the literature [2],[43],[64],[143],[183],[194]. Although the spectral transform method is well recognized and in principle leads to the general solutions of the underlying dynamical systems, nevertheless even in the simplest case, i.e. the KdV one, in practice the calculations are so complex that in the explicit form we can find only a restricted class of solutions. From the point of view of the applications the most important is the class of solutions connected with the discrete part of the spectrum of the spectral equation (1.3). These particular solutions are known as *multi-soliton solutions*. Thus, the nonlinear infinite dimensional dynamical system (1.1) which has Lax representation are called field (lattice) *soliton systems*.

Apart from the analytic approach to soliton systems, a very powerful one turns out to be the algebraic approach, intensely developed over the last several years. There exist several books dealing with algebraic aspects of integrable systems (see for example [58],[61],[77]). Of particular importance is a Hamiltonian formulation for the infinite dimensional dynamics (1.1), which is a natural extension of the Hamiltonian of classical mechanics onto field (lattice) systems. We assume that the reader is familiar with the basic concepts of theoretical mechanics so let us recollect some elementary information on the Hamiltonian formalism.

For a conservative system of N interacting particles, the fundamental problem is to solve its Newtonian equations of motion

$$(q_i)_{tt} = F_i = -\frac{\partial V(q_1, ..., q_N)}{\partial q_i}, \tag{1.17}$$

where $q_i = q_i(t), i = 1, ..., N$ are position coordinates, F_i are forces and $q_t = dq/dt$. Introducing momentum variables $p_i = (q_i)_t$, (1.17) may be written in the Hamiltonian form as a set of first order differential equations

$$(q_i)_t = \frac{\partial H}{\partial p_i}, \qquad (p_i)_t = -\frac{\partial H}{\partial q_i} \tag{1.18}$$

with the Hamiltonian function (energy) $H(q, p) = \frac{1}{2}\sum_{i=1}^{N} p_i^2 + V(q_1, ..., q_N)$.

Remark 1.2 In a generalized coordinate representation, (1.17) turn into the Lagrange equations and the Hamiltonian representation (1.18) is obtained through the well known Legendre transformation.

Now, (1.18) may be represented in the form

$$\begin{pmatrix} q \\ p \end{pmatrix}_t = \begin{pmatrix} \partial H/\partial p \\ -\partial H/\partial q \end{pmatrix} = \begin{pmatrix} 0 & I \\ -I & 0 \end{pmatrix} \nabla H(q, p), \tag{1.19}$$

where $q = (q_1, ..., q_N)^T, p = (p_1, ..., p_N)^T, \nabla = (\partial/\partial q_1, ..., \partial/\partial p_N)^T$ is the gradient operator and I is a unit matrix. Introducing the notation $u = (q, p)^T$ and $\theta = \begin{pmatrix} 0 & I \\ -I & 0 \end{pmatrix}$, (1.19) takes the more compact form

$$u_t = K(u) = \theta \nabla H(u) \tag{1.20}$$

which represents a Hamiltonian formulation of the finite dimensional dynamics (1.1). We call the form (1.19) of a Hamiltonian system a *canonical representation*; the variables (q, p) represent *canonical coordinates* and θ is the so called *canonical Poisson matrix*.

Example 1.4 A simple realization of the system (1.17) with $N = 1$ is the nonlinear oscillator (pendulum)

$$q_{tt} + \sin q = 0. \tag{1.21}$$

The Hamiltonian equation of motion takes the canonical form

$$\begin{pmatrix} q \\ p \end{pmatrix}_t = \begin{pmatrix} 0 & 1 \\ -1 & 0 \end{pmatrix} \nabla \left[\frac{1}{2}p^2 + (1 - \cos q) \right]. \tag{1.22}$$

This is a two-dimensional dynamical system on $M = \mathbb{R}^2$.

Besides the canonical Hamiltonian systems of finite dimension, classical mechanics offers us a *non-canonical Hamiltonian formulation* as well.

Example 1.5 Let us consider the Euler equations of motion of a rigid body

$$u_{1_t} = \frac{I_2 - I_3}{I_2 I_3} u_2 u_3 = K_1(u), \quad u_{2_t} = \frac{I_3 - I_1}{I_1 I_3} u_3 u_1 = K_2(u),$$

$$u_{3_t} = \frac{I_1 - I_2}{I_1 I_2} u_1 u_2 = K_3(u), \tag{1.23}$$

where I_1, I_2, I_3 are the moments of inertia about the coordinate axis and u_1, u_2, u_3 are the corresponding body angular momenta. The dynamics (1.23) can be presented in the following Hamiltonian form

$$u_t = K(u) = \begin{pmatrix} 0 & -u_3 & u_2 \\ u_3 & 0 & -u_1 \\ -u_2 & u_1 & 0 \end{pmatrix} \nabla \left(\frac{u_1^2}{2I_1} + \frac{u_2^2}{2I_2} + \frac{u_3^2}{2I_3} \right) = \theta(u) \nabla H(u),$$

$$\tag{1.24}$$

where $u = (u_1, u_2, u_3)^T$, $K(u) = (K_1, K_2, K_3)^T$, the Hamiltonian function is the kinetic energy of the body and $\theta(u)$ is a proper (i.e. skew-symmetric and satisfies the Jacobi identity) Poisson matrix. The dynamics take place on $M \subset \mathbb{R}^3$.

For a finite dimensional system, the Poisson matrix θ is invertible, if it is not degenerate, i.e. is of maximum rank everywhere. Moreover, as follows from the skew-symmetry, the rank of the Poisson matrix at any point is always an even integer. If the manifold is of dimension $m = 2n + k$ and the rank of θ is $2n$ everywhere, then there exist k functions of the dynamical variables $c_i = f_i(u), i = 1, ..., k$, called the *Casimir functions*, whose Poisson bracket with any variable u_i is always zero: $\{u_i, c_i\}_\theta = 0$. The Poisson bracket of two functions $F(u)$ and $G(u)$ is defined as follows

$$\{F(u), G(u)\}_\theta = \sum_{i,j} \frac{\partial F}{\partial u_i} \theta^{ij} \frac{\partial G}{\partial u_j}. \tag{1.25}$$

The Darbou'x theorem plays a crucial role in the theory of finite dimensional Poisson manifolds. This theorem says that at each point of M there exist local coordinates $(q_1, ..., q_n, p_1, ..., p_n, c_1, ..., c_k)$ in terms of which the Poisson matrix θ takes the canonical form

$$\theta = \begin{pmatrix} 0 & I & 0 \\ -I & 0 & 0 \\ \underbrace{0 \quad 0}_{2n} & \underbrace{0}_{k} \end{pmatrix}, \tag{1.26}$$

where q_i, p_i are canonical variables and c_i are Casimir variables. The estimation of values of Casimir variables, $c_i = \text{const}, i = 1, ..., k$, determines the particular leaf of the symplectic foliation of the Poisson manifold.

Example 1.6 Consider again the Euler equations of motion (1.23) and (1.24). The Casimir function of $\theta(u)$ has the form $u_1^2 + u_2^2 + u_3^2$. The map

$$q = \arccos \frac{u_1}{(u_1^2 + u_2^2)^{1/2}}, \quad p = u_3, \quad c = u_1^2 + u_2^2 + u_3^2, \qquad (1.27)$$

transforms $\theta(u)$ into the canonical form (1.26) and $H(u)$ into

$$H(q, p, c) = (c - p^2) \left(\frac{1}{2I_1} \cos^2 q + \frac{1}{2I_2} \sin^2 q \right) + \frac{p^2}{2I_3}. \qquad (1.28)$$

Classical Hamiltonian systems have a remarkable property that whenever dynamical equations are invariant under either translation or rotation then the system conserves the momentum or angular momentum, respectively. So there exists a close relation between the conservation laws of a system and its symmetry properties. Such a relation was revealed for the first time by Emmy Noether. But it suggests that by studying symmetry properties of a given system one can get information about conserved quantities. On the other hand if we are lucky enough to find a sufficient number of conserved quantities (integrals of motion), according to the classical theorem of Liouville/Arnold [15] the system under consideration is completely integrable. Actually a Hamiltonian system on a $2N$-dimensional manifold is called *completely integrable* if it has N integrals of motion $f_i, i = 1, ..., N$ called *action variables*, which do not depend explicitly on time, are functionally independent and are in involution w.r.t. the Poisson bracket: $\{f_i, f_j\}_\theta = 0$. In such a case the flow can be reduced to the N-dimensional submanifold

$$M_f = \{u \in M : f_1(u) = d_1, ..., f_N(u) = d_N\} \qquad (1.29)$$

with some constants $d_1, ..., d_N$. According to the Liouville/Arnold theorem, when the set (1.29) is compact, then it is homeomorphic to the N-dimensional torus. Otherwise, it can be shown to be a cylinder. Moreover, the theorem guarantees the existence of N, the so called *angle variables* φ_i being linear functions of time. Hence, in the action/angle variables (f_i, φ_i) the flow is linearized

$$\begin{aligned}
(f_i)_t &= 0, \quad i = 1, ..., N \\
(\varphi_i)_t &= \omega_i(f) = \text{const} \quad i = 1, ..., N
\end{aligned} \qquad (1.30)$$

and finding integral curves becomes a trivial task. We will call the scalar fields f_i, φ_i canonical (w.r.t. some θ) *action/angle variables* if the following Poisson bracket holds

$$\{f_i, f_j\}_\theta = \{\varphi_i, \varphi_j\} = 0, \quad \{f_i, \varphi_j\}_\theta = \delta_{ij}. \qquad (1.31)$$

Even this basic information demonstrates how important and powerful the algebraic methods might be when we analyse the properties and structure of the finite dynamical system (1.1). So, it seems very attractive to make an attempt at applying the methods from the Hamiltonian theory in classical mechanics to infinite dimensional dynamical systems. Indeed, as we will see, a correct Hamiltonian form of the KdV dynamics (1.9) is

$$u_t = u_{xxx} + auu_x = D_x\delta \int_R \left(-\frac{1}{2}u_x^2 + \frac{1}{6}au^3\right) dx, \qquad (1.32)$$

where now the role of the Poisson matrix is played by the operator of the derivative D_x, which acts on an arbitrary function $w(x)$ in the following way: $D_xw = w_x$; and the gradient of the Hamiltonian function is replaced by a variation of the Hamiltonian functional. In the same way the correct Hamiltonian form of the Toda system (1.12) is

$$\begin{pmatrix} v(n) \\ p(n) \end{pmatrix}_t = \begin{pmatrix} 0 & v(n)(1 - E^{-1}) \\ (E - 1)v(n) & 0 \end{pmatrix} \delta \sum_{n=-\infty}^{n=+\infty} \left[\frac{1}{2}p^2(n) + v(n)\right]$$

$$(1.33)$$

where E is a shift operation which acts on an arbitrary discrete function $w(n)$ in the following way: $Ew(n) = w(n + 1)$.

This book presents a theory which can be called a modern Hamiltonian theory of dynamical systems (1.1), in which systems of all types, i.e. finite, lattice and field systems, are treated in a unified way. Special attention will be paid to systems which are bi-Hamiltonian (multi-Hamiltonian). Roughly speaking this means that with respect to the same set of coordinates one can find more than one Hamiltonian formulation of a given dynamics. The importance of this brilliant and surprising idea of Magri [123] is related to the fact that once a bi-Hamiltonian formulation of a dynamics is found, a hierarchy (finite or infinite) of conserved quantities can be generated. Moreover, the multi-Hamiltonian property is closely connected with the existence of the Lax representation. This is not surprising as these are two different aspects of the same property, i.e. integrability.

Besides the multi-Hamiltonian theory of dynamical systems (1.1) we also present the multi-Hamiltonian theory of the Lax equations (1.2). Actually it was found that (1.2) can be considered as a dynamical Hamiltonian system on the appropriate dual Lie algebra (generally of infinite dimension), with a multi-Hamiltonian structure induced via the so called $\mathcal{R}$-matrix. For infinite systems, for example, this abstract approach has the advantage of covering simultaneously many particular cases of dynamics in the form given by (1.1) in one general scheme. Nowadays it seems to be the most general and systematic method for generating multi-Hamiltonian dynamical systems on M.

The book is composed of eight chapters and each chapter is divided into sections.

Chapter 2 presents a brief discussion of differential calculus, which is essential for our further considerations. We review the concept of tensor fields in finite and infinite dimensional cases. Then the transformation properties of various tensor fields via push-forward and pull-back given by an arbitrary C^∞– diffeomorphism are derived. Next, we define the directional derivative of an arbitrary tensor field and derive its explicit form for many particular cases. Also a brief discussion of differential forms, important in Hamiltonian theory, is given. Then, with special care, we present a theory of Lie transport and Lie derivatives, which are essential for understanding the contents of other chapters. Many formulas are derived step by step. All considerations of this chapter are performed for infinite dimensional cases (continuous and discrete) as well as for the finite dimensional cases.

In the first section of chap.3 the notion of a Lie algebra is introduced and it is shown that vector fields constitute such an algebra in a natural way. Then we prove the existence of Lie brackets for covector fields and scalar fields with respect to special bi-vectors (implectic operators) and other Lie brackets with respect to special two-forms (symplectic operators). These are followed by the proofs of some properties and relations between the algebras considered (subalgebras, homomorphisms, etc.).

In the second section we introduce the notion of Hamiltonian, inverse-Hamiltonian and bi-Hamiltonian vector fields and illustrate them by many examples of finite and infinite dimensional cases.

In the third section, the properties of Lie algebras of time-independent symmetries and conserved quantities of dynamical systems are introduced and discussed.

In the fourth section, by means of the Lie derivative, the concept of general tensor invariants for dynamical systems is introduced in such a way that the objects from the previous section are their special cases. We introduce such invariant objects as the Lie algebra of cosymmetries, Poisson (implectic) operators, symplectic operators, recursion operators for symmetries and recursion operators for cosymmetries. Everything is illustrated by many examples.

In the next section we examine many properties of tensor invariants and prove several important relations among them (the hereditary property, the compatibility property, commutativity, involutivity, etc.). We construct a Virasoro algebra (hereditary algebra) of symmetries and cosymmetries, and present the role of the Lie derivative in the generation of new invariant objects for dynamical systems. The concept of master tensor invariants is presented in a systematic way and their relations with time-dependent tensor invariants are discussed. As a consequence we construct Lie algebras of polynomial-in-time symmetries, cosymmetries and constants of motion. All considerations are illustrated by many examples.

In the last section we introduce a Miura transformation and illustrate its relevance to the theory.

In the first section of Chap. 4 it is shown that an arbitrary multi-Hamiltonian dynamical system can be related to one or more Lax representations (or equivalently zero-curvature representations), i.e. linear operators on some Hilbert space, whose spectrum stays invariant along the integral curves of a given dynamical system. Besides, the role of the so called nonisospectral Lax equations (or equivalently nonisospectral zero-curvature equations) in the theory is considered.

In the second section, a respective nonisospectral Lax representation or equivalently nonisospectral zero-curvature representations is shown for a given hierarchy of commuting symmetries, generated by a recursion operator.

In the last section a kind of algebraic structure of the space of the corresponding isospectral and nonisospectral Lax operators is introduced. Then we prove a Lie algebra isomorphism between algebras of Lax operators and algebras of time-independent and polynomial-in-time symmetries. Again many examples are given to illustrate the results.

It is well known that the majority of solvable field and lattice nonlinear evolution equations have the so called N-soliton solutions, which asymptotically decompose into a sum of single solitons, that is extended objects of permanent shape, moving at constant speed. In Chap. 5 we discuss the time independent decomposition of N-soliton solutions into a sum of extended objects (soliton particles) closely related to the eigenfunctions of the discrete part of the spectrum of a recursion operator. Then the analytical form of soliton particles, their equations of motion, the multi-Hamiltonian structure and other algebraic properties are proved. Finally, multi-soliton perturbation theory constructed in a pure algebraic way is presented.

Actually, the first section contains some basic information about multi-soliton solutions of dynamical systems and their equivalence to some linear systems.

In the second section the whole apparatus developed in the previous chapters is applied to suitable real and complex finite dimensional linear systems. For the linear cases the multi-Hamiltonian formulation of the dynamics is found and all important tensor invariants including the most general algebras of symmetries, cosymmetries and constants of motion are given. But of course we are interested in recovering the whole structure on the 'physical' N-soliton manifold in explicit form. This means that we are interested in expressing all quantities in terms of the original field variable. This is done in the third section, where, by applying a suitable diffeomorphism, we carry over the whole structure to the physical (multi-soliton) representation.

Finally, in the last section, on the basis of the previous considerations, the multi-soliton perturbation theory is formulated in a purely algebraic way. This follows from the fact that we have at our disposal a complete set of conserved quantities on the N-soliton manifold as well as a suitable set of vector fields forming the basis of a tangent bundle to the N-soliton flow. Again this chapter is furnished with many interesting examples.

One of the efficient ways of constructing new integrable bi-Hamiltonian finite dimensional systems is a restriction of infinite dimensional integrable systems to finite dimensional invariant submanifolds. Chap. 6 mainly concentrates on the multi-Hamiltonian aspects of such restrictions and the related problem of separability. A few different cases of restrictions are considered.

In the first section the main subject is stationary flows of field systems. We show that for a large class of bi-Hamiltonian field systems their finite dimensional stationary flows preserve the bi-Hamiltonian nature. A systematic method of finding two compatible Poisson matrices for finite systems in the so called Ostrogradsky generalized coordinates is outlined. The method is general but the representation is 'nonphysical' in the sense that the Hamiltonian function is not separable into kinetic and the potential parts.

Then, in the second section, a transformation to a new 'physical' representation is given and is henceforth called the Newton representation. In this new representation equations of motion take the form of Newtonian equations and Hamiltonian function separates into kinetic and potential parts. We perform explicit calculations for many representative examples.

In the third section we consider some finite dimensional dynamical systems obtained by the nonlinearization of the isospectral Lax equations under certain constraints between potentials and eigenfunctions. Such reductions lead to a certain class of bi-Hamiltonian mechanical system as well.

The forth section is devoted to the most general finite dimensional reductions of field systems, the so called restricted flows, which contain the previous ones as special cases. A systematic method is presented for finding a bi-Hamiltonian structure for such reductions.

A common feature of the systems considered in this chapter is that all Poisson matrices are degenerate and each bi-Hamiltonian structure leads to a bi-Hamiltonian chain of constants of motion sufficient for integration of a given dynamics according to the Arnold/Liouville theorem. Besides, each chain starts with a Casimir of a first Poisson matrix and terminates with a Casimir of a second Poisson matrix. For finite dimensional systems constructed in this way the evolution parameter is the space coordinate of the underlying field system.

In the fifth section it is proved that all bi-Hamiltonian systems constructed in Chap. 5 are equivalent to the so called quasi-bi-Hamiltonian systems and hence are separable. Actually, it is shown that for a special coordinate frame (the Nijenhuis coordinates), for each Hamiltonian from a hierarchy, the related Hamilton-Jacobi equation can be solved.

In the sixth section, we present a nonstandard algebraic description of some classes of integrable field systems, resulting from interchanging the role of the space and time variables. Bi- or multi-Hamiltonian hierarchies obtained this way reduce naturally to the stationary subspaces giving some insight into the recursive structure of stationary and restricted flows considered in previous sections.

Finally, in the last section we briefly outline the theory of bi-Hamiltonian dynamical systems on the so-called Poisson-Nijenhuis manifolds, i.e. for finite dimensional systems for which at least one Poisson matrix is invertible. This situation is different from that considered in previous sections of the current chapter. Two important examples of a nonperiodic Toda lattice and the Calogero-Moser system are discussed.

In the first section of Chap. 7 the notion of Lie groups and their algebras, adjoint mapping, adjoint action, co-adjoint action, etc., is briefly reviewed. On the basis of these considerations we finally demonstrate that the arbitrary Lax equations, themselves, can be considered as a Hamiltonian dynamics on a suitable Lie algebra.

After this introduction, the second section gives the reader some basic concepts of $\mathcal{R}$-matrix theory. In my opinion it is the most general and powerful theory of multi-Hamiltonian dynamical systems; it includes most examples which we have constructed in previous chapters, in a single general scheme. In fact the theory is still under development, but we have at our disposal enough significant results to present them in a compact and clear way. The famous Gelfand-Dikii and Adler-Kostant-Symes schemes appear here in a natural way as special cases of the theory. The application of the theory to the algebra of pseudo-differential operators (Sect. 7.3) and to the algebra of shift operators (Sect. 7.4) is discussed in detail. In the first case we obtain a unified theory of multi-Hamiltonian $(1 + 1)$-dimensional field systems; in the second case we get a unified multi-Hamiltonian theory of $(1+1)$-dimensional lattice systems.

Finally, in the last chapter, first, a relation between the Sato theory of $(2 + 1)$-dimensional field systems and the $\mathcal{R}$-matrix formalism is indicated. Then, we show the way the $\mathcal{R}$-matrix formalism, when applied to the algebra of pseudo-differential operators with coefficients being also operators instead of functions, leads to a multi-Hamiltonian theory of $(2 + 1)$-dimensional field systems, built up over noncommutative rings. The respective operators acting in such rings are named operands (operators that act in the operator space). The concept of operand invariants seems to be appropriate for overcoming the difficulties of multi-Hamiltonian theory for infinite systems with more then one space dimension.

2. Elements of Differential Calculus for Tensor Fields

In this chapter we briefly discuss some elements of differential calculus which are essential for our further considerations. For a more comprehensive treatment of the subject we refer the reader to the literature [3],[51],[52],[186].

2.1 Tensors

Let V be a linear space and V^* its dual. By definition, any element γ of V^* represents a real-valued function on V:

$$\gamma : v \to \gamma(v) \equiv \langle \gamma, v \rangle \in \mathbb{R}, \qquad v \in V. \tag{2.1}$$

The bilinear map

$$\langle .,. \rangle : V \times V^* \to \mathbb{R}, \tag{2.2}$$

called the natural pairing of V and V^*, is also known as a *duality map*. Now let us denote by

$$\mathcal{T}^{(r,s)} = \underbrace{V^* \otimes ... \otimes V^*}_{s} \ \underbrace{\otimes V \otimes ... \otimes V}_{r} \tag{2.3}$$

the set of all multilinear functions

$$\underbrace{V \times ... \times V}_{s} \ \underbrace{\times V^* \times ... \times V^*}_{r} \to \mathbb{R}. \tag{2.4}$$

The linear space structure of $\mathcal{T}^{(r,s)}$ is evident. Any element $T^{(r,s)}$ of $\mathcal{T}^{(r,s)}$ is called a *tensor* of type (r,s) over V (r-times contravariant and s-times covariant). Its action is as follows. Let

$$\mathcal{T}^{(r,s)} \ni T^{(r,s)} = \eta_1 \otimes ... \otimes \eta_s \otimes y_1 \otimes ... \otimes y_r, \qquad \eta_i \in V^*, \quad y_i \in V; \tag{2.5}$$

then

$$T^{(r,s)}(v_1, ..., v_s, \gamma_1, ..., \gamma_r)$$

$$= \langle \eta_1, v_1 \rangle ... \langle \eta_s, v_s \rangle \langle \gamma_1, y_1 \rangle ... \langle \gamma_r, y_r \rangle, \qquad v_i \in V, \gamma_i \in V^*. \tag{2.6}$$

Example 2.1

$$T^{(0,0)} = c \in \mathbb{R}, \qquad \text{scalar.}$$

Example 2.2

$$T^{(1,0)} = y \in V \Rightarrow y(\gamma) = \langle \gamma, y \rangle, \qquad y - \text{contravariant vector.}$$

Example 2.3

$$T^{(0,1)} = \eta \in V^* \Rightarrow \eta(v) = \langle \eta, v \rangle, \qquad \eta - \text{covariant vector.}$$

Example 2.4

$$T^{(2,0)} = y_1 \otimes y_2 \Rightarrow (y_1 \otimes y_2)(\gamma_1, \gamma_2) = \langle \gamma_1, y_1 \rangle \langle \gamma_2, y_2 \rangle.$$

To a tensor $T^{(2,0)}$ we can relate a linear operator

$$\theta : V^* \to V \tag{2.7}$$

through the relation

$$T^{(2,0)}(\gamma_1, \gamma_2) = \langle \gamma_1, \theta \circ \gamma_2 \rangle, \tag{2.8}$$

where the application of θ to γ_2 acts as a contraction

$$\theta \circ \gamma_2 = y_1 \langle \gamma_2, y_2 \rangle. \tag{2.9}$$

Example 2.5

$$T^{(0,2)} = \eta_1 \otimes \eta_2 \Rightarrow (\eta_1 \otimes \eta_2)(v_1, v_2) = \langle \eta_1, v_1 \rangle \langle \eta_2, v_2 \rangle$$

To a tensor $T^{(0,2)}$ we can relate a linear operator

$$J : V \to V^* \tag{2.10}$$

through the relation

$$T^{(0,2)}(v_1, v_2) = \langle J \circ v_1, v_2 \rangle, \tag{2.11}$$

where

$$J \circ v_1 = \langle \eta_1, v_1 \rangle \eta_2. \tag{2.12}$$

Example 2.6

$$T^{(1,1)} = \eta \otimes y \Rightarrow (\eta \otimes y)(v, \gamma) = \langle \eta, v \rangle \langle \gamma, y \rangle.$$

To a tensor $T^{(1,1)}$ we can relate two different linear operators, depending on the type of contraction. One is

$$\Phi : V \to V, \tag{2.13}$$

defined through the relation

$$T^{(1,1)}(v, \gamma) = \langle \gamma, \Phi \circ v \rangle, \tag{2.14}$$

where

$$\Phi \circ v = y \langle \eta, v \rangle, \tag{2.15}$$

and the other is

$$\Psi : V^* \to V^* \tag{2.16}$$

defined by the relation

$$T^{(1,1)}(v, \gamma) = \langle \Psi \circ \gamma, v \rangle, \tag{2.17}$$

where

$$\Psi \circ \gamma = \eta \langle \gamma, y \rangle. \tag{2.18}$$

When it does not lead to confusion, in our further considerations we will identify operators θ, J, Φ and Ψ with appropriate tensors $T^{(r,s)}$.

The V-space may be finite or infinite (linear function space) dimensional. In the finite case we can introduce a coordinate system (a basis). Let $\{e_j\}$ be a given basis on V, whose dual is denoted by $\{e^j\}$, so that

$$\langle \epsilon^k, e_j \rangle = \delta_j^k, \qquad k, j = 1, ..., n. \tag{2.19}$$

If for any pair $v \in V$ and $\gamma \in V^*$ we write

$$v = \sum_{k=1}^{n} v^k e_k, \quad \gamma = \sum_{j=1}^{n} \gamma_j e^j, \tag{2.20}$$

then it follows from (2.19) that

$$\langle \epsilon^k, v \rangle = v^k, \quad \langle \gamma, e_j \rangle = \gamma_j. \tag{2.21}$$

So, for any tensor $T^{(r,s)} \in \mathcal{T}^{(r,s)}$

$$T^{(r,s)}\left(e_{k_1}, ..., e_{k_s}, \epsilon^{j_1}, ..., \epsilon^{j_r}\right)$$

$$= (\eta_1 \otimes ... \otimes \eta_s \otimes y_1 \otimes ... \otimes y_r)\left(e_{k_1}, ..., e_{k_s}, \epsilon^{j_1}, ..., \epsilon^{j_r}\right) \tag{2.22}$$

$$= \langle \eta_1, e_{k_1} \rangle ... \langle \epsilon^{j_1}, y_r \rangle = \eta_{k_1} ... \eta_{k_s} y^{j_1} ... y^{j_r} = T^{j_1...j_r}{}_{k_1...k_s}$$

is a tensor component in our basis and the full form of $T^{(r,s)}$ reads

$$T^{(r,s)} = \sum_{k_1=1}^{n} ... \sum_{j_r=1}^{n} T^{j_1...j_r}{}_{k_1...k_s} \epsilon^{k_1} \otimes ... \otimes \epsilon^{k_s} \otimes e_{j_1} \otimes ... \otimes e_{j_r}. \tag{2.23}$$

Example 2.7

$$T^{(2,0)} = y_1 \otimes y_2 = \sum_{i,j=1}^{n} y_1^i y_2^j e_i \otimes e_j = \sum_{i,j=1}^{n} T^{ij} e_i \otimes e_j,$$

$$T^{(0,2)} = \eta_1 \otimes \eta_2 = \sum_{i,j=1}^{n} \eta_{1i} \eta_{2j} \epsilon^i \otimes \epsilon^j = \sum_{i,j=1}^{n} T_{ij} \epsilon^i \otimes \epsilon^j,$$

$$T^{(1,1)} = \eta \otimes y = \sum_{i,j=1}^{n} \eta_i y^j \epsilon^i \otimes e_j = \sum_{i,j=1}^{n} T_i^j \epsilon^i \otimes e_j.$$

2.2 Tensor Fields

To be able to perform certain differential calculus procedures we have to introduce a differential manifold M when we are concerned with a finite dimensional problem or an appropriate linear function space otherwise. Hence, our linear space V from the previous section may be identified with the tangent space $T_u M$ to M at a point $u \in M$, while the dual space V^* can be identified with an appropriate cotangent space $T_u^* M$. A tangent bundle $\mathcal{T}M$ over M is the union of all tangent spaces to M, that is, $\mathcal{T}M$ is the collection of all contravariant vectors (*vectors* for short) at each point $u \in M$. Analogously a cotangent bundle $\mathcal{T}^* M$ over M is the union of all cotangent spaces to M, i.e. the collection of all covariant vectors (*covectors* for short) at each point u of M. More generally, let us denote by $\mathcal{T}^{(r,s)}(u)$ the linear space of all (r, s) type tensors at u, constructed according to the method of the previous section, by means of the linear space $T_u M = V$ and its dual $T_u^* M = V^*$ (in the finite dimensional case, when $\dim M = n$ then $\dim \mathcal{T}^{(r,s)}(u) = n^{r+s}$). The union of all $\mathcal{T}^{(r,s)}(u)$ obtained by letting u run over M, keeping r and s fixed, is known as the tensor bundle of type (r, s) over M and is denoted by $\mathcal{T}^{(r,s)} M$. In particular $\mathcal{T}^{(1,0)} M$ is the tangent bundle $\mathcal{T}M$, $\mathcal{T}^{(0,1)} M$ is the cotangent bundle $\mathcal{T}^* M$, and $\mathcal{T}^{(0,0)} M$ we identify with $\mathbb{R}$. An (r, s) type *tensor field* $T^{(r,s)}$ on M is the mapping $T^{(r,s)} : M \to \mathcal{T}^{(r,s)} M$. Thus, for a given point u, $T^{(r,s)}(u)$ is a tensor from the previous section. Using the notation from the previous section we find

$$f : M \to \mathbb{R}, \qquad f(u) \in \mathbb{R},$$

$$v : M \to \mathcal{T}M, \qquad v(u) \in T_u M,$$

$$\gamma : M \to \mathcal{T}^* M, \qquad \gamma(u) \in T_u^* M,$$

$$\theta : M \to L(\mathcal{T}^* M, \mathcal{T}M), \qquad \theta(u) \in L(T_u^* M, T_u M),$$

$$J : M \rightarrow L(\mathcal{T}M, \mathcal{T}^*M), \qquad J(u) \in L(T_uM, T_u^*M),$$

$$\Phi : M \rightarrow L(\mathcal{T}M, \mathcal{T}M), \qquad \Phi(u) \in L(T_uM, T_uM), \tag{2.24}$$

$$\Psi : M \rightarrow L(\mathcal{T}M, \mathcal{T}M), \qquad \Psi(u) \in L(T_u^*M, T_u^*M),$$

where $L(A, B)$ means a linear map from A to B (see Fig 2.1).

Let us comment on the infinite dimensional case. Assume that M is a linear space of functions $u(x)$ which sufficiently rapidly vanishes at infinity $(x \rightarrow \pm\infty)$, the Schwartz space for example. Generally u can be a multicomponent function $u(x) = (u^1, ..., u^n)^{\mathrm{T}}$. The variable x can be identified with the index of the finite dimensional case which now can take a continuous value. With each point $u(x)$ we can associate a tangent space T_uM which in that case can be identified with M (as M is a linear space). The cotangent space T_u^*M (identified with M^*) can be introduced through the duality map in the form

$$\langle \gamma, v \rangle (u) = \langle \gamma(u), v(u) \rangle = \int_R \gamma^{\mathrm{T}}(u(x))v(u(x))\mathrm{d}x,$$
$$\tag{2.25}$$
$$v(u) \in T_uM \simeq M, \ \gamma(u) \in T_u^*M \simeq M^*,$$

which is a generalization of the scalar product of the finite dimensional case. For lattice systems the integral $\int_R$ is replaced by the infinite summation $\sum_{n=-\infty}^{n=+\infty}$, so

$$\langle \gamma, v \rangle (u) = \sum_{n=-\infty}^{n=+\infty} \gamma^{\mathrm{T}}(u(n))v(u(n)). \tag{2.26}$$

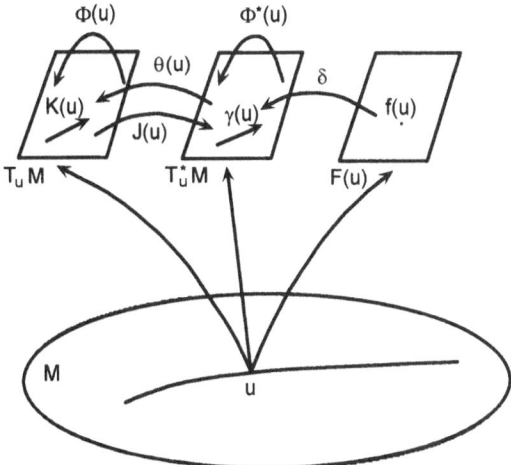

Fig. 2.1. Geometric scheme

The set of all tensors of the same rank (r, s) at each point $u \in M$ forms a linear space. Additionally, linear spaces $\mathcal{T}^{(0,0)}(u), \mathcal{T}^{(1,0)}(u)$ and $\mathcal{T}^{(0,1)}(u)$ can be endowed with a *Lie algebra* structure.

Definition 2.1 *A linear space V with a bilinear product $[.,.] : V \times V \to V$ which is antisymmetric*

$$[a, b] = -[b, a], \quad a, b \in V \tag{2.27}$$

and satisfies the so called Jacobi identity

$$[a, [b, c]] + [c, [a, b]] + [b, [c, a]] = 0 \qquad a, b, c \in V, \tag{2.28}$$

is called a Lie algebra.

We shall come back to this structure later on, as it plays an important role in the Hamiltonian theory.

We end this section by introducing the adjoint operation of some tensor fields with respect to the duality map.

Definition 2.2 *The tensor fields $\theta^\dagger, J^\dagger, \Phi^\dagger$ and $\Psi^\dagger$ are called the adjoint representatives of the tensor fields θ, J, Φ and Ψ with respect to the duality map if for arbitrary vector fields v, w and covector fields α, β the following equalities hold*

$$\langle \alpha, \theta \circ \beta \rangle (u) = \langle \beta, \theta^\dagger \circ \alpha \rangle (u),$$

$$\langle J \circ v, w \rangle (u) = \langle J^\dagger \circ w, v \rangle (u),$$

$$\langle \alpha, \Phi \circ v \rangle (u) = \langle \Phi^\dagger \circ \alpha, v \rangle (u), \tag{2.29}$$

$$\langle \Psi \circ \alpha, v \rangle (u) = \langle \alpha, \Psi^\dagger \circ v \rangle (u).$$

2.3 Transformation Properties of Tensor Fields

Let

$$\phi : M \to \overline{M} \tag{2.30}$$

be a $\mathbb{C}^\infty$-mapping. The *derivative* (differential) of ϕ at $u \in M$ is the linear map

$$\phi'(u) : T_u M \to T_{\phi(u)} \overline{M} \tag{2.31}$$

which is also smooth. If ϕ is a diffeomorphism, then for any vectors $v(u) \in T_u M$ and $\overline{v}(\overline{u}) = \overline{v}(\phi(u)) \in T_{\phi(u)} \overline{M}$, the *push-forward* (direct image) under ϕ of a vector field $v \in \mathcal{T} M$ is the vector field $\phi_* v \in \mathcal{T} \overline{M}$ which satisfies

$$\overline{v}(\overline{u}) = (\phi_* v)(\overline{u}) = \phi'(u) v(u) = (\phi' \circ v)(u) = (\phi' \circ v) \cdot \phi^{-1}(\overline{u}). \tag{2.32}$$

The quantity $\phi'(u)v(u)$ is most easily calculated from the directional derivative of ϕ

$$\phi'(u)v(u) \equiv \phi'(u)[v(u)] = \frac{d}{d\epsilon}\phi(u + \epsilon v(u))|_{\epsilon=0}. \tag{2.33}$$

The *pull-back* (inverse image) under ϕ of the vector field $\overline{v} \in \mathcal{T}\,\overline{M}$ is the
vector field $\phi^*\overline{v} \in \mathcal{T}\,M$, the direct image of $\overline{v}$ by ϕ^{-1}

$$v(u) \;= (\phi^*\overline{v})(u) = \left((\phi^{-1})_*v\right)(u) = \left(\phi^{-1}\right)'(\overline{u})\overline{v}(\overline{u})$$

$$= \phi'^{-1}(u)\overline{v}(\overline{u}) = (\phi'^{-1} \circ \overline{v} \cdot \phi)(u). \tag{2.34}$$

Now let us define the adjoint of $\phi'(u)$ with respect to the duality map, i.e.
$\phi'^\dagger(u)$ such that

$$\left\langle \overline{\gamma}, (\phi' \circ v) \cdot \phi^{-1} \right\rangle(\overline{u}) \;= \langle \overline{\gamma}(\overline{u}), (\phi' \circ v)(u) \rangle = \langle \overline{\gamma}(\overline{u}), \phi'(u)v(u) \rangle$$

$$= \langle \phi'^\dagger(u)\overline{\gamma}(\overline{u}), v(u) \rangle = \langle (\phi'^\dagger \circ \overline{\gamma} \cdot \phi)(u), v(u) \rangle$$

$$= \left\langle \phi'^\dagger \circ \overline{\gamma} \cdot \phi, v \right\rangle(u). \tag{2.35}$$

Hence, the *dual derivative* (dual differential) is a linear map

$$\phi'^\dagger(u) : T^*_{\phi(u)}\overline{M} \to T^*_u M \tag{2.36}$$

and defines a pull-back under ϕ of a covector field $\overline{\gamma}$

$$\gamma(u) = (\phi^*\overline{\gamma})(u) = \phi'^\dagger(u)\overline{\gamma}(\overline{u}) = (\phi'^\dagger \circ \overline{\gamma} \cdot \phi)(u). \tag{2.37}$$

The appropriate push-forward reads

$$\overline{\gamma}(\overline{u}) \;= (\phi_*\gamma)(\overline{u}) = (\phi'^\dagger)^{-1}(u)\gamma(u)$$

$$= ((\phi'^\dagger)^{-1} \circ \gamma)(u) = ((\phi'^\dagger)^{-1} \circ \gamma) \cdot \phi^{-1}(\overline{u}). \tag{2.38}$$

The pull-back of an arbitrary *scalar field* f can be defined as

$$f(u) = (\phi^*\overline{f})(u) = \overline{f}(\overline{u}) = (\overline{f} \cdot \phi)(u). \tag{2.39}$$

The present considerations allow us to define a push-forward and pull-back
for an arbitrary tensor field in the following way.

Definition 2.3 *If ϕ is a $\mathbb{C}^\infty$-diffeomorphism $M \to \overline{M}$, then for any ten-
sors $T^{(r,s)}(u) \in \mathcal{T}^{(r,s)}(u)$ and $\overline{T}^{(r,s)}(\phi(u)) = \overline{T}^{(r,s)}(\overline{u}) \in \overline{\mathcal{T}}^{(r,s)}(\phi(u))$
the push-forward under ϕ of a tensor field $T^{(r,s)} \in \mathcal{T}^{(r,s)}M$ is the ten-
sor field $\phi_*T^{(r,s)} \in \mathcal{T}^{(r,s)}\overline{M}$ and the pull-back under ϕ of a tensor field
$\overline{T}^{(r,s)} \in \overline{\mathcal{T}}^{(r,s)}\overline{M}$ is the tensor field $\phi^*\overline{T}^{(r,s)} \in \mathcal{T}^{(r,s)}M$ which satisfies*

$$\phi_*(T^{(r,s)}(u)) = (\phi_*T^{(r,s)})(\phi(u)) = \overline{T}^{(r,s)}(\overline{u}),$$

$$\phi^*(\overline{T}^{(r,s)}(\overline{u})) = (\phi^*\overline{T}^{(r,s)})(u) = T^{(r,s)}(u) \tag{2.40}$$

and

$$T^{(r,s)}(u)(v_1,..,v_s,\gamma_1,..,\gamma_r) = \overline{T}^{(r,s)}(\overline{u})(\overline{v}_1,..,\overline{v}_s,\overline{\gamma}_1,..,\overline{\gamma}_r), \tag{2.41}$$

where $\overline{v} = \phi_*v = (\phi' \circ v) \cdot \phi^{-1}, \gamma = \phi^*\overline{\gamma} = \phi'^\dagger \circ \overline{\gamma} \cdot \phi$.

From Definition 2.3 it is obvious that for arbitrary fixed values of r and s the $\phi_*\phi^*$ and $\phi^*\phi_*$ are the identities on $\overline{\mathcal{T}}^{(r,s)}\overline{M}$ and $\mathcal{T}^{(r,s)}M$, respectively. Now, equipped with the knowledge of this section we can find in explicit form transformation laws of various tensor fields under the map ϕ.

Example 2.8 For $f \in \mathcal{T}^{(0,0)}M$ and $\overline{f} \in \overline{\mathcal{T}}^{(0,0)}\overline{M}$ we have

$$\overline{f} = \phi_*f = f \cdot \phi^{-1} \Rightarrow \overline{f}(\overline{u}) = f(u),$$

$$f = \phi^*\overline{f} = \overline{f} \cdot \phi \Rightarrow f(u) = \overline{f}(\overline{u}). \tag{2.42}$$

The result follows from (2.39).

Example 2.9 For $K \in \mathcal{T}^{(1,0)}M$ and $\overline{K} \in \overline{\mathcal{T}}^{(1,0)}\overline{M}$ we have

$$\overline{K} = \phi_*K = (\phi' \circ K) \cdot \phi^{-1} \Rightarrow \overline{K}(\overline{u}) = \phi'(u)K(u),$$

$$K = \phi^*\overline{K} = \phi'^{-1} \circ \overline{K} \cdot \phi \Rightarrow K(u) = \phi'^{-1}(u)\overline{K}(\overline{u}). \tag{2.43}$$

The results follow from (2.32), (2.34).

Example 2.10 For $\gamma \in \mathcal{T}^{(0,1)}M$ and $\overline{\gamma} \in \overline{\mathcal{T}}^{(0,1)}\overline{M}$ we have

$$\overline{\gamma} = \phi_*\gamma = ((\phi'^\dagger)^{-1} \circ \gamma) \cdot \phi^{-1} \Rightarrow \overline{\gamma}(\overline{u}) = (\phi'^\dagger)^{-1}(u)\gamma(u),$$

$$\gamma = \phi^*\overline{\gamma} = \phi'^\dagger \circ \overline{\gamma} \cdot \phi^{-1} \Rightarrow \gamma(u) = \phi'^\dagger(u)\overline{\gamma}(\overline{u}). \tag{2.44}$$

The results follow from (2.37), (2.38).

Example 2.11 For $\theta \in \mathcal{T}^{(2,0)}M$ and $\overline{\theta} \in \overline{\mathcal{T}}^{(2,0)}\overline{M}$ we find

$$\overline{\theta} = \phi_*\theta = (\phi' \circ \theta \circ \phi'^\dagger \cdot \phi^{-1}) \Rightarrow \overline{\theta}(\overline{u}) = \phi'(u)\theta(u)\phi'^\dagger(u),$$

$$\theta = \phi^*\overline{\theta} = (\phi')^{-1} \circ \overline{\theta} \cdot \phi \circ (\phi'^\dagger)^{-1} \Rightarrow \theta(u) = (\phi')^{-1}(u)\overline{\theta}(\overline{u})(\phi'^\dagger)^{-1}(u). \tag{2.45}$$

The relations follow directly from the application of formulas (2.43) and (2.44). Actually as $K = \theta \circ \gamma$ we have

$$\overline{K} = (\phi' \circ K) \cdot \phi^{-1} = (\phi' \circ \theta \circ \gamma) \cdot \phi^{-1} = (\phi' \circ \theta \circ \overline{\gamma} \cdot \phi) \cdot \phi^{-1}$$

$$= (\phi' \circ \theta \circ \phi'^{\dagger}) \cdot \phi^{-1} \circ \overline{\gamma}, \tag{2.46}$$

but, on the other hand, $\overline{K} = \overline{\theta} \circ \overline{\gamma}$, so we find the result (2.45).

Example 2.12 For $J \in \mathcal{T}^{(0,2)}M$ and $\overline{J} \in \overline{\mathcal{T}}^{(0,2)}\overline{M}$ we find

$$\overline{J} = \phi_* J = \left[(\phi'^{\dagger})^{-1} \circ J \circ (\phi')^{-1} \right] \cdot \phi^{-1} \Rightarrow \overline{J}(\overline{u}) = (\phi'^{\dagger})^{-1}(u) J(u) (\phi')^{-1}(u),$$

$$J = \phi^* \overline{J} = \phi'^{\dagger} \circ \overline{J} \cdot \phi \circ \phi' \Rightarrow J(u) = \phi'^{\dagger}(u) \overline{J}(\overline{u}) \phi'(u). \tag{2.47}$$

The proof is similar to the one from Example 2.11, where now $\gamma = J \circ K$.

Example 2.13 For $\Phi \in \mathcal{T}^{(1,1)}M$ and $\overline{\Phi} \in \overline{\mathcal{T}}^{(1,1)}\overline{M}$ we find

$$\overline{\Phi} = \phi_* \Phi = (\phi' \circ \Phi \circ \phi'^{-1}) \cdot \phi^{-1} \Rightarrow \overline{\Phi}(\overline{u}) = \phi'(u) \Phi(u) \phi'^{-1}(u),$$

$$\Phi = \phi^* \overline{\Phi} = (\phi'^{-1} \circ \overline{\Phi} \cdot \phi \circ \phi') \cdot \phi^{-1} \Rightarrow \Phi(u) = \phi'^{-1}(u) \overline{\Phi}(\overline{u}) \phi'(u). \tag{2.48}$$

To derive formulas (2.48) we use the relation $K_2 = \Phi \circ K_1$.

The procedure can be continued in the same manner for higher order tensor fields. These examples illustrate the transformation properties of these tensor fields which are the most important for our considerations.

It is clear that for particular calculations we would like to know the explicit form of the maps $\phi'(u)$ and $\phi'^{\dagger}(u)$. Let us concentrate on the infinite dimensional case. We assume that each point of M, i.e. $u = (u^1, ..., u^m)^{\mathrm{T}}$ is an m-component function of the space variable x in the field case, or the discrete variable n in the lattice case. Hence, $\phi(u) = (\phi^1(u), ..., \phi^m(u))^T$ is an m-component differential (or difference) function of u. It means that in the continuous case, ϕ depends on $u(x)$ as well as on its x-derivatives u_x, u_{xx}, etc., and in the discrete case ϕ depends on $u(n)$ and its n-shifts $u(n+1), u(n+2)$, etc.

Observation 2.1 *The derivative and the dual derivative of a differential smooth mapping have the following form*

$$\left(\phi'(u) \right)^j_i = \sum_k \frac{\partial \phi^i}{\partial u^j_{kx}} D^k_x,$$

$$\left(\phi'^{\dagger}(u) \right)^j_i = \sum_k (-1)^k D^k_x \frac{\partial \phi^j}{\partial u^i_{kx}} \tag{2.49}$$

where $i, j = 1, ..., m$, *and* $u^j_{kx} = D^k_x u^j = \partial^k u^j / \partial x^k$. *The appropriate derivative and its dual of the difference smooth mapping reads*

$$\left(\phi'(u)\right)_i^j = \sum_k \frac{\partial \phi^i}{\partial u_k^j} E^k,$$

$$\left(\phi'^\dagger(u)\right)_i^j = \sum_k (E^k)^\dagger \frac{\partial \phi^j}{\partial u_k^i} = \sum_k E^{-k} \frac{\partial \phi^j}{\partial u_k^i}, \tag{2.50}$$

where now $u_k^i = E^k u^i(n) = u^i(n+k)$, E is the shift operation introduced in Example 1.2 and $E^\dagger$ its conjugate with respect to the duality map (2.26), where $E^\dagger = E^{-1}$.

Example 2.14 The field case with $m = 1$ and $\phi(u) = uu_x^2$. According to (2.49) we have

$$\phi'(u) = u_x^2 + 2uu_x D_x, \quad \phi'(u)[v] = u_x^2 v + 2uu_x v_x,$$

$$\phi'^\dagger(u) = u_x^2 - D_x 2uu_x = u_x^2 - 2(uu_x)_x - 2uu_x D_x = -u_x^2 - 2uu_{xx} - 2uu_x D_x.$$

We can verify the result by calculating the directional derivative from the definition (2.33)

$$\begin{aligned}
\phi'(u)[v] &= \tfrac{d}{d\epsilon}[(u+\epsilon v)(u_x + \epsilon v_x)^2]_{|\epsilon=0} \\
&= \tfrac{d}{d\epsilon}[(u+\epsilon v)(u_x^2 + 2\epsilon u_x v_x + \epsilon^2 v_x^2)]_{|\epsilon=0} \\
&= [v(u_x^2 + 2\epsilon u_x v_x + \epsilon^2 v_x^2) + (u+\epsilon v)(2u_x v_x + 2\epsilon v_x^2)]_{|\epsilon=0} \\
&= vu_x^2 + 2uu_x v_x.
\end{aligned}$$

The correctness of the dual derivative $\phi'^\dagger(u)$ can be verified by applying the explicit form of the duality map (2.25) and integrating by parts.

Example 2.15 The field case with $m = 2$. Let $(u^1, u^2) = (u, w)$ and

$$\phi(u, w) = \begin{pmatrix} u^2 w_{xx} \\ u_x w_x \end{pmatrix}.$$

Then we have

$$\phi'(u, w) = \begin{pmatrix} 2uw_{xx} & u^2 D_x^2 \\ w_x D_x & u_x D_x \end{pmatrix}$$

and

$$\phi'^\dagger(u, w) = \begin{pmatrix} 2uw_{xx} & -D_x w_x \\ D_x^2 u^2 & -D_x u_x \end{pmatrix}$$

$$= \begin{pmatrix} 2uw_{xx} & -w_{xx} - w_x D_x \\ 2u_x^2 + 2uu_{xx} + 4uu_x D_x + u^2 D_x^2 & -u_{xx} - u_x D_x \end{pmatrix}.$$

If $v = (q, p)^T$, then

$$\phi'(u)[v] = \begin{pmatrix} 2uw_{xx}q + u^2p_{xx} \\ w_xq_x + u_xp_x \end{pmatrix}.$$

Example 2.16 The lattice case with $m = 1$.
Let $\phi(u) = u(n-1)u(n)u(n+1)$. Then

$$\phi'(u) = u(n)u(n+1)E^{-1} + u(n-1)u(n+1) + u(n-1)u(n)E,$$

$$\begin{aligned} \phi'^\dagger(u) &= Eu(n)u(n+1) + u(n-1)u(n+1) + E^{-1}u(n-1)u(n) \\ &= u(n+1)u(n+2)E + u(n-1)u(n+1) + u(n-2)u(n-1)E^{-1}. \end{aligned}$$

In the case of a finite dimensional manifold, if in a local chart $u = x = (x^1, ..., x^m)^{\mathrm{T}}$ and $\bar{u} = \bar{x} = (\bar{x}^1, ..., \bar{x}^m)^{\mathrm{T}} = \phi(x) = \bar{x}(x)$, we find

$$\bar{x}'(x)^j_i = \frac{\partial \bar{x}^i}{\partial x^j}, \quad \bar{x}'^\dagger(x)^j_i = \frac{\partial \bar{x}^j}{\partial x^i} \Rightarrow \bar{x}'^\dagger(x) = \bar{x}'(x)^{\mathrm{T}} \qquad (2.51)$$

and, hence, we get standard transformation laws for an arbitrary tensor field

$$\bar{T}^{i_1...i_r}{}_{j_1...j_s}(\bar{x}) = \sum_{l_1...l_r}\sum_{k_1...k_s} \frac{\partial \bar{x}^{i_1}}{\partial x^{l_1}} \cdots \frac{\partial \bar{x}^{i_r}}{\partial x^{l_r}} \frac{\partial x^{k_1}}{\partial \bar{x}^{j_1}} \cdots \frac{\partial x^{k_s}}{\partial \bar{x}^{j_s}} T^{l_1...l_r}{}_{k_1...k_s}(x). \qquad (2.52)$$

2.4 Directional Derivative of Tensor Fields

Definition 2.4 *Let T represent an arbitrary tensor field and v be a given vector field. Then*

$$T'(u)[v] = \frac{\mathrm{d}}{\mathrm{d}\epsilon}T(u + \epsilon v)_{|\epsilon=0} \qquad (2.53)$$

is known as the directional derivative of $T(u)$ in the direction of $v(u)$.

From this definition it follows that: a directional derivative of an (r, s) type tensor is also an (r, s) type tensor, the directional derivative is a linear operator and the Leibniz chain rule holds for any kind of tensor compositions.

Let us pay some attention to these tensor fields $T^{(r,s)}$ which are important for our further considerations. Let us start from the simplest case $r + s = 0$, i.e. from a scalar field $T^{(0,0)} = f$ on M. In the finite dimensional case $f = f(x)$ is a function and in local coordinates the directional derivative has the well known form

$$f'(x)[v] = \sum_i \frac{\partial f}{\partial x^i}v^i = \langle \nabla f, v \rangle, \qquad (2.54)$$

where the covector field ∇f means the *gradient* of $f(x)$. More interesting for our considerations is the infinite dimensional case. Let u be an m-component function $u = (u^1, ..., u^m)^{\mathrm{T}}$. Then an arbitrary scalar field on M is represented by the formula

$$F(u) = \int_R f(u(x))dx, \tag{2.55}$$

where the density $f(u)$ is a differential function of the u^i components. In this case the gradient operator ∇ turns into a *variational operator* $\delta = (\delta/\delta u^1, ..., \delta/\delta u^m)^T$. Actually, we find

$$F'(u)[v] = \int_R f'(u)[v]dx = \int_R \sum_i \left(\sum_{k\geq 0} \frac{\partial f}{\partial u^i_{kx}} D^k_x v^i \right) dx$$

$$\overset{\text{by parts}}{=} \int_R \sum_i \left[\sum_{k\geq 0} (-D_x)^k \frac{\partial f}{\partial u^i_{kx}} \right] v^i dx = \langle \delta F, v \rangle, \tag{2.56}$$

where $\delta F = (\delta F/\delta u^1, ..., \delta F/\delta u^m)^T$ is a covector field and

$$\frac{\delta F}{\delta u^i} = \sum_{k\geq 0} (-D_x)^k \frac{\partial f}{\partial u^i_{kx}}. \tag{2.57}$$

In the lattice case $F(u)$ is represented by

$$F(u) = \sum_{n=-\infty}^{n=+\infty} f(u(n)), \tag{2.58}$$

where $f(u(n))$ is the difference function of the u^i components and

$$\frac{\delta F}{\delta u^i} = \sum_{k\in\mathbb{Z}} E^{-k} \frac{\partial f}{\partial u^i_k}, \qquad u^i_k = u^i(n+k). \tag{2.59}$$

Example 2.17

$$F(u, w) = \int_R \left(-\frac{1}{2}u^2_x + 2w^2_{xx} + \frac{1}{3}w^2 u_x + \frac{1}{6}u^3 \right) dx,$$

$$\delta F = \begin{pmatrix} \delta F/\delta u \\ \delta F/\delta w \end{pmatrix} = \begin{pmatrix} \frac{1}{2}u^2 - D_x \left(\frac{1}{3}w^2 - u_x\right) \\ \frac{2}{3}wu_x + D^2_x(4w_{xx}) \end{pmatrix} = \begin{pmatrix} \frac{1}{2}u^2 - \frac{2}{3}ww_x + u_{xx} \\ \frac{2}{3}wu_x + 4w_{4x} \end{pmatrix}.$$

Example 2.18

$$F(u, w) = \sum_{n=-\infty}^{n=+\infty} [u(n)w(n+2) + u^2(n+1)],$$

$$\delta F = \begin{pmatrix} w(n+2) + E^{-1}2u(n+1) \\ E^{-2}u(n) \end{pmatrix} = \begin{pmatrix} w(n+2) + 2u(n) \\ u(n-2) \end{pmatrix}.$$

Let $f(x, u(x))$ be a smooth differential function of x and field variables $u^i(x)$. The operation of D_x (or briefly D whenever it is not confusing) to f has the general form

$$Df = \frac{\partial f}{\partial x} + \sum_{i=1}^{m} \sum_{k \geq 0} \frac{\partial f}{\partial u_{kx}^i} u_{(k+1)x}^i, \tag{2.60}$$

known as the *total x-derivative* of $f(x)$. The formula (2.60) follows from a straightforward application of the chain rule. The important information is that $\delta D = 0$, which follows from the fact that [159]

$$\text{Ker } \delta = \text{Im } D. \tag{2.61}$$

But this means that the variation is defined up to a total derivative. We illustrate the above statement by a simple

Example 2.19

$$\delta \int_R D(uu_{xx}) \mathrm{d}x = \delta \int_R (u_x u_{xx} + uu_{3x}) \mathrm{d}x = u_{3x} - (u_{xx})_x + (u_x)_{xx} - u_{3x} = 0.$$

In the lattice case we can introduce a difference operation Δ

$$\Delta f(n) = f(n+1) - f(n) = (E - 1)f(u) \tag{2.62}$$

with the following properties

$$\Delta^\dagger = -E^{-1}\Delta, \quad \Delta^{-1}f(n) = \sum_{k=-\infty}^{n-1} f(k). \tag{2.63}$$

It is not difficult to check that $\delta\Delta = 0$.

Let us pass to the directional derivative of tensors of higher order than scalar fields. The case $r + s = 1$ contains two types of tensor fields, i.e. a vector field $T = T^{(1,0)}$ and a covector field $T = T^{(0,1)}$. Both types of vector fields have the form of an m-component differential function (the distinction between them lies in the different transformation properties rather than in their explicit forms), exactly like the m-component map $\phi(u)$. Hence, $T'(u)$ takes the form given by the formula (2.49) and the i-th component of $T'(u)[v]$ reads

$$(T'(u)[v])_i = \sum_{j,k} \frac{\partial(T(u))_i}{\partial u_{kx}^j} v_{kx}^j. \tag{2.64}$$

Example 2.20 The field case with $m = 1$.

$$T^{(1,0)}(u) \ni K(u) = u_{3x} + 6uu_x,$$

$$K'(u) = 6u_x + 6uD + D^3,$$

$$K'(u)[v] = 6u_x v + 6uv_x + v_{3x}.$$

Example 2.21 The two component field case $(u^1, u^2) = (u, w)$, $v = (p, q)$.

$$T^{(0,1)}(u) \ni \gamma(u, w) = \begin{pmatrix} w_x - \frac{1}{3}u^2 - \frac{2}{3}u_{xx} \\ 2w - u_x \end{pmatrix},$$

$$\gamma'(u, w) = \begin{pmatrix} -\frac{2}{3}u - \frac{2}{3}D^2 & D \\ -D & 2 \end{pmatrix}, \quad \gamma'(u, w)[v] = \begin{pmatrix} -\frac{2}{3}up - \frac{2}{3}p_{xx} + q_x \\ -p_x + 2q \end{pmatrix}.$$

The lattice case is analogous but now we apply formula (2.50).

Example 2.22 The two component lattice system, $u = (w, p)$, $v = (v_1, v_2)$.

$$T^{(1,0)}(u) \ni K(w, p) = \begin{pmatrix} w(n)[p(n) - p(n-1)] \\ w(n+1) - w(n) \end{pmatrix},$$

$$K'(w, p) = \begin{pmatrix} p(n) - p(n-1) & w(n) - w(n)E^{-1} \\ E - 1 & 0 \end{pmatrix},$$

$$K'(w, p)[v] = \begin{pmatrix} [p(n) - p(n-1)]v_1(n) + w(n)[v_2(n) - v_2(n-1)] \\ v_1(n+1) - v_1(n) \end{pmatrix}.$$

The case $r + s = 2$ covers four types of tensor fields which in our notation are θ, J, Φ and Ψ, respectively. In general, each field is represented by an $m \times m$ matrix differential operator whose elements can always be transformed to the form

$$(T(u))_{ij} = \sum_\alpha P_\alpha^{ij}(u)D^\alpha, \quad \alpha \in \mathbb{Z}, \tag{2.65}$$

where $P_\alpha^{ij}(u)$ are differential functions of u. The directional derivative (2.53) of such a tensor field T in the direction of the vector field $v = (v^1, ..., v^m)^{\mathrm{T}}$ has the following components

$$(T'(u)[v])_{ij} = \sum_\alpha \left(P_\alpha^{ij}\right)'(u)[v]D^\alpha, \tag{2.66}$$

where

$$\left(P_\alpha^{ij}\right)'(u)[v] = \sum_{k \geq 0} \sum_{l=1}^m \frac{\partial P_\alpha^{ij}(u)}{\partial u_{kx}^l} v_{kx}^l. \tag{2.67}$$

Example 2.23 One component tensor field.

$$T^{(1,1)}(u) \ni \Phi(u) = D^2 + u^2 + u_x D^{-1},$$

$$\Phi'(u)[v] = 2uv + v_x D^{-1}.$$

Example 2.24 The two component tensor field, $u = (u, w)^{\mathrm{T}}, v = (v_1, v_2)$.

$$T^{(2,0)}(u) \ni \theta(u, w) = \begin{pmatrix} u^2 D + w & w_x^2 D \\ u D^5 & D^3 \end{pmatrix},$$

$$\theta'(u, w)[v] = \begin{pmatrix} 2uv_1 D + v_2 & 2w_x v_{2_x} D \\ v_1 D^5 & 0 \end{pmatrix}.$$

The lattice case is similar to the field case but is now in matrix elements (2.65) of a second order tensor field $D \to E$, and P_α^{ij} are difference functions of u.

In the finite dimensional case, the directional derivative (2.53) of an arbitrary tensor field $T^{(r,s)}$ has the following form in the local coordinate frame

$$(T'(x)[v])^{j_1 \cdots j_r}{}_{k_1 \ldots k_s} = \sum_i \frac{\partial T^{j_1 \cdots j_r}{}_{k_1 \ldots k_s}(x)}{\partial x^i} v^i(x). \tag{2.68}$$

2.5 Differential k-Forms

A differential k-form $\omega_k(u)$ at a point $u \in M$ is a k-linear completely antisymmetric mapping

$$\omega_k(u) : \underbrace{T_u M \times \ldots \times T_u M}_{k} \to \mathbb{R}. \tag{2.69}$$

This means that we can identify a k-form with a completely antisymmetric tensor field $T^{(0,k)}$. Denote by $F^k(M)$ the set of all differential k-forms on M. We define zero-form as a scalar field on M and we identify one-forms with covector fields, respectively, so in these special cases $F^0(M) = \mathcal{T}^{(0,0)} M = \mathbb{R}$ and $F^1(M) = \mathcal{T}^{(0,1)} M = \mathcal{T}^* M$.

The *interior product* $i_v \omega$ of a k-form ω with a vector field v is a $(k-1)$-form defined by the equality

$$i_v \omega(u)(v_1, \ldots, v_{k-1}) = \omega(u)(v, v_1, \ldots, v_{k-1}). \tag{2.70}$$

In particular, the interior product of a two-form J with a vector field v yields a one-form γ

$$(i_v J)(v_1) = \langle i_v J, v_1 \rangle = J(v, v_2) = \langle J \circ v, v_1 \rangle = \langle \gamma, v_1 \rangle. \tag{2.71}$$

This means that $i_v J = J \circ v$. The interior product of a one-form γ with a vector field v yields a zero-form

$$i_v \gamma = \langle \gamma, v \rangle. \tag{2.72}$$

For a function $f(u)$ we define $i_v f = 0$. Moreover for an arbitrary k-form ω the following property holds: $i_v i_v \omega = 0$.

The interior product lowers the degree of a differential form. An operator which increases the degree of a differential form is an exterior differentiation. The *exterior derivative* d of a k-form ω is a $(k+1)$-form dω such that

$$\mathrm{d}\omega(u)(v_1, ..., v_{k+1}) = \sum_{i=1}^{k+1}(-1)^{i+1}\omega'(u)[v_i](v_1, ..., v_{i-1}, v_{i+1}, ..., v_{k+1}). \quad (2.73)$$

The exterior derivatives of a zero-form $F(u)$, a one-form $\gamma(u)$ and a two-form $J(u)$ are given by

$$\mathrm{d}F(u)(v) = F'(u)[v] = \langle \nabla F, v \rangle(u), \quad (2.74)$$

$$
\begin{aligned}
\mathrm{d}\gamma(u)(v_1, v_2) &= \gamma'(u)[v_1](v_2) - \gamma'(u)[v_2](v_1) \\
&= \langle \gamma'[v_1], v_2 \rangle(u) - \langle \gamma'[v_2], v_1 \rangle(u),
\end{aligned}
\quad (2.75)
$$

$$\mathrm{d}J(u)(v_1, v_2, v_3)$$

$$
\begin{aligned}
&= J'(u)[v_1](v_2, v_3) - J'(u)[v_2](v_1, v_3) + J'(u)[v_3](v_1, v_2) \\
&= J'(u)[v_1](v_2, v_3) + J'(u)[v_2](v_3, v_1) + J'(u)[v_3](v_1, v_2) \\
&= \langle J'[v_1] \circ v_2, v_3 \rangle(u) + \langle J'[v_2] \circ v_3, v_1 \rangle(u) + \langle J'[v_3] \circ v_1, v_2 \rangle(u).
\end{aligned}
\quad (2.76)
$$

Moreover, for an arbitrary k-form ω the following property holds: dd$\omega = 0$.

Definition 2.5 *A k-form ω with d$\omega = 0$ is called a closed k-form. A k-form ω is called an exact k-form if there exists $(k-1)$-form ζ such that $\omega = \mathrm{d}\zeta$.*

Since $\mathrm{d}(\mathrm{d}\zeta) = 0$ for all forms ζ, an exact form ω is always closed.

Lemma 2.1 *(Poincare).If M is a linear space, or more generally is of the star shape $\left(\bigwedge_{u \in M}\{\lambda u : 0 \le \lambda \le 1\} \subset M\right)$, each closed k-form is exact.*

Lemma 2.2 *A covariant vector field γ is exact one-form if*

$$\gamma' = \gamma'^\dagger. \quad (2.77)$$

Proof. If $\gamma = \mathrm{d}F$ then $\mathrm{d}\gamma = 0$ and hence

$$
\begin{aligned}
\mathrm{d}\gamma(v_1, v_2) &= \langle \gamma'[v_1], v_2 \rangle - \langle \gamma'[v_2], v_1 \rangle \\
&= \langle \gamma'[v_1], v_2 \rangle - \langle \gamma'^\dagger[v_1], v_2 \rangle \\
&= \langle (\gamma' - \gamma'^\dagger)[v_1], v_2 \rangle = 0.
\end{aligned}
\quad (2.78)
$$

From the arbitrariness of v_1, v_2 we get the condition (2.77) $\square$.

If M satisfies the condition of the Poincare Lemma, the following important problem arises. Given an exact k-form ω, how find a $(k-1)$-form ζ such that $d\zeta = \omega$. The general solution to this problem is connected with the theory of de Rham complexes and a suitable homotopy operator $h : h\omega = \zeta$ (see for example [159]), which goes beyond the scope of this book. Here, we only present a few useful results which are important for our applications.

Lemma 2.3 *Let M fulfil the condition of the Poincare Lemma.*

(i) For an exact one-form $\gamma(u)$

$$F(u) = h\gamma(u) = \int_0^1 \langle \gamma(\lambda u), u \rangle \, d\lambda \tag{2.79}$$

is a zero-form such that $dF(u) = \gamma(u)$.
(ii) For an exact two-form $J(u)$

$$\langle \gamma(u), v \rangle = \int_0^1 \langle J(\lambda u)\lambda u, v \rangle \, d\lambda, \qquad v(u) \in T_u M, \tag{2.80}$$

where $\gamma(u)$ is a one-form such that $d\gamma(u) = J(u)$.

Let us verify the case (i). For a scalar field (zero-form) $F(u)$ we have

$$\begin{aligned}
dF(u)(v) \quad &= F'(u)[v] = \int_0^1 \{\lambda \langle \gamma'(\lambda u)[v], u \rangle + \langle \gamma(\lambda u), v \rangle\} d\lambda \\
&= \int_0^1 \{\lambda \langle \gamma'^*(\lambda u)[u], v \rangle + \langle \gamma(\lambda u), v \rangle\} d\lambda \\
&= \int_0^1 \{\lambda \langle \gamma'(\lambda u)[u], v \rangle + \langle \gamma(\lambda u), v \rangle\} d\lambda \\
&= \int_0^1 \tfrac{d}{d\lambda} \langle \lambda\gamma(\lambda u), v \rangle \, d\lambda = \langle \gamma(u), v \rangle = \gamma(u)(v).
\end{aligned} \tag{2.81}$$

From the arbitrariness of v we have $dF(u) = \gamma(u)$. In our derivation we applied the result of Lemma 2.2 and the relation $\frac{d}{d\lambda}\gamma(\lambda u) = \gamma'(\lambda u)[u]$.

Example 2.25 Consider the one-form $\gamma(u) = u_{xx} + \tfrac{1}{2}au^2$. The $\gamma(u)$ is closed as $\gamma' = D^2 + au = \gamma'^*$. According to formula (2.79) we find

$$\begin{aligned}
F(u) \quad &= \int_0^1 \int_R \left(\lambda u_{xx} + \tfrac{1}{2}a\lambda^2 u^2\right) u dx d\lambda \\
&= \int_R \int_0^1 \left(\lambda u_{xx}u + \tfrac{1}{2}a\lambda^2 u^3\right) d\lambda dx = \int_R \left(\tfrac{1}{2}uu_{xx} + \tfrac{1}{6}au^3\right) dx.
\end{aligned}$$

Let us verify the result

$$\delta \int_R \left(\tfrac{1}{2}uu_{xx} + \tfrac{1}{6}au^3\right) dx = \tfrac{1}{2}u_{xx} + \tfrac{1}{2}au^2 + \tfrac{1}{2}u_{xx} = u_{xx} + \tfrac{1}{2}au^2.$$

Example 2.26 Consider the one-form $\gamma(u) = \frac{1}{2}u_x^2$. Then $J(u) = d\gamma(u) = \gamma' - \gamma'^\dagger = u_{xx} + 2u_x D$. Now we reconstruct $\gamma(u)$ from the formula (2.80)

$$
\begin{aligned}
\langle \gamma(u), v \rangle &= \int_0^1 \int_R (\lambda u_{xx} + 2\lambda u_x D)(\lambda u)v \, dx d\lambda \\
&= \int_0^1 \int_R (\lambda^2 u u_{xx} + 2\lambda^2 u_x^2)v \, dx d\lambda = \int_R \left(\tfrac{1}{3} u u_{xx} + \tfrac{2}{3} u_x^2 \right) v \, dx.
\end{aligned}
$$

So we have

$$
\gamma(u) = \frac{1}{3} u u_{xx} + \frac{2}{3} u_x^2 = \frac{1}{2} u_x^2 + \left(-\frac{1}{6} u_x^2 + \frac{1}{3} u_x^2 + \frac{1}{3} u u_{xx} \right) = \frac{1}{2} u_x^2 + dF(u),
$$

where $F(u) = \int_R \left(-\frac{1}{6} u u_x^2 \right) dx$, and hence

$$
J(u) = d\gamma(u) = d \left(\frac{1}{2} u_x^2 \right).
$$

For each $p \geq 0$ and $q \geq 0$ there exists an operation called an *exterior product*

$$
F^p(M) \times F^q(M) \to F^{p+q}(M),
$$

$$
(\omega, \eta) \to \omega \wedge \eta,
$$

$$(2.82)$$

defined by

$$
\omega \wedge \eta : u \to (\omega \wedge \eta)(u) = \omega(u) \wedge \eta(u) \in F^{p+q}(u), \qquad (2.83)
$$

where

$$
(\omega \wedge \eta)(u)(v_1, ..., v_p, v_{p+1}, ..., v_{p+q}) = \mathcal{A}\omega(u)(v_1, ..., v_p)\eta(u)(v_{p+1}, ..., v_{p+q})
$$
$$(2.84)$$

and $\mathcal{A}$ is the operator of antisymmetrization. The exterior product is anti-commutative, i.e. $\omega \wedge \eta = (-1)^{pq}\eta \wedge \omega$, associative, i.e. $(\omega \wedge \eta) \wedge \zeta = \omega \wedge (\eta \wedge \zeta)$ and bilinear, i.e. $\omega \wedge (\eta + \zeta) = \omega \wedge \eta + \omega \wedge \zeta$. Moreover, for an arbitrary q-form ω, the relations

$$
d(\omega \wedge \eta) = d\omega \wedge \eta + (-1)^q \omega \wedge d\eta,
$$

$$
i_v(\omega \wedge \eta) = i_v\omega \wedge \eta + (-1)^q \omega \wedge i_v\eta,
$$

$$(2.85)$$

hold. Note that given a collection of differential one-forms $\gamma_1, ..., \gamma_k$, we can form a differential k-form $\gamma_1 \wedge ... \wedge \gamma_k$ using the determinantal formula

$$
(\gamma_1 \wedge ... \wedge \gamma_k)(v_1, ..., v_k) = \det[\gamma_i(v_j)], \qquad 1 \leq i, j \leq k. \qquad (2.86)
$$

Endowed with the exterior product, $F(M) = \oplus_p F^p(M)$ is a covariant exterior algebra over $\mathbb{R}$.

Remark 2.1 The dual objects to differential forms on M are called *multivectors* and are defined as completely antisymmetric k-linear maps

$$V_k(u): \quad \underbrace{T_u^* M \times ... \times T_u^* M}_{k} \quad \to \mathbb{R}. \tag{2.87}$$

This means that we can identify k-vectors with completely antisymmetric tensor fields $T^{(k,0)}$. Uni-vectors are the same as vector fields. We can also define an exterior product $V \wedge W$ between multi-vectors in full analogy to the k-forms case and hence define a contravariant exterior algebra of multi-vector fields. Moreover, if V_k is a k-vector and V_l an l-vector, then there is a uniquely defined $(k + l - 1)$-vector $[V_k, V_l]$, called the Schouten bracket of V_k and V_l. It is a bilinear skew-symmetric $([V_k, V_l] = (-1)^{kl}[V_l, V_k])$ map identical to the ordinary Lie bracket in the case of vector fields (uni-vectors). Actually, let $v_1, ..., v_k, w_1, ..., w_l$ be vector fields over M. Then

$$[V_k, V_l] = [v_1 \wedge ... \wedge v_k, w_1 \wedge ... \wedge w_l]$$

$$= \sum (-1)^{i+j} v_1 \wedge ... \widehat{v}_i ... \wedge v_k \wedge [v_i, w_j] \wedge ... \widehat{w}_j ... \wedge w_l, \tag{2.88}$$

where $[.,.]$ is the Schouten bracket and $\widehat{v}$ stands for the omission of v. The Schouten bracket (2.88) is the unique extension of the Lie bracket of vector fields to the contravariant exterior algebra of multi-vector fields, making it a graded Lie algebra [119],[172],[184].

We end this section by giving standard information about the tensor representation in a particular basis when the finite dimensional case is considered. Let $(x^1, ..., x^n)$ be local coordinates and let us introduce the following notation (convention)

$$\langle \mathrm{d}f(x), v \rangle := v(f) = \sum_i v^i \frac{\partial f}{\partial x^i}. \tag{2.89}$$

This convention allow us to introduce the following basis of the tangent space $T_x M, x \in M$

$$\{e_j\}_{j=1}^n = \left\{ \frac{\partial}{\partial x^i} \right\}_{j=1}^n. \tag{2.90}$$

Hence, any tangent vector v at x can be written in the form

$$v(x) = \sum_i v^i(x) \frac{\partial}{\partial x^i} \tag{2.91}$$

and the application of v (2.91) to a function $f(x)$ is consistent with formula (2.89). The dual basis to $\left\{ \frac{\partial}{\partial x^i} \right\}_{j=1}^n$, i.e. the basis of the cotangent space $T_x^* M$, reads

$$\{\epsilon^i\}_{i=1}^n = \{\mathrm{d}x^i\}_{i=1}^n, \tag{2.92}$$

where $\mathrm{d}x^i$ is the exterior derivative (differential) of the coordinate x^i. Indeed, from (2.89) we have

$$\left\langle \mathrm{d}x^i, \frac{\partial}{\partial x^j} \right\rangle = \frac{\partial}{\partial x^j}(x^i) = \frac{\partial x^i}{\partial x^j} = \delta_{ij}. \tag{2.93}$$

Hence, for a vector (uni-vector) $v(x) = \sum_i v^i(x)\frac{\partial}{\partial x^i} \in T_x M$ and a covector (one-form) $\gamma(x) = \sum_i \gamma_i(x)\mathrm{d}x^i$ we find

$$\langle \gamma(x), v(x) \rangle = \sum_i \gamma_i(x)v^i(x). \tag{2.94}$$

Bases (2.90) and (2.92) give rise to the following representation of an arbitrary tensor $T^{r,s)}(x)$

$$T^{(r,s)}(x) = \sum_{k_1=1}^{n} \cdots \sum_{j_1=1}^{n} T^{j_1\ldots j_r}{}_{k_1\ldots k_s} \mathrm{d}x^{k_1} \otimes \ldots \otimes \mathrm{d}x^{k_s} \otimes \frac{\partial}{\partial x^{j_1}} \otimes \ldots \otimes \frac{\partial}{\partial x^{j_r}}. \tag{2.95}$$

In the special cases of the two-form $J(x)$ and the bi-vector $\theta(x)$ their representations have the following forms

$$J(x) = \sum_{i,j}^{n} J_{ij}(x)\mathrm{d}x^i \wedge \mathrm{d}x^j, \tag{2.96}$$

$$\theta(x) = \sum_{i,j}^{n} \theta^{ij}(x)\frac{\partial}{\partial x^i} \wedge \frac{\partial}{\partial x^j}, \tag{2.97}$$

where J_{ij} and θ^{ij} are $n \times n$ antisymmetric matrices.

2.6 Flows and Lie Transport

Let us introduce the notion of a *flow* on a manifold.

Definition 2.6 *A flow on a manifold M is defined as a $\mathbb{C}^\infty$-mapping*

$$\mathbb{R} \times M \to M : (t, u) \to \Psi(t, u) \tag{2.98}$$

such that for arbitrary $u \in M$ and all $t_1, t_2 \in \mathbb{R}$

$$\Psi(0, u) = u \quad \text{and} \quad \Psi(t_1, \Psi(t_2, u)) = \Psi(t_1 + t_2, u). \tag{2.99}$$

This implies that a flow represents a *one-parameter group* of diffeomorphisms on M and assigns to every $t \in \mathbb{R}$ some diffeomorphism ϕ_t

$$\phi_t : M \to M,$$
$$\bar{u} = \phi_t \cdot u = \phi_t(u) = \Psi(t, u), \quad u, \bar{u} \in M. \tag{2.100}$$

Notice that ϕ_t is a special case of transformation ϕ (2.30) where now $\overline{M} = M$ and $\phi_t^{-1} = \phi_{-t}$. The one parameter group is completely determined by its *infinitesimal generator*

$$K(u) = \frac{\mathrm{d}}{\mathrm{d}t}\Psi(t, u)_{|t=0}, \quad \text{or} \quad K = \frac{\mathrm{d}}{\mathrm{d}t}\phi_{t|t=0} \qquad (2.101)$$

which is a vector field on M. The t derivative of ϕ_t at arbitrary t is easily expressed by K:

$$\frac{\mathrm{d}}{\mathrm{d}t}\phi_t = K \circ \phi_t. \qquad (2.102)$$

Hence, ϕ_t is uniquely determined by the vector field $K(u)$. In other words, for any initial condition $u(0)$

$$u(t) = \phi_t \cdot u = \Psi(t, u(0)) \qquad (2.103)$$

is an integral curve of the dynamical system

$$u_t = K(u). \qquad (2.104)$$

We usually call t the *evolution parameter*.

Example 2.27 Finite dimensional case. Let us consider the one-parameter group of rotations in the plane $u = (x^1, x^2) = (x, y)$

$$\begin{pmatrix} x(t) \\ y(t) \end{pmatrix} = \phi_t \cdot \begin{pmatrix} x \\ y \end{pmatrix} = \begin{pmatrix} x\cos t - y\sin t \\ x\sin t + y\cos t \end{pmatrix} = \begin{pmatrix} \Psi_1(t, x, y) \\ \Psi_2(t, x, y) \end{pmatrix} = \Psi(t, x, y).$$

The infinitesimal generator of this flow has two components

$$K_1(x, y) = \frac{\mathrm{d}}{\mathrm{d}t}\Psi_{1|t=0} = -y, \quad K_2(x, y) = \frac{\mathrm{d}}{\mathrm{d}t}\Psi_{2|t=0} = x.$$

It is not difficult to verify that indeed $\Psi(t, x, y)$ is a flow of the dynamical system

$$u_t = \begin{pmatrix} x \\ y \end{pmatrix}_t = \begin{pmatrix} -y \\ x \end{pmatrix} = K(x, y) = K(u).$$

Example 2.28 Infinite dimensional case. We consider the one-parameter group of transformations in the form

$$u(t, x) = \phi_t \cdot u(x) = e^{2t}u(e^t x).$$

The infinitesimal generator reads $K(u) = 2x + xu_x$ and the reader can verify that $u(t, x)$ is a solution of the dynamical system

$$u_t = 2u + xu_x.$$

The *Lie transport* is a process of displacing a given geometric object along a flow. More precisely, for a given flow ϕ_t and an arbitrary tensor field $T^{(r,s)}$, the Lie transport of $T^{(r,s)}$ along ϕ_t is a push-forward $\phi_{t*}T^{(r,s)}$ and a pull-back $\phi_t^*T^{(r,s)}$. The fundamental problem to our further considerations is how to measure the rate of change of various tensor fields along a flow. The solution to the problem is the procedure of using the *Lie derivative*.

2.7 Lie Derivatives

As will be demonstrated in the next chapter, the Lie derivative plays a central role in the Hamiltonian theory of dynamical systems, so let us analyse this procedure in detail.

Definition 2.7 *Let ϕ_t be a flow, v its infinitesimal generator, $T^{(r,s)}$ an arbitrary tensor field and $\phi_{t*}T^{(r,s)}(\phi_t^*T^{(r,s)})$ its Lie transport. The Lie derivative of a tensor field along the flow ϕ_t is*

$$L_v T^{(r,s)}(u) = \frac{d}{dt}\left(\phi_t^* T^{(r,s)}\right)(u)_{|t=0} = \frac{d}{dt}\left(\phi_{-t*}T^{(r,s)}\right)(u)_{|t=0}. \quad (2.105)$$

Notice that the last equality follows from the fact that the Lie transports ϕ_t^* and ϕ_{t*} are interrelated in the following way

$$\phi_t^* = (\phi_{t*})^{-1} = \phi_{-t*}. \quad (2.106)$$

How do we understand this definition? Let us come back for a moment to Sect. 2.3 and its notation. Consider a flow ϕ_t and let for a particular t

$$\bar{u} = \phi_t(u) = u(t),$$
$$u = \phi_0(u) = u(0). \quad (2.107)$$

So $\bar{u}$ and u represent two different points on a curve. If t is infinitesimally small, then we have

$$\bar{u} = \phi_t(u) = u(0) + \frac{d}{dt}\left(\phi_t(u)\right)_{|t=0} \cdot t + O(t^2) \simeq u + t \cdot v(u), \quad (2.108)$$

where $v(u)$ is the infinitesimal generator of ϕ_t. On the other hand, according to the notation of Definition 2.3, formula (2.105) reads

$$
\begin{aligned}
L_v T^{(r,s)}(u) &= \frac{d}{dt}\left(\phi_{-t*}T^{(r,s)}\right)(u)_{|t=0} = -\frac{d}{dt}\left(\phi_{t*}T^{(r,s)}\right)(u)_{|t=0} \\
&= \lim_{t\to 0}\frac{T^{(r,s)}(u) - \overline{T}^{(r,s)}(u)}{t}.
\end{aligned} \quad (2.109)
$$

Now the definition (2.109) is clear. The recipe is as follows: push-forward a tensor $T^{(r,s)}(u)$ to the point $\bar{u}$ ($T^{(r,s)}(u) \to \overline{T}^{(r,s)}(\bar{u})$) according to a transformation law with respect to the transformation (2.108), pull-back its argument $\bar{u}$ to the point u ($\overline{T}^{(r,s)}(\bar{u}) \to \overline{T}^{(r,s)}(u)$) and compare the tensor before to the tensor after the transformation. Hence, the Lie derivative gives us about information how various tensor fields change locally along a given curve. From the definition it follows that the Lie derivative is a linear operation and that the Leibniz chain rule holds for any kind of tensor composition. Moreover the Lie derivative of a (r,s) type tensor is also an (r,s) type tensor.

Now we derive the explicit form of the Lie derivative for various tensor fields $T^{(r,s)}(u)$ applying the definition (2.109) and the transformation laws from Sect. 2.3, where now

$$\phi(u) = u + t \cdot v(u), \quad \phi'(u) = 1 + t \cdot v'(u), \quad \phi'^\dagger(u) = 1 + t \cdot v'^\dagger(u), \quad (2.110)$$

the relation

$$\overline{T}^{(r,s)}(\overline{u}) = \overline{T}^{(r,s)}(u) + t \cdot T'(u)[v(u)] + O(t^2) \quad (2.111)$$

and the fact that

$$\lim_{t \to 0} \overline{T}^{(r,s)}(\overline{u}) = \lim_{t \to 0} \overline{T}^{(r,s)}(u) = T^{(r,s)}(u). \quad (2.112)$$

Let us consider the cases most interesting with respect to our applications:

1. $T^{(0,0)}(u) = f(u)$:

$$f(u) = \overline{f}(\overline{u}) = \overline{f}(u + tv) = \overline{f}(u) + t \cdot \overline{f}'(u)[v(u)] + O(t^2),$$

$$f(u) - \overline{f}(u) = t \cdot \overline{f}'(u)[v(u)] + O(t^2),$$

$$L_v f(u) = f'(u)[v(u)] = \langle \nabla f, v \rangle (u). \quad (2.113)$$

2. $T^{(1,0)}(u) = K(u)$:

$$\overline{K}(\overline{u}) = \phi'(u)[K(u)] + t \cdot v'(u)[K(u)],$$

$$\overline{K}(\overline{u}) = \overline{K}(u) + t \cdot K'(u)[v(u)] + O(t^2),$$

$$K(u) - \overline{K}(u) = t \cdot K'(u)[v(u)] - t \cdot v'(u)[K(u)] + O(t^2),$$

$$L_v K(u) = K'(u)[v(u)] - v'(u)[K(u)] = [v, K](u). \quad (2.114)$$

3. $T^{(0,1)}(u) = \gamma(u)$:

$$\gamma(u) = \phi'^\dagger(u)\overline{\gamma}(\overline{u}) = \overline{\gamma}(\overline{u}) + t \cdot v'^\dagger(u)\overline{\gamma}(\overline{u}),$$

$$\overline{\gamma}(\overline{u}) = \overline{\gamma}(u) + t \cdot \overline{\gamma}'(u)[v(u)] + O(t^2),$$

$$\gamma(u) - \overline{\gamma}(u) = t \cdot \overline{\gamma}'(u)[v(u)] + t \cdot v'^\dagger(u)\overline{\gamma}(\overline{u}) + O(t^2),$$

$$L_v \gamma(u) = \gamma'(u)[v(u)] + v'^\dagger(u)\gamma(u). \quad (2.115)$$

4. $T^{(1,1)}(u) = \Phi(u)$:

$$\overline{\Phi}(\overline{u})\phi'(u) = \phi'(u)\Phi(u) \Rightarrow \overline{\Phi}(\overline{u})(1 + t \cdot v'(u)) = (1 + t \cdot v'(u))\Phi(u)$$

$$\Rightarrow \overline{\Phi}(\overline{u}) + t \cdot \overline{\Phi}(\overline{u})v'(u) = \Phi(u) + t \cdot v'(u)\Phi(u),$$

$$\overline{\Phi}(\overline{u}) = \overline{\Phi}(u) + t \cdot \overline{\Phi}'(u)[v(u)] + O(t^2),$$

$$L_v \Phi(u) = \Phi'(u)[v(u)] + \Phi(u)v'(u) - v'(u)\Phi(u). \quad (2.116)$$

Similarly we can find

5. $T^{(2,0)}(u) = \theta(u)$:

$$L_v\theta(u) = \theta'(u)[v(u)] - \theta(u)v'^\dagger(u) - v'(u)\theta(u). \tag{2.117}$$

6. $T^{(0,2)}(u) = J(u)$:

$$L_v J(u) = J'(u)[v(u)] + J(u)v'(u) + v'^\dagger(u)J(u). \tag{2.118}$$

In the u-free representation we have

$$L_v f = f' \circ v = \langle \nabla f, v \rangle,$$

$$L_v K = K' \circ v - v' \circ K = [v, K],$$

$$L_v \gamma = \gamma' \circ v + v'^\dagger \circ \gamma,$$

$$L_v \Phi = \Phi' \circ v + \Phi \circ v' - v' \circ \Phi, \tag{2.119}$$

$$L_v \theta = \theta' \circ v - \theta \circ v'^\dagger - v' \circ \theta,$$

$$L_v J = J' \circ v + J \circ v' + v'^\dagger \circ J,$$

$$L_v \Phi^\dagger = L_v \Psi = \Psi' \circ v + v'^\dagger \circ \Psi - \Psi \circ v'^\dagger,$$

where

$$\left(L_v T^{(r,s)}\right)(u) = L_v T^{(r,s)}(u). \tag{2.120}$$

Formulas (2.119) can be derived from the definition

$$L_v T^{(r,s)} = \frac{\mathrm{d}}{\mathrm{d}t}\left(\phi_t^* T^{(r,s)}\right)\Big|_{t=0} = \lim_{t\to 0}\frac{T^{(r,s)} - \overline{T}^{(r,s)}}{t}, \tag{2.121}$$

appropriate formulas for ϕ^* from Sect. 2.3 and the relation

$$\overline{T}^{(r,s)} \cdot \phi = T^{(r,s)} + t \cdot T^{(r,s)'} \circ v + O(t^2). \tag{2.122}$$

Notice that from the (2.105), (2.121) we immediately find that

$$\frac{\mathrm{d}}{\mathrm{d}t}\phi_t^* = L_v \phi_t^*, \tag{2.123}$$

which is clearly a linear differential equation, contrary to (2.102) as in general a flow ϕ_t is far from being linear.

Remark 2.2 The Lie derivative can be introduced in a more formal and coordinate free way. Actually an L_v–derivative of an arbitrary tensor field $T^{(r,s)}$ in the direction of a vector field v is called a linear map

$$L_v : \mathcal{T}^{(r,s)}M \to \mathcal{T}^{(r,s)}M,$$

such that

$$L_v f = \langle \nabla f, v \rangle, \qquad f \in \mathcal{T}^{(0,0)} M,$$

$$L_v K = [v, K], \qquad K \in \mathcal{T}^{(1,0)} M$$

and the Leibniz chain rule holds for any kind of tensor composition. Indeed, to find $L_v \gamma$, for example, we apply the L_v-derivative to a scalar field $\langle \gamma, K \rangle$ and use the chain rule

$$\langle L_v \gamma, K \rangle = L_v \langle \gamma, K \rangle - \langle \gamma, [v, K] \rangle.$$

To recover the explicit form (2.119) is an easy task:

$$\langle L_v \gamma, K \rangle = (\langle \gamma, K \rangle)' \circ v - \langle \gamma, K' \circ v \rangle + \langle \gamma, V' \circ K \rangle$$

$$= \langle \gamma' \circ v, K \rangle + \langle \gamma, K' \circ v \rangle - \langle \gamma, K' \circ v \rangle + \langle \gamma, v' \circ K \rangle$$

$$= \langle \gamma' \circ v, K \rangle + \langle \gamma, v' \circ K \rangle$$

$$= \langle \gamma' \circ v, K \rangle + \langle v'^\dagger \circ \gamma, K \rangle$$

$$= \langle (\gamma' \circ v + v'^\dagger \circ \gamma), K \rangle.$$

Although this method of introducing the Lie derivative is elegant, nevertheless it is devoid of a 'physical' interpretation.

In the finite dimensional case, with the canonical basis (2.90), (2.92), the components of the appropriate Lie derivatives are the following

$$L_v f(x) = \frac{\partial f}{\partial x^k} v^k,$$

$$(L_v K(x))^i = \frac{\partial K^i}{\partial x^l} v^l - \frac{\partial v^i}{\partial x^l} K^l, \quad L_v e_i = -\frac{\partial v^k}{\partial x^i} e_k,$$

$$(L_v \gamma(x))_i = \frac{\partial \gamma_i}{\partial x^k} v^k + \gamma_k \frac{\partial v^k}{\partial x^i}, \quad L_v \epsilon^i = \frac{\partial v^i}{\partial x^k} \epsilon^k,$$

$$(L_v \Phi(x))^i_j = \frac{\partial \Phi^i_j}{\partial x^k} v^k - \Phi^k_j \frac{\partial v^i}{\partial x^k} + \Phi^i_k \frac{\partial v^k}{\partial x^j},$$

$$(L_v \theta(x))^{ij} = \frac{\partial \theta^{ij}}{\partial x^k} v^k - \theta^{kj} \frac{\partial v^i}{\partial x^k} - \theta^{ik} \frac{\partial v^j}{\partial x^k},$$

$$(L_v J(x))_{ij} = \frac{\partial J_{ij}}{\partial x^k} v^k + J_{kj} \frac{\partial v^k}{\partial x^i} + J_{ik} \frac{\partial v^k}{\partial x^j}, \qquad (2.124)$$

where the summation convention over repeating indices is used. The general rule can be guessed now very easily but it is not difficult to derive it by applying the chain rule

$$L_v \left(T^{i_1 \ldots i_r}_{k_1 \ldots k_s} e_1 \otimes \ldots \otimes e_r \otimes \epsilon^1 \otimes \ldots \otimes \epsilon^s \right)$$

$$= L_v \left(T^{i_1 \ldots i_r}_{k_1 \ldots k_s} \right) e_1 \otimes \ldots \otimes \epsilon^s$$

$$+ T^{i_1 \ldots i_r}_{k_1 \ldots k_s} L_v(e_1) \otimes \ldots \otimes \epsilon^s + \ldots + T^{i_1 \ldots i_r}_{k_1 \ldots k_s} e_1 \otimes \ldots \otimes L_v(\epsilon^s).$$

$$(2.125)$$

Hence from the relations (2.124) we find

$$\left(L_v T^{(r,s)}(x) \right)^{j_1 \ldots j_r}_{l_1 \ldots l_s} = \frac{\partial T^{j_1 \ldots j_r}_{l_1 \ldots l_s}}{\partial x^k} v^k - \sum_{\alpha=1}^r T^{j_1 \ldots j_{\alpha-1} k j_{\alpha+1} \ldots j_r}_{l_1 \ldots l_s} \frac{\partial v^{j_\alpha}}{\partial x^k}$$

$$+ \sum_{\beta=1}^s T^{j_1 \ldots j_r}_{l_1 \ldots l_{\beta-1} k l_{\beta+1} \ldots l_s} \frac{\partial v^k}{\partial x^{l_\beta}}. \qquad (2.126)$$

We finish this section with various properties of the L_v–derivative which are useful in the further considerations.

Lemma 2.4 *For the Lie derivative L_v, the exterior derivative d, the interior product i_v and for arbitrary k-form ω, the following relations hold*

(i)
$$L_v \omega = (d \circ i_v + i_v \circ d)\omega, \qquad (2.127)$$

(ii)
$$d \circ L_v \omega = L_v \circ d\omega, \qquad (2.128)$$

(iii)
$$i_v \circ L_v \omega = L_v \circ i_v \omega. \qquad (2.129)$$

Proof.

(i)   We proceed by induction on k. For $k = 0$ we have $\omega = f$, $i_v f = 0$ and $L_v f = \langle df, v \rangle = i_v f$. Now assume that (i) holds for k. Then a $(k+1)$-form may be written as $\alpha \wedge \omega$, where ω is a k-form and α is a one-form. Hence

$$i_v \circ d(\alpha \wedge \omega) + d \circ i_v(\alpha \wedge \omega)$$

$$= i_v \circ (d\alpha \wedge \omega) - i_v \circ (\alpha \wedge d\omega) + d \circ (i_v \alpha \wedge \omega) - d \circ (\alpha \wedge i_v \omega)$$

$$= i_v \circ d\alpha \wedge \omega + d\alpha \wedge i_v \omega - i_v \alpha \wedge d\omega + \alpha \wedge i_v \circ d\omega + d \circ i_v \alpha \wedge \omega$$

$$+ i_v \alpha \wedge d\omega - d\alpha \wedge i_v \omega + \alpha \wedge d \circ i_v \omega = \alpha \wedge L_v \omega + L_v \alpha \wedge \omega$$

$$= L_v(\alpha \wedge \omega)$$

by our inductive assumption and formulas (2.85). $\square$

(ii) Applying the result (i) and the fact that $d \circ d = 0$ we find

$$L_v \circ d = d \circ i_v \circ d + i_v \circ d \circ d = d \circ i_v \circ d = d \circ (L_v - d \circ i_v)$$

$$= d \circ L_v - d \circ d \circ i_v = d \circ L_v. \square$$

(iii) Applying the result (i) and the fact that $i_v \circ i_v = 0$ we find

$$L_v \circ i_v = (d \circ i_v + i_v \circ d) \circ i_v = i_v \circ d \circ i_v$$

$$= i_v \circ (L_v - i_v \circ d) = i_v \circ L_v. \square$$

3. The Theory of Hamiltonian and Bi-Hamiltonian Systems

This chapter is devoted to the standard results of the algebraic theory of bi-Hamiltonian systems, developed during the last two decades. The crucial concepts like the one of the recursion operator introduced by Olver [158], the bi-Hamiltonian property formulated by Magri [123] as well as the hereditary property introduced by Fuchssteiner [82] gave a background for the modern algebraic theory of integrable systems.

It is well known that for a given Hamiltonian dynamical system, the crucial objects are constants of the motion. For finite systems, for example, the existence of a sufficient number of such objects leads directly to integrability of the dynamics considered. Hence, it is of great importance to have at our disposal some algebraic method for recursive generation of a hierarchy of constants of motion, if such exists. On the other hand, it is known that with each constant of motion one can relate an appropriate symmetry generator (via Noether's theorem), i.e. an infinitesimal generator of a symmetry group of a system with the property that it transforms solutions of the system to other solutions. So, alternatively, one can search for a recursive algorithm for the generation of the hierarchy of Hamiltonian symmetries. Thus, the recognition in 1977 by P. Olver of the recursion operator, which realized to the algorithmic scheme for certain soliton field systems was the first milestone of the algebraic theory of integrable systems. The second milestone of the theory was the observation made by F. Magri in 1978, that the existence of a recursion operator is closely related to the existence of two different Hamiltonian formulations for a given dynamics, i.e. the existence of the so called implectic-symplectic factorization of the recursion operator. Next, in 1979, B. Fuchssteiner introduced the concept of the hereditary property, which guarantees the commutativity of Hamiltonian symmetries and involutivity of related constants of motion. Then, together with A.S. Fokas [70], they related this property to the notion of compatibility [92] between two Poisson operators of the Magri scheme. Finally, in 1981, Fokas and Fuchssteiner [71] introduced the concept of master symmetries which revealed the importance of time-dependent symmetries as well as non-Hamiltonian symmetries in the theory of integrable systems. On the basis of these fundamental results, the algebraic theory of tensor invariants of nonlinear dynamical systems started to develop. What is interesting is that it was first developed for field systems (soliton systems); then it was extended to lattice systems. Finally, in the last decade the theory was applied to finite dimensional systems, i.e. it has been implemented to classical mechanics.

Although historically, many basic theorems about the bi-Hamiltonian formulation, hereditariness, non-Hamiltonian symmetries, etc., presented in this Chapter, were formulated for the first time by Gelfand and Dorfman [93], Fokas and Fuchssteiner [70] and Fokas [73], nevertheless, here we present either the later proofs or new ones, which are consistent with our Lie-derivative formalism.

3.1 Lie Algebras

In Definition 2.1 we introduced the notion of a Lie algebra which, together with the Lie derivative, is crucial for the Hamiltonian theory. Vector fields constitute a Lie algebra in a very natural way.

Theorem 3.1 *Vector fields on M constitute a Lie algebra with respect to the Lie bracket (commutator) defined as follows*

$$[v, w](u) = (L_v w)(u) = [v(u), w(u)] = w'(u)[v(u)] - v'(u)[w(u)], \qquad (3.1)$$

where $v, w \in T^{(1,0)} M$.

Proof. Antisymmetry follows from definition (3.1), and the Jacobi identity is a consequence of the fact that the second directional derivative $v''[a, b] = v''[b, a]$ is a symmetric bilinear form. $\square$

Remark 3.1 Lie derivatives constitute a vector space. The commutator

$$[L_{v_1}, L_{v_2}] = L_{v_1} \circ L_{v_2} - L_{v_1} \circ L_{v_1}$$

endows this vector space in a natural way with a Lie algebra structure. The map $v \to L_v$ is a Lie algebra isomorphism from the Lie algebra of vector fields onto the Lie derivatives

$$[L_{v_1}, L_{v_2}] = L_{[v_1, v_2]}. \qquad (3.2)$$

Lemma 3.1 *For the Lie derivative L_{v_1}, the inner product i_{v_2} and an arbitrary k-form ω, the following relation holds*

$$(L_{v_1} \circ i_{v_2} - i_{v_2} \circ L_{v_1}) \omega = i_{[v_1, v_2]} \omega. \qquad (3.3)$$

Proof. Applying relation (2.127) to relation (3.2) the reader can verify the validity of relation (3.3).

For covector fields and scalar fields there is no natural Lie bracket; nevertheless we can introduce it via appropriate θ or J tensor fields. First let us concentrate on scalar fields. Having applied the exterior derivative d to a scalar field F we get $dF = \nabla F \, (\delta F) = F'$. Hence, for convenience we will use hereafter the symbol ∇ to stand for mapping scalar fields onto covector fields for all cases, that is finite, lattice and field, respectively. Now, as the total derivative D (difference Δ) lies in the kernel of ∇ (2.61), let us denote the quotient space $\mathcal{T}^{(0,0)}/\partial\mathcal{T}^{(0,0)}(\mathcal{T}^{(0,0)}/D\mathcal{T}^{(0,0)})$ by $\mathcal{F}$, which is a set of equivalence classes of differential (difference) functions with respect to the total derivative D (difference Δ). Now for a given $\theta \in \mathcal{T}^{(2,0)}$ we define a bracket $\mathcal{F} \times \mathcal{F} \to \mathcal{F}$ among scalar fields by

$$\{F_1, F_2\}_\theta(u) = \langle \nabla F_2, \theta \circ \nabla F_1 \rangle (u) = L_{\theta \circ \nabla F_1} F_2(u). \qquad F_1, F_2 \in \mathcal{F}. \quad (3.4)$$

An important question arises: when is the bracket (3.4) a Lie bracket?

Theorem 3.2 *The bracket* $\{.,.\}_\theta$ *defines a Lie algebra on* $\mathcal{F}$ *if*

(i) θ *is a bi-vector, i.e. is skew-symmetric with respect to the duality map:* $\theta = -\theta^\dagger$.
(ii) for arbitrary covector fields α, β, γ *the following identity holds*

$$\langle \alpha, \theta'[\theta \circ \beta] \circ \gamma \rangle + \langle \gamma, \theta'[\theta \circ \alpha] \circ \beta \rangle + \langle \beta, \theta'[\theta \circ \gamma] \circ \alpha \rangle = 0. \quad (3.5)$$

Proof. The skew-symmetry of θ guarantees that the bracket (3.4) is antisymmetric. Computation of the double bracket gives

$$\{F_1, \{F_2, F_3\}_\theta\}_\theta$$

$$= \langle \nabla \{F_2, F_3\}_\theta, \theta \circ \nabla F_1 \rangle = \langle \nabla \langle \nabla F_3, \theta \circ \nabla F_2 \rangle, \theta \circ \nabla F_1 \rangle$$

$$= F_3''[\theta \circ \nabla F_2, \theta \circ \nabla F_1] + \langle \nabla F_3, \theta'[\theta \circ \nabla F_1] \circ \nabla F_2 \rangle - F_2''[\theta \circ \nabla F_3, \theta \circ \nabla F_1].$$
$$(3.6)$$

Since second derivatives are symmetric with respect to their entries then all of them cancel when we take $\{F_1, \{F_2, F_3\}_\theta\}_\theta +$ cyclic permutations and the condition (3.6) guarantees that the bracket (3.4) satisfies the Jacobi identity.
□

An operator $\theta(u) : T_u^* M \to T_u M$ of the properties described in Theorem 3.2 is called a *Poisson operator* or an *implectic operator*, the corresponding bracket (3.4) is called a *Poisson bracket* and a manifold M equipped with a Lie algebra $\mathcal{F}_\theta$ is called a *Poisson manifold (implectic manifold)*.

Lemma 3.2 [148] *For an implectic operator* θ *the following identity holds*

$$L_{\theta \circ \gamma} \theta = -\theta \circ d\gamma \circ \theta, \qquad \gamma \in \mathcal{T}^{(0,1)}. \quad (3.7)$$

Proof. For arbitrary $\alpha, \beta \in \mathcal{T}^{(0,1)}$ we have

$$\langle \alpha, ((L_{\theta \circ \gamma}\theta) + (\theta \circ d\gamma \circ \theta)) \beta \rangle$$

$$= \langle \alpha, \theta'[\theta \circ \gamma] \circ \beta \rangle - \langle \alpha, \theta \circ (\theta \circ \gamma)'^{\dagger} \circ \beta \rangle - \langle \alpha, (\theta \circ \gamma)'[\theta \circ \beta] \rangle$$

$$\quad + \langle \alpha, (\theta \circ \gamma' \circ \theta) \circ \beta \rangle - \langle \alpha, (\theta \circ \gamma'^{\dagger} \circ \theta) \circ \beta \rangle$$

$$= \langle \alpha, \theta'[\theta \circ \gamma] \circ \beta \rangle + \langle \alpha, ((\theta \circ \gamma)' \circ \theta)^{\dagger} \circ \beta \rangle - \langle \alpha, (\theta \circ \gamma)' \circ \theta \circ \beta \rangle$$

$$\quad + \langle \alpha, \theta \circ \gamma'[\theta \circ \beta] \rangle + \langle \alpha, (\gamma' \circ \theta)^{\dagger} \circ \theta \circ \beta \rangle \tag{3.8}$$

$$= \langle \alpha, \theta'[\theta \circ \gamma] \circ \beta \rangle + \langle \beta, (\theta \circ \gamma)'[\theta \circ \alpha] \rangle - \langle \alpha, (\theta \circ \gamma)'[\theta \circ \beta] \rangle$$

$$\quad + \langle \alpha, \theta \circ \gamma'[\theta \circ \beta] \rangle - \langle \beta, \theta \circ \gamma'[\theta \circ \alpha] \rangle$$

$$= \langle \alpha, \theta'[\theta \circ \gamma] \circ \beta \rangle + \langle \beta, \theta'[\theta \circ \alpha] \circ \gamma \rangle - \langle \alpha, \theta'[\theta \circ \beta] \circ \gamma \rangle$$

$$= \langle \alpha, \theta'[\theta \circ \gamma] \circ \beta \rangle + \langle \beta, \theta'[\theta \circ \alpha] \circ \gamma \rangle + \langle \gamma, \theta'[\theta \circ \beta] \circ \alpha \rangle$$

$$= 0. \quad \square$$

The bracket (3.4) is not the unique bracket which can be defined on $\mathcal{F}$. For a given $J \in \mathcal{T}^{(0,2)}$ and for an arbitrary $F_1, F_2 \in \mathcal{F}$, let the vector fields v_1 and v_2 be such that $J \circ v_1 = \nabla F_1$ and $J \circ v_2 = \nabla F_2$. Then we define a bracket $\mathcal{F} \times \mathcal{F} \to \mathcal{F}$ among scalar fields by

$$\{F_1, F_2\}^J(u) = \langle J \circ v_2, v_1 \rangle (u) = L_{v_1} F_2(u). \tag{3.9}$$

Theorem 3.3 *The bracket* $\{.,.\}^J$ *defines a Lie algebra on* $\mathcal{F}$ *if*

(i) J *is a two-form, i.e. is skew-symmetric with respect to the duality map:* $J = -J^{\dagger}$.

(ii) J *is a closed two-form, i.e. for arbitrary vector fields* a, b, c *the following identity holds*

$$\langle J'[a] \circ b, c \rangle + \langle J'[b] \circ c, a \rangle + \langle J'[c] \circ a, b \rangle = 0. \tag{3.10}$$

Proof. The skew-symmetry of J guarantees that the bracket (3.9) is antisymmetric. Calculation of the double bracket gives

$$\{F_1, \{F_2, F_3\}^J\}^J$$

$$= \{F_1, \langle J \circ v_3, v_2 \rangle\}^J$$

$$= \langle \nabla \langle J \circ v_3, v_2 \rangle, v_1 \rangle \tag{3.11}$$

$$= \langle J'[v_1] \circ v_3, v_2 \rangle + \langle J \circ v_3'[v_1], v_2 \rangle - \langle J \circ v_2'[v_1], v_3 \rangle.$$

Taking $\{F_1,\{F_2,F_3\}^J\}^J$ + c.p., the second and third terms of all double brackets cancel due to the property $\langle J \circ v_a'[v_b], v_c\rangle = \langle J \circ v_a'[v_c], v_b\rangle$ which again follows from the fact that a second derivative is symmetric with respect to its entries. Hence, the condition (3.10) is sufficient for the bracket (3.9) to fulfil the Jacobi identity. $\square$

An operator $J(u) : T_u M \to T_u^* M$ of the properties described in Theorem 3.3 is called a *symplectic operator* and a manifold M equipped with a Lie algebra $\mathcal{F}^J$ is called a *symplectic manifold*. Of course in a particular case, when θ is invertible, then $J = \theta^{-1}$, a Poisson manifold becomes a symplectic manifold and both brackets (3.4) and (3.9) are equivalent.

Lemma 3.3 *For a symplectic operator J the following identity holds*

$$L_v J = \mathrm{d}(J \circ v) \qquad v \in \mathcal{T}^{(1,0)}. \tag{3.12}$$

Proof. Applying relation (2.128) from Lemma 2.4 we immediately have

$$L_v J = (\mathrm{d}\circ i_v + i_v \circ \mathrm{d})\circ J = \mathrm{d}(i_v \circ J) + i_v \circ \mathrm{d}J = \mathrm{d}(i_v \circ J) = \mathrm{d}(J\circ v).\square \tag{3.13}$$

Theorem 3.4 *The map $\theta \circ \nabla : \mathcal{F}_\theta \to \mathcal{L}$ is a Lie algebra homomorphism of the scalar field Lie algebra $\mathcal{F}_\theta$ into the vector field Lie algebra $\mathcal{L}$*

$$\theta \circ \nabla\{F_1, F_2\}_\theta = [\theta \circ \nabla F_1, \theta \circ \nabla F_2]. \tag{3.14}$$

Proof. Notice that from Lemma 3.2 it follows that for closed $\gamma : L_{\theta\circ\gamma}(\theta) = 0$. Then we have

$$\theta \circ \nabla\{F_1, F_2\}_\theta \;=\; \theta \circ \nabla L_{\theta\circ\nabla F_1} F_2 = \theta \circ L_{\theta\circ\nabla F_1} \nabla F_2$$

$$= \theta \circ L_{\theta\circ\nabla F_1} \nabla F_2 + L_{\theta\circ\nabla F_1}(\theta) \circ \nabla F_2$$

$$= L_{\theta\circ\nabla F_1}(\theta \circ \nabla F_2) = [\theta \circ \nabla F_1, \theta \circ \nabla F_2].\square$$

Now we can pass to covector fields. Let θ be an implectic operator. We define a bracket $\mathcal{T}^{(0,1)} \times \mathcal{T}^{(0,1)} \to \mathcal{T}^{(0,1)}$ in the following way [84]

$$\{\gamma_1, \gamma_2\}_\theta(u) = L_{\theta\circ\gamma_1}\gamma_2(u) - i_{\theta\circ\gamma_2}\mathrm{d}\gamma_1(u), \qquad \gamma_1, \gamma_2 \in \mathcal{T}^{(0,1)} \tag{3.15}$$

and denote the $\mathcal{T}^{(0,1)}$ equipped with the bracket (3.15) by $\mathcal{L}_\theta^*$.

Theorem 3.5 *θ is a homomorphism from $\mathcal{L}_\theta^*$ into the vector field Lie algebra $\mathcal{L}$.*

Proof. Applying the property (3.7) of an implectic operator we find

$$\theta \circ \{\gamma_1, \gamma_2\}_\theta = \theta \circ (L_{\theta \circ \gamma_1} \gamma_2 - i_{\theta \circ \gamma_2} \circ d\gamma_1) = \theta \circ L_{\theta \circ \gamma_1} \gamma_2 - \theta \circ d\gamma_1 \circ \theta \circ \gamma_2$$

$$= \theta \circ L_{\theta \circ \gamma_1} \gamma_2 + (L_{\theta \circ \gamma_1} \theta) \gamma_2 = L_{\theta \circ \gamma_1} (\theta \circ \gamma_2)$$

$$= [\theta \circ \gamma_1, \theta \circ \gamma_2]. \qquad \square \tag{3.16}$$

Theorem 3.6 $\mathcal{L}_\theta^*$ *is the covector field Lie algebra.*

Proof. When θ is invertible then the proof is immediate as the skew-symmetry of θ guarantees the antisymmetry of the bracket (3.15) and θ^{-1} lifts the Jacobi identity from $\mathcal{L}$ to $\mathcal{L}_\theta^*$. A bit more complex is the case when θ is not invertible, i.e. when $\ker \theta \neq 0$, because then, from Theorem 3.5 we only know that the Jacobi identity in $\mathcal{L}_\theta^*$ is satisfied up to an element of the $\ker \theta$. So, firstly notice that $\ker \theta$ is an ideal in $\mathcal{L}_\theta^*$. Indeed, if $\bar\gamma \in \ker \theta$ then for an arbitrary γ

$$\begin{aligned} 0 &= L_{\theta \circ \gamma}(\theta \circ \bar\gamma) = (L_{\theta \circ \gamma}\theta) \circ \bar\gamma + \theta \circ L_{\theta \circ \gamma}\bar\gamma \\ &= -(\theta \circ d\gamma \circ \theta) \circ \bar\gamma + \theta \circ L_{\theta \circ \gamma}\bar\gamma \\ &= \theta \circ L_{\theta \circ \gamma}\bar\gamma \quad \Rightarrow \quad L_{\theta \circ \gamma}\bar\gamma \in \ker \theta. \end{aligned} \tag{3.17}$$

On the other hand

$$\{\gamma, \bar\gamma\}_\theta = L_{\theta \circ \gamma}\bar\gamma \Rightarrow \{\gamma, \bar\gamma\}_\theta \in \ker \theta. \tag{3.18}$$

Secondly, the following lemma holds.

Lemma 3.4 *Let* γ_1, γ_2 *and* γ_3 *be such that at least one of them belongs to* $\ker \theta$. *Then*

$$\{\gamma_1, \{\gamma_2, \gamma_3\}_\theta\}_\theta + \text{cyclic permutations} = 0. \tag{3.19}$$

Proof. Let $\gamma_1 = \bar\gamma \in \ker \theta$. Then

$$\begin{aligned} \{\bar\gamma, \{\gamma_2, \gamma_3\}\} &= \{\bar\gamma, L_{\theta \circ \gamma_2}\gamma_3\} - \{\bar\gamma, i_{\theta \circ \gamma_3} \circ d\gamma_2\} \\ &= \{\bar\gamma, L_{\theta \circ \gamma_2}\gamma_3\} - \{\bar\gamma, (d\gamma_2) \circ \theta \circ \gamma_3\} \\ &= -i_{\theta \circ L_{\theta \circ \gamma_2}\gamma_3}d\bar\gamma + i_{\theta \circ d\gamma_2 \circ \theta \circ \gamma_3}d\bar\gamma \\ &= -d\bar\gamma \circ \theta \circ L_{\theta \circ \gamma_2}\gamma_3 + d\bar\gamma \circ \theta \circ d\gamma_2 \cdot \theta \cdot \gamma_3 \\ &= -d\bar\gamma \circ \theta \circ L_{\theta \circ \gamma_2}\gamma_3 - d\bar\gamma(L_{\theta \circ \gamma_2}\theta) \circ \gamma_3 \\ &= -d\bar\gamma \circ L_{\theta \circ \gamma_2}(\theta \circ \gamma_3) = d\bar\gamma \circ L_{\theta \circ \gamma_3}(\theta \circ \gamma_2), \end{aligned}$$

$$\{\gamma_3, \{\overline{\gamma}, \gamma_2\}\} = \{\gamma_3, -i_{\theta\circ\gamma_2}\circ d\overline{\gamma}\} = -\{\gamma_3, (d\overline{\gamma})\circ\theta\circ\gamma_2\}$$

$$= -L_{\theta\circ\gamma_3}(d\overline{\gamma}\circ\theta\circ\gamma_2) + i_{\theta\circ(d\overline{\gamma})\circ\theta\circ\gamma_2}d\gamma_3$$

$$= -L_{\theta\circ\gamma_3}(d\overline{\gamma}\circ\theta\circ\gamma_2) - d\gamma_3\circ\theta\circ d\overline{\gamma}\circ\theta\circ\gamma_2$$

$$= -L_{\theta\circ\gamma_3}(d\overline{\gamma}\circ\theta\circ\gamma_2) - d\gamma_3\circ(L_{\theta\circ\overline{\gamma}}\theta)\circ\gamma_2$$

$$= -L_{\theta\circ\gamma_3}(d\overline{\gamma}\circ\theta\circ\gamma_2),$$

$$\{\gamma_2, \{\gamma_3, \overline{\gamma}\}\} = \{\gamma_2, L_{\theta\circ\gamma_3}\overline{\gamma}\} = L_{\theta\circ\gamma_2}L_{\theta\circ\gamma_3}\overline{\gamma} - i_{\theta\circ L_{\theta\circ\gamma_3}\overline{\gamma}}\circ d\gamma_2$$

$$= (d\circ i_{\theta\circ\gamma_2} + i_{\theta\circ\gamma_2}\circ d)\circ L_{\theta\circ\gamma_3}\overline{\gamma} - d\gamma_2\circ\theta\circ L_{\theta\circ\gamma_3}\gamma_3$$

$$= d(i_{\theta\circ\gamma_2}(L_{\theta\circ\gamma_3}\overline{\gamma})) + i_{\theta\circ\gamma_2}(L_{\theta\circ\gamma_3}d\overline{\gamma}) - d\gamma_2\circ\theta\circ L_{\theta\circ\gamma_3}\overline{\gamma}$$

$$= d\langle L_{\theta\circ\gamma_3}\overline{\gamma}, \theta\circ\gamma_2\rangle + (L_{\theta\circ\gamma_3}d\overline{\gamma})\circ\theta\circ\gamma_2 - d\gamma_2\circ\theta\circ L_{\theta\circ\gamma_3}\overline{\gamma}$$

$$= -d\langle\gamma_2, \theta\circ(L_{\theta\circ\gamma_3}\overline{\gamma})\rangle + (L_{\theta\circ\gamma_3}d\overline{\gamma})\circ\theta\circ\gamma_2$$

$$= (L_{\theta\circ\gamma_3}d\overline{\gamma})\circ\theta\circ\gamma_2,$$

where we have used relations (2.127), (2.128) and (3.7) respectively. Hence the sum of double brackets gives zero. □

Lemma 3.4 and the fact that $\ker\theta$ is an ideal in $\mathcal{L}_\theta^*$ are sufficient conditions for preserving the Jacobi identity by the bracket (3.15) and hence $\mathcal{L}_\theta^*$ is a Lie algebra. □

Let J be a symplectic operator and for arbitrary $\gamma_1, \gamma_2 \in \mathcal{T}^{(0,1)}$ the vector fields v_1 and v_2 are such that $\gamma_1 = Jv_1$ and $\gamma_2 = Jv_2$, then we can define a bracket $\mathcal{T}^{(0,1)} \times \mathcal{T}^{(0,1)} \to \mathcal{T}^{(0,1)}$ in the following way

$$\{\gamma_1, \gamma_2\}^J(u) = L_{v_1}\gamma_2(u) - i_{v_2}d\gamma_1(u) \tag{3.20}$$

and denote the $\mathcal{T}^{(0,1)}$ equipped with the bracket (3.18) by $\mathcal{L}^{*J}$.

Theorem 3.7

(i) J is a homomorphism from the vector field Lie algebra $\mathcal{L}$ into $\mathcal{L}^{*J}$.
(ii) $\mathcal{L}^{*J}$ is a covector field Lie algebra.

Proof.

(i) $\{Jv_1, Jv_2\}^J = \{\gamma_1, \gamma_2\}^J = L_{v_1}\gamma_2 - i_{v_2}d\gamma_1$

$\qquad = L_{v_1}(J \circ v_2) - i_{v_2} \circ d(J \circ v_1)$

$\qquad = (L_{v_1}J) \circ v_2 + J \circ L_{v_1}v_2 - d(J \circ v_1) \circ v_2$

$\qquad = (L_{v_1}J) \circ v_2 + J \circ L_{v_1}v_2 - (L_{v_1}J) \circ v_2 = J \circ L_{v_1}v_2$
(3.21)

$\qquad = J[v_1, v_2].\square$

(ii) Antisymmetry of the bracket (3.18) follows from the skew-symmetry of J and the Jacobi identity is satisfied as, according to (i), the operator J transports this property from the $\mathcal{L}$ algebra. $\square$

In the special case when J is invertible, the bracket (3.18) is equivalent to the bracket (3.15) as $J^{-1} = \theta$. Generally, when $\ker J \neq 0$ and we postulate that $\{0, \gamma\}^J = 0$ for arbitrary γ, $\ker J$ must be an ideal in $\mathcal{L}$.

Let $^{\mathrm{cl}}\mathcal{T}^{(0,1)}$ be a subspace of closed covector fields, i.e. $d\gamma = 0$ for $\gamma \in {}^{\mathrm{cl}}\mathcal{T}^{(0,1)}$, and let M fulfil the condition for the Poincare Lemma 2.1. Then the following theorem holds.

Theorem 3.8

(i) $^{\mathrm{cl}}\mathcal{T}^{(0,1)}$ is a Lie subalgebra with respect to the brackets $\{.,.\}_\theta$ and $\{.,.\}^J$.
(ii) Let us denote the subalgebras from (i) by $^{\mathrm{cl}}\mathcal{L}^*_\theta$ and $^{\mathrm{cl}}\mathcal{L}^{*J}$, respectively. Then ∇ is an isomorphism between $\mathcal{F}_\theta$ and $^{\mathrm{cl}}\mathcal{L}^*_\theta$ as well as between $\mathcal{F}_\theta$ and $^{\mathrm{cl}}\mathcal{L}^{*J}$.

Proof.

(i) For $\gamma_1, \gamma_2 \in {}^{\mathrm{cl}}\mathcal{L}^*_\theta$ we have $\{\gamma_1, \gamma_2\}_\theta = L_{\theta \circ \gamma_1}\gamma_2$. On the other hand

$$dL_{\theta \circ \gamma_1}\gamma_2 = L_{\theta \circ \gamma_1}d\gamma_2 = 0 \Rightarrow \{\gamma_1, \gamma_2\}_\theta \in {}^{\mathrm{cl}}\mathcal{L}^*_\theta. \qquad (3.22)$$

For $\gamma_1, \gamma_2 \in {}^{\mathrm{cl}}\mathcal{L}^{*J}$ we have $\{\gamma_1, \gamma_2\}^J = L_{v_1}\gamma_2$, where $Jv_1 = \gamma_1$. Moreover

$$dL_{v_1}\gamma_2 = L_{v_1}d\gamma_2 = 0 \Rightarrow \{\gamma_1, \gamma_2\}^J \in {}^{\mathrm{cl}}\mathcal{L}^{*J}.\square \qquad (3.23)$$

(ii) ∇ is a homomorphism as

$$\nabla\{F_1, F_2\}_\theta = \nabla L_{\theta \circ \nabla F_1}F_2 = L_{\theta \circ \nabla F_1}\nabla F_2 = \{\nabla F_1, \nabla F_2\}_\theta,$$

$$\nabla\{F_1, F_2\}^J = \nabla L_{v_1}F_2 = L_{v_1}\nabla F_2 = \{\nabla F_1, \nabla F_2\}^J. \qquad (3.24)$$

On the other hand ∇ is a bijection with inverse equal to the homotopy operator h (see Lemma 2.3(i)). $\square$

3.2 Hamiltonian and Bi-Hamiltonian Vector Fields

We are now prepared to introduce the concept of Hamiltonian vector fields.

Definition 3.1

(i) *An element $K \in \mathcal{L}$ is called a Hamiltonian vector field with respect to the Poisson (implectic) operator θ if there exists an element $H \in \mathcal{F}$ such that*

$$K = \theta \circ \nabla H. \tag{3.25}$$

The scalar field H is called a Hamiltonian function (functional).

(ii) *An element $K \in \mathcal{L}$ is called an inverse Hamiltonian vector field with respect to the symplectic operator J if there exists an element $G \in \mathcal{F}$ such that*

$$J \circ K = \nabla G. \tag{3.26}$$

Definition 3.2

(i) *An element $K \in \mathcal{L}$ is called a bi-Hamiltonian with respect to the Poisson operators θ_0 and θ_1 if there exist elements $H, G \in \mathcal{F}$ such that*

$$K = \theta_0 \circ \nabla H = \theta_1 \circ \nabla G. \tag{3.27}$$

(ii) *An element $K \in \mathcal{L}$ is called a bi-Hamiltonian with respect to the Poisson operator θ and the symplectic operator J (such that $J^{-1} \neq \theta$) if there exist elements $H, G \in \mathcal{F}$ such that*

$$K = \theta \circ \nabla H \qquad \text{and} \qquad J \circ K = \nabla G. \tag{3.28}$$

Notice that if θ_0 is invertible $\theta_0^{-1} = J$, then case (i) is equivalent to case (ii).

The dynamical system (2.1) with $K(u)$ being an appropriate Hamiltonian vector field is called a *Hamiltonian dynamical system*.

Example 3.1 The Korteweg-de Vries (KdV) equation [123]

$$u_t = u_{3x} + auu_x = K(u) = D \nabla \int_R \left(-\frac{1}{2}u_x^2 + \frac{1}{6}au^3 \right) dx = \theta_0 \nabla H(u)$$

$$= \left(D^3 + \frac{1}{3}auD + \frac{1}{3}aDu \right) \nabla \int_R \frac{1}{2}u^2 \, dx = \theta_1(u)\nabla G(u). \tag{3.29}$$

Example 3.2 The modified Korteweg-de Vries (MKdV) equation [123]

$$u_t = u_{3x} - au^2 u_x = K(u) = (-D) \nabla \int_R \left(\frac{1}{2} u_x^2 + \frac{1}{12} au^4 \right) dx = \theta_0 \nabla H(u)$$

$$= \left(-D^3 + \frac{2}{3} a D u D^{-1} u D \right) \nabla \int_R \left(-\frac{1}{2} u^2 \right) dx = \theta_1(u) \nabla G(u). \qquad (3.30)$$

Example 3.3 The Harry–Dym (HD) equation [110],[123]

$$w_t = w^3 w_{3x} = (-w^2 D w^2) \nabla \int_R \frac{1}{2} w^{-1} w_x^2 dx = \theta_0(w) \nabla H(w)$$

$$= (w^3 D^3 w^3) \nabla \int_R (-w^{-1}) dx = \theta_1(w) \nabla G(w), \qquad (3.31)$$

which is also known in the form

$$u_t = \frac{1}{4} \left(u^{-1/2} \right)_{3x} = \left(-\frac{1}{2} u D - \frac{1}{2} D u \right) \nabla \int_R \left(-\frac{1}{16} u_x^2 u^{-5/2} \right) dx$$

$$= \left(\frac{1}{4} D^3 \right) \nabla \int_R 2u^{1/2} dx. \qquad (3.32)$$

The reader can verify, using the transformation properties from Examples 2.8-2.11, that both representations are related through the map $w = \phi(u) = -\frac{1}{2} u^{-1/2}$.

Example 3.4 The nonlinear Schrödinger equation (NLS)

$$\begin{pmatrix} u \\ u^* \end{pmatrix}_t = \begin{pmatrix} iu_{xx} + 2iu |u|^2 \\ -iu_{xx}^* - 2iu^* |u|^2 \end{pmatrix}$$

$$= \begin{pmatrix} 0 & -i \\ i & 0 \end{pmatrix} \nabla \int_R \left(|u_x|^2 - |u|^4 \right) dx$$

$$= \begin{pmatrix} 2uD^{-1}u & -D - 2uD^{-1}u^* \\ -D - 2u^* D^{-1}u & 2u^* D^{-1}u^* \end{pmatrix} \nabla \int_R \frac{1}{2} i \left(u_x^* u - u_x u^* \right) dx \qquad (3.33)$$

Example 3.5 The dispersive water waves (DWW) equation [113]

$$\begin{pmatrix} q \\ r \end{pmatrix}_t = \begin{pmatrix} -\frac{1}{2} q_{xx} + (qr)_x \\ \frac{1}{2} r_{xx} + rr_x + q_x \end{pmatrix}$$

$$= \begin{pmatrix} 0 & D \\ D & 0 \end{pmatrix} \nabla \int_R \left(-\frac{1}{2} rq_x + \frac{1}{2} qr^2 + \frac{1}{2} q^2 \right) dx$$

$$= \frac{1}{2} \begin{pmatrix} qD + Dq & -D^2 + rD \\ D^2 + Dr & 2D \end{pmatrix} \nabla \int_R qr dx, \qquad (3.34)$$

which is also known in the form [9]

$$
\begin{pmatrix} u \\ v \end{pmatrix}_t = \begin{pmatrix} \frac{1}{4}v_{3x} + uv_x + \frac{1}{2}vu_x \\ u_x + \frac{3}{2}vv_x \end{pmatrix}
$$

$$
= \begin{pmatrix} -\frac{1}{2}vD - \frac{1}{2}Dv & D \\ D & 0 \end{pmatrix} \nabla \int_R \left(-\frac{1}{8}v_x^2 + \frac{1}{2}u^2 + \frac{3}{4}uv^2 + \frac{5}{32}v^4 \right) dx
$$

$$
= \begin{pmatrix} \frac{1}{4}D^3 + \frac{1}{2}uD + \frac{1}{2}Du & 0 \\ 0 & D \end{pmatrix} \nabla \int_R \left(uv + \frac{1}{4}v^3 \right) dx. \tag{3.35}
$$

Both representations are related through the map: $q = u + \frac{1}{4}v^2 - \frac{1}{2}v_x, r = v$.

Example 3.6 The Toda Lattice equation [149]

$$
\begin{pmatrix} v(n) \\ p(n) \end{pmatrix}_t = \begin{pmatrix} v(n)[p(n) - p(n-1)] \\ v(n+1) - v(n) \end{pmatrix}
$$

$$
= \begin{pmatrix} 0 & v(n)(1 - E^{-1}) \\ (E-1)v(n) & 0 \end{pmatrix} \nabla \sum_{n=-\infty}^{n=+\infty} \left[\frac{1}{2}p^2(n) + v(n) \right]
$$

$$
= \begin{pmatrix} v(n)(E - E^{-1})v(n) & v(n)(1 - E^{-1})p(n) \\ p(n)(E-1)v(n) & Ev(n) - v(n)E \end{pmatrix} \nabla \sum_{n=-\infty}^{n=+\infty} [-p(n)].
$$

$$\tag{3.36}$$

Example 3.7 Finite dimensional system [22]

$$
\begin{pmatrix} q \\ p \\ c \end{pmatrix}_t = \begin{pmatrix} 2pq^2 \\ 2cq - 2qp^2 \\ 0 \end{pmatrix} = \begin{pmatrix} 0 & 1 & 0 \\ -1 & 0 & 0 \\ 0 & 0 & 0 \end{pmatrix} \nabla \left(p^2 q^2 - cq^2 - c \right)
$$

$$
= \begin{pmatrix} 0 & 1+q^2 & 2pq^2 \\ -1-q^2 & 0 & 2cq - 2qp^2 \\ -2pq^2 & -2cq + 2qp^2 & 0 \end{pmatrix} \nabla(c). \tag{3.37}
$$

Example 3.8 Generalized Henon–Heiles (gHH) system [14]

$$
\begin{pmatrix} q_1 \\ q_2 \\ p_1 \\ p_2 \\ \alpha \end{pmatrix}_t = \begin{pmatrix} p_1 \\ p_2 \\ -3q_1^2 - \frac{1}{2}q_2^2 \\ -q_1 q_2 + \alpha/q_2^3 \\ 0 \end{pmatrix}
$$

$$= \begin{pmatrix} 0 & 0 & 1 & 0 & 0 \\ 0 & 0 & 0 & 1 & 0 \\ -1 & 0 & 0 & 0 & 0 \\ 0 & -1 & 0 & 0 & 0 \\ 0 & 0 & 0 & 0 & 0 \end{pmatrix} \nabla \left(\frac{1}{2}p_1^2 + \frac{1}{2}p_2^2 + q_1^3 + \frac{1}{2}q_1 q_2^2 + \frac{1}{2}\alpha/q_2^2 \right)$$

$$= \begin{pmatrix} 0 & 0 & 0 & -4/q_2 & 8q_2 p_2 \\ 0 & 0 & -4/q_2 & 8q_1/q_2^2 & 8p_1 q_2 - 16p_2 q_1 \\ 0 & 4/q_2 & 0 & -4p_2/q_2^2 & -8q_1 q_2^2 + 8p_2^2 + 8\alpha/q_2^2 \\ 4/q_2 & -8q_1/q_2^2 & 4p_2/q_2^2 & 0 & A \\ * & * & * & 0 \end{pmatrix}$$

$$\times \nabla \left(\frac{1}{4}q_1 p_2^2 - \frac{1}{4}q_2 p_1 p_2 - \frac{1}{8}q_1^2 q_2^2 - \frac{1}{32}q_2^4 + \frac{1}{4}\alpha q_1/q_2^2 \right), \tag{3.38}$$

where $A = -8p_1 p_2 - 8q_1^2 q_2 - 4q_2^3 - 16\alpha q_1/q_2^3$ and stars in the lowest row mean matrix elements which make the whole matrix skew-symmetric.

All the above examples represent various bi-Hamiltonian dynamical systems. A few natural questions arise when considering our examples. First: is the bi-Hamiltonian property a common feature of all Hamiltonian systems? Second: what is the relevance of the bi-Hamiltonian formulation? Third: how do we find the bi-Hamiltonian representation? The answer to the first question is negative. The bi-Hamiltonian property is a feature of some class of Hamiltonian systems only. The answer to the second question is given in this chapter and the answer to the third question the reader will also find, at least partially, later in this book.

3.3 Symmetries and Conserved Quantities of Dynamical Systems

Before we define a general tensor invariant of a given dynamical system, let us make some basic considerations. We again consider an arbitrary dynamical system $u_t = K(u)$. A scalar field $F(u(t))$ is said to be a *conserved quantity* (1.1) if

$$\frac{d}{dt}F(u) = F'(u)[u_t] = F'(u)[K(u)] = \langle \nabla F, K \rangle (u) = 0, \tag{3.39}$$

where t is the evolution parameter of the system. Hence, (3.39) guarantees that $F(u)$ is constant along the integral curve of (1.1). If $K(u)$ is a Hamiltonian vector field, i.e. $K = \theta \circ \nabla H$, then the condition (3.39) reads

$$\frac{d}{dt}F(u) = \langle \nabla F, \theta \circ \nabla H \rangle (u) = \{H, F\}_\theta(u) = 0, \tag{3.40}$$

which means that the Poisson bracket of F with the Hamiltonian function H is equal to zero. We can also generalize the notion of conserved quantity to a scalar field which depends explicitly on the evolution parameter $t : F(u(t), t)$. Then, a scalar field $F(u(t), t)$ is said to be a time dependent conserved quantity for (1.1) if

$$\frac{d}{dt} F(u(t), t) = \frac{\partial F}{\partial t} + \langle \nabla F, K \rangle (u) = \frac{\partial F}{\partial t} + \{H, F\}_\theta(u) = 0. \qquad (3.41)$$

Remark 3.2 Note that the angle variables (2.29) of a completely integrable system are closely related to the time dependent constants of motion, i.e. obviously the functions

$$\varphi_i - \omega_i(f) \cdot t \qquad (3.42)$$

are conserved along the flow. In other words, for an integrable system on $M \subset \mathbb{R}^{2N}$ we can always construct N time independent and N linear in time constants of motion, i.e. scalar fields fulfilling the condition (3.41).

Theorem 3.9 *Time dependent conserved quantities constitute a subalgebra of the scalar field Lie algebra $\mathcal{F}_\theta$.*

Proof. Let F_1 and F_2 be time dependent conserved fields of a Hamiltonian dynamical system $u_t = \theta \circ \nabla H$ and let us assume that θ does not depend explicitly on time (this assumption will be justified later). Then

$$\frac{d}{dt}\{F_1, F_2\}_\theta$$

$$= \frac{\partial}{\partial t}\{F_1, F_2\}_\theta + \{H, \{F_1, F_2\}_\theta\}_\theta$$

$$= \left\langle \nabla \frac{\partial F_2}{\partial t}, \theta \circ \nabla F_1 \right\rangle + \left\langle \nabla F_2, \theta \circ \nabla \frac{\partial F_1}{\partial t} \right\rangle$$

$$\quad - \{F_2, \{H, F_1\}_\theta\}_\theta - \{F_1, \{F_2, H\}_\theta\}_\theta$$

$$= \left\{ \frac{\partial F_1}{\partial t}, F_2 \right\}_\theta + \left\{ F_1, \frac{\partial F_2}{\partial t} \right\}_\theta + \{\{H, F_1\}_\theta, F_2\}_\theta + \{F_1, \{H, F_2\}_\theta\}_\theta$$

$$= \left\{ \frac{\partial F_1}{\partial t} + \{H, F_1\}_\theta, F_2 \right\}_\theta + \left\{ F_1, \frac{\partial F_2}{\partial t} + \{H, F_1\}_\theta \right\}_\theta$$

$$= 0. \ \square \qquad (3.43)$$

Theorem 3.9 shows us that the Poisson bracket of two time dependent conserved quantities gives another conserved quantity. Thus, we have at our disposal a powerful method for the generation of constants of motion for a given Hamiltonian system. Of course we can get the same results for an inverse Hamiltonian system and the Poisson bracket $\{.,.\}^J$. Now we pass to the concept of a *symmetry* of the dynamical system (1.1).

Definition 3.3 *A symmetry group of the dynamical system (1.1) is a one-parameter group of transformations ϕ_ϵ (a flow), such that if $u = f(x,t)$ is an arbitrary solution of (1.1) then $\phi_\epsilon \cdot f(x,t)$ is also a solution of the same equation.*

In other words, for an arbitrary initial condition $u(0)$ and a fixed value of ϵ, ϕ_ϵ maps one integral curve $u(t)$ onto another integral curve $u_\epsilon(t)$ of (1.1). If ϕ_t is a given flow and ϕ_ϵ is its symmetry group, then

$$\phi_t \cdot (\phi_\epsilon \cdot) = \phi_\epsilon \cdot (\phi_t \cdot). \tag{3.44}$$

On the level of infinitesimal generators the condition (3.44) is equivalent to

$$[K, \sigma] = 0, \tag{3.45}$$

where $K(u), \sigma(u)$ stand for the symmetry generators of the flows ϕ_t and ϕ_ϵ, respectively. Indeed, taking the dual derivative of (3.44) we get

$$\phi_t^* \cdot (\phi_\epsilon \cdot) = \phi_\epsilon^* \cdot (\phi_t \cdot). \tag{3.46}$$

Then, differentiating (3.46) with respect to t and ϵ at $t = \epsilon = 0$, according to results (2.102) and (2.123) we find

$$L_K \sigma = L_\sigma K \Longleftrightarrow [K, \sigma] = 0.$$

In the case when $\sigma = \sigma(u,t)$ depends explicitly on time, the condition (3.45) turns into

$$\frac{\partial \sigma}{\partial t} + [K, \sigma] = 0. \tag{3.47}$$

As in our further considerations we will stay on the level of vector fields rather then on the level of flows, so hereafter we will use the notion of symmetry for an infinitesimal symmetry generator σ.

Theorem 3.10 *The time dependent symmetries of a given dynamical system constitute a subalgebra of the vector field Lie algebra $\mathcal{L}$.*

Proof. Let $\sigma_1, \sigma_2 \in \mathcal{L}$ be time dependent symmetries of a dynamical system $u_t = K(u)$. Then

$$\frac{\partial}{\partial t}[\sigma_1, \sigma_2] + [K, [\sigma_1, \sigma_2]]$$

$$= \left[\frac{\partial \sigma_1}{\partial t}, \sigma_1\right] + \left[\sigma_1, \frac{\partial \sigma_2}{\partial t}\right] - [\sigma_2, [K, \sigma_1]] - [\sigma_1, [\sigma_2, K]]$$

$$= -\left[\sigma_2, \frac{\partial \sigma_1}{\partial t} + [K, \sigma_1]\right] + \left[\sigma_1, \frac{\partial \sigma_2}{\partial t} + [K, \sigma_2]\right]$$

$$= -[\sigma_2, 0] + [\sigma_1, 0]$$
$$= 0, \tag{3.48}$$

hence the commutator $[\sigma_1, \sigma_2]$ is again a symmetry. $\square$

When $\sigma = \theta \circ \nabla F$ is a Hamiltonian vector field, then F is an appropriate conserved quantity satisfying (3.41), as $\theta \circ \nabla$ is a Lie algebra homomorphism (Theorem 3.4). In that sense the map $\theta \circ \nabla$ can be called a *Noether map* as it relates the Lie algebra of conserved quantities to the corresponding Lie algebra of symmetries. Because the Noether map is a homomorphism, the inverse may not exist and we can find symmetries which are non-Hamiltonian and hence are not related to any conserved quantity. As we will see later, the non-Hamiltonian symmetries play an important role in the theory of dynamical systems.

Example 3.9 Lie symmetries of the KdV.
The KdV equation (2.9) is invariant under the following four so called point transformations

1. $\bar{u}(\bar{x}, \bar{t}) = \phi_\epsilon \cdot u(x, t) = u(x + \epsilon, t)$ space translation,
2. $\bar{u}(\bar{x}, \bar{t}) = \phi_\epsilon \cdot u(x, t) = u(x, t + \epsilon)$ time translation,
3. $\bar{u}(\bar{x}, \bar{t}) = \phi_\epsilon \cdot u(x, t) = u(x + \epsilon a t, t) + \epsilon$ Galilean transformation,
4. $\bar{u}(\bar{x}, \bar{t}) = \phi_\epsilon \cdot u(x, t) = e^{2\epsilon} u(e^\epsilon x, e^{3\epsilon} t)$ conformal transformation.

with the following generators

$$\sigma_1 = u_x,$$

$$\sigma_2 = u_t = u_{3x} + a u u_x,$$

$$\sigma_3 = 1 + a t u_x,$$

$$\sigma_4 = 2u + x u_x + 3t u_t = 2u + x u_x + 3t(u_{3x} + a u u_x),$$

called Lie symmetries. For more information on the problem of finding Lie symmetries we recommend the book [159]. Other symmetries of the KdV which are not connected with any point transformation are the so called Lie-Bäcklund symmetries and will be considered later. For these generalized symmetries we cannot present the flow ϕ_ϵ in an explicit form so we stay on the level of infinitesimal generators $\sigma(u(x,t))$. Note that with respect to the Poisson operator $\theta = D$, the symmetries σ_1, σ_2 and σ_3 are Hamiltonian with the corresponding scalar fields (conserved quantities)

$$F_1 = \int_R \frac{1}{2} u^2 \mathrm{d}x, \quad F_2 = \int_R \left(-\frac{1}{2} u_x^2 + \frac{1}{6} a u^3 \right) \mathrm{d}x, \quad F_3 = \int_R \left(xu + \frac{1}{2} a t u^2 \right) \mathrm{d}x,$$

while σ_4 is non-Hamiltonian. On the other hand, with respect to the Poisson operator $\theta = D^3 + \frac{1}{3} auD + \frac{1}{3} aDu$, the symmetries σ_1, σ_2 and σ_4 are Hamiltonian with appropriate scalar fields

$$F_1 = \int_R \frac{3}{a} u \mathrm{d}x, \quad F_2 = \int_R \frac{1}{2} u^2 \mathrm{d}x, \quad F_4 = \int_R \left(\frac{3}{a} xu + \frac{3}{2} t u^2 \right) \mathrm{d}x$$

and σ_3 is a non-Hamiltonian scalar field.

Lemma 3.5 *The symmetries $\sigma(u)$ of the dynamical system $u_t = K(u)$ are solutions of the linearized (tangent) equation*

$$v_t = K'[v], \qquad v \in \mathcal{L}. \tag{3.49}$$

Proof. As a tangent trajectory $\sigma(t) = \sigma(u(t))$ is determined by the curve $u(t)$, so we have

$$\sigma_t = \frac{\partial \sigma}{\partial t} + \sigma'[u_t] = \frac{\partial \sigma}{\partial t} + \sigma'[K(u)]. \tag{3.50}$$

On the other hand, for the solutions $v = \sigma(u(t))$ of (3.49) $\sigma_t = K'[\sigma(u)]$, hence

$$\frac{\partial \sigma}{\partial t} + \sigma'[K] = K'[\sigma] \Leftrightarrow \frac{\partial \sigma}{\partial t} + [K, \sigma] = 0. \ \square \tag{3.51}$$

3.4 Tensor Invariants of Dynamical Systems

Let us translate the conditions (3.41) and (3.47) for a conserved quantity and a symmetry into the language of Lie derivatives, then we find

$$\left(\frac{\partial}{\partial t} + L_K \right) F = 0 \quad \text{and} \quad \left(\frac{\partial}{\partial t} + L_K \right) \sigma = 0. \tag{3.52}$$

The result (3.52) is not a surprise as a Lie derivative is a proper operation to measure changes of various tensor fields along a given flow. So it suggests the general definition of *tensor invariant* for a given dynamical system in the form [181]:

Definition 3.4 *A tensor field T is an invariant of a dynamical system $u_t = K(u)$, if*

$$\left(\frac{\partial}{\partial t} + L_K\right) T = 0. \tag{3.53}$$

In that sense invariant scalar fields (constants of motion) and invariant vector fields (symmetries) are only examples of various invariant objects of a dynamical system. So let us analyse other possible invariant fields.

We start from invariant covector fields γ defined by the equality

$$\left(\frac{\partial}{\partial t} + L_K\right)\gamma = \frac{\partial \gamma}{\partial t} + \gamma'[K] + K'^\dagger[\gamma] = 0. \tag{3.54}$$

Let us call them *cosymmetries* as they are covectors and are dual objects to symmetries (in the literature they are also called *adjoint symmetries* or *conserved covariants*).

Lemma 3.6 *The cosymmetries $\gamma(u)$ of a dynamical system $u_t = K(u)$ are solutions of the adjoint linear equation*

$$\alpha_t = -K'^\dagger[\alpha], \qquad \alpha \in \mathcal{T}^{(0,1)}. \tag{3.55}$$

Proof. As a cotangent trajectory $\gamma(t) = \gamma(u(t))$ is determined by the integral curve $u(t)$, so we have

$$\gamma_t = \frac{\partial \gamma}{\partial t} + \gamma'[u_t] = \frac{\partial \gamma}{\partial t} + \gamma'[K]. \tag{3.56}$$

On the other hand, for the solutions $\alpha = \gamma(u(t))$ of (3.55) we have $\gamma_t = -K'^\dagger[\gamma]$, hence

$$\frac{\partial \gamma}{\partial t} + \gamma'[K] = -K'^\dagger[\gamma] \Rightarrow \left(\frac{\partial}{\partial t} + L_K\right)\gamma = 0. \,\square \tag{3.57}$$

Theorem 3.11

(i) *For a Hamiltonian dynamical system $u_t = K(u) = \theta \circ \nabla H$, the time dependent cosymmetries constitute a subalgebra of the covector field Lie algebra $\mathcal{L}_\theta^*$.*

(ii) *For an inverse-Hamiltonian system $u_t = K(u)$, $J \circ K = \nabla H$, the time dependent cosymmetries constitute a subalgebra of the covector field Lie algebra $\mathcal{L}^{*J}$.*

Proof. We only prove the case (i) as the proof of (ii) is fully analogous to the case (i) so we leave it to the reader as an exercise.

We assume that θ does not depend explicitly on t. As

$$\frac{\partial}{\partial t} L_G = L_{\frac{\partial G}{\partial t}} + L_G \frac{\partial}{\partial t} \quad \text{and} \quad \frac{\partial}{\partial t} i_G = i_{\frac{\partial G}{\partial t}} + i_G \frac{\partial}{\partial t},$$

we find that

$$\frac{\partial}{\partial t}\{\gamma_1, \gamma_2\}_\theta = \left\{\frac{\partial \gamma_1}{\partial t}, \gamma_2\right\}_\theta + \left\{\gamma_1, \frac{\partial \gamma_2}{\partial t}\right\}_\theta.$$

On the other hand

$$L_K\{\gamma_1, \gamma_2\}_\theta = L_K L_{\theta \circ \gamma_1} \gamma_2 - L_K \, i_{\theta \circ \gamma_2} \circ d\gamma_1$$

$$= L_{\theta \circ \gamma_1} L_K \gamma_2 + L_{[K, \theta \circ \gamma_1]} \gamma_2 - i_{[K, \theta \circ \gamma_2]} \circ d\gamma_1 - i_{\theta \circ \gamma_2} \circ L_K d\gamma_1$$

$$= L_{\theta \circ \gamma_1} L_K \gamma_2 + L_{\theta \circ L_K \gamma_1} \gamma_2 - i_{\theta \circ L_K \gamma_2} \circ d\gamma_1 - i_{\theta \circ \gamma_2} \circ dL_K \gamma_1$$

$$= \{L_K \gamma_1, \gamma_2\}_\theta + \{\gamma_1, L_K \gamma_2\}_\theta, \tag{3.58}$$

where we have used relations (2.128), (3.4), (3.5) and the fact that $L_K \theta = L_{\theta \circ \nabla H} \theta = 0$. Now, let $\gamma_1, \gamma_2 \in \mathcal{L}_\theta^*$ be cosymmetries of a dynamical system $u_t = K(u) = \theta \circ \nabla H$. Then

$$\left(\frac{\partial}{\partial t} + L_K\right)\{\gamma_1, \gamma_2\}_\theta = \left\{\frac{\partial \gamma_1}{\partial t} + L_K \gamma_1, \gamma_2\right\}_\theta + \left\{\gamma_1, \frac{\partial \gamma_2}{\partial t} + L_K \gamma_2\right\}_\theta = 0. \; \square$$

$$\tag{3.59}$$

Let us pass to higher order tensor invariants.

Lemma 3.7

(i) *A Poisson operator θ which maps cosymmetries onto symmetries of a dynamical system $u_t = K(u)$ is a tensor invariant of that system.*

(ii) *A symplectic operator J which maps symmetries onto cosymmetries of a dynamical system $u_t = K(u)$ is a tensor invariant of that system.*

Proof.

(i) Let σ and γ be a symmetry and a cosymmetry, respectively and $\sigma = \theta \circ \gamma$. Then

$$0 = \frac{\partial \sigma}{\partial t} + L_K \sigma = \frac{\partial(\theta \circ \gamma)}{\partial t} + L_K(\theta \circ \gamma)$$

$$= \frac{\partial \theta}{\partial t} \circ \gamma + \theta \circ \frac{\partial \gamma}{\partial t} + (L_K \theta) \circ \gamma + \theta \circ L_K \gamma$$

$$= \left(\frac{\partial \theta}{\partial t} + L_K \theta\right) \circ \gamma + \theta \circ \left(\frac{\partial \gamma}{\partial t} + L_K \gamma\right) = \left(\frac{\partial \theta}{\partial t} + L_K \theta\right) \circ \gamma$$

$$\Rightarrow \left(\frac{\partial \theta}{\partial t} + L_K \theta\right) = 0. \; \square \tag{3.60}$$

(ii) The proof is similar to the case (i) when $\gamma = J \circ \sigma$.

Lemma 3.8

(i) *If a vector field $K(u)$ is Hamiltonian with respect to its invariant Poisson tensor θ, then θ does not depend explicitly on time.*

(ii) *If a vector field $K(u)$ is inverse-Hamiltonian with respect to its invariant symplectic operator J, then J does not depend explicitly on time.*

Proof.

(i) We find

$$0 = \frac{\partial \theta}{\partial t} + L_K \theta = \frac{\partial \theta}{\partial t} + L_{\theta \circ \nabla H} \theta = \frac{\partial \theta}{\partial t}, \tag{3.61}$$

which follows from Lemma 3.2 and the fact that ∇H is a closed one-form.
□

(ii) We have

$$0 = \frac{\partial J}{\partial t} + L_K J = \frac{\partial J}{\partial t} + d(J \circ K) = \frac{\partial J}{\partial t}, \tag{3.62}$$

which follows from Lemma 3.3 and the fact that $J \circ K$ is a closed one-form. □

Lemma 3.8 explains why in all examples 3.1-3.8 the Poisson operators are time independent. The example with time dependent θ and J operators (cylindrical KdV) will be given later.

Of special importance are invariant tensor fields Φ and $\Psi = \Phi^\dagger \in \mathcal{T}^{(1,1)}$. They are known as the *recursion operator* for symmetries [158] and the recursion operator for cosymmetries, respectively. The adequacy of the name follows from the following

Lemma 3.9

(i) *The recursion operator Φ acting on one symmetry produces another symmetry of a given dynamics.*

(ii) *The recursion operator Ψ acting on one cosymmetry produces another cosymmetry of a given dynamics.*

Proof. Let σ be a symmetry and γ a cosymmetry of a dynamical system $u_t = K(u)$.

(i) As the derivative $\left(\frac{\partial}{\partial t} + L_K \right)$ obeys the Leibniz rule for an arbitrary product of tensor fields, then

$$\left(\frac{\partial}{\partial t} + L_K \right)(\Phi \circ \sigma) = \left(\frac{\partial \Phi}{\partial t} + L_K \Phi \right) \sigma + \Phi \circ \left(\frac{\partial \sigma}{\partial t} + L_K \sigma \right) = 0. \square \tag{3.63}$$

(ii)

$$0 = \left(\frac{\partial}{\partial t} + L_K\right) \langle \gamma, \Phi \circ \sigma \rangle = \left(\frac{\partial}{\partial t} + L_K\right) \langle \Phi^\dagger, \gamma \circ \sigma \rangle$$

$$= \left\langle \left(\frac{\partial}{\partial t} + L_K\right)(\Phi^\dagger \circ \gamma), \sigma \right\rangle + \left\langle \Phi^\dagger \circ \gamma, \left(\frac{\partial}{\partial t} + L_K\right)\sigma \right\rangle$$

$$= \left\langle \left(\frac{\partial}{\partial t} + L_K\right)(\Phi^\dagger \circ \gamma), \sigma \right\rangle \quad \Rightarrow \quad \left(\frac{\partial}{\partial t} + L_K\right)\Psi = 0. \ \Box \quad (3.64)$$

Lemma 3.9 reveals to us the power of the recursion operator. That is, when for some dynamical system we have one symmetry σ_0 and the corresponding cosymmetry γ_0, and the system has the recursion operator Φ, then we can immediately generate the whole hierarchy of symmetries $\sigma_n = \Phi^n \circ \sigma_0$ and the corresponding hierarchy of cosymmetries $\gamma_n = (\Phi^\dagger)^n \circ \gamma_0$. Moreover, because the composition of the θ and J invariants gives $\theta \circ J = \Phi$ and $J \circ \theta = \Phi^\dagger = \Psi$ respectively, this relationship clarifies the importance of a bi-Hamiltonian representation of a dynamical system. That is, a bi-Hamiltonian (J, θ) or $(J = \theta_0^{-1}, \theta = \theta_1)$ formulation of a dynamics guarantees the existence of a hierarchy of symmetries and the corresponding hierarchy of cosymmetries. In such a case the recursion operator does not depend explicitly on time.

Example 3.10 The KdV equation.
Both θ_0 and θ_1 are tensor invariants and moreover $\theta_0^{-1} = J = D_x^{-1}$. Hence the recursion operator Φ reads

$$\Phi(u) = \theta_1 \theta_0^{-1} = D^2 + \frac{2}{3}au + \frac{1}{3}au_x D^{-1}.$$

Applying the operator Φ to the Lie symmetries from Example 3.9 we find two hierarchies of symmetries (called also the Lie-Bäcklund symmetries)

$$K_n = \Phi^n \circ K_0, \quad \sigma_n = \Phi^{n+1} \circ \sigma_{-1}, \quad K_0(u) = u_x, \ \sigma_{-1}(u) = \frac{3}{a}(1 + atu_x),$$

where $K_0, K_1, \frac{a}{3}\sigma_{-1}, \sigma_0$ are equal to the symmetries from Example 3.9. Notice that the K_n symmetries are time independent and the σ_n ones are linear in time. The recursion operator Ψ for cosymmetries has the form

$$\Psi(u) = D^2 + \frac{2}{3}au - \frac{1}{3}aD^{-1}u_x$$

and generates two hierarchies of cosymmetries

$$\gamma_n = \Psi^n \circ \gamma_0, \quad \zeta_n = \Psi^{n+1} \circ \zeta_{-1},$$

$$\gamma_0(u) = JK_0(u) = u, \quad \zeta_{-1}(u) = J\sigma_{-1}(u) = \frac{3}{a}(x + atu).$$

Example 3.11 The MKdV equation.

$$\Phi(u) = \theta_1(u)\theta_0^{-1}(u) = D^2 - \frac{2}{3}aDuD^{-1}u, \qquad \Psi(u) = D^2 - \frac{2}{3}aD^{-1}uD,$$

$$K_n = \Phi^n \circ K_0, \quad \sigma_n = \Phi^n \circ \sigma_0, \quad K_0(u) = u_x, \quad \sigma_0(u) = (xu)_x + 3t(u_{3x} - au^2u_x).$$

Example 3.12 The HD equation.

$$\Phi(w) = \theta_1(w)\theta_0^{-1}(w) = -w^3 D^3 w D^{-1} w^{-2}, \qquad \Psi(w) = -w^{-2} D^{-1} w D^3 w^3,$$

$$K_n = \Phi^n \circ K, \quad \sigma_n = \Phi^n \circ \sigma_0, \quad K(w) = w^3 w_{3x}, \quad \sigma_0(w) = (xw)_x + 6tw^3 w_{3x}.$$

Notice that $\Phi w_x = 0$.

Example 3.13 The Sawada–Kotera (SK) equation [171]

$$u_t = u_{5x} + \frac{5}{2}auu_{3x} + \frac{5}{2}au_x u_{xx} + \frac{5}{4}a^2 u^2 u_x = K(u) \qquad (3.65)$$

has an invariant Poisson operator θ and an invariant symplectic operator J of the form [85]

$$\theta(u) = D^3 + auD + aDu, \quad J(u) = D^{-1}\left(D^2 + \frac{1}{2}au\right)D\left(D^2 + \frac{1}{2}au\right)D^{-1}$$

and hence, the recursion operator

$$\Phi(u) = (D^2 + 2au + auD^{-1})\left(D^2 + \frac{1}{2}au\right)D\left(D^2 + \frac{1}{2}au\right)D^{-1}.$$

The equation (3.65) has three hierarchies of symmetries

$$K_n(u) = \Phi^n(u)u_x, \quad \tilde{K}_n(u) = \Phi^n(u)K(u), \quad \sigma(u) = \Phi^n(u)\sigma_0(u),$$

where u_x is a generator of space translations, K serves as a generator of time translations and $\sigma_0(u) = 2u + xu_x + 5tK(u)$ is a scaling symmetry (generator of a conformal transformation). First two hierarchies are time independent whereas the third one depends linearly on time. Two separate time independent hierarchies appear because the recursion operator Φ does not relate u_x to $K(u)$ so both symmetries can initiate different hierarchies.

Example 3.14 The Ablowitz, Kaup, Newell and Segur (AKNS) dynamical system [1], [2]

$$\begin{pmatrix} q \\ r \end{pmatrix}_t = \begin{pmatrix} iq_{xx} - 2iq^2 r \\ -ir_{xx} + 2iqr^2 \end{pmatrix} = \begin{pmatrix} 0 & i \\ -i & 0 \end{pmatrix}\nabla\int_R(-q_x r_x - q^2 r^2)\,dx$$

$$= \begin{pmatrix} -2qD^{-1}q & -D + 2qD^{-1}r \\ -D + 2rD^{-1}q & -2rD^{-1}r \end{pmatrix}\nabla\int_R \frac{1}{2}i(r_x q - q_x r)\,dx$$

$$(3.66)$$

has the recursion operator of the form

$$\Phi(q,r) = \theta_1(q,r)\theta_0^{-1} = \begin{pmatrix} iD - 2iqD^{-1}r & -2iqD^{-1}q \\ 2irD^{-1}r & -iD + 2irD^{-1}q \end{pmatrix}$$

and two hierarchies of symmetries

$$K_n = \Phi^{n+1} \circ K_{-1}, \quad \sigma_n = \Phi^{n+1} \circ \sigma_{-1},$$

$$K_{-1}(q,r) = \begin{pmatrix} -iq \\ ir \end{pmatrix}, \quad \sigma_{-1}(q,r) = \begin{pmatrix} ixq - itq \\ -ixr + itr \end{pmatrix},$$

where the AKNS vector field (3.66) is equal to K_1. Note that for the special case when $r = q^*$, the dynamics (3.66) reduce to the NLS equation (Example 3.4).

Example 3.15 The cylindrical KdV equation [133]

$$u_t = u_{3x} + auu_x - \frac{u}{2t} = K(u,t) \tag{3.67}$$

is of special interest as it is an example of a vector field which explicitly depends on the t variable. Equation (3.67) has two invariant Poisson operators [147]

$$\theta_0(t) = \frac{1}{t}D, \quad \theta_1(u,t) = D^3 + \frac{1}{3}auD + \frac{1}{3}aDu + \frac{1}{3}\frac{x}{t}D + \frac{1}{6t}$$

and the recursion operator

$$\Phi(u,t) = \theta_1(u,t)\theta_0^{-1}(t) = t\left(D^2 + \frac{2}{3}au + \frac{1}{3}au_xD^{-1}\right) + \frac{1}{3}x + \frac{1}{6}D^{-1}.$$

All these operators depend explicitly on time. Applying the recursion operator to Lie symmetries we again find two hierarchies of symmetries

$$K_n = \Phi^n \circ K_0, \quad \sigma_n = \Phi^{n+1} \circ \sigma_{-1}, \quad K_0(u) = a\sqrt{t}u_x + \frac{1}{2\sqrt{t}}, \quad \sigma_{-1}(u) = 3u_x.$$

Contrary to the previous examples, here both hierarchies depend on the time variable in a nonlinear way. The vector field $K(u,t)$ is non-Hamiltonian with respect to θ_0 and θ_1, which is in excellent agreement with Lemma 3.8, and does not belong to any of the hierarchies.

Example 3.16 The Toda lattice.
Inverting the first Poisson tensor from Example 3.6 we get the following recursion operator

$$\Phi(p,v) = \begin{pmatrix} vE^{-1}\Delta pE\Delta^{-1}(1/v) & v(1+E^{-1}) \\ (Ev - vE^{-1})E\Delta^{-1}(1/v) & p \end{pmatrix}.$$

If for a constant sequence $c(n) = 1$ we formally define $(\Delta^{-1}c)(n) = n - 1$, the two hierarchies of symmetries are the following

$$K_n = \Phi^n \circ K, \quad \sigma_n = \Phi^n \circ \sigma_0, \quad \sigma_0(n) = \begin{pmatrix} 2v(n) \\ p(n) \end{pmatrix} + tK(v(n), p(n)),$$

where K is the Toda vector field and σ_0 is a conformal symmetry.

Example 3.17 Consider the following lattice system

$$\begin{pmatrix} q(n) \\ r(n) \end{pmatrix}_t = \begin{pmatrix} q(n+1) - q^2(n)r(n) \\ -r(n-1) + q(n)r^2(n) \end{pmatrix}$$

$$= \begin{pmatrix} 0 & 1 \\ -1 & 0 \end{pmatrix} \nabla \sum_{n=-\infty}^{n=+\infty} \left[r(n-1)q(n) - \frac{1}{2}q^2(n)r^2(n) \right]. \quad (3.68)$$

The recursion operator of the dynamics (3.68) was found [98] in the form

$$\Phi(q, r) = \begin{pmatrix} E - 2qE\Delta^{-1}r & q^2 - 2qE\Delta^{-1}q \\ -r^2 + 2rE\Delta^{-1}r & E^{-1} - 2qr + 2rE\Delta^{-1}q \end{pmatrix}$$

and two hierarchies of symmetries are

$$K_n = \Phi^n \circ K, \quad \sigma_n = \Phi^n \circ \sigma_0, \quad \sigma_0(n) = \begin{pmatrix} (n+1)q(n) \\ -nr(n) \end{pmatrix} + tK(q(n), r(n)),$$

where K is the vector field (3.68) and σ_0 is a conformal symmetry.

3.5 Algebraic Properties of Tensor Invariants

In the previous section we introduced various tensor invariants of a given dynamical system. Now, we examine some of their properties and find a few important relations among them.

Definition 3.5 *Let $\Phi \in \mathcal{T}^{(1,1)}$ be a tensor field such that for an arbitrary vector field $v \in \mathcal{L}$*

$$L_{\Phi \circ v}\Phi = \Phi \circ L_v\Phi. \quad (3.69)$$

Then $\Phi(u)$ is called a hereditary operator [82] (regular operator [93], operator with vanishing Nijenhuis torsion [181] or just Nijenhuis operator).

Remark 3.3 *For every tensor field $\Phi \in \mathcal{T}^{(1,1)}$ there exists a tensor field $N_\Phi \in \mathcal{T}^{(1,2)}$, dependent only on Φ, such that for all vector fields $v \in \mathcal{L}$*

$$L_{\Phi \circ v}\Phi - \Phi \circ L_v\Phi = N_\Phi \circ v. \quad (3.70)$$

The alternative definition is given by the formula

$$N_\Phi(v,w) = [\Phi \circ v, \Phi \circ w] - \Phi \circ [\Phi \circ v, w] - \Phi \circ [v, \Phi \circ w] + \Phi^2 \circ [v,w], \quad (3.71)$$

obtained from the application of (3.70) to (a contraction with) a vector field w. The tensor field N_Φ is called the *Nijenhuis torsion* of Φ [144]. For finite systems, in local coordinates the Nijenhuis torsion has the following entries

$$(N_\Phi)^i_{jk} = \frac{\partial \Phi^i_k}{\partial x^l}\Phi^l_j - \frac{\partial \Phi^i_j}{\partial x^l}\Phi^l_k + \frac{\partial \Phi^l_j}{\partial x^k}\Phi^i_l - \frac{\partial \Phi^l_k}{\partial x^j}\Phi^i_l. \quad (3.72)$$

Theorem 3.12 *Let Φ be a hereditary operator and K be a vector field such that $L_K\Phi = 0$. Then*

(i) the vector fields $K_n = \Phi^n \circ K$, constitute an Abelian subalgebra of $\mathcal{L}$;
(ii) for an arbitrary $n \in N$, $L_{K_n}\Phi = 0$.

Proof.

(i)

$$[K_i, K_j] = L_{K_i}K_j = L_{\Phi^i \circ K}(\Phi^j \circ K) = L_{\Phi^i \circ K}(\Phi^j) \circ K + \Phi^j \circ L_{\Phi^i \circ K}K$$

$$= \sum_{n=0}^{j-1} \Phi^n \circ (L_{\Phi^i \circ K}\Phi) \circ \Phi^{j-n-1} \circ K + \Phi^i \circ \Phi^j \circ L_K K$$

$$= \sum_{n=0}^{j-1} \Phi^n \circ (\Phi^i \circ L_K\Phi) \circ \Phi^{j-n-1} \circ K = 0. \quad \Box \quad (3.73)$$

(ii) follows immediately from the hereditary property (3.69).

For each dynamical system from Examples 3.10-3.16 the recursion operator Φ is hereditary and the first vector field K from the hierarchy K_n has the desired property $L_K\Phi = 0$. This means that the K_n-symmetries do commute. Moreover, besides the cylindrical KdV, in all examples the vector field K belongs to the hierarchy K_n and all K_n do not depend explicitly on time. As a consequence, Theorem 3.12 is equivalent to the statement that for an arbitrary evolution equation

$$u_{t_m} = K_m(u) \quad (3.74)$$

all vector fields K_m as well as the operator Φ are tensor invariants of the dynamics (3.74).

Definition 3.6 *Let θ and $\bar{\theta} \in \mathcal{T}^{(0,2)}$ be implectic operators. The pair $(\theta, \bar{\theta})$ is said to be compatible if $\theta + \bar{\theta}$ is also implectic. Then, $\theta_\epsilon = \theta + \epsilon\bar{\theta}$ for all real numbers ϵ is called a Poisson pencil.*

Definition 3.7 *A bi-Poisson manifold is a manifold M endowed with a pair $(\theta, \bar{\theta})$ of compatible implectic operators.*

Definition 3.8 *Let $\theta \in \mathcal{T}^{(0,2)}$ be an implectic operator and $J \in \mathcal{T}^{(2,0)}$ be a symplectic operator. The pair (θ, J) is said to be compatible if $J \circ \theta \circ J$ is also symplectic.*

Definition 3.9 *A Poisson–Nijenhuis manifold is a manifold M endowed with a pair (θ, J) of compatible implectic and symplectic operators.*

Lemma 3.10 *If $(\theta, \bar{\theta})$ is a compatible pair and $\bar{\theta}$ is invertible $\bar{\theta}^{-1} = J$, then the pair (θ, J) is also compatible.*

Proof. Let γ be an arbitrary covector and $\gamma = J \circ \tau$. According to Lemma 3.2 we find

$$0 = L_{(\theta + \bar{\theta}) \circ \gamma}(\theta + \bar{\theta}) + (\theta + \bar{\theta}) \circ d\gamma \circ (\theta + \bar{\theta})$$

$$= L_{\theta \circ \gamma} \bar{\theta} + L_{\bar{\theta} \circ \gamma} \theta + \theta \circ d\gamma \circ \bar{\theta} + \bar{\theta} \circ d\gamma \circ \theta$$

$$= L_{\theta \circ J \circ \tau} \bar{\theta} + L_\tau \theta + \theta \circ (L_\tau J) \circ \bar{\theta} + \bar{\theta} \circ (L_\tau J) \circ \theta.$$

As

$$L_{\theta \circ J \circ \tau} \bar{\theta} = L_{\bar{\theta}} \bar{\theta} = L_{\bar{\theta} \circ \gamma} \bar{\theta} = -\bar{\theta} \circ d\bar{\gamma} \circ \bar{\theta} = -\bar{\theta} \circ (L_{\bar{\tau}} J) \circ \bar{\theta}$$

$$= -\bar{\theta} \circ (L_{\theta \circ J \circ \tau} J) \circ \bar{\theta}$$

we have

$$0 = -\bar{\theta} \circ (L_{\theta \circ J \circ \tau} J) \circ \bar{\theta} + L_\tau \theta + \theta \circ (L_\tau J) \circ \bar{\theta} + \bar{\theta} \circ (L_\tau J) \circ \theta.$$

Multiplying from the left and right by $J = \bar{\theta}^{-1}$ we get

$$0 = -(L_{\theta \circ J \circ \tau} J) + J \circ (L_\tau \theta) \circ J + J \circ \theta \circ (L_\tau J) + (L_\tau J) \circ \theta \circ J$$

$$= -(L_{\theta \circ J \circ \tau} J) + L_\tau (J \circ \theta \circ J) = -d(J \circ \theta \circ J \circ \tau) + L_\tau (J \circ \theta \circ J)$$

and hence $J \circ \theta \circ J$ is symplectic. □

Theorem 3.13 [83], [148] *Let (θ, J) be a compatible pair and let K be a vector field such that $L_K \theta = 0$ and $L_K J = 0$.*

(i) *The operator $\Phi = \theta \circ J$ is hereditary.*
(ii) *All operators $J_n = J \circ \Phi^n = \Psi^n \circ J$ are symplectic.*
(iii) *All operators $\theta_n = \Phi^n \circ \theta = \theta \circ \Psi^n$ are implectic.*
(iv) *All covectors $\gamma_n = J \circ K_n = J \circ \Phi^n \circ K = J_n \circ K = \Psi^n \circ J \circ K = \Psi^n \circ \gamma$ are closed.*

Proof. Let $\gamma = J \circ \tau$, so $L_{\theta \circ J \circ \tau} \theta = -\theta \circ (L_\tau J) \circ \theta$ and then

(i)

$$
\begin{aligned}
L_{\Phi \circ \tau} \Phi - \Phi \circ L_\tau \Phi &= L_{\Phi \circ \tau}(\theta \circ J) - \Phi \circ L_\tau \Phi \\
&= (L_{\theta \circ J \circ \tau} \theta) \circ J + \theta \circ L_{\Phi \circ \tau} J - \Phi \circ L_\tau \Phi \\
&= -\theta \circ (L_\tau J) \circ \theta \circ J + \theta \circ L_{\Phi \circ \tau} J - \Phi \circ L_\tau \Phi \\
&= \theta \circ [-(L_\tau J) \circ \Phi + L_{\Phi \circ \tau} J - J \circ L_\tau \Phi] \\
&= \theta \circ [L_{\Phi \circ \tau} J - L_\tau(J \circ \Phi)] \\
&= \theta \circ [\mathrm{d}(J \circ \Phi \circ \tau) - L_\tau(J \circ \Phi)] = 0.
\end{aligned}
$$

$$(3.75)$$

The last equality follows from the fact that $J \circ \Phi = J \circ \theta \circ J$ is symplectic as (θ, J) is compatible. □

(ii) According to Lemma 3.3 one has to show that $L_\tau(J \circ \Phi^n) = \mathrm{d}(J \circ \Phi^n \circ \tau)$. The proof is inductive. Assuming that (ii) is true for n, we get

$$
\begin{aligned}
\mathrm{d}(J \circ \Phi^{n+1} \circ \tau) &= L_{\Phi^{n+1} \circ \tau} J = L_{\Phi^n \circ \Phi \circ \tau} J = L_{\Phi \circ \tau}(J \circ \Phi^n) \\
&= (L_{\Phi \circ \tau} J) \circ \Phi^n + J \circ (L_{\Phi \circ \tau} \circ \theta) \circ J \circ \Phi^{n-1} \\
&\quad + J \circ \theta \circ L_{\Phi \circ \tau}(J \circ \Phi^{n-1}) = L_\tau(J \circ \Phi) \circ \Phi^n \\
&\quad - J \circ \theta \circ (L_\tau J) \circ \theta \circ J \circ \Phi^{n-1} + J \circ \theta \circ L_\tau(J \circ \Phi^n) \\
&= L_\tau(J \circ \Phi) \circ \Phi^n + J \circ \Phi \circ L_\tau \Phi^n = L_\tau(J \circ \Phi^{n+1}). \square
\end{aligned}
$$

$$(3.76)$$

(iii) According to Lemma 3.2 one has to show that

$$
\begin{aligned}
L_{\Phi^n \circ \theta \circ \gamma}(\Phi^n \circ \theta) &= -\Phi^n \circ \theta \circ \mathrm{d}\gamma \circ \Phi^n \circ \theta \\
&= -\Phi^n \circ \theta \circ \mathrm{d}\gamma \circ \theta \circ J \circ \Phi^{n-1} \circ \theta \qquad (3.77) \\
&= \Phi^n(L_{\theta \circ \gamma} \theta) \circ J \circ \Phi^{n-1} \circ \theta.
\end{aligned}
$$

Assuming that (3.77) is true and applying the hereditary property of Φ, the inductive step is as follows

$$L_{\Phi^{n+1} \circ \theta \circ \gamma}(\Phi^{n+1} \circ \theta) \quad = (L_{\Phi^{n+1} \circ \theta \circ \gamma} \Phi) \circ \Phi^n \circ \theta +$$

$$\Phi \circ (L_{\Phi^n \circ \Phi \circ \theta \circ \gamma} \Phi^n \circ \theta)$$

$$= \Phi^{n+1} \circ (L_{\theta \circ \gamma} \Phi) \circ \Phi^n \circ \theta$$

$$+ \Phi^{n+1} \circ (L_{\theta \circ J \circ \theta \circ \gamma} \theta) \circ J \circ \Phi^{n-1} \circ \theta \quad (3.78)$$

$$= \Phi^{n+1} \circ (L_{\theta \circ \gamma} \theta \circ J) \circ \Phi^n \circ \theta$$

$$- \Phi^{n+1} \circ \theta \circ (L_{\theta \circ \gamma} J) \circ \Phi^n \circ \theta$$

$$= \Phi^{n+1} \circ (L_{\theta \circ \gamma} \theta) \circ J \circ \Phi^n \circ \theta,$$

where we have used the relation

$$L_{\theta \circ J \circ \theta \circ \gamma} \theta = -\theta \circ \mathrm{d}(J \circ \theta \circ \gamma) \circ \theta = -\theta \circ (L_{\theta \circ \gamma} J) \circ \theta. \square$$

(iv) From (ii) we know that all J_n are closed. On the other hand $L_K J_n = 0$ which follows from the assumption that $L_K J = 0$, $L_K \theta = 0$ and the application of the Leibniz product rule to the L_K–derivative. Then from Lemma 2.4(ii) we find that for J_n

$$0 = \mathrm{d}(i_K \circ J_n) = \mathrm{d}(J_n \circ K) = \mathrm{d}\gamma_n. \square$$

An important consequence of Theorem 3.13 is that vector fields $K_n = \Phi_n \circ K$ are multi-Hamiltonian vector fields. Let us introduce a hierarchy of scalar fields

$$H_n = \int_0^1 \langle \gamma_n(\lambda u), u \rangle \, \mathrm{d}\lambda, \qquad \nabla H_n = \gamma_n, \qquad (3.79)$$

and let $K = \theta \circ \gamma = \theta \circ \nabla H$. Then

$$K_n = \Phi^n \circ \theta \circ \nabla H \quad = \theta \circ \Psi^n \circ \nabla H = \theta \circ \nabla H_n$$

$$= \theta \circ \Psi \circ \nabla H_{n-1} = \theta_1 \circ \nabla H_{n-1}$$

$$= \theta_1 \circ \Psi \circ \nabla H_{n-2} = \theta_2 \circ \nabla H_{n-2} \qquad (3.80)$$

$$\vdots$$

As Examples 3.9–3.15 satisfy the conditions of Theorem 3.13, hence $\{K_n\}$ is a commuting hierarchy of multi-Hamiltonian symmetries and $\{H_n\}$ is a hierarchy of constants of motion. Moreover, from Theorems 3.7 and 3.8 we know that $\theta \circ \nabla$ is a homomorphism from $\mathcal{F}_\theta$ to $\mathcal{L}$ and $h \circ J$ is a homomorphism from a subalgebra of inverse Hamiltonian vector fields to $\mathcal{F}^J$, hence

$$\{H_i, H_j\}_\theta = 0 \quad \text{and} \quad \{H_i, H_j\}^J = 0, \tag{3.81}$$

so the constants of motion are in involution with respect to both Poisson brackets.

Lemma 3.11 *The conserved quantities $\{H_n\}$ are in involution with respect to an arbitrary Poisson bracket $\{.,.\}_{\theta_n}$ and $\{.,.\}^{J_n}$.*

Proof.

$$\begin{aligned}
\{H_i, H_j\}_{\theta_n} &= \langle \nabla H_j, \theta_n \circ \nabla H_i \rangle = \langle \nabla H_j, \theta \circ \Psi^n \circ \nabla H_i \rangle \\
&= \langle \nabla H_j, \theta \circ \nabla H_{i+n} \rangle = \{H_{i+n}, H_j\}_\theta = 0,
\end{aligned} \tag{3.82}$$

$$\begin{aligned}
\{H_i, H_j\}^{J_n} &= \langle J_n \circ K_j, K_i \rangle = \langle J\Phi^n \circ K_j, K_i \rangle \\
&= \langle J \circ K_{j+n}, K_i \rangle = \{H_i, H_{j+n}\}^J = 0. \square
\end{aligned} \tag{3.83}$$

For a detailed discussion of Poisson–Nijenhuis manifolds we refer the reader to the literature [124], [125].

Example 3.18 The KdV hierarchy.

With the KdV equation (2.9) we can relate the whole hierarchy of dynamical systems

$$u_{t_n} = K_n(u) = \Phi^n(u)u_x = \left(D^2 + \frac{2}{3}au + \frac{1}{3}au_x D^{-1} \right)^n u_x, \tag{3.84}$$

where symmetries K_n are

$$K_0 = u_x,$$

$$K_1 = u_{3x} + auu_x,$$

$$K_2 = u_{5x} + \frac{10}{3}au_x u_{xx} + \frac{5}{3}auu_{3x} + \frac{5}{6}a^2 u^2 u_x,$$

$$K_3 = u_{7x} + \frac{21}{3}au_x u_{4x} + \frac{7}{3}auu_{5x} + \frac{35}{3}au_{2x}u_{3x}$$

$$+ \frac{70}{9}a^2 uu_x u_{xx} + \frac{35}{18}a^2 u^2 u_{3x} + \frac{35}{18}a^2 u_x^3 + \frac{35}{54}a^3 u^3 u_x,$$

$$\vdots$$

related cosymmetries γ_n are

$$\gamma_0 \;=\; u,$$

$$\gamma_1 \;=\; u_{xx} + \tfrac{1}{2}au^2,$$

$$\gamma_2 \;=\; u_{4x} + \tfrac{5}{3}auu_{xx} + \tfrac{5}{6}au_x^2 + \tfrac{5}{18}a^2u^3,$$

$$\gamma_3 \;=\; u_{6x} + \tfrac{7}{3}auu_{4x} + \tfrac{14}{3}au_xu_{3x} + \tfrac{7}{2}au_{xx}^2$$

$$+ \tfrac{35}{18}a^2u^2u_{xx} + \tfrac{35}{18}a^2uu_x^2 + \tfrac{35}{6\cdot36}a^3u^4,$$

$$\vdots$$

respective conserved quantities H_n are

$$H_0 \;=\; \int_R \tfrac{1}{2}u^2 \mathrm{d}x,$$

$$H_1 \;=\; \int_R \left(-\tfrac{1}{2}u_x^2 + \tfrac{1}{6}au^3\right) \mathrm{d}x,$$

$$H_2 \;=\; \int_R \left(\tfrac{1}{2}u_{xx}^2 + \tfrac{5}{12}au^2u_{xx} + \tfrac{5}{72}a^2u^4\right) \mathrm{d}x,$$

$$H_3 \;=\; \int_R \left(-\tfrac{1}{2}u_{3x}^2 + \tfrac{7}{6}auu_{2x}^2 - \tfrac{35}{36}a^2u^2u_x^2 + \tfrac{7}{6\cdot36}a^3u^5\right) \mathrm{d}x,$$

$$\vdots$$

and $\theta_n = \Phi^n \circ D$, $J_n = D^{-1} \circ \Phi^n$. For an arbitrary equation from the hierarchy (3.84) the sets $\{K_i\}, \{\gamma_i\}, \{H_i\}, \{\theta_i\}$ and $\{J_i\}$ are the sets of suitable invariant tensor fields.

In our present considerations we have not taken into account one important case. Actually we have to discuss a situation when a dynamical system is bi-Hamiltonian with respect to noninvertible Poisson operators θ_0 and θ_1. This is a typical situation for finite dimensional systems. In such a case there is no recursion operator undergoing implectic-symplectic factorization and we cannot apply our previous results. So, we have to consider this case separately.

Theorem 3.14 *Let a vector field K be a bi-Hamiltonian with respect to the compatible, noninvertible pair (θ_0, θ_1)*

$$K = \theta_0 \circ \nabla H_1 = \theta_1 \circ \nabla H_0 \tag{3.85}$$

and let H_0 be a Casimir of θ_0: $\theta_0 \circ \nabla H_0 = 0$. Then the hierarchy of vector fields $\{K_n\}$ generated by the pair (θ_0, θ_1)

$$\theta_0 \circ \gamma_0 \quad = 0$$

$$\theta_0 \circ \gamma_1 \quad = K_1 = \theta_1 \circ \gamma_0$$

$$\theta_0 \circ \gamma_2 \quad = K_2 = \theta_1 \circ \gamma_1 \tag{3.86}$$

$$\vdots \qquad \vdots$$

$$\theta_0 \circ \gamma_n \quad = K_n = \theta_1 \circ \gamma_{n-1},$$

$$\vdots \qquad \vdots$$

where $K = K_1, \gamma_0 = \nabla H_0, \gamma_1 = \nabla H$, is a Hamiltonian and Abelian subalgebra of $\mathcal{L}$.

Proof. Our proof is inductive. From the assumption that γ_n is closed we find that γ_{n+1} is also closed. Actually, from the compatibility of (θ_0, θ_1) we find

$$0 = L_{(\theta_0 + \theta_1)\gamma}(\theta_0 + \theta_1) + (\theta_0 + \theta_1) \circ d\gamma \circ (\theta_0 + \theta_1)$$

$$= L_{\theta_0 \circ \gamma}\theta_1 + L_{\theta_1 \circ \gamma}\theta_0 + \theta_0 \circ d\gamma \circ \theta_1 + \theta_1 \circ d\gamma \circ \theta_0.$$

For $\gamma = \gamma_n$ we have

$$0 = L_{\theta_0 \circ \gamma_n}\theta_1 + L_{\theta_1 \circ \gamma_n}\theta_0 + \theta_0 \circ d\gamma_n \circ \theta_1 + \theta_1 \circ d\gamma_n \circ \theta_0$$

$$= L_{\theta_1 \circ \gamma_{n-1}}\theta_1 + L_{\theta_0 \circ \gamma_{n+1}}\theta_0$$

$$= -\theta_1 \circ d\gamma_{n-1} \circ \theta_1 - \theta_0 \circ d\gamma_{n+1} \circ \theta_0$$

$$= -\theta_0 \circ d\gamma_{n+1} \circ \theta_0 \qquad \Longrightarrow d\gamma_{n+1} = 0, \tag{3.87}$$

Hence, all γ_n are closed. Now we pass to commutativity. Because all K_n are Hamiltonian, then $L_{K_n}\theta_0 = L_{K_n}\theta_1 = 0$. Our proof is again inductive. For fixed K_n we assume that $[K_n, K_m] = 0$. This means that

$$0 = L_{K_n}K_m = L_{K_n}(\theta_0 \circ \gamma_m) = (L_{K_n}\theta_0) \circ \gamma_m + \theta_0 \circ L_{K_n}\gamma_m$$

$$= \theta_0 \circ L_{K_n}\gamma_m \Longrightarrow L_{K_n}\gamma_m = 0.$$

Now we have

$$[K_n, K_{m+1}] = L_{K_n}K_{m+1} = L_{K_n}(\theta_1 \circ \gamma_m)$$

$$= (L_{K_n}\theta_1) \circ \gamma_m + \theta_1 \circ L_{K_n}\gamma_m = 0. \; \Box \tag{3.88}$$

When the vector field K represents a finite dimensional dynamical system, the hierarchy (3.86) terminates or is infinite. In this second case certainly only a finite number of integrals of motion are linearly independent.

Example 3.19 Let us consider the following Newtonian equations of motion

$$r_{1_{tt}} = r_2 - \tfrac{5}{2}r_1^2,$$

$$r_{2_{tt}} = c - 5r_1r_2 + \tfrac{5}{2}r_1^3. \tag{3.89}$$

Introducing the conjugate momenta $s_1 = r_{2_t}$ and $s_2 = r_{1_t}$, the system (3.89) has the following bi-Hamiltonian representation [30]

$$
\begin{pmatrix} r_1 \\ r_2 \\ s_1 \\ s_2 \\ c \end{pmatrix}_t
=
\begin{pmatrix} s_2 \\ s_1 \\ c - 5r_1r_2 + \tfrac{5}{2}r_1^3 \\ r_2 - \tfrac{5}{2}r_1^2 \\ 0 \end{pmatrix}
= K_1 = \theta_0 \circ \nabla H_1 = \theta_1 \circ \nabla H_0, \tag{3.90}
$$

where

$$
\theta_0 =
\begin{pmatrix}
0 & 0 & 1 & 0 & 0 \\
0 & 0 & 0 & 1 & 0 \\
-1 & 0 & 0 & 0 & 0 \\
0 & -1 & 0 & 0 & 0 \\
0 & 0 & 0 & 0 & 0
\end{pmatrix},
$$

$$
\theta_1 =
\begin{pmatrix}
0 & 0 & r_1 & -1 & 2s_2 \\
0 & 0 & 2r_2 & r_1 & 2s_1 \\
-r_1 & -2r_2 & 0 & s_2 & -10r_1r_2 + 5r_1^3 + 2c \\
1 & -r_1 & -s_2 & 0 & 2r_2 - 5r_1^2 \\
-2s_2 & -2s_1 & 10r_1r_2 - 5r_1^3 - 2c & -2r_2 + 5r_1^2 & 0
\end{pmatrix},
$$

$$H_0 = \tfrac{1}{2}c, \quad H_1 = s_1s_2 + \tfrac{5}{2}r_1^2r_2 - \tfrac{1}{2}r_2^2 - \tfrac{5}{8}r_1^4 - cr_1.$$

The vector field K_1 belongs to the bi-Hamiltonian chain

$$\theta_0 \circ \nabla H_0 = 0$$

$$\theta_0 \circ \nabla H_1 = K_1 = \theta_1 \circ \nabla H_0$$

$$\theta_0 \circ \nabla H_2 = K_2 = \theta_1 \circ \nabla H_1$$

$$0 = \theta_1 \circ \nabla H_2,$$

$$H_2 = -\tfrac{1}{2}s_1^2 - r_1s_1s_2 + s_2^2r_2 - 2r_1r_2^2 + \tfrac{1}{2}r_1^5 + cr_2 + \tfrac{1}{2}cr_1^2,$$

which starts with the Casimir of θ_0 and terminates with the Casimir of θ_1. One can verify that indeed θ_0 and θ_1 are compatible.

Example 3.20 Let us take two implectic operators [22]

$$
\theta_0 = \begin{pmatrix}
0 & 0 & 0 & 0 & 0 \\
0 & 0 & 0 & e_1 & 0 \\
0 & 0 & 0 & 0 & e_1 \\
0 & -e_1 & 0 & 0 & 0 \\
0 & 0 & -e_1 & 0 & 0
\end{pmatrix},
$$

$$
\theta_1 = \begin{pmatrix}
0 & 2e_1 & -e_2 & e_5 & 0 \\
-2e_1 & 0 & 2e_3 & e_4 & -e_5 \\
e_2 & -2e_3 & 0 & 0 & e_4 \\
-e_5 & -e_4 & 0 & 0 & 0 \\
0 & e_5 & -e_4 & 0 & 0
\end{pmatrix},
$$

which are not compatible. Starting from the Casimir $C_0 = f(e_1)$ of θ_0 we get the following chain

$$
\theta_0 \circ \nabla H_0 = 0
$$

$$
\theta_0 \circ \nabla H_1 = K_1 = \theta_1 \circ \nabla H_0
$$

$$
\theta_0 \circ \nabla H_2 = K_2 = \theta_1 \circ \nabla H_1
$$

$$
K_3 = \theta_1 \circ \nabla H_2,
$$

where

$$
H_0 = \tfrac{1}{3}e_1^3,
$$

$$
H_1 = e_1 e_2 e_5 - 2e_1^2 e_4,
$$

$$
H_2 = -e_3 e_5^2 + \tfrac{1}{2}e_1^{-1}e_2^2 e_5^2 - 3e_2 e_4 e_5 + 3e_1 e_4^2.
$$

Fortunately the vector fields K_1, K_2 are Hamiltonian with respect to the implectic operator θ_0 but K_3 is not. This truncates the chain. The flows of K_1 and K_2 are completely integrable since we have three commuting (with respect to θ_0 and θ_1) constants of motion H_0, H_1 and H_2. The implectic operator θ_1 has the Casimir function $C_1 = e_1 e_4^2 - e_2 e_4 e_5 - e_3 e_5^2 = H_2 - \tfrac{1}{2}H_1^2(3H_0)^{-1}$.

Let us come back again to systems with a recursion operator, represented by Examples 3.10-3.16. We still miss some information about the second hierarchy of symmetries, i.e. $\sigma_n = \Phi^n \circ \sigma_0$. In our investigations the following theorem is of help.

Theorem 3.15 *Let $u_t = K(u)$ be a dynamical system, σ its symmetry and T its arbitrary tensor invariant. The $L_\sigma T$ is again a tensor invariant of the same type.*

Proof.

$$\frac{\partial}{\partial t}(L_\sigma T) + L_K(L_\sigma T) = L_{\frac{\partial \sigma}{\partial t}}T + L_\sigma \frac{\partial T}{\partial t} + L_\sigma L_K T + L_{[K,\sigma]}T$$

$$= L_{\frac{\partial \sigma}{\partial t}+[K,\sigma]}T + L_\sigma(\frac{\partial T}{\partial t} + L_K T) = 0. \;\square \quad (3.91)$$

Generally Hamiltonian symmetries K_n are useless when applied to Theorem 3.15, because the time independent tensor fields T are tensor invariants of an arbitrary K_n and hence $L_{K_n}T = 0$. Quite a different situation occurs when non-Hamiltonian symmetries σ_n are considered.

Theorem 3.16 *Let Φ be a recursion hereditary operator for the vector field K and let σ be its symmetry which is a scaling vector field for K and Φ*

$$L_\sigma K = \rho K, \quad L_\sigma \Phi = \alpha \Phi, \quad \rho, \alpha = \text{const.}$$

and moreover,

$$K_n = \Phi^n \circ K, \quad \sigma_n = \Phi^n \circ \sigma.$$

Then,

$$L_{K_n}\Phi = 0, \quad L_{\sigma_n}\Phi = \alpha \Phi^{n+1}$$

and

$$[K_n, K_m] = 0, \quad [\sigma_n, K_m] = (\rho + \alpha m)K_{m+n},$$

$$(3.92)$$

$$[\sigma_n, \sigma_m] = \alpha(m - n)\sigma_{n+m}.$$

The algebra (3.92) is called a hereditary algebra or a Virasoro algebra of the dynamical system $u_t = K(u)$.

Proof. From the hereditary property of Φ we have

$$L_{K_n}\Phi = L_{\Phi^n \circ K}\Phi = \Phi^n \circ L_K \Phi = 0,$$

$$L_{\sigma_n}\Phi = L_{\Phi^n \circ \sigma}\Phi = \Phi^n \circ L_\sigma \Phi = \alpha \Phi^{n+1},$$

and hence

$$[K_n, K_m] = L_{K_n}(\Phi^m \circ K) = \Phi^m \circ L_{K_n}K$$
$$= -\Phi^m \circ L_K(\Phi^n \circ K) = -\Phi^{n+m} \circ L_K K$$
$$= 0,$$

$$[\sigma_n, K_m] = -L_{K_m}(\Phi^n \circ \sigma) = -\Phi^n \circ L_{K_m}\sigma = \Phi^n \circ L_\sigma K_m$$
$$= \Phi^n \circ L_\sigma(\Phi^m \circ K) = \Phi^n \circ [(L_\sigma \Phi^m) \circ K + \Phi^m \circ L_\sigma K]$$
$$= (\rho + \alpha m)K_{m+n},$$

$$\begin{aligned}
[\sigma_n, \sigma_m] &= L_{\sigma_n}\sigma_m = L_{\sigma_n}(\varPhi^m \circ \sigma) = L_{\sigma_n}(\varPhi^m) \circ \sigma + \varPhi^m \circ L_{\sigma_n}\sigma \\
&= m\alpha\varPhi^{n+m} \circ \sigma - \varPhi^m \circ L_\sigma(\varPhi^n \circ \sigma_0) \\
&= m\alpha\varPhi^{n+m} \circ \sigma_0 - n\alpha\varPhi^{n+m} \circ \sigma_0 \\
&= \alpha(m-n)\sigma_{n+m}. \qquad \Box
\end{aligned}$$

Summarizing the results of Theorem 3.16 we find that if for a given dynamical system $u_t = K(u)$ there exists a hereditary recursion operator $\varPhi$, then we can relate to that system the following infinite hierarchy of dynamical systems

$$u_{t_m} = K_m(u), \quad u_{\epsilon_n} = \sigma(u,t), \tag{3.93}$$

where the vector fields $K_m(u)$ and $\sigma_n(u,t)$ are symmetries of a given dynamical system and satisfy the hereditary algebra (3.92).

If now the recursion operator is factorized, $\varPhi = \theta \circ J$, $J \circ K_n = \gamma_n$ and $J \circ \sigma_n = \zeta_n$, then $K_n = \theta \circ \gamma_{n-1}$, $\sigma_n = \theta \circ \zeta_{n-1}$ and we can construct two hereditary algebras of cosymmetries

$$\{\gamma_n, \gamma_m\}^J = 0, \quad \{\zeta_n, \gamma_m\}^J = (\rho + \alpha m)\gamma_{n+m},$$

$$\{\zeta_n, \zeta_m\}^J = \alpha(m-n)\zeta_{n+m}, \tag{3.94}$$

$$\{\gamma_n, \gamma_m\}_\theta = 0, \quad \{\zeta_n, \gamma_m\}_\theta = (\rho + \alpha m)\gamma_{n+m+1},$$

$$\{\zeta_n, \zeta_m\}_\theta = \alpha(m-n)\zeta_{n+m+1}. \tag{3.95}$$

We leave to the reader the proof of the statement that $\varPsi$ is a Lie algebra homomorphism of the hereditary algebra $\{.,.\}_\theta$ into the hereditary algebra $\{.,.\}^J$.

The importance of non-Hamiltonian symmetries is revealed in the following theorem [148]

Theorem 3.17 *For a vector field K let $\varPhi$ be a recursion hereditary operator with implectic-symplectic factorization $\varPhi = \theta \circ J$. Then, let σ be a scaling (conformal) symmetry of K and simultaneously a scaling vector field for J, θ and hence for $\varPhi$ as well:*

$$L_\sigma K = \rho K, \quad L_\sigma \theta = \lambda \theta, \quad L_\sigma J = \beta J, \quad L_\sigma \varPhi = (\lambda + \beta)\varPhi = \alpha\varPhi.$$

Moreover let

$$K_n = \varPhi^n \circ K, \quad \sigma_n = \varPhi^n \circ \sigma, \quad \theta_n = \varPhi^n \circ \theta, \quad J_n = J \circ \varPhi^n,$$

$$J \circ K_n = \gamma_n = \nabla H_n, \quad K_n = \theta \circ \gamma_{n-1}, \quad J \circ \sigma_n = \zeta_n, \quad \sigma_n = \theta \circ \zeta_{n-1}.$$

Then

$$\text{1.} \qquad L_{\sigma_n} K_m = (\rho + m\alpha) K_{n+m}$$

$$\text{2.} \qquad L_{\sigma_n} \sigma_n = (m-n)\alpha \; \sigma_{n+m}$$

$$\text{3.} \qquad L_{\sigma_n} \Phi = \alpha \; \Phi^{n+1}$$

$$\text{4.} \qquad L_{\sigma_n} \theta_m = (-\beta + (m-n+1)\alpha)\theta_{n+m}$$

$$\text{5.} \qquad L_{\sigma_n} J_m = (-\beta + (n+m)\alpha) J_{n+m}$$

$$\text{6.} \qquad L_{\sigma_n} \gamma_m = (\rho + \beta + (n+m)\alpha)\gamma_{n+m}$$

$$\text{7.} \qquad L_{\sigma_n} \zeta_m = (\beta + m\alpha)\zeta_{n+m}$$

$$\text{8.} \qquad L_{\sigma_n} H_m = (\rho + \beta + (n+m)\alpha) H_{n+m}.$$

$$(3.96)$$

Proof. We proved 1–3 in the previous theorem. For 4–5 we have

$$L_{\sigma_n} J = \mathrm{d}(J \circ \sigma_n) = L_\sigma J_n = L_\sigma (J \circ \Phi^n) = (\beta + n\alpha) J_{n+1},$$

$$L_{\sigma_n} \theta = L_{\theta \circ J_n \circ \sigma}\theta = -\theta \circ \mathrm{d}(J_n \circ \sigma) \circ \theta = -\theta \circ (L_\sigma J_n) \circ \theta = -(\beta + (n-1)\alpha)\theta_n.$$

Now application of the Leibniz chain rule and 3 gives the desired result. Eq. 6-8 follow again from the application of the chain rule and results 1-5. $\square$

Obviously, we can always rescale formulas (3.96) through the choice $\bar{\sigma} = \frac{1}{\alpha}\sigma$. This gives a new hierarchy $\bar{\sigma}_n = \frac{1}{\alpha}\sigma_n$ and new scaling constants $\bar{\alpha} = 1$, $\bar{\rho} = \rho/\alpha$, $\bar{\beta} = \beta/\alpha$ and $\bar{\lambda} = \lambda/\alpha$.

Theorem 3.17 assures us that if $L_\sigma \Phi \neq 0$, i.e. $\alpha \neq 0$, then an arbitrary symmetry σ_n of a system $u_t = K(u)$ is a non-Hamiltonian vector field with respect to any θ_m except at most one. Indeed, let $\sigma_n = \theta_m \circ \zeta$. Then

$$L_{\sigma_n}\theta_m = L_{\theta_m \circ \zeta}\theta_m = -\theta_m \circ \mathrm{d}\zeta \circ \theta_m = 0 \Longrightarrow \zeta = \mathrm{d}F.$$

On the other hand, from Theorem 3.17(4), for fixed n we have at most one solution of equation $-\beta + (m-n)\alpha = 0$ with respect to m such that $m \in N_0$. So if for $n = 0$, $m = \frac{\beta}{\alpha} - 1 = r \in N_0$, there exists one linear in time integral of motion $G(u,t)$ such that $\sigma = \theta_r \circ \nabla G$ and hence,

$$\sigma_n = \theta_{r+n} \circ \nabla G. \qquad (3.97)$$

Before we pass to examples we introduce some gradation in the algebra of symmetries. For a given vector field v, let the set $\{K_n\}$ be an Abelian subalgebra of its symmetries. Then a symmetry σ of a vector field v is called a *symmetry of degree* d if for an arbitrary $K_i \in \{K_n\}$

$$(L_{K_i})^{d+1}\sigma = 0. \qquad (3.98)$$

For the hereditary algebra (3.92) we find that the symmetries $\{K_n\}$ are of degree 0 and the symmetries $\{\sigma_n\}$ are of degree 1.

Assume now that $v = K$ is a time independent vector field and $K \in \{K_n\}$. In this case we can introduce a gradation for all types of tensor invariants and relate it to the explicit dependence of the invariants on time. First, we notice that for an arbitrary time independent tensor field T_0, the formal series

$$\exp(-t \cdot L_K)T_0 = \sum_{n=0}^{\infty} \frac{(-t)^n}{n!}(L_K)^n T_0 = T(u,t) \qquad (3.99)$$

is a tensor invariant of the dynamical system $u_t = K(u)$ as

$$\frac{\partial T}{\partial t} + L_K T = 0.$$

In general, of course, such a series does not make sense as it might not converge. So, of special interest are these T_0 for which the series (3.99) truncates, as they generate polynomial in time tensor invariants.

Definition 3.10

(i)   The time independent tensor field T_0 is called a master tensor invariant of degree d with respect to a dynamical system $u_t = K(u)$ if

$$(L_K)^{d+1}T_0 = 0. \qquad (3.100)$$

(ii) We call the time dependent tensor invariant T of a dynamical system $u_t = K(u)$ a tensor invariant of degree d if

$$(L_K)^{d+1}T = 0. \qquad (3.101)$$

The formula (3.99) gives a one to one correspondence between a master tensor invariant of degree d and an invariant tensor field being a polynomial of degree d with respect to t. Of course, the master tensor invariant of degree 0 is equal to a time independent tensor invariant. In the case of vector fields, the idea of master symmetries was introduced by Fokas and Fuchssteiner [71], [86]. If τ is a time independent scaling vector field for K, then in a natural way it is a master symmetry of degree 1 for K. If, moreover, τ is a scaling vector field for a recursion operator Φ then τ is a scaling vector field for the whole hierarchy $\{K_n\}$ and hence a master symmetry of degree 1 for all members of the hierarchy. As a result, applying the formula (3.99), for an arbitrary $K \in \{K_n\}$

$$\sigma = \tau + \rho t K \qquad (3.102)$$

is a scaling symmetry of K and ρ is a scaling constant $L_\tau K = \rho K$. Applying the recursion operator Φ we get the whole hierarchy of symmetries

$$\sigma(u,t) = \Phi^n \tau(u) + \rho t \Phi^n K(u) = \tau_n(u) + \rho t K_n(u), \qquad (3.103)$$

where τ_n are suitable master symmetries of degree 1. Notice that a replacement of τ_n for σ_n in the hereditary algebra (3.92) does not change the algebra. This means that we have a one to one correspondence between the algebra of symmetries and the algebra of master symmetries. The same exchange can be made in formulas (3.96) of Theorem 3.17. Finally, according to (3.97) there exists at most one θ_r such that

$$\tau = \theta_r \circ \nabla F \quad \text{and} \quad G(u) = F(u) + \rho t H(u), \qquad (3.104)$$

where $\theta_r \circ \nabla H = K$.

Remark 3.4 Note that if for a Hamiltonian system $u_t = K(u) = \theta(u)\nabla H(u)$, with invertible θ, we can find in any way (by computer algebra, for example) one nontrivial non-Hamiltonian master symmetry, let say τ_1, then we can generate the whole hereditary algebra as well as all tensor invariants, applying formulas (3.96). The recursion operator, for example, can be constructed by the formula

$$\Phi = (L_{\tau_1}\theta) \circ \theta^{-1}.$$

Example 3.21 The KdV hierarchy.
Let us supplement Example 3.18 by additional information. We have

$$\tau_{-1} \;=\; \tfrac{3}{a},$$

$$\tau_0 \;=\; 2u + xu_x,$$

$$\tau_1 \;=\; xK_1 + 4u_{xx} + \tfrac{4}{3}au^2 + \tfrac{1}{3}au_x D^{-1}u,$$

$$\tau_2 \;=\; xK_2 + 6u_{4x} + 6au_x^2 + 8auu_{xx} + \tfrac{1}{3}au_{3x}D^{-1}u$$

$$+\tfrac{8}{9}a^2 u^3 + 2a^2 uu_x D^{-1}u + \tfrac{1}{6}a^2 u_x D^{-1}u^2,$$

$$\vdots$$

and for $K_0 = u_x$ we find $\rho = 1, \beta = 2, \lambda = 0, \alpha = 2, T = \tfrac{3}{a}\int_R xu\,dx$. Notice that in general, the vector fields τ_n are non-local objects.

Example 3.22
MKdV: $K_0 = u_x, \rho = 1, \lambda = 0, \beta = 2, \alpha = 2.$
SK: $K_0 = u_x, \overline{K}_0 =$ the SK vector field, $\rho = 1, \overline{\rho} = 5, \lambda = 0, \ \beta = 2, \alpha = 2.$
AKNS: $K_0 = \begin{pmatrix} q \\ r \end{pmatrix}_x, \rho = 1, \lambda = 1, \beta = 0, \alpha = 1.$
Toda: $K_0 =$ the Toda vector field $\rho = 1, \lambda = 0, \beta = 1, \alpha = 1.$

Assume now that for a given vector field K, the set $\{K_n\} := \tau_0$ forms a hierarchy of its time independent commuting symmetries (i.e. master symmetries of degree 0). Let τ_s and τ_r be master symmetries of degree s and r, respectively. Then $[\tau_s, \tau_r]$ is a master symmetry of degree $s + r - 1$. Indeed

$$(L_K)^{r+s}[\tau_s, \tau_r] = \sum_{i=0}^{r+s} \binom{r+s}{i} [(L_K)^i \tau_s, (L_K)^{r+s-i} \tau_r], \qquad (3.105)$$

where $\binom{:}{:}$ denotes the binomial coefficient, but as $(L_K)^{s+1}\tau_s = 0$ and $(L_K)^{r+s}\tau_r = 0$, the only term different from zero can be the one with $i = s$, i.e. $[(L_K)^s \tau_s, (L_K)^r \tau_r]$. On the other hand $(L_K)^s \tau_s, (L_K)^r \tau_r \in \tau_0$, hence (3.105) is equal to zero. Henceforth we shall use the notation $\tau_{r,n}$ for the nth master symmetry of degree r.

Theorem 3.18 *Assume that for a given vector field K we have master symmetries $\tau_{0,0}, \tau_{1,0}, \tau_{1,1}$, and $\tau_{2,0}$ such that*

$$[\tau_{1,0}, \tau_{0,0}] = \rho \tau_{0,0}, \quad [\tau_{1,0}, \tau_{1,1}] = \tau_{1,1}, \quad [\tau_{2,0}, \tau_{0,0}] = 2\rho \tau_{1,0},$$

$$[\tau_{2,0}, \tau_{1,0}] = \rho \tau_{2,0}, \qquad \rho = \text{const.} \qquad (3.106)$$

Moreover all $\tau_{0,n}$, defined by $[\tau_{1,1}, \tau_{0,n}] = (n + \rho)\tau_{0,n+1}$, commute in pairs. Then, the following Lie algebra of master symmetries exists [24]

$$[\tau_{s,m}, \tau_{r,n}] = (s(n + \rho) - r(m + \rho))\tau_{s+r-1,n+m}, \qquad (3.107)$$

where for $r \geq 1$ $\tau_{r,n}$ are defined recursively by $[\tau_{2,0}, \tau_{r,n}] = (2(n + \rho) - r\rho)\tau_{r+1,n}$.

Proof. From the commutativity of $\tau_{0,n}$ the commutator of master symmetries of degree s and r gives indeed a master symmetry of degree $s + r - 1$. The rest of the proof consists of the following inductive steps
  1. $[\tau_{1,0}, \tau_{0,n}] = (n + \rho)\tau_{0,n}$.
  2. $[\tau_{1,0}, \tau_{r,n}] = ((n + \rho) - r\rho)\tau_{r,n}$.
  3. $[\tau_{0,0}, \tau_{r,n}] = -r\rho\tau_{r-1,n}$.
  4. $[\tau_{0,m}, \tau_{r,n}] = -r(m + \rho)\tau_{r-1,m+n}$.
  5. $[\tau_{s,m}, \tau_{r,n}] = (s(n + \rho) - r(m + \rho))\tau_{s+r-1,m+n}$.
In each of the steps we use the results of the previous ones, as well as the Jacobi identity. $\square$

Let $\sigma_{0,1} := (1 + \rho)K$. Then with arbitrary master symmetry $\tau_{r,n}$ we can relate a time dependent symmetry $\sigma_{r,n}$ of degree r in the form

$$\sigma_{r,n} = \exp(-tL_K)\tau_{r,s} = \sum_{i=0}^{r} \binom{r}{i} t^i \tau_{r-i,n+i}, \qquad (3.108)$$

which can be verified by direct calculations.

Lemma 3.12 *The symmetries $\sigma_{r,n}$ form the same Lie algebra (3.107) as the master symmetries.*

Proof. From (2.106) and (2.123) the push-forward ϕ_{t*} has a formal representation $\phi_{t*} = \exp(-tL_K)$. But the push-forward is a Lie algebra homomorphism and $\phi_{t*}\tau_{r,n} = \sigma_{r,n}$.$\square$

Summarizing the results we find that if a given dynamical system $u_t = K(u)$ satisfies the conditions of Theorem 3.18 then we can relate to it an infinite number of dynamical systems

$$u_{t_{r,s}} = \sigma_{r,s}(u,t), \qquad r,m \in N_0, \tag{3.109}$$

where the vector fields $\sigma_{r,s}(u,t)$ represent symmetries of a given dynamical system and satisfy the *general algebra* (3.107). Moreover, for a fixed r all symmetries $\sigma_{r,s}(u,t)$ are t-polynomials of order r. The symmetries $\sigma_{0,n}, \sigma_{1,m}$ (respective master symmetries $\tau_{0,n}, \tau_{1,m}$) constitute a subalgebra:

$$[\sigma_{0,n}, \sigma_{0,m}] = 0, \qquad [\sigma_{1,n}, \sigma_{0,m}] = (\rho + m)\sigma_{0,n+m},$$

$$[\sigma_{1,n}, \sigma_{1,m}] = (m - n)\sigma_{1,n+m}, \tag{3.110}$$

which remains the hereditary algebra (3.92). Nevertheless, the reader should notice that for a hereditary algebra, $\Phi\sigma_n = \sigma_{n+1}$ and for the general algebra (3.107) $[\sigma_{2,1}, \sigma_{0,n}] = 2(n + \rho)\sigma_{1,n+1}$, hence, besides the scaling symmetry $\sigma_0 = \sigma_{1,0}$, $\{\sigma_n\}$ and $\{\sigma_{1,n}\}$ are completely different sets of symmetries of degree 1. On the other hand, as will be verified later on, a proper relation between the symmetries $\sigma_{0,n}$ and K_n is the following one

$$\sigma_{0,n} = (n + \rho)K_n,$$

so, in the $(K_n, \sigma_{1,m})$ representation, the algebra (3.110) reads

$$[K_n, K_m] = 0, \quad [\sigma_{1,n}, K_m] = (\rho + n + m)K_{n+m},$$

$$[\sigma_{1,n}, \sigma_{1,m}] = (m - n)\sigma_{1,n+m}. \tag{3.111}$$

For reasons which will be more clear later on let us call the algebra (3.111) the *non-canonical action/angle algebra* [25]. Notice that if vector fields K and $\tau_{2,0}$ are Hamiltonian with respect to some Poisson operator θ, then the whole algebra (3.107) is the algebra of Hamiltonian vector fields. As a matter of fact this is the case for all known infinite dimensional examples. Consequently, one can find respective algebras of covectors $\xi_{r,n}$ (master cosymmetries) and scalar fields $T_{r,n}$ (master constants of motion)

$$\{\xi_{s,m}, \xi_{r,n}\}_\theta = (s(n + \rho) - r(m + \rho))\xi_{s+r-1,m+n},$$

$$\{T_{s,m}, T_{r,n}\}_\theta = (s(n + \rho) - r(m + \rho))T_{s+r-1,m+n}, \tag{3.112}$$

related to the time dependent cosymmetries $\gamma_{r,n}$ and conserved quantities $H_{r,n}$ by the relations

$$\gamma_{r,n} = \exp(-tL_K)\xi_{r,n} = \sum_{i=0}^{r} \binom{r}{i} t^i \xi_{r-i,n+i},$$

$$\hspace{8cm}(3.113)$$

$$H_{r,n} = \exp(-tL_K)T_{r,n} = \sum_{i=0}^{r} \binom{r}{i} t^i T_{r-i,n+i},$$

as θ and ∇ commute with L_K. Unfortunately, for all examples we have considered so far, there is no master symmetry of degree 2 (at least in the form of a differential function of the u variable with respect to the x-coordinate) and hence we cannot generate algebra (3.107). We will do it later on for the finite dimensional restrictions of infinite systems. Let us pass to examples of infinite systems with the Hamiltonian algebra (3.107).

Example 3.23 The Benjamin-Ono (BO) equation [160] of motion

$$u_t = 4auu_x + \mathcal{H}\,u_{xx} = K(u)$$

$$= D\,\nabla \int_R \left(\frac{1}{2}u\mathcal{H}\,u_x + \frac{2}{3}au^3\right) dx = \theta\nabla H(u), \quad a = \text{const. (3.114)}$$

where $\mathcal{H}$ means the Hilbert transform

$$(\mathcal{H}\,f)(x) = P\,\frac{1}{\pi} \int_R \frac{f(y)}{y-x}\,dy \qquad \text{(principal value integral),} \qquad (3.115)$$

is an excellent example of a dynamical system with a general Hamiltonian algebra (3.107) [20],[24]. The first few master symmetries are of the form [71]

$$T_{0,0} = u_x, \quad T_{0,1} = 2K = 8auu_x + 2\mathcal{H}\,u_{xx}, \quad \dots$$

$$T_{1,-1} = \frac{1}{4a}, \quad T_{1,0} = xu_x + u, \quad T_{1,1} = x\sigma_{0,1} + 4au^2 + 3\mathcal{H}\,u_x, \quad \dots$$

$$T_{2,-1} = \frac{x}{a}, \quad T_{2,0} = x^2 u_x + 2xu, \quad \dots$$

$$\dots$$

$$\hspace{8cm}(3.116)$$

and generate the algebra (3.107) with $\rho = 1$. The master symmetry $T_{2,0}$ is the Hamiltonian vector field $T_{2,0} = D\,\nabla \int_R \frac{1}{2}x^2 u^2 dx$, and the whole algebra (3.107) then consists of Hamiltonian vector fields as well. Because we have

$$\{T_{r,n+1}, T_{1,-1}\}_\theta = \langle \nabla T_{1,-1}, T_{r,n+1} \rangle = -(n+2)T_{r,n}$$

and $\nabla T_{1,-1} = \frac{x}{4a}$, we find

$$T_{r,n} = -\frac{1}{4a}\frac{1}{n+2} \int_R xT_{r,n+1} dx. \hspace{3cm} (3.117)$$

Finally, we mention some properties of the Hilbert transform, useful for the calculation procedure. If $u, v \in S(\mathbb{R})$, then

(i) $\int_R u\mathcal{H}\, v \mathrm{d}x = - \int_R v\mathcal{H}\, u \mathrm{d}x,$
(ii) $\mathcal{H}\mathcal{H}\, u = -u,$
(iii) $D\mathcal{H}\, u = \mathcal{H}\, Du,$
(iv) $(\mathcal{H}\, u)(\mathcal{H}\, v) = uv + \mathcal{H}\,(u\mathcal{H}\, v) + \mathcal{H}\,(v\mathcal{H}\, u),$
(v) $x\mathcal{H}\, u = -\frac{1}{\pi} \int_R u(x)\mathrm{d}x + \mathcal{H}\,(xu).$

Contrary to the previous examples, for the BO equation non-Hamiltonian master symmetries have not been found in the form of differential functions (with the Hilbert transform operator) of the variable u with respect to the x-coordinate. Hence, there is no bi-Hamiltonian formulation and no recursion operator, at least in the framework of the theory which has been presented till now.

Before we pass to another example, the reader should be supplied with some information about the dimensionality of infinite systems. So far our theory has been developed for $(1+1)$-dimensional field and lattice systems, that is the systems described by one space variable x (or discrete space variable n) and one time variable t. In principle, it does not seem difficult to extend all our differential formulas from one space variable to their greater number. Actually, whenever we have the summation over x-derivatives we change it into the multi-index summation. For example, the total x_i derivative of a differential function $f(x,u)$ where $x = (x_1, ..., x_m)$ and $u = (u^1, ..., u^n)$, has the general form

$$D_{x_i} f = \frac{\partial f}{\partial x_i} + \sum_{j=1}^{n} \sum_{|\alpha|} (u_{\alpha x}^j)_{x_i} \frac{\partial f}{\partial u_{\alpha x}^j}, \tag{3.118}$$

and the variational derivative (gradient) reads

$$\frac{\delta}{\delta u^j} \int_\Omega f(u)\mathrm{d}x_1...\mathrm{d}x_m = \sum_{|\alpha|} (-D)^{|\alpha|} \frac{\partial f}{\partial u_{\alpha x}^j}, \tag{3.119}$$

where the summation is extended over all multi-indices $\alpha = (\alpha_1, ..., \alpha_m)$, $|\alpha| = \sum_{i=1}^{m} \alpha_i$ and

$$u_{\alpha x}^j = u_{\alpha_1 x...\alpha_m x_m}^j = \frac{\partial^{\alpha_1 + ... + \alpha_m} u^j}{\partial x_1^{\alpha_1}...\partial x_m^{\alpha_m}}, \quad (-D)^{|\alpha|} = (-1)^{\alpha_1 + ... + \alpha_m} D_{x_1}^{\alpha_1}...D_{x_m}^{\alpha_m}.$$

When $m = 2, n = 1, x_1 = x, x_2 = y$ we find

$$D_x f = \frac{\partial f}{\partial x} + u_x \frac{\partial f}{\partial u} + u_{xx} \frac{\partial f}{\partial u_x} + u_{xy} \frac{\partial f}{\partial u_y} + u_{3x} \frac{\partial f}{\partial u_{xx}} + ...,$$

$$\frac{\delta}{\delta u} \int_\Omega f(u)\, \mathrm{d}x\mathrm{d}y = \frac{\partial f}{\partial u} - \partial_x \frac{\partial f}{\partial u_x} - \partial_y \frac{\partial f}{\partial u_y} + \partial_x^2 \frac{\partial f}{\partial u_{xx}} + \partial_x \partial_y \frac{\partial f}{\partial u_{xy}} + ... \tag{3.120}$$

The problem arises when one tries to find a bi-Hamiltonian formulation and then a recursion operator. Since the bi-Hamiltonian theory was introduced, there has been no evidence of bi-Hamiltonian field systems in $(2+1)$-dimensions. Then it was proved in [195] that bi-Hamiltonian structures in the form of operators (2.65) do not exist for $(n+1)$-dimensional infinite systems when $n \geq 2$. So, to overcome this difficulty, for multi-dimensional systems we need a more general concept of a bi-Hamiltonian structure. We will come back to this problem later on. Now we shall consider our first $(2+1)$-dimensional field system.

Example 3.24 The Kadomtsev–Petviashvili (KP) dynamical system [100], [196]

$$u_t = u_{3x} + 6uu_x + 3D_x^{-1}u_{yy} = K(u)$$

$$= D_x \nabla \int_\Omega \left(-\frac{1}{2}u_x^2 + u^3 + \frac{3}{2}(D_x^{-1}u_y)^2 \right) dxdy, \qquad (3.121)$$

is one of the most important $(2+1)$-dimensional field systems, with the Lax representation, which will appear in our further considerations many times. In this example we are interested in its algebra of symmetries and conserved quantities. The first few master symmetries [49],[146] can be chosen in the following form

$$\tau_{0,-1} = u_x, \ \tau_{0,0} = 2u_y, \ \tau_{0,1} = 3K = 3u_{3x} + 18uu_x + 9D_x^{-1}u_{yy}, \ ...$$

$$\tau_{1,-1} = yu_x, \ \tau_{1,0} = 2yu_y + xu_x + 2u, \ \tau_{1,1} = 2xu_y + y(u_{3x} + 6uu_x$$

$$+3D_x^{-1}u_{yy}) + 4D_x^{-1}u_y, \ ...$$

$$\tau_{2,-1} = x + y^2u_x, \ \tau_{2,0} = 2y^2u_y + 2xyu_x + 4yu, \ ...$$

$$\vdots$$

$$(3.122)$$

which generate the algebra (3.107) with $\rho = 2$. The master symmetry $\tau_{2,0}$ is the Hamiltonian vector field

$$\tau_{2,0} = D_x \nabla \int_\Omega y^2 u D_x^{-1} u_y \, dxdy,$$

hence the whole algebra (3.107) consists of Hamiltonian vector fields. Indeed, as

$$\{\tau_{r+1,n+1}, T_{0,-1}\}_\theta = \langle \nabla T_{0,-1}, \tau_{r+1,n+1} \rangle = (r+1)T_{r,n}$$

and $\nabla T_{0,-1} = u$, we find

$$T_{r,n} = \frac{1}{r+1} \int_\Omega u\tau_{r+1,n+1} \, dxdy. \qquad (3.123)$$

3.6 The Miura Transformation

Miura presented his famous transformation almost 30 years ago [136]. He showed that if

$$u = M(v) = -v_x - v^2 \qquad (3.124)$$

and v satisfies the MKdV equation (3.30)

$$v_t = v_{3x} - 6v^2 v_x, \qquad (3.125)$$

then u satisfies the KdV equation (3.29)

$$u_t = u_{3x} + 6uu_x. \qquad (3.126)$$

Indeed, applying the transformation formula (2.43) we find

$$
\begin{aligned}
M' \circ v_t &= (-D - 2v)(v_{3x} - 6v^2 v_x) \\
&= -v_{4x} - 6(v^2 v_x)_x - 2vv_{3x} + 12v^3 v_x \\
&= (-v_x - v^2)_{3x} + 6(-v_x - v^2)(-v_x - v^2)_x \\
&= u_{3x} + 6uu_x.
\end{aligned} \qquad (3.127)
$$

Although the map (3.124) is nonlinear and not invertible, and involves v and v_x , it 'magically' in the correct combination gives (3.126), i.e. again the local equation. The most important property of the transformation (3.124) is that it could be used to construct the second Hamiltonian structure of the KdV equation out of the first Hamiltonian structure of the MKdV equation. Actually, applying the transformation law (2.45) to the first Poisson operator $(-D)$ of the MKdV we find

$$
\begin{aligned}
M'(v)(\partial^{-1})M'(v) &= (-D - 2v)(-D)(-D - 2v) \\
&= D^3 + 2(-v_x - v^2)D + 2D(-v_x - v^2) \\
&= D^3 + 2uD + 2Du = D^3 + 4uD + 2u_x.
\end{aligned} \qquad (3.128)
$$

The implectic nature of the operator (3.128) follows from that of $(-D)$. On the other hand the transformation (3.124) if applied to the first symplectic operator of the KdV equation produces the second symplectic operator of the MKdV equation

$$
\begin{aligned}
M'^{\dagger}(v)(D^{-1})M'(v) &= (-D - 2v)D^{-1}(-D - 2v) \\
&= -D + 4vD^{-1}v = (-D)\varPhi(v).
\end{aligned} \qquad (3.129)
$$

In a more general algebraic setting, let $u = (u^1, ..., u^n)^{\mathrm{T}}$ and $v = (v^1, ..., v^n)^{\mathrm{T}}$ be the respective unmodified and modified variables. Then:

Definition 3.11

(i) *For an infinite dimensional system $v_t = K(v)$ the mapping $u = M(u)$ is a Miura map, if M is not invertible and, when $K(v)$ is a Hamiltonian vector field with respect to some local Poisson operator $\theta(v)$, then $\theta(u) = M'(v)\theta(v)M'^{\dagger}(v)$ is also a local operator with respect to the u variable.*

(ii) *For a finite dimensional system $v_t = K(v)$ the mapping $u = M(v)$ is a Miura map if M is nonlinear, and when $K(v)$ is a Hamiltonian vector field, M is a noncanonical transformation.*

The Miura map will appear many times in our further considerations. Here, we only give a few more examples.

Example 3.25 The modified dispersive water wave (MDWW) equation [113]

$$
\begin{pmatrix} v \\ w \end{pmatrix}_t = \frac{1}{2} \begin{pmatrix} -v_x + 2vw - v^2 \\ w_x - 2v_x - 2v^2 + 2vw + w^2 \end{pmatrix}_x
$$

$$
= \frac{1}{2} \begin{pmatrix} 0 & D \\ D & 2D \end{pmatrix} \nabla \int_R (-v_x w - v^2 w + vw^2)\mathrm{d}x
$$

$$
= \frac{1}{4} \begin{pmatrix} vD + Dv & -D^2 + wD + 2Dv \\ D^2 + Dw + 2vD & 2wD + 2Dw \end{pmatrix} \nabla \int_R (2vw - 2v^2)\mathrm{d}x
$$

$$(3.130)$$

is related to the DWW equation (3.34) by the noninvertible Miura map

$$
q = -v_x - v^2 + vw, \qquad r = w. \tag{3.131}
$$

The reader can verify that the image of the first Poisson operator of the MDWW equation is equal to the second Poisson operator of the DWW system

$$
M' \begin{pmatrix} 0 & \frac{1}{2}D \\ \frac{1}{2}D & D \end{pmatrix} M'^{\dagger} = \frac{1}{2} \begin{pmatrix} qD + Dq & -D^2 + rD \\ D^2 + Dr & 2D \end{pmatrix}.
$$

Example 3.26 The modified Kadomtsev–Petviashvili (MKP) equation

$$
v_t = v_{3x} - 6v^2 v_x + 3D_x^{-1}v_{yy} - 6v_x D_x^{-1}v_y \tag{3.132}
$$

is our second example of (2+1)-dimensional field systems. The Miura map relating the MKP equation (3.132) with the KP equation (3.121) is the following

$$
u = -v_x - v^2 - D_x^{-1}v_y \tag{3.133}
$$

and can be considered as a (2+1)-generalization of the MKdV-KdV Miura transformation (3.124). But contrary to the (1+1)-case, (3.132) does not have a Hamiltonian formulation in the framework of the present formalism. Nevertheless, MKP is an inverse-Hamiltonian system with a symplectic operator $J(v)$ being the Miura image of the KP symplectic operator $J = \theta^{-1} = D^{-1}$

$$J(v) = M'^{\dagger}(D^{-1})M' = -D_x + 4vD_x^{-1}v + 2vD_yD_x^{-2} + 2D_yD_x^{-2}v + D_y^2D_x^{-3}$$
$$(3.134)$$

and might be considered as a (2+1)-dimensional version of (3.129). The appropriate Hamiltonian is $H(v) = H(u) \circ M$.

A more general and systematic treatment of the Miura transformation and related modified systems will be given in the frames of the $\mathcal{R}$-matrix formalism in Chap. 7.

4. Lax Representations of Multi-Hamiltonian Systems

As mentioned in the Introduction, the existence of nontrivial, i.e. not coming from a local conservation law [44], Lax representation or zero-curvature representation is a principal condition necessary for the solution of the Cauchy problem for the underlying dynamical system. It suggests that a Lax representation also carries important information about algebraic properties of related evolution equations. The recursion relations for differential equations in Lax form first appeared in the context of the inverse scattering method. The infinite family of equations, which leave the eigenvalues of the KdV spectral problem and the AKNS ones invariant in time, were found by Gardner et al. [90] and Ablowitz at al. [1] in 1974. As mentioned in the previous Chapter, the real importance of the recursion scheme was recognized by Olver in 1977. Then, Adler [4] in 1979 and Kupershmidt and Wilson [111] in 1981 derived the Hamiltonian structure of some dynamical systems directly from their Lax representations. Finally, in 1982, Fokas and Anderson [72] proved that the recursion operators constructed from Lax equations have the hereditary property.

All these basic results on relations between Lax representation and the algebraic structure of dynamical systems were the background of further investigations. There have been two main directions of such investigations. The first direction concentrates on the derivation of the proper Lax formulation on the basis of the information on the algebraic structure of the dynamical system, i.e. using the bi-Hamiltonian formalism. The following chapter is devoted to the solution of this problem. The reader can find different frameworks for such constructions in the literature [18],[120],[121],[122],[139],[109]. All these papers reveal the basic fact that complete information on the algebraic structure of solvable dynamical systems is hidden in their Lax (zero-curvature) representation.

The second direction of investigations is opposite to the first one. It concentrates on the construction of the whole algebraic structure of underlying dynamical systems from the Lax representation. The reader can find a typical example of this method in a series of papers by Antonowicz and Fordy [6],[7],[8]. Nevertheless, the most powerful tool for the investigation of the passage from a spectral problem to the bi-Hamiltonian representation seems nowadays to be the $\mathcal{R}$-matrix formalism, which will be presented in the last two chapters of this book.

4.1 Lax Operators and Their Spectral Deformations

In Chap. 1 we mentioned that the so called Lax spectral equation can be found for some nonlinear dynamical systems, that is, a linear operator on some Hilbert space, depending on the point of the manifold M in such a way that its spectrum stays invariant along the integral curves of the dynamical system. In fact, one can find a Lax representation for an arbitrary multi-Hamiltonian dynamical system. Further more, there are a several different Lax representations for each such system. Let us begin from the system $u_t = K(u)$ with a hereditary algebra. Obviously a recursion operator is such a Lax operator, since the invariance condition $L_K \Phi = 0$ means that $\Phi'[K] = K' \circ \Phi - \Phi \circ K'$ and on the other hand $\Phi'[K] = \Phi'[u_t] = \Phi_t$, so

$$\Phi_t = [K', \Phi], \tag{4.1}$$

which is just the Lax equation (1.2), where now $L = \Phi$ and $A = K'$. Equation (4.1) is the compatibility condition for the isospectral problem (1.3)

$$\Phi\psi = \mu\psi, \qquad \mu_t = 0$$
$$\psi_t = K'\psi, \tag{4.2}$$

hence eigenfunctions of the recursion operator Φ are symmetries of the underlying dynamical system, not necessarily equal to those known from the hereditary algebra. This important observation will be referred to later on. Another Lax pair is the recursion operator $\Psi \doteq \Phi^\dagger$ for cosymmetries and the operator $-K'^\dagger$. Actually, the invariance condition $L_K \Psi = 0$ reads (2.119)

$$\Psi_t = [-K'^\dagger, \Psi], \tag{4.3}$$

which is another Lax equation (1.2) with $L = \Psi$ and $A = -K'^\dagger$. Equation (4.3) is the compatibility condition for another isospectral problem

$$\Psi\psi^* = \mu\psi^*, \qquad \mu_t = 0$$
$$\psi_t^* = -K'^\dagger\psi^*, \tag{4.4}$$

in some Hilbert space. Again the second equation of (4.4) is the adjoint linear equation from Lemma 3.6, hence eigenfunctions of the recursion operator Ψ are cosymmetries of the underlying dynamics. Of course the following relation holds:

$$\psi = \theta\psi^*. \tag{4.5}$$

But as recursion operators are usually integro-differential operators, one prefers equivalent operators for the Lax representation, which are purely differential.

Remark 4.1 In order to avoid confusion in our notation we observe that the composition of a differential operator $A(\partial) = a_0 + a_1\partial + ... + a_N\partial^N$ with a function ϕ is of the form

$$A(\partial)\phi = (a_0\phi + a_1\phi_x + ... + a_N\phi_{x...x}) + (a_1\phi + a_2\phi_x + ...)\partial + ... + (...)\partial^N.$$

Hence, the eigenvalue problem

$$a_0\phi + a_1\phi_x + ... + a_N\phi_{x...x} = \lambda\phi$$

can be described as the zero order term $(A\phi)_0$ of the operator $A(\partial)\phi$:

$$(A(\partial)\phi)_0 = A(D_x)\phi = \lambda\phi.$$

Example 4.1 The KdV equation.
We know from Chap. 1 that the KdV equation is related to the isospectral deformation of the Schrodinger operator $4\partial_x^2 + \frac{2}{3}au = L(\partial_x)$. Of course the spectrum of the Schrodinger operator coincides with the spectrum of the recursion operator $\Phi = D^2 + \frac{2}{3}au + \frac{1}{3}au_x D^{-1}$ as they are two different representations of the same isospectral flow. The simple way of showing it is as follows. Let

$$(L(D_x) - \lambda)\varphi \equiv P\varphi = 0 \tag{4.6}$$

and

$$(\Phi - \mu)\psi \equiv R\psi = 0. \tag{4.7}$$

Then

$$\frac{1}{2}\varphi(P\varphi)_x + \frac{3}{2}(P\varphi)\varphi_x = R\psi \tag{4.8}$$

if

$$\psi = (\varphi^2)_x \quad \text{and} \quad \mu = \lambda. \tag{4.9}$$

This means that square eigenfunctions $\varphi^2 = \psi^*$ of L are cosymmetries and $(\varphi^2)_x$ the respective symmetries of the KdV.

Remark 4.2 The literature very often gives different versions of the same dynamical system which are related to each other through an appropriate rescaling of x, t and u. For the KdV system the most popular version is that with $a = 6$. Then the spectral operator reads $L = 4\partial^2 + 4u$. Sometimes it is more convenient to use the spectral operator in the form $L = \partial^2 + u$, whose spectrum coincides with the rescaled recursion operator $\Phi \to \frac{1}{4}\Phi = \frac{1}{4}D^2 + u + \frac{1}{2}u_x D^{-1}$. So, in this representation, the KdV vector field reads $K(u) \to \frac{1}{4}K(u) = \frac{1}{4}u_{3x} + \frac{3}{2}uu_x$, its second Poisson operator takes the form $\theta(u) = \frac{1}{4}D^3 + \frac{1}{2}uD + \frac{1}{2}Du$ and generally $K_m(u) \to 4^{-m}K_m(u), \sigma(u,t) \to$

$4^{-m}\sigma_m(u,t)$. The same transformation can be applied the MKdV system if we do not change the form of the Miura map.

Example 4.2 The Harry–Dym equation.
The isospectral problem for the representation (3.32) takes the form

$$(D^2 + u\lambda)\varphi = 0 \tag{4.10}$$

and coincides with the spectrum of the recursion operator

$$(\Phi + \mu)\psi = 0, \tag{4.11}$$

where

$$\Phi(u) = \theta_1(u)\theta_0^{-1}(u) = -\frac{1}{4}D^3 u^{-1/2}D^{-1}u^{-1/2}, \tag{4.12}$$

if

$$\psi = (\varphi^2)_{3x}, \qquad \mu = 2\lambda. \tag{4.13}$$

Example 4.3 The AKNS equation.
The isospectral problem for (3.66) takes the form

$$\begin{pmatrix} \varphi_1 \\ \varphi_2 \end{pmatrix}_x = \begin{pmatrix} -i\lambda & q \\ r & i\lambda \end{pmatrix}\begin{pmatrix} \varphi_1 \\ \varphi_2 \end{pmatrix} \tag{4.14}$$

and coincides with the spectrum of the recursion operator $\Phi(q,r)$ (Example 3.14)

$$\Phi(q,r)\psi = \mu\psi, \qquad \psi = (\psi_1, \psi_2)^{\mathrm{T}}, \tag{4.15}$$

if

$$\psi_1 = i\varphi_1^2, \quad \psi_2 = i\varphi_2^2, \quad \mu = 2\lambda. \tag{4.16}$$

With each hierarchy of commuting flows $u_{t_m} = K_m(u)$ we can relate a hierarchy of isospectral Lax equations

$$L_{t_m} = [A_m, L] \Longleftrightarrow u_{t_m} = K_m(u), \tag{4.17}$$

being the compatibility condition of the isospectral problems

$$L(u)\varphi = \lambda\varphi, \qquad\qquad \lambda_{t_m} = 0$$
$$\varphi_{t_m} = A_m(u)\varphi. \tag{4.18}$$

Now the question arises: what does the Lax representation look like for the hierarchy of dynamical systems $u_{\epsilon_n} = \sigma(u,t)$? The Lie derivative of the recursion operator Φ in the direction of σ_n has been established in Theorem 3.17 in the following form:

$$L_{\sigma_n}\Phi = \alpha\Phi^{n+1} \Longleftrightarrow \Phi'; [\sigma_n] = \Phi_{t_n} = [\sigma_n', \Phi] + \alpha\Phi^{n+1}. \tag{4.19}$$

Hence a natural guess is that a similar situation holds for the Lax operator L, i.e. there should be operators $B_n(u)$ such that

$$L_{\epsilon_n} = [B_n, L] + \alpha L^{n+1}. \tag{4.20}$$

Indeed (4.20) is the compatibility condition of the *nonisospectral problem*

$$L(u)\varphi = \lambda\varphi, \qquad\qquad \lambda_{\epsilon_m} = \alpha\lambda^{n+1}, \tag{4.21}$$

$$\varphi_{\epsilon_m} = B_m(u)\varphi,$$

hence can be called the *nonisospectral Lax equation*.

Remark 4.3 Sometimes the nonisospectral problem takes the form

$$\varphi_x = U(u; \lambda)\varphi, \qquad\qquad \lambda_{\epsilon_n} = \alpha\lambda^{n+1}, \tag{4.22}$$

$$\varphi_{\epsilon_n} = W_n(u; \lambda)\varphi,$$

and then the nonisospectral Lax equation (4.20) takes the form of the so called *nonisospectral zero-curvature equation*

$$U_{\epsilon_n} + \alpha U_\lambda - W_{n_x} + [U, W_n] = 0. \tag{4.23}$$

There are many methods of generation of the operators A_n and B_n [4],[25],[143], [148]. Here, for our purpose, we first adopt the method of Ma [120],[121], [122].

4.2 Lax Representations of Isospectral and Nonisospectral Hierarchies of Equations

In the following we give some basic symbols and notation. Let the independent variables $x \in \mathbb{R}^p, t \in \mathbb{R}$, and the dependent variables $u^i(x,t)$, $1 \le i \le q$, belong to the Schwartz space over $\mathbb{R}^p$ for any fixed t i.e. $M = S^q(\mathbb{R}^p, \mathbb{R})$. For the multi-index $\alpha = (\alpha_1, .., \alpha_p)$, $\alpha_i \in \mathbb{Z}$, we write $\partial^\alpha = \partial_{x_1}^{\alpha_1}...\partial_{x_p}^{\alpha_p}$, $|\alpha| = \sum_{i=1}^p \alpha_i$. Let $\mathcal{L}^r$ denote the Lie algebra of r-component vector fields on M and $\mathcal{T}^r$ denote all linear operators $T^r(x, t, u) : \mathcal{L}^r \to \mathcal{L}^r$ which are C^∞–differentiable with respect to x and t and C^∞–Gateaux differentiable (in the sense of definition (2.53)) with respect to u. That is, for an operator $T^r \in \mathcal{T}^r$, let us define its Gateaux derivative operator $T^{r\prime} : \mathcal{L}^q \to \mathcal{T}^r$ as

$$T'[K]S = \frac{\mathrm{d}}{\mathrm{d}\epsilon}T^r(u + \epsilon K)S|_{\epsilon=0}, \qquad K \in \mathcal{L}^q, \ S \in \mathcal{L}^r. \tag{4.24}$$

Notice that the linear space $\mathcal{T}^r$ equipped with the commutator

$$[A, B] := AB - BA, \qquad A, B \in \mathcal{T}^r \tag{4.25}$$

is an operator Lie algebra. Finally, let $\mathcal{T}_0^r (\subseteq \mathcal{T}^r)$ denote all matrix differential operators $\mathcal{T}_0^r : \mathcal{L}^r \to \mathcal{L}^r$ of the special form

$$\mathcal{T}_0^r = \sum_{|\alpha|}^{\infty} T_{(\alpha)}^r, \quad T_{(\alpha)}^r = (P^{ij}[u]\partial^\alpha)_{r \times r}, \tag{4.26}$$

where $P^{ij}[u]$ are differential functions which are $\mathbb{C}^\infty$–differentiable with respect to x and t, and $\mathbb{C}^\infty$–Gateaux differentiable (in the sense of (4.24)) with respect to u. We assume that the spectral operator $L = L(x, u)$ belongs to $\mathcal{T}_0^r$ and that its Gateaux derivative operator $L' : \mathcal{L}^q \to \mathcal{T}_0^r$ is an injective homomorphism.

First suppose that we have a hierarchy of evolution equations

$$u_{t_m} = K_m(u) = \Phi^m K_0, \quad K_m \in \mathcal{L}^q, \ m \geq 0, \tag{4.27}$$

where Φ stands for a recursion operator, and let $L \in \mathcal{T}_0^r$ be a spectral operator connected with the hierarchy. For any fixed vector field $X \in \mathcal{L}^q$, we construct the following *characteristic operator equation* with respect to $V \in \mathcal{T}_0^r$

$$[V, L] = L'[\Phi X] - L'[X]L. \tag{4.28}$$

We assume that the operator equation (4.28) has solutions and that $V = V(X)$ denotes one of the solutions.

Theorem 4.1 *Let the differential operator $V_0 \in \mathcal{T}_0^r$ satisfy $[V_0, L] = L'(K_0)$, and suppose the set $\{A_m\}$ is defined in the following way*

$$A_m = \sum_{i=0}^{m} V_i L^{m-i} = V_0 L^m + \sum_{i=1}^{m} V(K_{i-1})L^{m-i}, \quad m \geq 0. \tag{4.29}$$

Then we have

$$[A_m, L] = L'[K_m], \quad m \geq 0 \tag{4.30}$$

and thus for any $m \geq 0$, the evolution equation $u_{t_m} = K_m(u)$ has the isospectral Lax representation $L_{t_m} = [A_m, L]$.

Proof. By assumption, we know the equality (4.30) holds when $m = 0$. In the following let $m \geq 1$. From (4.28) we obtain

$$[V_i, L] = L'[K_i] - L'[K_{i-1}]L, \quad i \geq 1. \tag{4.31}$$

Therefore we have

$$[A_m, L] = \left[\sum_{i=0}^{m} V_i L^{m-i}, L\right] = \sum_{i=0}^{m}[V_i, L]L^{m-i}$$

$$= L'[K_0]L^m + \sum_{i=1}^{m}(L'[K_i] - L'[K_{i-1}]L)L^{m-i} \tag{4.32}$$

$$= L'[K_m].$$

So, the equality (4.30) holds when $m \geq 1$. But as $L'[K_m] = L'[u_{t_m}] = L_{t_m}$ (4.30) is the isospectral Lax representation for the evolution equation $u_{t_m} = K_m(u)$.□

Next, suppose that we have another hierarchy of evolution equations

$$u_{\epsilon_n} = \sigma_n(u) = \Phi^n \sigma_0, \qquad \sigma_n \in \mathcal{L}^q, \ n \geq 0, \tag{4.33}$$

where the operator Φ is the same as that of the hierarchy (4.27).

Theorem 4.2 *Let the differential operator $W_0 \in \mathcal{T}_0^r$ satisfy $[W_0, L] = L'[\sigma_0] - \alpha L$ and define the set $\{B_n\}$ in the following way*

$$B_n = \sum_{j=0}^{n} W_j L^{n-j} = W_0 L^n + \sum_{j=1}^{n} V(\sigma_{j-1}) L^{n-j}, \qquad n \geq 0. \tag{4.34}$$

Then we have

$$[B_n, L] = L'[\sigma_n] - \alpha L^{n+1}, \qquad n \geq 0 \tag{4.35}$$

and thus for any $n \geq 0$ the evolution equation $u_{\epsilon_n} = \sigma_n(u)$ has the non-isospectral Lax representation

$$L_{\epsilon_n} = [B_n, L] + \alpha L^{n+1}. \tag{4.36}$$

Proof. When $n = 0$, $B_0 = W_0$. Thus by hypothesis, the equality (4.35) holds when $n = 0$. For $n \geq 1$ we see that

$$[W_j, L] = L'[\sigma_j] - L'[\sigma_{j-1}]L, \qquad j \geq 1. \tag{4.37}$$

Therefore we have

$$[B_n, L] = \left[\sum_{j=0}^{n} W_j L^{n-j}, L \right] = \sum_{j=0}^{n} [W_j, L] L^{n-j}$$

$$= (L'[\sigma_0] - \alpha L) L^n + \sum_{j=1}^{n} (L'[\sigma_j] - L'[\sigma_{j-1}]L) L^{n-j} \tag{4.38}$$

$$= L'[\sigma_n] - \alpha L^{n+1},$$

which implies that the equality (4.35) holds for arbitrary n. On the other hand as $L'[\sigma_n] = L'[u_{t_n}] = L_{\epsilon_n}$ then ,(4.36) is the nonisospectral Lax representation of the hierarchy (4.33). □

Remark 4.4 According to our convention, if a given system has space translational symmetry with the generator u_x we always choose the latter as the $K_0(u)$. Moreover as the $\sigma_0(u, t)$ we choose the scaling symmetry. On the other hand, for various examples of dynamical systems one can find lower order symmetries such as $K_{-1}(u)$ or $\sigma_{-1}(u, t)$. In such cases in Theorem 4.1 $m \geq -1$ and we assume that the operator $V_{-1} \in \mathcal{T}_0^r$ satisfies $[V_{-1}, L] = L'[K_{-1}]$ and in Theorem 4.2 $n \geq -1$ and we assume that the operator $W_{-1} \in \mathcal{T}_0^r$ satisfies $[W_{-1}, L] = L'[\sigma_{-1}] - \alpha I$, where I is the identity operator in $\mathcal{T}_0^r$.

Example 4.4 The KdV isospectral and nonisospectral Lax hierarchies. Here we consider the KdV dynamical system in the form

$$u_t = \frac{1}{4}u_{3x} + \frac{2}{3}uu_x = K_1(u) \tag{4.39}$$

and the two related hierarchies of evolution equations

$$u_{t_m} = K_m(u) = \Phi^m u_x, \qquad m \geq 0, \tag{4.40}$$

$$u_{\epsilon_n} = \sigma_n(u, t) = \Phi^{n+1}(1 + tu_x), \qquad n \geq -1, \tag{4.41}$$

where Φ is the hereditary recursion operator

$$\Phi(u) = \frac{1}{4}D^2 + u + \frac{1}{2}u_x D^{-1} \tag{4.42}$$

and $\alpha = 1$ as $L_{\sigma_0}\Phi = \Phi$. These two hierarchies correspond to the following spectral problem

$$(L\varphi)_0 = \lambda\varphi, \qquad L = \partial^2 + u. \tag{4.43}$$

Evidently, we have $L'[X] = X$, $X \in \mathcal{L}^1$, and thus L' is injective. The corresponding operator equation (4.28) has one special solution

$$V = V(X) = -\frac{1}{4}X + \frac{1}{2}(D^{-1}X)\partial. \tag{4.44}$$

In this case $V_0 = \partial$ and $W_{-1} = t\partial$ so that $[V_0, L] = L'[K_0]$ and $[W_{-1}, L] = L'[\sigma_{-1}] - 1$. Then we obtain by Theorems 4.1 and 4.2 that the isospectral and nonisospectral KdV hierarchies have the following isospectral and nonisospectral Lax representations, respectively,:

$$L_{t_m} = [A_m, L], \qquad m \geq 0$$

$$A_m = \sum_{i=0}^{m} V_i L^{m-i} = \partial L^m + \sum_{i=1}^{m} \left(-\frac{1}{4}K_{i-1} + \frac{1}{2}\gamma_{i-1}\partial\right) L^{m-i}, \tag{4.45}$$

and

$$L_{\epsilon_n} = [B_n, L] + L^{n+1}, \qquad n \geq -1$$

$$B_n = \sum_{j=-1}^{n} W_j L^{n-j} = t\partial L^{n+1} + \sum_{j=0}^{n} \left(-\frac{1}{4}\sigma_{j-1} + \frac{1}{2}\zeta_{j-1}\partial\right) L^{n-j}, \tag{4.46}$$

where $\gamma_l = D^{-1}K_l$ and $\zeta_l = D^{-1}\sigma_l$ are respective cosymmetries of the KdV system. Notice that if we put $t = 0$ then

$$L_{\epsilon_n} = [B_n(t = 0), L]L^{n+1} \tag{4.47}$$

is the nonisospectral Lax representation for the hierarchy

$$u_{\epsilon_n} = \tau_n(u), \tag{4.48}$$

where $\tau_n(u)$ are master symmetries of degree 1 for the KdV equation. For (4.39) the operator A_1 takes the form

$$A_1 = \partial L - \frac{1}{4}K_0 + \frac{1}{2}\gamma_0\partial = \partial^3 + \frac{3}{2}u\partial + \frac{3}{4}u_x.$$

Example 4.5 The AKNS isospectral and nonisospectral Lax hierarchies. Here, we consider the AKNS dynamical system in the form

$$\begin{pmatrix} q \\ r \end{pmatrix}_t = \begin{pmatrix} iq_{xx} - 2iq^2r \\ -ir_{xx} + 2ir^2q \end{pmatrix} = K_1(q,r) \tag{4.49}$$

and the two related hierarchies of evolution equations

$$\begin{pmatrix} q \\ r \end{pmatrix}_{t_m} = K_m(q,r) = \Phi^{m+1}(q,r)\begin{pmatrix} -4iq \\ 4ir \end{pmatrix}, \qquad m \geq -1, \tag{4.50}$$

$$\begin{pmatrix} q \\ r \end{pmatrix}_{\epsilon_n} = \sigma_n(q,r) = \Phi^{n+1}(q,r)\begin{pmatrix} 4ixq - 8itq \\ -4ixr + 8itr \end{pmatrix}, \qquad n \geq -1, \tag{4.51}$$

where Φ is the hereditary recursion operator

$$\Phi(q,r) = i\begin{pmatrix} \frac{1}{2}D - qD^{-1}r & -qD^{-1}q \\ rD^{-1}r & -\frac{1}{2}D + rD^{-1}q \end{pmatrix} \tag{4.52}$$

and $\alpha = 2$ as $L_{\sigma_0}\Phi = 2\Phi$. Those two hierarchies correspond to the following spectral problem

$$(L\varphi)_0 = \zeta\varphi, \quad L = i\begin{pmatrix} \partial & -q \\ r & -\partial \end{pmatrix}, \quad \varphi = \begin{pmatrix} \varphi_1 \\ \varphi_2 \end{pmatrix} \tag{4.53}$$

and thus we have

$$L'[X] = i\begin{pmatrix} 0 & -X_1 \\ X_2 & 0 \end{pmatrix}, \quad X = \begin{pmatrix} X_1 \\ X_2 \end{pmatrix} \in \mathcal{L}^2, \tag{4.54}$$

which implies that L' is injective. One can verify that the operator equation (4.28) of this case has the following special solution

$$V(X) = \frac{1}{2}i\begin{pmatrix} -D^{-1}(rX_1 + qX_2) & X_1 \\ -X_2 & D^{-1}(rX_1 + qX_2) \end{pmatrix}. \tag{4.55}$$

For this case we find

$$V_{-1} = \begin{pmatrix} -2i & 0 \\ 0 & 2i \end{pmatrix}, \quad W_{-1} = \begin{pmatrix} -2i(x-2t) & 0 \\ 0 & 2i(x-2t) \end{pmatrix}. \tag{4.56}$$

Thus we obtain by Theorem 4.1 and Remark 4.4 that the isospectral and non-isospectral AKNS hierarchies have the following isospectral and nonisospectral Lax operators, respectively,:

$$A_m = \sum_{i=-1}^{m} V_i L^{m-i} = 2\mathrm{i} \begin{pmatrix} -1 & 0 \\ 0 & 1 \end{pmatrix} L^{m+1}$$

$$+ \sum_{i=0}^{m} \frac{1}{2}\mathrm{i} \begin{pmatrix} -D^{-1}(rK_{i-1}^1 + qK_{i-1}^2) & K_{i-1}^1 \\ -K_{i-1}^2 & D^{-1}(rK_{i-1}^1 + qK_{i-1}^2) \end{pmatrix} L^{m-i},$$

$$(4.57)$$

$$B_n = \sum_{j=-1}^{n} W_j L^{n-j} = 2\mathrm{i} \begin{pmatrix} -x + 2t & 0 \\ 0 & x - 2t \end{pmatrix} L^{n+1}$$

$$+ \sum_{j=0}^{n} \frac{1}{2}\mathrm{i} \begin{pmatrix} -D^{-1}(r\sigma_{j-1}^1 + q\sigma_{j-1}^2) & \sigma_{j-1}^1 \\ -\sigma_{j-1}^2 & D^{-1}(r\sigma_{j-1}^1 + q\sigma_{j-1}^2) \end{pmatrix} L^{n-j},$$

$$(4.58)$$

where $K_i = (K_i^1, K_i^2)^{\mathrm{T}}$ and $\sigma_j = (\sigma_j^1, \sigma_j^2)^{\mathrm{T}}$. Again if one puts $t = 0$, we get the Lax representation for the hierarchy

$$u_{\epsilon_n} = \tau_n(u),$$
$$(4.59)$$

where $\tau_n(u) = \Phi^{n+1}\tau_{-1}(u)$ is the hierarchy of master symmetries of degree 1. For (4.49) the operator A_1 takes the form

$$A_1 = \mathrm{i} \begin{pmatrix} 2\partial^2 - qr & -2q\partial - q_x \\ 2r\partial + r_x & -2\partial^2 + qr \end{pmatrix}.$$
$$(4.60)$$

Now we can extend our results to include the isospectral and nonisospectral zero curvature representations of dynamical systems. First let us introduce the affine extension of $\mathcal{T}_0^r \to \widetilde{\mathcal{T}}_0^r = \mathcal{T}_0^r \otimes C[\lambda, \lambda^{-1}]$ being the set of formal power series in λ, with coefficients in $\mathcal{T}_0^r$ such that for arbitrary $\widetilde{T}_0^r \in \widetilde{\mathcal{T}}_0^r$ we have $\widetilde{T}_0^r = T_0^r \otimes C[\lambda, \lambda^{-1}] = \sum_{|\alpha|}^{\infty} \widetilde{T}_{(\alpha)}^r$, $\widetilde{T}_{(\alpha)}^r = T_{(\alpha)}^r \otimes C[\lambda, \lambda^{-1}]$. In order to discuss the zero curvature representations (1.8) and (4.23) associated with the linear problems (1.7) and (4.22), we confine ourselves to $(1+1)$-dimensional systems, and assume that the operators $U = U(u; \lambda), V = V(u; \lambda)$ and $W = W(u; \lambda)$ belong to the space $\widetilde{\mathcal{T}}_{(0)}^r$. Moreover, we require that the Gateaux derivative operator $U' : \mathcal{L}^q \to \widetilde{\mathcal{T}}_{(0)}^r$ is injective.

Now, let us suppose that the isospectral ($\lambda_{t_m} = 0$) zero curvature equations

$$U_{t_m} - V_{m_x} + [U, V] = 0, \qquad V_m \in \widetilde{\mathcal{T}}_{(0)}^r, \; m \geq 0, \qquad (4.61)$$

determine a hierarchy of isospectral dynamical systems

$$u_{t_m} = K_m(u) = \Phi^m(u)K_0(u), \quad \Phi \in \mathcal{T}^q, \quad K_0 \in \mathcal{L}^q, \ m \geq 0. \quad (4.62)$$

Here, the operator Φ is usually a recursion operator. For a given vector field $X \in \mathcal{L}^q$ we construct the following operator equation with respect to $\Omega = \tilde{\Omega}(u, x, t; \lambda) \in \tilde{\mathcal{T}}^r_{(0)}$:

$$[\Omega, U] + \Omega_x = U'[\Phi X] - \lambda U'[X], \quad (4.63)$$

and call it the characteristic operator equation for the associated zero curvature representation. Assume that (4.63) has solutions and that $\Omega = \Omega(X)$ denotes the solution corresponding to $X \in \mathcal{L}^q$.

Theorem 4.3 *Let the operator* $V_0 \in \tilde{\mathcal{T}}^r_{(0)}$ *satisfy the operator equation*

$$U'[K_0] - V_{0_x} + [U, V_0] = 0 \quad (4.64)$$

and set

$$V_m = \lambda^m V_0 + \sum_{i=1}^{m} \lambda^{m-i} \Omega(K_{i-1}), \quad m \geq 0, \quad (4.65)$$

then we have

$$U'[K_m] - V_{m_x} + [U, V_m] = 0, \quad m \geq 0 \quad (4.66)$$

and thus for any $m \geq 0$*, the evolution equations* $u_{t_m} = K_m(u)$ *have the isospectral zero curvature representation*

$$U_{t_m} - V_{m_x} + [U, V_m] = 0. \quad (4.67)$$

Proof. As follows from equality (4.66), applying representation (4.65) and the characteristic equation (4.63):

$$
\begin{aligned}
[U, V_m] &= [U, \lambda^m V_0 + \sum_{i=1}^{m} \lambda^{m-i} \Omega(K_{i-1})] \\
&= \lambda^m [U, V_0] + \sum_{i=1}^{m} \lambda^{m-i} [U, \Omega(K_{i-1})] \\
&= \lambda^m (V_{0_x} - U'[K_0]) + \sum_{i=1}^{m} \lambda^{m-i} (\Omega(K_{i-1})_x - U'[K_i] + \lambda U'[K_{i-1}]) \\
&= \lambda^m V_{0_x} + \sum_{i=1}^{m} \lambda^{m-i} \Omega(K_{i-1})_x - U'[K_m] \\
&= V_{m_x} - U'[K_m]. \quad (4.68)
\end{aligned}
$$

As $U'[K_m] = U'[u_{t_m}] = U_{t_m}$, (4.67) is the isospectral zero curvature representation for the evolution equation $u_{t_m} = K_m(u)$.$\square$

Now suppose that the nonisospectral $(\lambda_{\epsilon_n} = \alpha\lambda^{n+1})$ zero curvature equations

$$U_{\epsilon_n} + \alpha U_\lambda - W_{n_x} + [U, W_n] = 0, \quad W_n \in \tilde{\mathcal{T}}_{(0)}^r, \ n \geq 0, \tag{4.69}$$

determine a hierarchy of nonisospectral dynamical systems

$$u_{\epsilon_n} = \sigma_n(u, t) = \Phi^n \sigma_0, \qquad \Phi \in \mathcal{T}^q, \ \sigma_0 \in \mathcal{L}^q, \ n \geq 0, \tag{4.70}$$

where the operator Φ is the same as that of the hierarchy (4.62).

Theorem 4.4 *Let the operator $W_0 \in \tilde{\mathcal{T}}_{(0)}^r$ satisfy the operator equation*

$$U'[\sigma_0] + \alpha\lambda U_\lambda - W_{0_x} + [U, W_0] = 0 \tag{4.71}$$

and set

$$W_n = \lambda^n W_0 + \sum_{j=1}^{n} \lambda^{n-j} \Omega(\sigma_{j-1}). \tag{4.72}$$

Then we have

$$U'[\sigma_n] + \alpha\lambda^{n+1} U_\lambda - W_{n_x} + [U, W_n] = 0, \quad n \geq 0 \tag{4.73}$$

and hence for any $n \geq 0$, the evolution equation $u_{\epsilon_n} = \sigma_n(u, t)$ has a nonisospectral zero curvature representation

$$U_{\epsilon_n} + \alpha\lambda^{n+1} U_\lambda - W_{n_x} + [U, W_n] = 0. \tag{4.74}$$

Proof. As follows from equality (4.73), applying representation (4.72) and the characteristic equation (4.61):

$$[U, W_n] = [U, \lambda^n W_0 + \sum_{j=1}^{n} \lambda^{n-j} \Omega(\sigma_{j-1})]$$

$$= \lambda^n [U, W_0] + \sum_{j=1}^{n} \lambda^{n-j} [U, \Omega(\sigma_{j-1})]$$

$$= \lambda^n (W_{0_x} - \alpha\lambda U_\lambda - U'[\sigma_0])$$

$$+ \sum_{j=1}^{n} \lambda^{n-j} \left(\Omega(\sigma_{j-1})_x - U'[\sigma_j] + \lambda U'[\sigma_{j-1}] \right)$$

$$= \lambda^n W_{0_x} + \sum_{j=1}^{n} \lambda^{n-j} \Omega(\sigma_{j-1})_x - \alpha\lambda^{n+1} U_\lambda - U'[\sigma_n]$$

$$= W_{n_x} - \alpha\lambda^{n+1} U_\lambda - U'[\sigma_n]. \tag{4.75}$$

As $U'[\sigma_n] = U'[u_{\epsilon_n}] = U_{\epsilon_n}$, equation (4.74) is the nonisospectral zero curvature representation for the evolution equation $u_{\epsilon_n} = \sigma(u,t).\square$

Remark 4.3 is also applicable to the zero curvature representation. In this case the operator equation (4.71) from Theorem 4.4 takes the new form

$$U'[\sigma_{-1}] + \alpha U_\lambda - W_{-1_x} + [U, W_{-1}] = 0. \tag{4.76}$$

We illustrate Theorems 4.3 and 4.4 on the same examples of the KdV and the AKNS hierarchies of evolution equations in order to compare the Lax representation and the zero curvature one.

Example 4.6 The KdV isospectral and nonisospectral zero curvature hierarchies.
We again consider the KdV dynamical system (4.39) and related hierarchies of evolution equations (4.40), (4.41). The spectral problem $(\partial^2 + u)\varphi = \lambda\varphi$ is equivalent to the following:

$$\psi_x = U(u;\lambda)\psi, \quad U = \begin{pmatrix} 0 & 1 \\ \lambda - u & 0 \end{pmatrix}, \quad \psi = (\psi_1, \psi_2)^T, \quad \psi_1 = \varphi, \psi_2 = \varphi_x. \tag{4.77}$$

The corresponding characteristic equation (4.63) has a solution

$$\Omega(X) = \begin{pmatrix} -\frac{1}{4}X & \frac{1}{2}(D^{-1}X) \\ -\frac{1}{4}(DX) + \frac{1}{2}(D^{-1}X)(\lambda - u) & \frac{1}{4}X \end{pmatrix} \tag{4.78}$$

and in this case $V_0 = U$ and $W_{-1} = tU$. Then we obtain by Theorems 4.3 and 4.4 that the isospectral and nonisospectral KdV hierarchies have the following isospectral and nonisospectral zero curvature representations, respectively:

$$U_{t_m} - V_{m_x} + [U, V_m] = 0,$$

$$V_m = \lambda^m \begin{pmatrix} 0 & 1 \\ \lambda - u & 0 \end{pmatrix}$$

$$+ \sum_{i=1}^{m} \lambda^{m-i} \begin{pmatrix} -\frac{1}{4}K_{i-1} & \frac{1}{2}\gamma_{i-1} \\ -\frac{1}{4}(K_{i-1})_x + \frac{1}{2}\gamma_{i-1}(\lambda - u) & \frac{1}{4}K_{i-1} \end{pmatrix} \tag{4.79}$$

and

$$U_{\epsilon_n} + \alpha\lambda^{n+1}U_\lambda - W_{n_x} + [U, W_n] = 0,$$

$$W_n = \lambda^{n+1} \begin{pmatrix} 0 & t \\ (\lambda - u)t & 0 \end{pmatrix}$$

$$+ \sum_{j=1}^{n} \lambda^{n-i} \begin{pmatrix} -\frac{1}{4}\sigma_{j-1} & \frac{1}{2}\zeta_{j-1} \\ -\frac{1}{4}(\sigma_{j-1})_x + \frac{1}{2}\zeta_{j-1}(\lambda - u) & \frac{1}{4}\sigma_{j-1} \end{pmatrix}. \tag{4.80}$$

For the KdV equation (4.39) the operator V takes the form

$$V_1 = \begin{pmatrix} -\frac{1}{4}u_x & \lambda + \frac{1}{2}u \\ \lambda^2 - \frac{1}{2}\lambda u - \frac{1}{2}u^2 - \frac{1}{4}u_{xx} & \frac{1}{4}u_x \end{pmatrix}. \tag{4.81}$$

Example 4.7 The AKNS isospectral and nonisospectral zero curvature hierarchies.

We again consider the AKNS dynamical system (4.49) and the related hierarchies of evolution equations (4.50), (4.51). The spectral problem (4.53) is equivalent to the following:

$$\varphi_x = U(q, r; \lambda)\varphi, \quad U = \begin{pmatrix} -i\lambda & q \\ r & i\lambda \end{pmatrix}, \quad \varphi = (\varphi_1, \varphi_2)^{\mathrm{T}}. \tag{4.82}$$

In this case the solution $\Omega(X)$ of the characteristic equation (4.63) is equal to the solution $V(X)$ (4.55) of the characteristic equation (4.28), i.e.

$$\Omega(X) = \frac{1}{2}i \begin{pmatrix} -D^{-1}(rX_1 + qX_2) & X_1 \\ -X_2 & D^{-1}(rX_1 + qX_2) \end{pmatrix}, \tag{4.83}$$

and V_{-1} and W_{-1} are equal to their counterparts (4.55) from Example 4.5. Then we obtain by Theorems 4.3 and 4.4 that the isospectral and nonisospectral AKNS hierarchies have the following isospectral and nonisospectral zero curvature representations, respectively:

$$U_{t_m} - V_{m_x} + [U, V_m] = 0,$$

$$V_m = \lambda^{m+1} \begin{pmatrix} -2i & 0 \\ 0 & 2i \end{pmatrix}$$

$$+ \sum_{i=0}^{m} \frac{1}{2}i \begin{pmatrix} -D^{-1}(rK_{i-1}^1 + qK_{i-1}^2) & K_{i-1}^1 \\ -K_{i-1}^2 & D^{-1}(rK_{i-1}^1 + qK_{i-1}^2) \end{pmatrix} \lambda^{m-i}$$

$$\tag{4.84}$$

and

$$U_{\epsilon_n} + 2\lambda^{n+1}U_\lambda - W_{n_x} + [U, W_n] = 0,$$

$$W_n = 2i \begin{pmatrix} -x + 2t & 0 \\ 0 & x - 2t \end{pmatrix} \lambda^{n+1}$$

$$+ \sum_{j=0}^{n} \frac{1}{2}i \begin{pmatrix} -\partial^{-1}(r\sigma_{j-1}^1 + q\sigma_{j-1}^2) & \sigma_{j-1}^1 \\ -\sigma_{j-1}^2 & \partial^{-1}(r\sigma_{j-1}^1 + q\sigma_{j-1}^2) \end{pmatrix} \lambda^{n-j}.$$

$$\tag{4.85}$$

For the AKNS equation (4.49) the operator V takes the form

$$V_1 = \begin{pmatrix} -2i\lambda^2 - iqr & 2\lambda q + iq_x \\ 2\lambda r - ir_x & 2i\lambda^2 + iqr \end{pmatrix}. \tag{4.86}$$

4.3 The Lax Operator Algebra

In this section, for the spectral operator $L : \mathcal{L}^r \to \mathcal{L}^r$, we establish a kind of algebraic structure of the space of the corresponding isospectral and nonisospectral Lax operators and further derive the Lie algebraic structure of a quotient algebra of the Lax operator algebra.

Definition 4.1

(i) Let $A \in \mathcal{T}^r$. If there exists a vector field $X \in \mathcal{L}^q$ such that $[A, L] = L'[X]$, then A is called an isospectral Lax operator, or Lax operator for short, and X is called an eigenvector field of the Lax operator A.

(ii) Let $B \in \mathcal{T}^r$. If there exists a vector field $Y \in \mathcal{L}^q$ such that $[B, L] + \alpha L^k = L'[Y]$, $\alpha = $ const and $k \in N$, then B is called a nonisospectral Lax operator and Y is called an eigenvector field of the operator B.

Notice that a Lax operator only has an eigenvector field if L' is injective. We denote by $\mathcal{M}(L)$ all Lax operators related to the spectral operator L, by $E(\mathcal{M})$ all eigenvector fields of Lax operators and by $\mathcal{N}(L)$ and $E(\mathcal{N})$ the respective subsets of nonisospectral Lax operators and their eigenvector fields.

Definition 4.2 Let two Lax operators $A, B \in \mathcal{M}(L)$ have eigenvector fields $X, Y \in E(\mathcal{M})$, respectively. Then let us define the product operator of two Lax operators A, B as follows:

$$[|A, B|] = B'[X] - A'[Y] + [A, B]. \qquad (4.87)$$

We shall show that this product operator $[|A, B|]$ corresponds to the commutator $[X, Y]$, but we first need the following basic results.

Lemma 4.1 Let $P(x, t, u) \in \mathcal{L}$, $K(x, t, u), S(x, t, u) \in \mathcal{L}^q$. Then we have the relation

$$(P'[K])'[S] - (P'[S])'[K] = P'[R], \quad R = [S, K]. \qquad (4.88)$$

Proof. By the definition of the directional (Gateaux) derivative, we have

$$(P'[K])'[S] = \left(\frac{\partial}{\partial \epsilon} P(u + \epsilon K)_{|\epsilon=0} \right)' [S]$$

$$= \frac{\partial^2}{\partial \delta \partial \epsilon} P(u + \delta S + \epsilon K(u + \delta S))_{|\delta, \epsilon = 0}$$

$$= \frac{\partial^2}{\partial \delta \partial \epsilon} P(u + \delta S + \epsilon K)_{|\delta, \epsilon = 0} + \frac{\partial}{\partial \mu} P(u + \mu K'[S])_{|\mu=0}.$$

At the same time, we similarly have

$$(P'[S])'[K] = \frac{\partial^2}{\partial\delta\partial\epsilon}P(u + \delta K + \epsilon S)|_{\delta,\epsilon=0} + \frac{\partial}{\partial\mu}P(u + \mu S'[K])|_{\mu=0}.$$

Thus we obtain

$$(P'[K])'[S] - (P'[S])'[K]$$

$$= \frac{\partial}{\partial\mu}\left(P(u + \mu K'[S]) - P(u + \mu S'[K])\right)|_{\mu=0}$$

$$= P'[K'[S]] - P'[S'[K]] = P'[R].\square$$

From the above Lemma we can easily deduce the following:

Lemma 4.2 *Let $T \in \mathcal{T}_0^r$, $K, S \in \mathcal{L}^q$. Then we have*

$$(T'[K])'[S] - (T'[S])'[K] = T'[R], \qquad R = [S, K]. \tag{4.89}$$

Theorem 4.5 *Suppose that two isospectral Lax operators $A, B \in \mathcal{N}(L)$ have eigenvector fields $X, Y \in E(\mathcal{N})$, respectively. Then we have the equality*

$$[[\|A, B\|], L] = L'[Z], \qquad Z = [X, Y], \tag{4.90}$$

which shows that the product operator $[\|A, B\|] \in \mathcal{T}^r$ is a Lax operator, too, and that its eigenvector field is the commutator $[X, Y]$.

Proof. Since $(\mathcal{T}^r; [., .])$ is an operator Lie algebra, we have

$$[[B, A], L] = [B, [A, L]] - [A, [B, L]] = [B, L'[X]] - [A, L'[Y]].$$

Therefore,

$$[[\|A, B\|], L] = [B'[X], L] + [B, L'[X]] - [A'[Y], L] - [A, L'[Y]]$$

$$= [B, L]'[X] - [A, L]'[Y] = (L'[Y])'[X] - (L'[X])'[Y] \tag{4.91}$$

$$= L'[Z],$$

by Lemma 3.2. $\square$

Evidently we see by (4.87) that the multiplication operation $[\|A, B\|]$ is bilinear and antisymmetric. Therefore noticing that the multiplication operation $[., .]$ given by (4.25) satisfies the Jacobi identity we obtain the following results by the above theorem.

Lemma 4.3

(i) $(\mathcal{N}(L); [|.,.|])$ is an antisymmetric algebra and $(E(\mathcal{N}); [.,.])$ is a Lie
      subalgebra of $\mathcal{L}^q$.

(ii) Let $A, B, C \in \mathcal{N}(L)$. Then we have

$$[[| [|A, B|], C|] + \text{ cyclic perm}(A, B, C), \ L] = 0. \tag{4.92}$$

(iii) Set $K(L) = \{A \in \mathcal{T}^r, \ [A, L] = 0\}$. Thus $K(L)$ generates an equivalence
      relation $\sim$ of $\mathcal{T}^r$

$$A \sim B \ \Leftrightarrow \ [A, L] = [B, L], \qquad A, B \in \mathcal{T}^r. \tag{4.93}$$

Then $([|.,.|])$ is an ideal subalgebra of $(\mathcal{N}(L); [|.,.|])$.

Let us denote by $C(A)$ the equivalence class to which $A \in \mathcal{T}^r$ belongs.
We can generate a quotient algebra $C\mathcal{M}(L) = \mathcal{M}(L)/K(L)$, whose multiplication operation is as follows

$$[|C(A), C(B)|] = C([|A, B|]), \qquad A, B \in \mathcal{N}(L). \tag{4.94}$$

Theorem 4.6 *The quotient algebra* $(C\mathcal{N}(L), [|.,.|])$ *is a Lie algebra and is
isomorphic to the Lie algebra* $(E(\mathcal{M}); [.,.])$. *Moreover the following mapping*

$$\pi : \ C\mathcal{N}(L) \to E(\mathcal{N}), \qquad C(A) \to X \ \ ([A, L] = L^{'}[X]) \tag{4.95}$$

is a Lie algebra isomorphism.

Proof. Obviously, π is a linear isomorphism. If the Lax operators $A, B \in
\mathcal{N}(L)$ have the eigenvector fields $X, Y \in E(\mathcal{N})$, respectively, then we have

$$\pi([|C(A), C(B)|]) = \pi(C([|A, B|])) = [X, Y] = [\pi C(A), \pi C(B)]. \tag{4.96}$$

Thus by Lemma 4.3 (i) we obtain the result that $(C\mathcal{N}(L), [|.,.|])$ is a Lie
algebra and further we see that π is a Lie algebra isomorphism. $\square$

From Theorem 4.6 follows the important observation that if a given vector
field K is related to a spectral operator L, and K has a general Lie algebra
of Hamiltonian master symmetries (3.107) then there exists a related Lie
algebra of isospectral Lax operators $A_{r,n}$

$$[|A_{s,m}, A_{r,n}|] = (s(n + \rho) - r(m + \rho))A_{s+r-1,n+m}, \tag{4.97}$$

where

$$L'[\tau_{r,n}] = [A_{r,n}, L]. \tag{4.98}$$

Obviously the Lax operators of polynomial in t symmetries are

$$\tilde{A}_{r,n}(t) = \sum_{k=0}^{r} t^k A_{r-k,n+k},$$ (4.99)

$$L'[\sigma_{r,n}(t)] = [\tilde{A}_{r,n}(t), L]$$ (4.100)

and form the same Lie algebra (4.97) as the Lax operators of master symmetries.

Example 4.8 Isospectral Lax algebra of the KP system.
The KP dynamical system (3.121) is equivalent to the isospectral Lax equation (1.2) where

$$L = \partial_y + \partial_x^2 + u, \quad A = 3\partial_x^3 + 6u\partial_x + 3u_x - 3\partial_x^{-1}u_y.$$ (4.101)

The first few Lax operators of the master symmetries (3.122) are

$$A_{0,-1} = \partial_x, \ A_{0,0} = -2\partial_x^2 - 2u, \ A_{0,1} = 3A = 12\partial_x^3 + 18u\partial_x + 9u_x - 9\partial_x^{-1}u_y, \dots$$

$$A_{1,-1} = y\partial_x - \frac{1}{2}x, \ A_{1,0} = -2y\partial_x^2 + x\partial_x - 2y^2u_y - yu + \frac{1}{2},$$

$$A_{1,1} = xA_{0,0} + \frac{1}{3}yA_{0,1} - 2\partial_x - \partial_x^{-1}u, \dots$$

$$A_{2,0} = -2y^2\partial_x^2 + 2xy\partial_x - 2y^2u - \frac{1}{2}x^2 + y, \dots$$ (4.102)

$$\vdots$$

Other operators can be generated by applying formulas (4.97) and (4.87). Notice that because the operators $A_{0,-1}, A_{1,1}$ and $A_{2,0}$ are differential polynomial operators in ∂_x only, non of the others contain the operator ∂_y either.

Now we can pass to the Lax representation of the hereditary algebra. Suppose we consider the hierarchies of evolution equations (4.27) and (4.33) respectively, with appropriate isospectral (4.30) and nonisospectral (4.35) Lax representations.

Theorem 4.7 *Let $A_m \in \mathcal{M}(L)$ and $B_n \in \mathcal{N}(L)$ be isospectral and non-isospectral Lax operators with respective eigenvector fields $K_m \in E(\mathcal{M})$, $\sigma_n \in E(\mathcal{N})$. Then we have*

(i)

$$[[|A_m, A_n|], L] = L'[[K_m, K_n]],$$ (4.103)

(ii)

$$[[|B_m, A_n|], L] = L'[[\sigma_m, K_n]],$$ (4.104)

(iii)

$$[[|B_m, B_n|], L] = L'[[\sigma_m, \sigma_n]] - \alpha(n - m)L^{n+m+1}.$$ (4.105)

Proof. (i) was proved in Theorem 4.5. Here we only prove the equality (ii). The proof of (iii) is completely similar. Because $(\mathcal{T}^r; [., .])$ is a Lie algebra of operators, we obtain

$$[[A_n, B_m], L] = [A_n, [B_m, L]] - B_m, [A_n, L]].$$

Therefore we have

$[[|B_m, A_n|], L]$

$$= [A_n'[\sigma_m] - B_m'[K_n] + [A_n, B_m], L]$$

$$= [A_n'[\sigma_m], L] - [B_m'[K_n], L] + [A_n, [B_m, L]] - [B_m, [A_n, L]]$$

$$= [A_n'[\sigma_m], L] - [B_m'[K_n], L] + [A_n, L'[\sigma_m]] - \alpha[A_n, L^{m+1}] - [B_m, L'[K_n]]$$

$$= [A_n, L]'[\sigma_m] - [B_m, L]'[K_n] - \alpha[A_n, L^{m+1}]$$

$$= (L'[K_n])'[\sigma_m] - (L'[\sigma_m])'[K_n] + \alpha(L^{m+1})'[K_n] - \alpha[A_n, L^{m+1}]$$

$$= L'[[\sigma_m, K_n]] + \alpha(L^{m+1})'[K_n] - \alpha[A_n, L^{m+1}].$$

$$(4.106)$$

In addition we have

$$(L^{m+1})'[K_n] = \sum_{k=0}^{m} L^k L'[K_n] L^{m-k} = \sum_{k=0}^{m} L^k (A_n L - L A_n) L^{m-k}$$

$$= A_n L^{m+1} - L^{m+1} A_n = [A_n, L^{m+1}].$$

Thus we arrive at

$$[[|B_m, A_n|], L] = L'[[\sigma_m, K_n]]. \qquad \Box$$

The results of Theorem 4.7 allow us to generalize the results of Lemma 4.3 and Theorem 4.6 to include the case of $\mathcal{M}(L)$. This means that the quotient algebra $(C\mathcal{M}(L), [|., .|])$ of Lax operators is a Lie algebra and is isomorphic to the Lie algebra $(E(\mathcal{M}), [., .])$ of eigenvector fields. Hence the Lax operators from Theorem 4.1 and 4.2 constitute a hereditary algebra

$$[|A_n, A_m|] = 0, \quad [|B_n, A_m|] = (\rho + \alpha m) A_{n+m}, \quad [|B_n, B_m|] = \alpha(m - n) B_{n+m}.$$

$$(4.107)$$

Now we can see the main difference between symmetries of degree 1 from the noncanonical action/angle algebra (3.111) and symmetries of degree 1 from the hereditary algebra (3.92), i.e. the first ones are eigenvector fields of isospectral Lax operators and the second ones are eigenvector fields of nonisospectral Lax operators.

There are two main reasons why we pay so much attention to the Lax representations of multi-Hamiltonian dynamical systems. Obviously, they are the key objects for solving the Cauchy problem in the framework of the analytic approach. Besides, it occurs that the isospectral Lax equations (1.2) themselves can be considered as Hamiltonian dynamical systems on the appropriate algebra, with the multi-Hamiltonian structure. The advantage of such an abstract approach is the possibility of finding in a very systematic way multi-Hamiltonian dynamical systems on M. The theory of isospectral Lax dynamics will be developed in Chap. 7. In the next two chapters we shall apply the theory presented so far to particular finite dimensional dynamical systems.

5. Soliton Particles

5.1 General Aspects

It is well known that the majority of solvable field and lattice nonlinear evolution equations have the so called *N-soliton solutions* u_N, which asymptotically, i.e. for $t \to \pm\infty$, decompose into a sum of single *solitons* s_i, that is extended objects of permanent shape, moving at a constant speed. Their dynamic behaviour has been studied extensively and solitons have been found to be stable against mutual collisions and to behave like particles. These useful properties make them attractive for a description of not only a wide class of physical phenomena [41],[114],[115],[173],[193], but also biological [57] and others [192],[174]. In this chapter we discuss the time independent decomposition of N-soliton solutions into a sum of extended objects being closely related to the eigenfunctions of the discrete part of the spectrum of a recursion hereditary operator. These objects will be called *soliton particles (interacting solitons)* [19],[21],[87]. Moreover, we present the analytic form of soliton particles, their equations of motion with the multi-Hamiltonian structure and other algebraic properties. Finally we present multisoliton perturbation theory constructed in a purely algebraic way.

As we have mentioned before, for real field systems, the N-soliton solutions decompose asymptotically, for $t \to \pm\infty$, into 1-solitons of the form

$$u_N \simeq \sum_{i=1}^{N} s_i(x + c_i t + q_i). \tag{5.1}$$

If the asymptotic speeds c_i and the phases q_i are considered as variables, then the set of these solutions forms a $2N$-dimensional invariant submanifold M_N of M. In the case of complex fields, u_N is parametrized by some complex variables $\tilde{c}_i$ and $\tilde{q}_i$ and the N-soliton solutions decompose asymptotically for $t \to \pm\infty$ into 1-solitons of the usual form

$$u_N \simeq \sum_{k=1}^{N} s_k(\operatorname{im} \tilde{c}_k x + \operatorname{im} \tilde{c}_k^2 t + \operatorname{im} \tilde{q}_k) \exp i(\operatorname{re} \tilde{c}_k x + \operatorname{re} \tilde{c}_k^2 t + \operatorname{re} \tilde{q}_k). \tag{5.2}$$

If the asymptotic data $(\operatorname{re} \tilde{c}_k, \operatorname{im} \tilde{c}_k)$, $(\operatorname{re} \tilde{q}_k, \operatorname{im} \tilde{q}_k)$ are considered as variables, then the set of these solutions forms a $4N$-dimensional invariant submanifold M_N of M.

In the submanifold M_N we are now going to give a new parametrization in the following way [23],[145]. Define a map Π which assigns to each u_N the set of asymptotic data $(q_1, ..., q_N, c_1, ..., c_N)$ or $(\operatorname{re}\tilde{q}_1, ..., \operatorname{im}\tilde{c}_N)$. Of course, this map cannot be written down explicitly; however this is not necessary for our further considerations. All algebraic properties of the objects we are interested in are invariant w.r.t. the choice of a special chart of the manifold M_N, so what we need are only the appropriate transformation laws for the objects, found in Chap. 2. The general procedure of finding new coordinates is very simple in principle. Take an arbitrary $u = u(0)$ out of this manifold as an initial condition for the flow given by $u_t = K(u)$, solve this equation and get quantities (q_i, c_i) by comparing the solution with (5.1) or (5.2), respectively. Observe that although we use the asymptotic behaviour of the N-solitons this new parametrization is defined for arbitrary time. The quantities q_i, c_i are scalar fields on the submanifold M_N and we now define their time dependence by setting

$$q_i(t) = q_i(u(t)),$$
$$c_i(t) = c_i(u(t)).$$

Lemma 5.1

(i) For real N-solitons:

$$\frac{\partial}{\partial t}q_i(t) = c_i, \qquad \frac{\partial}{\partial t}c_i(t) = 0, \qquad i = 1, ..., N. \tag{5.3}$$

(ii) For complex N-solitons:

$$\frac{\partial}{\partial t}\operatorname{re}\tilde{q}_i(t) = \operatorname{re}\tilde{c}_i^2, \qquad \frac{\partial}{\partial t}\operatorname{re}\tilde{c}_i(t) = 0, \qquad i = 1, ..., N$$

$$\frac{\partial}{\partial t}\operatorname{im}\tilde{q}_i(t) = \operatorname{im}\tilde{c}_i^2, \qquad \frac{\partial}{\partial t}\operatorname{im}\tilde{c}_i(t) = 0, \qquad i = 1, ..., N \tag{5.4}$$

Proof. We only prove the case (i) as the proof of (ii) is the same. Let $\bar{u}(t) \simeq \sum_{i=1}^{N} s_i(x + c_i(0)t + q_i(0))$ be a solution of a given system for the initial condition $\bar{u}(0) = u_0$. For a fixed t_0 we take now $\bar{u}(t_0)$ as a new initial condition. Solving the given equation with that new initial condition we get

$$\bar{u}(t) \simeq \sum_{i=1}^{N} s_i(x + c_i(t_0)(t - t_0) + q_i(t_0)).$$

Now, a comparison of the two solutions yields the desired result

$$q_i(t_0) = c_i t_0 + q_i(0),$$
$$c_i(t_0) = c_i(0) = c_i. \square$$

Lemma 5.1 shows that the flow of $u_t = K(u)$ on the submanifold M_N is linearized in our new coordinates. Because of this linear structure a given dynamics on M_N is trivial with respect to this parametrization. Of course at this stage these considerations do not give any practical information about the problem with respect to its original coordinates x, t. Let us for convenience call the manifold M_N parametrized in the x, t-coordinates, a physical representation and the manifold M_N endowed with the coordinates (q_i, c_i), a linear representation. Lemma 5.1 gives us an effective tool for the construction of the algebraic structure on M_N with respect to physical representation. The method is as follows. Find the algebraic structure of the linear system (5.3) respectively (5.4). Relate the known algebraic objects in physical representations with the respective ones of the linear representation via the map Π. Construct the unknown objects in the physical representation applying the transformation laws to the respective known objects in the linear representation. Notice that the multi-soliton manifold M_N is not a linear space at all.

We consider all $(1 + 1)$-dimensional systems with N-soliton solutions for which there exists a hierarchy of vector fields $K_n(u)$ (symmetries), a scaling vector field $\tau_0(u)$ and the Poisson operator θ so that $K_n(u)$ and $\tau_0(u)$ are Hamiltonian vector fields with respect to θ

$$K_n(u) = \theta(u)\nabla H_n(u), \quad \tau_0(u) = \theta(u)\nabla F(u), \tag{5.5}$$

where $H_n(u)$ and $F(u)$ are suitable scalar fields on the manifold under consideration, and they satisfy the following commutation relations in the Lie algebra of vector fields

$$[K_n, K_m] = 0, \quad [\tau_0, K_n] = (n + \rho)K_n, \quad \rho = \text{const.} \tag{5.6}$$

When the field variable u is a complex function we always consider the extended system

$$\mathbf{u}_t = \begin{pmatrix} u \\ u^* \end{pmatrix}_t = \begin{pmatrix} K(u, u^*) \\ K^*(u, u^*) \end{pmatrix} = \mathbf{K}(\mathbf{u}). \tag{5.7}$$

If the vector fields K_n are generated by the recursion operator Φ, this means that $L_{\tau_0}\Phi = \Phi$.

5.2 Algebraic Structure of Linear Systems

We now restrict our considerations for a while to the linearized systems (5.3) and (5.4) to examine their structure. Starting from the real system (5.3) on $\mathbb{R}^{2N}$, one observes that this system is given in a Hamiltonian form. Let $q = (q_1, ..., q_N)^T$, $c = (c_1, ..., c_N)^T$, $\bar{u} = (q, c)^T \in \mathbb{R}^{2N}$, $\{\partial/\partial q_1, ..., \partial/\partial q_N, \partial/\partial c_1, ..., \partial/\partial c_N\}$ be a basis of $T_{\bar{u}}\mathbb{R}^{2N}$ and $\{dq_1, ..., dq_N, dc_1, ..., dc_N\}$ be a basis of $T_{\bar{u}}^*\mathbb{R}^{2N}$. Then (5.3) can be expressed as

$$\bar{u}_t = \begin{pmatrix} q \\ c \end{pmatrix}_t = \begin{pmatrix} c \\ 0 \end{pmatrix} = \begin{pmatrix} 0 & I \\ -I & 0 \end{pmatrix} \begin{pmatrix} 0 \\ c \end{pmatrix} = \begin{pmatrix} 0 & I \\ -I & 0 \end{pmatrix} \nabla \left(\frac{1}{2} \sum_{i=1}^{N} c_i^2 \right),$$

(5.8)

i.e. in the canonical Hamiltonian representation. But in fact (5.3) admits many different Hamiltonian formulations. For every $p \neq 2$ we have

$$\bar{u}_t = \begin{pmatrix} q \\ c \end{pmatrix}_t = \begin{pmatrix} c \\ 0 \end{pmatrix} = \begin{pmatrix} 0 & \Lambda_p \\ -\Lambda_p & 0 \end{pmatrix} \nabla \left(\frac{1}{2-p} \sum_{i=1}^{N} c_i^{2-p} \right), \qquad (5.9)$$

where $\Lambda_p = \mathrm{diag}(c_1^p, ..., c_N^p)$. To make the following compatible with formulas (5.5) and (5.6) we choose the Poisson operators in the following form

$$\bar{\theta}_{-1} = \begin{pmatrix} 0 & \Lambda_{-\rho} \\ -\Lambda_{-\rho} & 0 \end{pmatrix}, \qquad \bar{\theta}_0 = \begin{pmatrix} 0 & \Lambda_{1-\rho} \\ -\Lambda_{1-\rho} & 0 \end{pmatrix}, \qquad (5.10)$$

where ρ is the scaling factor from (5.6). The sum of these two implectic operators is again an implectic operator, hence they give rise to the hereditary recursion operator

$$\bar{\Phi} = \bar{\theta}_0 \circ \bar{\theta}_{-1}^{-1} = \begin{pmatrix} \Lambda_1 & 0 \\ 0 & \Lambda_1 \end{pmatrix}. \qquad (5.11)$$

Thus, $\mathbb{R}^{2N}$ equipped with compatible Poisson structures (5.10) turns into a bi-Poisson manifold, which due to nondegeneracy of $\bar{\theta}_{-1}$ is simultaneously the Poisson-Nijenhuis one. Since $\bar{\Phi}$ is in diagonal form the eigenvalues of $\bar{\Phi}$ are $c_1, ..., c_N$ and each of them occurs twice. For every $i = 1, ..., N$, the eigenvectors $\bar{A}_i$ and $\bar{B}_i$ of $\bar{\Phi}$ with respect to c_i are given by the partial derivatives

$$\bar{A}_i = \frac{\partial \bar{u}}{\partial q_i}, \quad \bar{B}_i = \frac{\partial \bar{u}}{\partial c_i}, \qquad \bar{\Phi}\,\bar{A}_i(\bar{B}_i) = c_i \bar{A}_i(\bar{B}_i). \qquad (5.12)$$

Lemma 5.2

(i)   With respect to the Poisson bracket given by $\bar{\theta}_0$ the coordinates c_i^p, q_i for all $i, j = 1, ..., N$ satisfy the following relations

$$\{c_i^p, q_j\}_{\bar{\theta}_0} = \rho \delta_{ij}, \quad \{c_i, c_j\}_{\bar{\theta}_0} = \{q_i, q_j\}_{\bar{\theta}_0} = 0. \qquad (5.13)$$

Hence $(\rho^{-1} c_i^p, q_i)$ are the canonical coordinates corresponding to $\bar{\theta}_0$. They are called canonical action/angle variables.

(ii)   For every $i = 1, ..., N$, the canonical action/angle vector fields are of the form

$$\bar{A}_i = \frac{\partial \bar{u}}{\partial q_i} = \bar{\theta}_0 \circ \nabla \left(\frac{1}{\rho} c_i^p \right),$$

$$c_i^{1-p} \bar{B}_i = \rho \frac{\partial \bar{u}}{\partial c_i^p} = \bar{\theta}_0 \circ \nabla(-q_i). \qquad (5.14)$$

(iii) The symmetries $\overline{K}_n = \overline{\Phi}(c,0)^T = (c^n, 0)^T$, where $c^n = (c_1^n, ..., c_N^n)$, are bi-Hamiltonian vector fields with

$$\overline{K}_n = \begin{pmatrix} c^n \\ 0 \end{pmatrix} = \overline{\theta}_0 \circ \nabla \left(\frac{1}{n+\rho} \sum_{i=1}^{N} c_i^{n+\rho} \right)$$

$$= \overline{\theta}_{-1} \circ \nabla \left(\frac{1}{n+1+\rho} \sum_{i=1}^{N} c_i^{n+1+\rho} \right) \tag{5.15}$$

and commute in pairs: $[\overline{K}_n, \overline{K}_m] = 0$.

(iv) The Hamiltonian vector field

$$\overline{\tau}_0 = \begin{pmatrix} -\rho q \\ c \end{pmatrix} = \overline{\theta}_0 \circ \nabla \left(-\sum_{i=1}^{N} c_i^{\rho} q_i \right) = \overline{\theta}_0 \circ \nabla \overline{F} \tag{5.16}$$

is a scaling vector field for $\overline{K}_n$, i.e.

$$[\overline{\tau}_0, \overline{K}_n] = (n+\rho)\overline{K}_n. \tag{5.17}$$

Observe that the angle vector fields $c_i^{1-\rho}\overline{B}_i$ as well as the scaling one are Hamiltonian vector fields only with respect to $\overline{\theta}_0$.

Now, for the sake of further considerations, including the recovery of the full structure in the Lie algebra of the vector fields for the nonlinear soliton systems, let us define the *fundamental algebra $\overline{A}$* [24],[25] consisting of the following vector fields in the linear representation

$$\overline{P}_{r,n} = \overline{\Phi}^n \sum_{i=1}^{N} q_i^r \overline{A}_i, \qquad \overline{M}_{s,n} = \overline{\Phi}^{n+1} \sum_{i=1}^{N} q_i^s \overline{B}_i. \tag{5.18}$$

These vector fields satisfy the commutator relations

$$[\overline{P}_{r,n}, \overline{P}_{s,m}] = (s-r)\overline{P}_{r+s-1,n+m},$$

$$[\overline{M}_{r,n}, \overline{M}_{s,m}] = (m-n)\overline{M}_{r+s,n+m},$$

$$[\overline{P}_{r,n}, \overline{M}_{s,m}] = s\overline{M}_{s+r-1,n+m} - n\overline{P}_{s+r,n+m}. \tag{5.19}$$

In general $\overline{P}_{r,n}$ and $\overline{M}_{s,n}$ are non-Hamiltonian vector fields, hence our first goal is to construct a subalgebra of vector fields which are Hamiltonian with respect to the Poisson operator

$$\overline{\theta}_p(\rho) = \overline{\Phi}^p \circ \overline{\theta}_0(\rho) = \begin{pmatrix} 0 & \Lambda_{p+1-\rho} \\ -\Lambda_{p+1-\rho} & 0 \end{pmatrix}. \tag{5.20}$$

For a fixed p we define the following vector fields

$$\overline{\tau}_{r,n}^p = (-1)^r [(n+\rho-p)\overline{P}_{r,n} - r\overline{M}_{r-1,n}], \tag{5.21}$$

satisfying the commutator relations

$$[\overline{T}^p_{r,n}, \overline{T}^p_{s,m}] = ((r(m + \rho - p) - s(n + \rho - p))\overline{T}^p_{s+r-1,n+m}. \tag{5.22}$$

Notice that $\overline{T}^0_{1,0} = \overline{T}_0$. Each vector field $\overline{T}^p_{r,n}$ is a Hamiltonian vector field with respect to the Poisson operator $\bar{\theta}_p(\rho)$

$$\overline{T}^p_{r,n} = \bar{\theta}_p \circ \nabla \overline{T}_{r,n-p}, \tag{5.23}$$

where

$$\overline{T}_{r,n} = (-1)^r \sum_{i=1}^{N} q_i^r c_i^{n+\rho} \tag{5.24}$$

are some scalar fields. The following commutator relations between these scalar fields holds:

$$\{\overline{T}_{r,n}, \overline{T}_{s,m}\}_{\bar{\theta}_p} = ((r(m + \rho) - s(n + \rho))\overline{T}_{r+s-1,n+m+p}. \tag{5.25}$$

For our linear system (5.3) the vector fields $\overline{T}^p_{r,n}$ are Hamiltonian, with respect to $\bar{\theta}_p$, the master symmetries of degree r and respective scalar fields $\overline{T}_{r,n}$ are master integrals of motion. Moreover, the vector fields

$$\overline{\sigma}^p_{r,n}(t) = \sum_{m=0}^{r} \binom{r}{n} t^m \overline{T}^p_{r-m,n+m} \tag{5.26}$$

are Hamiltonian w.r.t. $\bar{\theta}_p$, polynomial in t variable, symmetries of the system (5.3) and

$$\overline{H}^p_{r,n}(t) = \sum_{m=0}^{r} \binom{r}{n} t^m \overline{T}_{r-m,n+m-p} = \sum_{i=1}^{N}(c_i t - q_i)^r c_i^{n-p+\rho} \tag{5.27}$$

are suitable time dependent conserved quantities. Notice that for $p = 0$ (5.22) and (5.25) recover the general algebra from Chap. 3. On the other hand vector fields

$$\overline{K}_n = \overline{P}_{0,n} = \overline{\Phi}^n \circ \overline{K}_0, \qquad \overline{T}_n = \overline{M}_{0,n} - \rho \overline{P}_{1,n} = \overline{\Phi}^n \circ \overline{T}_0 \tag{5.28}$$

constitute the hereditary algebra

$$[\overline{K}_n, \overline{K}_m] = 0, \; [\overline{T}_n, \overline{K}_m] = (m + \rho)\overline{K}_{n+m}, \; [\overline{T}_n, \overline{T}_m] = (m - n)\overline{T}_{n+m}. \tag{5.29}$$

Note that the $\overline{K}_n$ vector fields are Hamiltonian with respect to arbitrary $\bar{\theta}_p(\rho)$ while an arbitrary $\overline{T}_n$ vector field is Hamiltonian only with respect to the Poisson operator $\bar{\theta}_p(\rho)$

$$\overline{T}_n = \bar{\theta}_p(\rho) \circ \nabla \overline{F}. \tag{5.30}$$

In the convention of this chapter we denote by $\bar{\theta}_0$ the Poisson operator θ_r of (3.104). Of course for a given vector field $\overline{K}_m$ the hierarchy

$$\bar{\sigma}_n(t) = \bar{\tau}_n + (\rho + m)t\overline{K}_{n+m} \tag{5.31}$$

represents its linear in time non-Hamiltonian symmetries.

Finally let us reconstruct the non-canonical action/angle algebra. For $\overline{K}_n = \frac{1}{n+\rho-p}\,\bar{\sigma}_{0,n}^p$ and $\bar{\sigma}_{1,n}^p$ we find

$$[\overline{K}_n, \overline{K}_m] = 0, \qquad [\bar{\sigma}_{1,n}^p, \overline{K}_m] = (n + m + \rho - p)\overline{K}_{n+m},$$

$$[\bar{\sigma}_{1,n}^p, \bar{\sigma}_{1,m}^p] = (m - n)\bar{\sigma}_{1,n+m}^p. \tag{5.32}$$

Notice that for $p = 0$ we recover the algebra (3.111). On the level of scalar fields

$$\overline{H}_n \equiv \frac{1}{n+\rho}\overline{T}_{0,n} = \frac{1}{n+\rho}\sum_{i=1}^{N} c_i^{n+\rho}, \qquad \overline{T}_m \equiv \overline{T}_{1,n} = -\sum_{i=1}^{N} q_i c_i^{n+\rho} \tag{5.33}$$

one gets the following noncanonical action/angle algebra

$$\{\overline{H}_n, \overline{H}_m\}_{\bar{\theta}_p} = 0, \qquad \{\overline{T}_n, \overline{H}_m\}_{\bar{\theta}_p} = (n + m + \rho + p)\overline{H}_{n+m+p},$$

$$\{\overline{T}_n, \overline{T}_m\}_{\bar{\theta}_p} = (n - m)\overline{T}_{n+m+p}. \tag{5.34}$$

We will come back to the representation (5.10) with $\rho = 1$ in the next chapter as it plays a crucial role in the theory of all finite dimensional integrable systems.

Now we shall pass to complex field systems (5.7) and examine the algebraic structure of the linear representation of $4N$ complex soliton scattering data $(q_1, ..., q_N, q_1^*, ..., q_N^*, c_1, ..., c_N, c_1^*, ..., c_N^*) := \bar{u}$. We confine ourselves to the case $\rho = 1$, because it is a common feature for such systems, but of course generalization to other ρ does not pose special difficulties. Let $q = (q_1, ..., q_N)^{\mathrm{T}}, q^* = (q_1^*, ..., q_N^*)^{\mathrm{T}}, c = (c_1, ..., c_N)^{\mathrm{T}}$ and $c^* = (c_1^*, ..., c_N^*)^{\mathrm{T}}$ and let us construct the hierarchy of commuting Hamiltonian systems of the form

$$\bar{u}_t = \overline{K}_m = \bar{\theta}_0 \circ \nabla \overline{H}_m, \tag{5.35}$$

where the field variable $\bar{u} = (q, c, q^*, c^*)^{\mathrm{T}}$ and

$$\overline{K}_m = (c^m, 0, c^{*m}, 0)^{\mathrm{T}},$$

$$\bar{\theta}_0 = \mathrm{i}\begin{pmatrix} 0 & I & 0 & 0 \\ -I & 0 & 0 & 0 \\ 0 & 0 & 0 & I \\ 0 & 0 & -I & 0 \end{pmatrix}, \qquad \overline{H}_m = \frac{1}{\mathrm{i}(m+1)}\sum_{k=1}^{N} (c_k^{m+1} - c_k^{*m+1}).$$

$$\tag{5.36}$$

The Poisson operator $\bar{\theta}_0$ is canonical for our variables in the sense that the only nonvanishing Poisson brackets are the following

$$\{q_k, c_l\}_{\bar{\theta}_0} = \mathrm{i}\delta_{kl}, \qquad \{q_k^*, c_l^*\}_{\bar{\theta}_0} = -\mathrm{i}\delta_{kl}, \tag{5.37}$$

hence, the coordinates (c_k, q_k) and (c_k^*, q_k^*) are called the canonical action/angle variables. The hierarchy (5.35) admits the bi-Hamiltonian formulation

$$\overline{K}_m = \overline{\theta}_0 \circ \nabla \overline{H}_m = \overline{\theta}_1 \circ \nabla \overline{H}_{m-1}, \qquad (5.38)$$

where

$$\overline{\theta}_1 = \mathrm{i} \begin{pmatrix} 0 & \Lambda & 0 & 0 \\ -\Lambda & 0 & 0 & 0 \\ 0 & 0 & 0 & -\Lambda^* \\ 0 & 0 & \Lambda^* & 0 \end{pmatrix}, \quad \Lambda = \mathrm{diag}(c_1, ..., c_N). \qquad (5.39)$$

The recursion operator $\overline{\Phi}$ takes the diagonal form

$$\overline{\Phi} = \overline{\theta}_1 \circ \overline{\theta}_0^{-1} = \begin{pmatrix} \Lambda & & & 0 \\ & \Lambda & & \\ & & \Lambda^* & \\ 0 & & & \Lambda^* \end{pmatrix}, \qquad (5.40)$$

so, for every $k = 1, ..., N$ the eigenvectors $\overline{A}_k(\overline{A}_k^\dagger)$ and $\overline{B}_k(\overline{B}_k^\dagger)$ w.r.t. $c_k(c_k^*)$ are given by the partial derivatives

$$\overline{A}_k = \frac{\partial \overline{u}}{\partial q_k}, \quad \overline{B}_k = \frac{\partial \overline{u}}{\partial c_k}, \quad \overline{\Phi} \overline{A}_k(\overline{B}_k) = c_k \overline{A}_k(\overline{B}_k),$$

$$\overline{A}_k^\dagger = \frac{\partial \overline{u}}{\partial q_k^*}, \quad \overline{B}_k^\dagger = \frac{\partial \overline{u}}{\partial c_k^*}, \quad \overline{\Phi}\, \overline{A}_k^\dagger(\overline{B}_k^\dagger) = c_k \overline{A}_k^\dagger(\overline{B}_k^\dagger) \qquad (5.41)$$

and may represent the basis of the tangent space to the considered phase space. Moreover, they are Hamiltonian action/angle vector fields w.r.t. $\overline{\theta}_0$ as

$$\overline{A}_k = \overline{\theta}_0 \circ \nabla(-\mathrm{i}c_k), \qquad \overline{A}_k^\dagger = \overline{\theta}_0 \circ \nabla(\mathrm{i}c_k^*),$$

$$\overline{B}_k = \overline{\theta}_0 \circ \nabla(\mathrm{i}q_k), \qquad \overline{B}_k^\dagger = \overline{\theta}_0 \circ \nabla(-\mathrm{i}q_k^*). \qquad (5.42)$$

This means that all results obtained for real systems with $\rho = 1$ are valid for the complex case. What we have to do on the level of vector fields is to take any vector field $V(q, c)$ from the real case with $\rho = 1$, treat the coordinates q and c as complex numbers and substitute

$$V(q, c) \Longrightarrow \begin{pmatrix} V(q, c) \\ V^*(q, c) \end{pmatrix}. \qquad (5.43)$$

All commutator relations of the real case hold. For example

$$\overline{\tau}_0 = \begin{pmatrix} -q \\ c \\ -q^* \\ c^* \end{pmatrix}, \qquad \overline{\tau}_n = \overline{\Phi}^n \circ \overline{\tau}_0,$$

$$\overline{P}_{r,n} = \overline{\Phi}^n \sum_{i=1}^{N} \left(q_i^r \overline{A}_i + q_i^{*r} \overline{A}_i^\dagger \right), \quad \overline{M}_{s,n} = \overline{\Phi}^{n+1} \sum_{i=1}^{N} \left(q_i^s \overline{B}_i + q_i^{*s} \overline{B}_i^\dagger \right). \quad (5.44)$$

On the level of scalar fields, $\overline{H}_m$ is given by (5.37) and

$$\overline{T}_{r,n} = (-1)^r \frac{1}{i} \sum_{k=1}^{N} \left(q_k^r c_k^{n+1} - q_k^{*r} c_k^{*n+1} \right), \qquad \overline{F} = \overline{T}_{1,0}. \quad (5.45)$$

From the physical point of view we definitely prefer a real linear representation of the complex system. Let

$$q_k = \gamma_k + i\delta_k, \qquad c_k = a_k + i b_k \qquad (5.46)$$

and perform the linear change of variables

$$\phi: \quad \gamma_k = \tfrac{1}{2}(q_k + q_k^*),$$

$$\delta_k = \tfrac{1}{2i}((q_k - q_k^*),$$

$$a_k = \tfrac{1}{2}(c_k + c_k^*), \qquad (5.47)$$

$$b_k = \tfrac{1}{2i}(c_k - c_k^*) \qquad k = 1, ..., N.$$

Hence, according to the transformation rules from Chap. 2 we find

$$\overline{\theta}_0 = \frac{1}{2} \begin{pmatrix} & 0 & & I \\ & & I & \\ & -I & & 0 \\ -I & & & \end{pmatrix}, \quad \overline{\theta}_1 = \frac{1}{2} \begin{pmatrix} 0 & 0 & -\Lambda_b & \Lambda_a \\ 0 & 0 & \Lambda_a & \Lambda_b \\ \Lambda_b & -\Lambda_a & 0 & 0 \\ -\Lambda_a & -\Lambda_b & 0 & 0 \end{pmatrix}$$

$$\overline{\Phi} = \begin{pmatrix} \Lambda_a & -\Lambda_b & 0 & 0 \\ \Lambda_b & \Lambda_a & 0 & 0 \\ 0 & 0 & \Lambda_a & -\Lambda_b \\ 0 & 0 & \Lambda_b & \Lambda_a \end{pmatrix}, \quad \Lambda_y = \mathrm{diag}(y_1, .., y_N), \quad y = a, b,$$

$$(5.48)$$

where now

$$\overline{A}_k = \frac{1}{2} \left(\frac{\partial \overline{u}}{\partial \gamma_k} - i \frac{\partial \overline{u}}{\partial \delta_k} \right), \qquad \overline{A}_k^\dagger = \frac{1}{2} \left(\frac{\partial \overline{u}}{\partial \gamma_k} + i \frac{\partial \overline{u}}{\partial \delta_k} \right),$$

$$\overline{B}_k = \frac{1}{2} \left(\frac{\partial \overline{u}}{\partial a_k} - i \frac{\partial \overline{u}}{\partial b_k} \right), \qquad \overline{B}_k^\dagger = \frac{1}{2} \left(\frac{\partial \overline{u}}{\partial a_k} + i \frac{\partial \overline{u}}{\partial b_k} \right). \quad (5.49)$$

Our new variables are canonical action/angle variables w.r.t. the new Poisson operator $\overline{\theta}_0$

$$\{a_k, \delta_l\}_{\overline{\theta}_0} = \delta_{kl}, \qquad \{b_k, \gamma_l\}_{\overline{\theta}_0} = \delta_{kl}, \qquad (5.50)$$

and the respective canonical action/angle Hamiltonian vector fields are

$$\frac{\partial \overline{u}}{\partial \delta_k} = \mathrm{i}(\overline{A}_k - \overline{A}_k^\dagger) = \overline{\theta}_0 \circ \nabla a_k, \quad \frac{\partial \overline{u}}{\partial a_k} = \overline{B}_k + \overline{B}_k^\dagger = \overline{\theta}_0 \circ \nabla(-\delta_k),$$

$$\frac{\partial \overline{u}}{\partial \gamma_k} = \overline{A}_k + \overline{A}_k^\dagger = \overline{\theta}_0 \circ \nabla b_k, \quad \frac{\partial \overline{u}}{\partial b_k} = \mathrm{i}(\overline{B}_k - \overline{B}_k^\dagger) = \overline{\theta}_0 \circ \nabla(-\gamma_k). \quad (5.51)$$

For the hierarchy of commuting vector fields (5.35) we have

$$\overline{K}_m = \overline{\Phi}^m \circ \overline{K}_0 = \overline{\Phi}^m (1,0,0,0)^{\mathrm{T}}, \quad \overline{H}_m = \frac{2}{m+1} \sum_{k=1}^{N} \mathrm{im}\, c_k^{m+1}, \quad (5.52)$$

so the first few conserved functions and the corresponding vector fields in our new variables are as follows

$$\overline{H}_0 = 2 \sum_{k=1}^{N} b_k, \quad \overline{H}_1 = 2 \sum_{k=1}^{N} a_k b_k, \quad \overline{H} = \frac{2}{3} \sum_{k=1}^{N} (3a_k^2 b_k - b_k^3), \dots$$

$$\overline{K}_0 = (1,0,0,0)^{\mathrm{T}}, \quad \overline{K}_1 = (a,b,0,0)^{\mathrm{T}}, \quad \overline{K}_2 = (a^2 - b^2, 2ab, 0, 0)^{\mathrm{T}}, \dots$$

Moreover, we have the following relation between Hamiltonian master symmetries $\overline{\tau}_{r,n}^p$ and non-Hamiltonian master symmetries $\overline{\tau}_n$

$$\overline{\tau}_{r,n}^p = \overline{\theta}_p \circ \nabla \overline{T}_{r,n-p}, \quad \overline{\theta}_p = \overline{\Phi}^p \circ \overline{\theta}_0,$$

$$\overline{T}_{r,n} = (-1)^r 2 \sum_{k=1}^{N} \left(\mathrm{re}\, q_k^r \mathrm{im}\, c_k^{n+1} + \mathrm{im}\, q_k^r \mathrm{re}\, c_k^{n+1} \right),$$

$$\overline{\tau}_0 = (-\gamma, -\delta, a, b)^T = \overline{\theta}_0 \nabla \overline{F}, \quad \overline{F} = \overline{T}_{1,0} = -2 \sum_{k=1}^{N} (a_k \delta_k + b_k \gamma_k),$$

$$\overline{\tau}_n = \overline{\Phi}^n \circ \overline{\tau}_0 = \overline{\theta}_n \circ \nabla \overline{F}. \quad (5.53)$$

In order to compare the results for complex systems of the present section with the literature we perform another change of coordinates given by

$$\phi : \qquad \widetilde{\gamma}_k = a_k \gamma_k - b_k \delta_k,$$

$$\widetilde{\delta}_k = a_k \delta_k + b_k \gamma_k,$$

$$\qquad\qquad\qquad\qquad\qquad (5.54)$$

$$\widetilde{a}_k = 2a_k,$$

$$\widetilde{b}_k = 2b_k.$$

In terms of this parametrization, if $\widetilde{u} = (\widetilde{\gamma}, \widetilde{\delta}, \widetilde{a}, \widetilde{b})^T$ then

$$\widetilde{H}_n(\widetilde{u}) = 2^{-(n+1)} \overline{H}_n(\widetilde{u}), \quad \widetilde{K}_n(\widetilde{u}) = 2^{-(n+1)} \overline{K}_{n+1}(\widetilde{u}), \quad \widetilde{\theta}_n(\widetilde{u}) = \overline{\theta}_{n+1}(\widetilde{u}),$$

$$\widetilde{\tau}_0 = (0,0,\widetilde{a},\widetilde{b})^T = \widetilde{\theta}_0 \circ \nabla \widetilde{F}, \quad \widetilde{F} = \widetilde{T}_{1,-1} = -2 \sum_{k=1}^{N} \widetilde{\delta}_k. \quad (5.55)$$

5.3 Algebraic Structure of Multi-Soliton Representation

In the previous section we found the algebraic structure for the linear system
(5.3) in the real and complex cases, respectively. But of course, the natural
question arises of whether it is possible to recover the whole structure on the
physical N-soliton manifold in explicit form. By 'explicit' we mean that it
must be possible to express all quantities in terms of the original field variable
u.

As we have seen in the last section, the structure of the linear representation is more or less trivial. Now, applying the Π transformation, introduced
in Sect. 5.1, we try to carry over the whole structure to the physical representation. Although we do not know the explicit form of our variable transformation Π, we know how tensor fields behave under a change of coordinates.
In other words, Π induces the push-forward

$$\Pi' : T_u M_N \to T_{\bar{u}} \mathbb{R}^{2N}, \tag{5.56}$$

which maps vector fields of the nonlinear space onto vector fields on the linear
space. Furthermore we obtain the pull-back

$$\Pi'^\dagger : T_{\bar{u}}^* \mathbb{R}^{2N} \to T_u^* M_N, \tag{5.57}$$

which is the conjugate of Π' with respect to the duality between the tangent
and the cotangent bundle.

Let us first concentrate on the real multisoliton case.

Theorem 5.1

(i) $\Phi(u) = (\Pi')^{-1} \bar{\Phi}(\bar{u}) \Pi' : T_u M_N \to T_u M_N$ *is a hereditary recursion
operator in the physical representation.*

(ii) *The eigenvalues of Φ are $c_1, ..., c_N$; each of them occurs twice. At each
point $u \in M_N$ the tangent space $T_u M_N$ is the span of the eigenvectors
of Φ.*

(iii) *The eigenvectors $A_i, B_i, i = 1, ..., N$, of Φ are given by the derivative of
the field variable u with respect to the coordinates $q_i, c_i, i = 1, ..., N$.*

(iv) *The eigenvectors*

$$A_i = \frac{\partial u}{\partial q_i} \quad \text{and} \quad c_i^{1-\rho} B_i = c_i^{1-\rho} \frac{\partial u}{\partial c_i} \tag{5.58}$$

*of $\Phi(u)$ are Hamiltonian vector fields with respect to the implectic oper-
ator $\theta_{|\text{red}}(u)$ determined by*

$$\tau_0 = \theta(u) \nabla F(u).$$

*The reduced implectic operator $\theta_{|\text{red}}(u)$ has the following representation
on M_N:*

$$\theta_{|\text{red}}(u) = \theta_0 = (\Pi')^{-1} \circ \bar{\theta}_0 \circ (\Pi'^\dagger)^{-1}.$$

(v)   The Hamiltonians E_i and Ω_ι of the eigenvectors A_i and $c_i^{1-\rho}B_i$

$$A_i = \frac{\partial u}{\partial q_i} = \theta_0 \circ \nabla\left(\frac{1}{\rho}E_i\right),$$

$$c_i^{1-\rho}B_i = c_i^{1-\rho}\frac{\partial u}{\partial c_i} = \rho\frac{\partial u}{\partial(c_i^\rho)} = \theta_0\nabla(-\Omega_i), \qquad (5.59)$$

are given by the partial derivatives

$$E_i = -\frac{\partial F}{\partial q_i}, \qquad \Omega_i = -\frac{1}{\rho}c_i^{1-\rho}\frac{\partial F}{\partial c_i} = -\frac{\partial F}{\partial(c_i^\rho)}. \qquad (5.60)$$

(vi) $\frac{1}{\rho}E_i$ and Ω_i are canonical coordinates with respect to θ_0, i.e. for all $i, j = 1, ..., N$,

$$\{E_i, E_j\}_{\theta_0} = \{\Omega_i, \Omega_j\}_{\theta_0} = 0, \quad \{E_i, \Omega_j\}_{\theta_0} = \rho\delta_{ij}. \qquad (5.61)$$

Proof.

(i) and (ii) are obvious consequences of Lemma 5.2, the definition of Π' and the fact that Π' is a Lie algebra isomorphism, i.e.

$$\Pi'[A, B] = [\Pi'A, \Pi'B], \qquad A, B \in T_u M_N.$$

(iii) By definition of the eigenvectors of $\overline{\Phi}$ we obtain for example that

$$
\begin{aligned}
B_i &= (\Pi')^{-1}\overline{B}_i = (\Pi')^{-1}\frac{\partial}{\partial c_i}\overline{u} \\
&= (\Pi')^{-1}\frac{\partial}{\partial c_i}\Pi(u(q_1, ..., q_N, c_1, ..., c_N)) \\
&= (\Pi')^{-1}\Pi'\frac{\partial u}{\partial c_i} = \frac{\partial}{\partial c_i}u. \qquad (5.62)
\end{aligned}
$$

is an eigenvector of Φ for the eigenvalue c_i.

(iv) One has to keep in mind that the pull-back conserves the Hamiltonian structure. Then (iv) is a direct consequence of the fact that in the linear space the corresponding eigenvectors and the scaling vector field are Hamiltonian only with respect to the same implectic operator $\overline{\theta}_0$.

(v) From (iv) we have that there are scalar fields E_i and Ω_i with

$$\frac{\partial u}{\partial q_i} = \theta_0 \circ \nabla\left(\frac{1}{\rho}E_i\right) \qquad \text{and} \qquad c_i^{1-\rho}\frac{\partial u}{\partial c_i} = \theta_0 \circ \nabla(-\Omega_i). \qquad (5.63)$$

Since these scalar fields are conserved quantities it follows that

$$
\begin{aligned}
E_i &= L_{\tau_0}\left(\frac{1}{\rho}E_i\right) = \left\langle\frac{1}{\rho}\nabla E_i, \tau_0\right\rangle = \left\langle\frac{1}{\rho}\nabla E_i, \theta_0 \circ \nabla F\right\rangle \\
&= -\left\langle\nabla F, \frac{1}{\rho}\nabla E_i\right\rangle = -\left\langle\nabla F, \frac{\partial u}{\partial q_i}\right\rangle = -\frac{\partial F}{\partial q_i}, \qquad (5.64)
\end{aligned}
$$

$$-\rho\Omega_i = L_{\tau_0}\Omega_i = \langle\nabla\Omega_i, \theta_0 \circ \nabla F\rangle = \langle\nabla F, \theta_0 \circ \nabla(-\Omega_i)\rangle = c_i^{1-\rho}\frac{\partial F}{\partial c_i}.$$

(vi) is a direct consequence of (v) together with Lemma 5.2 (i),(ii).$\square$

Now, via the inverse of the push-forward Π' we define the soliton fundamental algebra $\mathcal{A}$ as the image of $\overline{\mathcal{A}}$ under $(\Pi')^{-1}$. Then the basic vector fields have the form

$$(\Pi')^{-1}\overline{P}_{r,n} = P_{r,n}(u_N) = \Phi^n \sum_{k=1}^{N} q_k^r A_k = \sum_{k=1}^{N} q_k^r c_k^n A_k,$$

$$(\Pi')^{-1}\overline{M}_{s,n} = M_{s,n}(u_N) = \Phi^{n+1} \sum_{k=1}^{N} q_k^s B_k = \sum_{k=1}^{N} q_k^s c_k^{n+1} B_k. \tag{5.65}$$

Since the push-forward is a Lie algebra isomorphism the commutator relations (5.19) of $\overline{\mathcal{A}}$ are also valid for $\mathcal{A}$. The same holds for all subalgebras. Hence, we have found the representation of commuting symmetries $K_n(u_N)$, non-Hamiltonian master symmetries $\tau_n(u_N) = M_{0,n}(u_N) - \rho P_{1,n}(u_N)$ and the Hamiltonian master symmetries $\tau_{r,n}^p(u_N) = (-1)^r[(n + \rho - p)P_{r,n}(u_N) - rM_{r-1,n}(u_N)]$. This means that, at least on the N-soliton manifold M_N, each dynamical system (5.5),(5.6) contains the hereditary algebra (5.30) as well as the general algebra (5.22).

Lemma 5.3 *On the N-soliton manifold the Hamiltonian vector fields*

$$K_n(u_N) = \theta_p \circ \nabla H_{n-p}, \qquad \tau_{r,n}^p(u_N) = \theta_p \circ \nabla T_{n-p} \tag{5.66}$$

have the following representation

$$K_n(u_N) = \sum_{i=1}^{N} \Phi^n A_i = \sum_{i=1}^{N} c_i^n A_i, \tag{5.67}$$

$$\tau_{r,n}^p(u_N) = (-1)^r \left[(n + \rho - p) \sum_{i=1}^{N} q_i^r c_i^n A_i - r \sum_{i=1}^{N} q_i^{r-1} c_i^{n+1} B_i\right]. \tag{5.68}$$

The corresponding scalar fields are given by

$$H_n(u_N) = -\frac{1}{n+\rho} \int_R \nabla F \cdot K_n \mathrm{d}x = \frac{1}{n+\rho} \sum_{i=1}^{N} c_i^n E_i, \tag{5.69}$$

$$\begin{aligned} T_{r,n}(u_N) &= -\frac{1}{n+\rho(1-r)} \int_R \nabla F \cdot \tau_{r,n}^0 \mathrm{d}x \\ &= (-1)^r \left[\frac{n+\rho}{n+\rho(1-r)} \sum_{i=1}^{N} q_i^r c_i^n E_i - \frac{r\rho}{n+\rho(1-r)} \sum_{i=1}^{N} q_i^{r-1} c_i^{n+\rho} \Omega_i.\right] \end{aligned}$$

$$\tag{5.70}$$

Proof. Since the formulas for the vector fields are obvious by construction, we only prove the representation of the scalar fields.

$$H_n(u_N) = \frac{1}{n+\rho} L_{\tau_0} H_n = -\frac{1}{n+\rho} \langle \nabla F, K_n \rangle = -\frac{1}{n+\rho} \int_R \nabla F \cdot K_n \mathrm{d}x$$

$$= -\frac{1}{n+\rho} \sum_{i=1}^{N} c_i^n \langle \nabla F, A_i \rangle = \frac{1}{n+\rho} \sum_{i=1}^{N} c_i^n E_i, \qquad (5.71)$$

$$T_{r,n}(u_N)$$

$$= \frac{1}{n+\rho(1-r)} L_{\tau_0} T_{r,n}(u_N) = -\frac{1}{n+\rho(1-r)} \langle \nabla F, \tau_{r,n}^0 \rangle$$

$$= -\frac{1}{n+\rho(1-r)} \int_R \nabla F \cdot \tau_{r,n}^0 \mathrm{d}x$$

$$= -\frac{1}{n+\rho(1-r)} \left\langle \nabla F, (-1)^r \left[(n+\rho) \sum_{i=1}^{N} q_i^r c_i^n A_i - r \sum_{i=1}^{N} q_i^{r-1} c_i^{n+1} B_i \right] \right\rangle$$

$$= -\frac{(-1)^r}{n+\rho(1-r)} \left[(n+\rho) \sum_{i=1}^{N} q_i^r c_i^n \langle \nabla F, A_i \rangle - r\rho \sum_{i=1}^{N} q_i^{r-1} c_i^{n+1} \langle \nabla F, B_i \rangle \right]$$

$$= \frac{(-1)^r}{n+\rho(1-r)} \left[(n+\rho) \sum_{i=1}^{N} q_i^r c_i^n E_i - r\rho \sum_{i=1}^{N} q_i^{r-1} c_i^{n+\rho} \Omega_i \right]. \qquad (5.72)$$

where we have used the relations

$$\langle \nabla F, A_i \rangle = \frac{\partial F}{\partial q_i} = -E_i \quad \text{and} \quad \langle \nabla F, B_i \rangle = \frac{\partial F}{\partial c_i} = -\rho c_i^{\rho-1} \Omega_i. \square$$

According to the theory presented, one can treat solitons as models of *extended noninteracting particles* or in other words as field representatives of non-interacting point particles. Indeed each point particle is represented by canonical functional coordinates

$$E_i = \int_R p_i(x,t) \mathrm{d}x, \qquad \Omega_i = \int_R q_i(x,t) \mathrm{d}x \qquad (5.73)$$

where

$$p_i(x,t) = -\frac{\partial f(u_N)}{\partial q_i}, \qquad q_i(x,t) = -\frac{\partial f(u_N)}{\partial (c_i^\rho)}, \qquad (5.74)$$

are canonical momentum and position densities related to vector fields $\frac{\partial u_N}{\partial q_i}$ and $\frac{\partial u_N}{\partial (c_i^\rho)}$ where $F = \int_R f(u_N) \mathrm{d}x$. Of course the evaluation of the integrals

(5.67), (5.68) gives $E_i = \beta c_i^\rho$ and $\Omega_i = \beta q_i$, where β is some constant depending on the appropriate equation.

The extension of the above considerations to include complex soliton particles is straightforward. One can formulate the analogs of Theorem 5.1 and Lemma 5.3. Here we only present the final results. The vector fields $\frac{\partial u_N}{\partial \gamma_i}, \frac{\partial u_N}{\partial \delta_i}, \frac{\partial u_N}{\partial a_i}$ and $\frac{\partial u_N}{\partial b_i}, i = 1, ..., N$, are Hamiltonian vector fields

$$\frac{\partial u_N}{\partial \delta_i} = \theta_0 \circ \nabla E_i, \qquad \frac{\partial u_N}{\partial a_i} = \theta_0 \circ \nabla(-\Omega_i),$$

$$\frac{\partial u_N}{\partial \gamma_i} = \theta_0 \circ \nabla G_i, \qquad \frac{\partial u_N}{\partial b_i} = \theta_0 \circ \nabla(-\Delta_i), \qquad (5.75)$$

where the implectic operator $\theta_0 = \theta_{|\mathrm{red}}(u)$ is determined by $\tau_0 = \theta(u)\nabla F(u)$, and the Hamiltonian functionals E_i, Ω_i, G_i and Δ_i are given by the partial derivatives

$$E_i = -\frac{\partial F(u_N)}{\partial \delta_i}, \qquad \Omega_i = -\frac{\partial F(u_N)}{\partial a_i},$$

$$G_i = -\frac{\partial F(u_N)}{\partial \gamma_i}, \qquad \Delta_i = -\frac{\partial F(u_N)}{\partial b_i}. \qquad (5.76)$$

The functionals (Ω_i, E_i) and (G_i, Δ_i) are canonical coordinates w.r.t. θ_0, i.e. for all $i, j = 1, ..., N$ the only Poisson brackets different from zero are the following

$$\{E_i, \Omega_j\}_{\theta_0} = \delta_{ij}, \qquad \{G_i, \Delta_j\}_{\theta_0} = \delta_{ij}. \qquad (5.77)$$

The Hamiltonian vector fields

$$K_n(u_N) = \theta_p \circ \nabla H_{n-p}, \qquad \tau_{r,n}^p(u_N) = \theta_p \circ \nabla T_{n-p} \qquad (5.78)$$

have the following representation

$$K_n(u_N) = \sum_{i=1}^{N} \left(\mathrm{re}\, c_i^n \frac{\partial u_N}{\partial \gamma_i} + \mathrm{im}\, c_i^n \frac{\partial u_N}{\partial \delta_i} \right),$$

$$\tau_{r,n}^p(u_N) = (-1)^r \left\{ (n+1-p) \sum_{i=1}^{N} \left[(\mathrm{re}\, q_i^r \mathrm{re}\, c_i^n - \mathrm{im}\, q_i^r \mathrm{im}\, c_i^n) \frac{\partial u_N}{\partial \gamma_i} \right. \right.$$

$$\left. + (\mathrm{re}\, q_i^r \mathrm{im}\, c_i^n + \mathrm{im}\, q_i^r \mathrm{re}\, c_i^n) \frac{\partial u_N}{\partial \delta_i} \right]$$

$$- r \sum_{i=1}^{N} \left[(\mathrm{re}\, q_i^{r-1} \mathrm{re}\, c_i^{n+1} - \mathrm{im}\, q_i^{r-1} \mathrm{im}\, c_i^{n+1}) \frac{\partial u_N}{\partial a_i} \right.$$

$$\left. \left. + (\mathrm{re}\, q_i^{r-1} \mathrm{re}\, c_i^{n+1} - \mathrm{im}\, q_i^{r-1} \mathrm{im}\, c_i^{n+1}) \frac{\partial u_N}{\partial b_i} \right] \right\}. \qquad (5.79)$$

The corresponding scalar fields are given by

$$H_n(u_N) = -\frac{1}{n+1} \int_R \nabla F(u_N) \cdot K_n(u_N) dx$$

$$= \frac{1}{n+1} \sum_{-=1}^{N} (\operatorname{re} c_i^n G_i + \operatorname{im} c_i^n E_i),$$

$$T_{r,n}(u_N) = -\frac{1}{n+1-1} \int_R \nabla F(u_N) \cdot \tau_{r,n}^0(u_N) dx$$

$$= \frac{n+1}{n+1-r}(-1)^r \sum_{i=1}^{N} [(\operatorname{re} q_i^r \operatorname{re} c_i^n - \operatorname{im} q_i^r \operatorname{im} c_i^n)G_i$$

$$+ (\operatorname{re} q_i^r \operatorname{im} c_i^n + \operatorname{im} q_i^r \operatorname{re} c_i^n)E_i]$$

$$- \frac{r}{n+1-r}(-1)^r \sum_{i=1}^{N} [(\operatorname{re} q_i^{r-1} \operatorname{re} c_i^{n+1} - \operatorname{im} q_i^{r-1} \operatorname{im} c_i^{n+1})\Omega_i$$

$$+ (\operatorname{re} q_i^{r-1} \operatorname{re} c_i^{n+1} - \operatorname{im} q_i^{r-1} \operatorname{im} c_i^{n+1})\Delta_i].$$

$$(5.80)$$

The following recursion relations are useful for particular calculations

$$\operatorname{re} q^{r+1} = \gamma \operatorname{re} q^r - \delta \operatorname{im} q^r, \quad \operatorname{im} q^{r+1} = \gamma \operatorname{im} q^r + \delta \operatorname{re} q^r,$$

$$\operatorname{re} c^{n+1} = a \operatorname{re} c^n - b \operatorname{im} c^n, \quad \operatorname{im} c^{n+1} = a \operatorname{im} c^n + b \operatorname{re} c^n. \qquad (5.81)$$

Example 5.1 The KdV soliton manifold.
Here we consider the KdV equation in the form $u_t = u_{3x} + auu_x$, with the hierarchy of commuting symmetries

$$K_n(u) = \Phi^n(u)K_0(u) = \left(D^2 + \frac{a}{3}uD^{-1} + \frac{a}{3}u\right)^n u_x,$$

the hierarchy of master symmetries

$$\tau_n(u) = \Phi^n(u)\tau_0(u) = \left(D^2 + \frac{a}{3}uD^{-1} + \frac{a}{3}u\right)^n \left(\frac{1}{2}xu_x + u\right)$$

and the hereditary algebra

$$[K_n, K_m] = 0, \quad [\tau_n, K_m] = \left(m + \frac{1}{2}\right)K_{n+m}, \quad [\tau_n, \tau_m] = (m-n)\tau_{n+m}.$$

The scaling master symmetry τ_0 is a Hamiltonian vector field with respect to the second implectic operator

$$\frac{1}{2}xu_x + u = \left(D^3 + \frac{1}{3}aDu + \frac{1}{3}auD\right) \cdot \nabla \int_R \frac{3}{2a}xu\,dx = \theta(u)\nabla F(u). \quad (5.82)$$

The N-soliton solution decomposing at $\pm\infty$ into 1-soliton solutions is given by [2]

$$u_N \cong \sum_{i=1}^{N} \frac{3}{a}c_i\text{sech}^2\left[\frac{1}{2}\sqrt{c_i}(x + c_it + q_i)\right] \quad (5.83)$$

where c_i are the eigenvalues of the recursion operator Φ. Since $\rho = \frac{1}{2}$ Theorem 5.1 gives the action/angle variables of the soliton particles explicitly as

$$E_i = -\frac{\partial F(u_N)}{\partial q_i} = -\frac{3}{2a}\int_R x(u_N)_{q_i}\,dx,$$

$$\Omega_i = -\frac{\partial F(u_N)}{\partial(c_i^{1/2})} = -\frac{3}{a}c_i^{1/2}\int_R x(u_N)_{c_i}\,dx. \quad (5.84)$$

and suitable action/angle vector fields

$$A_i = \frac{\partial u_N}{\partial q_i}, \quad B_i = \frac{\partial u_N}{\partial c_i}. \quad (5.85)$$

The Hamiltonian vector fields $H_n(u_N)$ and $\tau_{r,n}(u_N)$, with respect to the Poisson operator $\theta_0 = \theta_{|red}$, are the following

$$K_n(u_N) = \Phi^n \sum_{i=1}^{N}(u_N)_{q_i} = \sum_{i=1}^{N} c_i^n(u_N)_{q_i},$$

$$\tau_{r,n}(u_N) = \tau_{r,n}^0(u_N) = (-1)^r[(n + \frac{1}{2})P_{r,n} - rM_{r-1,n}]$$

$$= (-1)^r\Phi^n \sum_{i=1}^{N} q_i^{r-1}[(n + \frac{1}{2})q_i(u_N)_{q_i} - rc_i(u_N)_{c_i}]$$

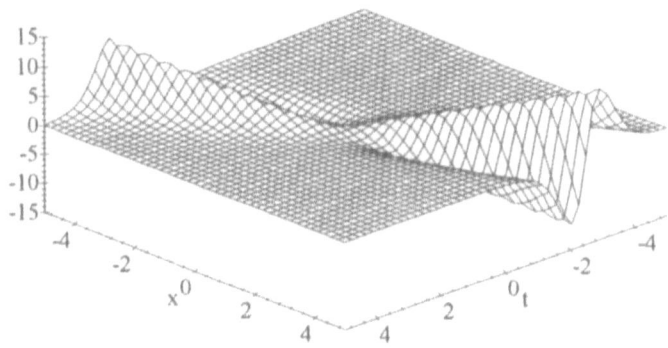

Fig. 5.1. Action density of the KdV one-soliton with $c = 2$

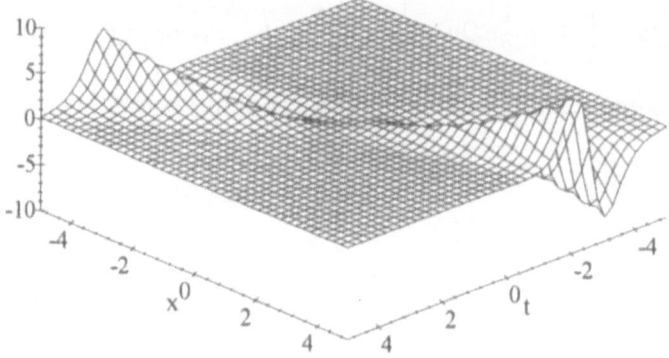

Fig. 5.2. Angle density of the one-soliton with $c = 2$

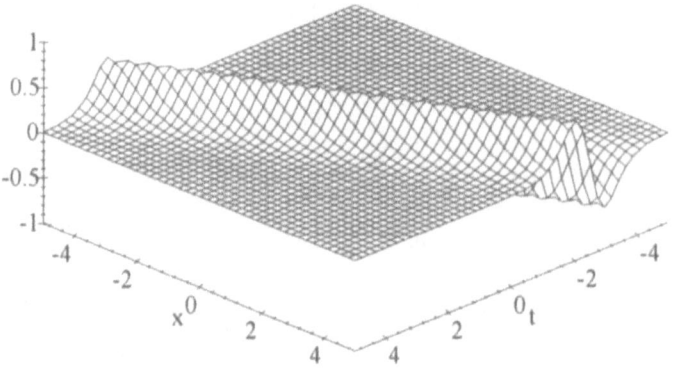

Fig. 5.3. Action field A for the KdV one-soliton with $c = 2$

$$= (-1)^r \sum_{i=1}^{N} q_i^{r-1} c_i^n [(n + \frac{1}{2}) q_i(u_N)_{q_i} - r c_i(u_N)_{c_i}], \qquad (5.86)$$

and the related scalar fields are

$$H_n(u_N) = -\frac{3}{a} \frac{1}{2n+1} \int_R x K_n(u_N) \mathrm{d}x = \frac{2}{2n+1} \sum_{i=1}^{N} c_i^n E_i,$$

$$T_{r,n}(u_N) = -\frac{3}{a} \frac{1}{2n+1-r} \int_R x T_{r,n}(u_N) \mathrm{d}x$$

$$= \frac{(-1)^r}{2n+1-r} \left[(2n+1) \sum_{i=1}^{N} q_i^r c_i^n E_i - r \sum_{i=1}^{N} q_i^{r-1} c_i^{n+\frac{1}{2}} \Omega_i \right].$$

$$(5.87)$$

Notice that the vector fields $T_{r,n}(u_N)$ constitute the general algebra (3.107). Additionally, the soliton non-Hamiltonian master symmetries $\tau_n(u)$ are as

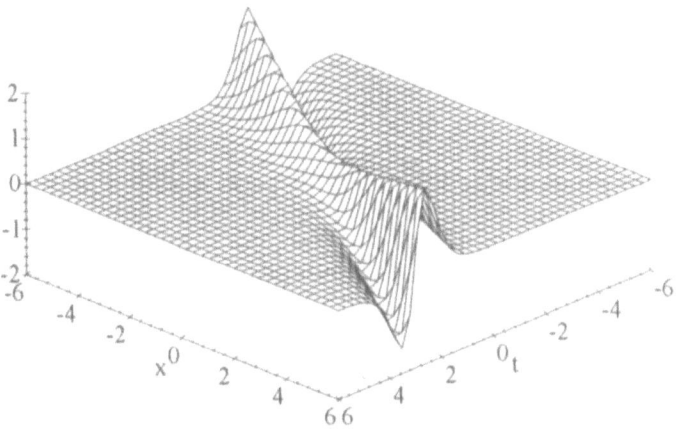

Fig. 5.4. Angle field B of the KdV one-soliton with $c = 2$

follows

$$\tau_n(u_N) = \Phi^n \sum_{i=1}^{N} \left[c_i(u_N)_{c_i} - \frac{1}{2} q_i(u_N)_{q_i} \right].$$ (5.88)

Evaluating the integrals from (5.84) we find

$$E_i = \beta c_i^{\frac{1}{2}}, \quad \Omega_i = \beta q_i, \quad \beta = \frac{18}{a^2}$$

and hence

$$H_n = \frac{2\beta}{2n+1} \sum_{i=1}^{N} c_i^{n+\frac{1}{2}} = \beta \overline{H}_n,$$

$$T_{r,n} = \beta(-1)^r \sum_{i=1}^{N} q_i^r c_i^{n+\frac{1}{2}} = \beta \overline{T}_{r,n}.$$

One should notice that on M_N

$$\sum_{i=1}^{N} (u_N)_{q_i} = (u_N)_x, \quad \sum_{i=1}^{N} \left[c_i(u_N)_{c_i} - \frac{1}{2} q_i(u_N)_{q_i} \right] = u + \frac{1}{2} x u_x$$

holds, hence $K_n(u_N)$ and $\tau_n(u_N)$ are expressible in terms of u_N and its x-derivatives (integrals) contrary to $\tau_{r,n}(u_N)$ which additionally contains derivatives of the field variable u_N with respect to the asymptotic data. On the other hand, applying relations (5.86) and (5.88), $(u_N)_{q_i}$ and $(u_N)_{c_i}$ can be expressed as a linear combination of $K_n(u_N)$ and $\tau_n(u_N)$, but still the coefficients of the differential functions depend in an explicit way on c_i and q_i.

Example 5.2 The MKdV soliton particles.
We consider the MKdV equation in the form $u_t = u_{3x} + au^2 u_x$. The related

hierarchies of commuting symmetries $K_n(u)$ and master symmetries $\tau_n(u)$ are

$$K_n(u) = \Phi^n(u)K_0(u) = \left(D^2 + \frac{2}{3}auD^{-1}u\right)^n u_x,$$

$$\tau_n(u) = \Phi^n(u)\tau_0(u) = \left(D^2 + \frac{2}{3}auD^{-1}u\right)^n \left(\frac{1}{2}xu_x + \frac{1}{2}u\right), \qquad (5.89)$$

which constitute the hereditary algebra with $\rho = \frac{1}{2}$. The scaling master symmetry $\tau_0(u)$ is a Hamiltonian vector field with respect to the implectic operator $\theta = D$

$$\frac{1}{2}(xu_x + u) = D \ \nabla\frac{1}{4}\int_R xu^2\mathrm{d}x = \theta_0\nabla F(u).$$

The N-soliton solutions decomposing into 1-solitons for $t \to \infty$ are given by [114]

$$u_N \cong \sum_{i=1}^{N} \left(\frac{6}{a}c_i\right)^{\frac{1}{2}} \mathrm{sech}[c_i^{\frac{1}{2}}(x + c_it + q_i)]. \qquad (5.90)$$

Since $\rho = \frac{1}{2}$, Theorem 5.1 determines the action/angle variables of soliton particles

$$E_i = -\frac{\partial F(u_N)}{\partial q_i} = -\frac{1}{2}\int_R x(u_N)_{q_i}\mathrm{d}x,$$

$$\Omega_i = -\frac{\partial F(u_N)}{\partial(c_i^{1/2})} = -c_i^{1/2}\int_R x(u_N)_{c_i}\mathrm{d}x. \qquad (5.91)$$

Example 5.3 The BO soliton particles.
In Example 3.22 we presented the Benjamin–Ono equation and its general Hamiltonian algebra of symmetries (master symmetries). The scaling vector field is a Hamiltonian vector field w.r.t. $\theta = D$

$$\tau_0(u) := \tau_{1,0}(u) = xu_x + u = D \ \nabla \int_R \frac{1}{2}xu^2\mathrm{d}x.$$

The N-soliton solutions, decomposing at $t \to \pm\infty$ into 1-solitons, are given by [132]

$$u_N \cong \sum_{i=1}^{N} \frac{1}{a}\frac{c_i}{c_i^2(x + c_it + q_i)^2 + 1}, \qquad (5.92)$$

where c_i are the eigenvalues of the recursion operator Φ which on our level of considerations exists in an implicit form.

As $\rho = 1$, Theorem 5.1 gives the canonical action/angle variables of soliton particles

$$E_i = -\frac{\partial F(u_N)}{\partial q_i} = -\int_R x(u_N)_{q_i}\mathrm{d}x,$$

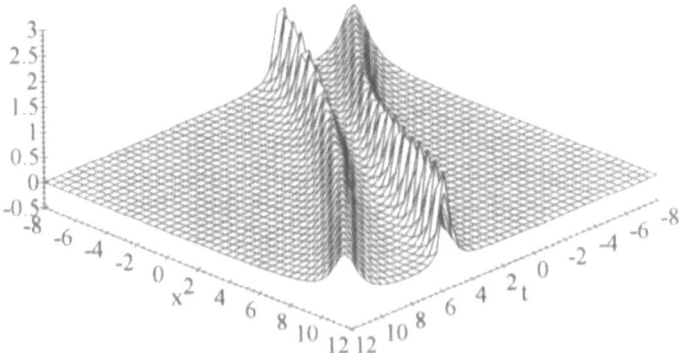

Fig. 5.5. BO two-soliton solution with $c_1 = 1$ and $c_2 = 2$

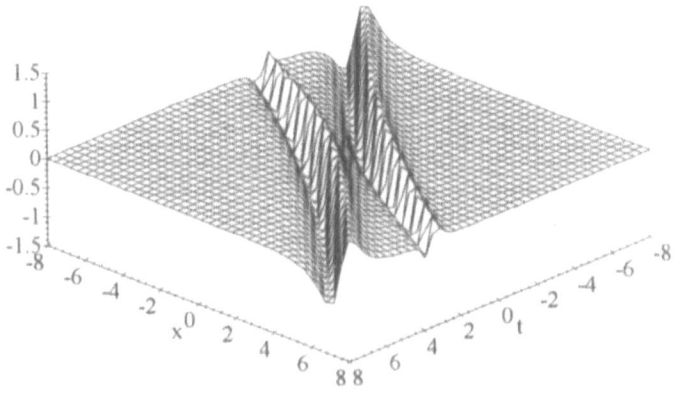

Fig. 5.6. Action vector field A_1 of BO two-soliton

$$\Omega_i = -\frac{\partial F(u_N)}{\partial(c_i)} = -\int_R x(u_N)_{c_i} dx, \qquad (5.93)$$

and suitable Hamiltonian vector fields

$$A_i = \frac{\partial u_N}{\partial q_i}, \quad B_i = \frac{\partial u_N}{\partial c_i}. \qquad (5.94)$$

Additionally, on M_N there exist non-Hamiltonian master symmetries

$$\tau_n(u_N) = \sum_{i=1}^{N} \left[c_i^{n+1}(u_N)_{c_i} - q_i(u_N)_{q_i} \right] \qquad (5.95)$$

and the vector fields $K_n(u_N) = \frac{1}{n+1}\tau_{0,1}(u_N)$ and $\tau_m(u_N)$ constitute the hereditary algebra

$$[K_n(u_N), K_m(u_N)] = 0, \quad [\tau_n(u_N), K_m(u_N)] = \left(m + \frac{1}{2} \right) K_{n+m}(u_N),$$

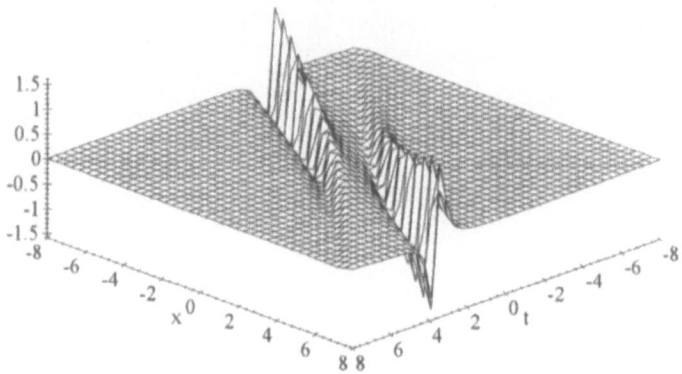

Fig. 5.7. Action vector field A_2 of BO two-soliton

$$[\tau_n(u_N), \tau_m(u_N)] = (m - n)\tau_{n+m}(u_N). \tag{5.96}$$

The appropriate values of integrals (5.93) are

$$E_i = \beta c_i, \quad \Omega_i = \beta q_i, \quad \beta = \frac{\pi}{4a^2}.$$

Example 5.4 The NLS soliton particles.
Here we consider the nonlinear Schrödinger equation from Example 3.4. The hierarchy of commuting symmetries

$$\begin{pmatrix} K_n \\ K_n^* \end{pmatrix} = \Phi^{n+1}\mathbf{K}_{-1} = i\begin{pmatrix} -D - 2uD^{-1}u^* & -2uD^{-1}u \\ 2u^*D^{-1}u^* & +D + 2u^*D^{-1}u \end{pmatrix} \begin{pmatrix} -iu \\ iu^* \end{pmatrix} \tag{5.97}$$

and the hierarchy of master symmetries

$$\boldsymbol{\tau}(u, u^*) = \Phi^{n+1}\boldsymbol{\tau}_{-1}(u, u^*) = \Phi^{n+1}\begin{pmatrix} ixu \\ -ixu^* \end{pmatrix}, \tag{5.98}$$

constitute the hereditary algebra with $\rho = 1$. The scaling vector field τ_0 has the following Hamiltonian representation

$$\begin{aligned} \boldsymbol{\tau}_0 &= \begin{pmatrix} xu_x + u \\ -xu_x^* + u^* \end{pmatrix} = \theta \circ \nabla F \\ &= \begin{pmatrix} 2uD^{-1}u & -D - 2uD^{-1}u^* \\ -D - 2u^*D^{-1}u & 2u^*D^{-1}u^* \end{pmatrix} \nabla \int_R (-x)|u|^2 \, dx. \end{aligned} \tag{5.99}$$

The N-soliton solutions decompose asymptotically with $t \to \infty$ into 1-soliton solutions as follows [64]

$$u_N \cong$$

$$\sum_{k=1}^{N} \frac{1}{2} \tilde{b}_k \operatorname{sech} \left(\frac{1}{2} \tilde{b}_k x + \frac{1}{2} \tilde{a}_k \tilde{b}_k t + \tilde{\delta}_k \right) \exp \left[i \left(\frac{1}{2} \tilde{a}_k x + \frac{1}{4} \left(\tilde{a}_k^2 - \tilde{b}_k^2 \right) t + \tilde{\gamma}_k \right) \right].$$

$$(5.100)$$

Unfortunately, with respect to the coordinates $(\tilde{a}_k, \tilde{b}_k), (\tilde{\gamma}_k, \tilde{\delta}_k)$ we have

$$F(u_N) = T_{-1,1}(u_N) = \overline{T}_{-1,1} = -2 \sum_{k=1}^{N} \tilde{\delta}_k \qquad (5.101)$$

and hence it cannot be used as a generation function for the canonical action/angle variables. Instead, we have to use a proper $\tilde{T}_{1,0}$ functional on M_N, that is

$$\tilde{T}_{1,0}(u_N) = -2 \sum_{k=1}^{N} \left(\tilde{\gamma}_k \frac{\partial H_0}{\partial \tilde{a}_k} + \tilde{\delta}_k \frac{\partial H_0}{\partial \tilde{b}_k} \right)$$

$$= i \sum_{k=1}^{N} \int_{R} [\tilde{\gamma}_k (u_x u^* - u_x^* u)_{\tilde{a}_k} + \tilde{\delta}_k (u_x u^* - u_x^* u)_{\tilde{b}_k}] dx.$$

$$(5.102)$$

The canonical representation of soliton particles reads [23]

$$\frac{\partial u_N}{\partial \tilde{\delta}_k} = \tilde{\theta}_0 \nabla \tilde{E}_k, \quad \frac{\partial u_N}{\partial \tilde{a}_k} = \tilde{\theta}_0 \nabla (-\tilde{\Omega}_k), \quad \tilde{E}_k = -\frac{\partial \tilde{T}_{1,0}(u_N)}{\partial \tilde{\delta}_k}, \quad \tilde{\Omega}_k = -\frac{\partial \tilde{T}_{1,0}(u_N)}{\partial \tilde{a}_k},$$

$$\frac{\partial u_N}{\partial \tilde{\gamma}_k} = \tilde{\theta}_0 \nabla \tilde{G}_k, \quad \frac{\partial u_N}{\partial \tilde{b}_k} = \tilde{\theta}_0 \nabla (-\tilde{\Delta}_k), \quad \tilde{G}_k = -\frac{\partial \tilde{T}_{1,0}(u_N)}{\partial \tilde{\gamma}_k}, \quad \tilde{\Delta}_k = -\frac{\partial \tilde{T}_{1,0}(u_N)}{\partial \tilde{b}_k},$$

$$(5.103)$$

where $\tilde{\theta}_0$ is reduced to M_N, the second Poisson operator of the NLS equation. The other possibility is to pass to new coordinates $(\gamma_k, \delta_k), (a_k, b_k)$ through the inverse of the transformation (5.54). In these new coordinates

$$F(u_N) = T_{1,0}(u_N) = \overline{T}_{1,0} = -2 \sum_{k=1}^{N} (\delta_k a_k + \gamma_k b_k) \qquad (5.104)$$

and another canonical representation of soliton particles reads

$$\frac{\partial u_N}{\partial \delta_k} = \theta_0 \nabla E_k, \quad \frac{\partial u_N}{\partial a_k} = \theta_0 \nabla (-\Omega_k), \quad E_k = -\frac{\partial F(u_N)}{\partial \delta_k}, \quad \Omega_k = -\frac{\partial F(u_N)}{\partial a_k},$$

$$\frac{\partial u_N}{\partial \gamma_k} = \theta_0 \nabla G_k, \quad \frac{\partial u_N}{\partial b_k} = \theta_0 \nabla (-\Delta_k), \quad G_k = -\frac{\partial F(u_N)}{\partial \gamma_k}, \quad \Delta_k = -\frac{\partial F(u_N)}{\partial b_k},$$

$$(5.105)$$

where now θ_0 is reduced to M_N, the first Poisson operator of the NLS equation.

The reader can find other examples of soliton particle representations in [23],[145].

5.4 Multi-Soliton Perturbation Theory

On the basis of the previous considerations for the soliton systems, we are now able to formulate the multi-soliton perturbation theory in a purely algebraic way. This follows from the fact that we have at our disposal a complete set of conserved quantities on the N-soliton manifold as well as a suitable set of vector fields, forming the basis of a tangent bundle to N-soliton flow.

We consider the evolution equation $u_t = K(u)$ with N-soliton solutions admitting asymptotic behaviour in the form (5.1). As we proved in the previous Section, such an equation always has a recursion operator Φ on M_N in an explicit or implicit form, with the following eigenstates

$$A_i = \frac{\partial u}{\partial q_i}, \quad B_i = \frac{\partial u}{\partial c_i} \qquad \Phi A_i(B_i) = c_i \Phi, \quad A_i, B_i \in T_u M_N. \qquad (5.106)$$

Provided that there is an implectic-symplectic factorization of Φ, i.e. $\Phi = \theta_0 \circ J$, we find

$$A_i^* = J A_i = c_i \theta_0^{-1} A_i, \quad \Psi A_i^* = c_i A_i^*,$$

$$B_i^* = J B_i = c_i \theta_0^{-1} B_i, \quad \Psi B_i^* = c_i B_i^*, \qquad A_i^*, B_i^* \in T_u^* M_N, \qquad (5.107)$$

where $\Psi = J \circ \theta_0$ is the recursion operator for cosymmetries.

As the complete set of conserved quantities we choose the noncanonical action/angle variables

$$H_n = \frac{1}{n+\rho} \sum_{i=1}^{N} c_i^n E_i = \frac{1}{n+\rho} \sum_{i=1}^{N} c_i^{n+\rho},$$

$$T_n := T_{1,n} = -\frac{n+\rho}{n} \sum_{i=1}^{N} q_i c_i^n E_i + \frac{\rho}{n} \sum_{i=1}^{N} c_i^{n+\rho} \Omega_i = -\beta \sum_{i=1}^{N} c_i^{n+\rho} q_i. \qquad (5.108)$$

The suitable gradients and vector fields are as follows

$$\nabla H_n = \sum_{i=1}^{N} c_i^{n-1} A_i^*, \quad K_n = \theta_0 \circ \nabla H_n = \sum_{i=1}^{N} c_i^n A_i, \qquad (5.109)$$

$$\nabla T_n = \sum_{i=1}^{N} [c_i^n B_i^* - (n+\rho) c_i^{n-1} q_i A_i^*],$$

$$S_n := \tau_{1,n}^0 = \theta_0 \circ \nabla T_n = \sum_{i=1}^{N} [c_i^{n+1} B_i - (n+\rho) c_i^n q_i A_i], \qquad (5.110)$$

and satisfy the commutator relations

$$[K_n, K_m] = 0, \quad [S_n, K_m] = (n+m+\rho) K_{n+m},$$

$$[S_n, S_m] = (m - n)S_{n+m}. \tag{5.111}$$

Now let us consider the perturbed m-th equation from the hierarchy (5.4)

$$u_t = K_m(u) + \epsilon R(u), \tag{5.112}$$

where ϵ is the smallness parameter. In our considerations we confine our considerations to a perturbation on M_N, so we perform all calculations in the so called *adiabatic approximation*. What we mean by this approximation can be explained as follows. It is well known that perturbations do not remain invariant on a soliton submanifold. This means that in general the first order multi-soliton perturbation theory, besides the deformation of N-soliton dynamics, induces radiational effects. This is because perturbations mix discrete and continuous parts of the spectrum of the Lax operator of a given system. The adiabatic approximation neglecting radiational effects is restricted only to the soliton's deformation under various perturbations. Although this approximation weakens the power of the theory, nevertheless, in the 1-soliton case, it has been the most popular perturbation approach with well recognized limitations. Here, we apply this approach to the multi-soliton case [25].

The time evolution of H_n and T_n along the perturbed flow (5.111) are as follows

$$\frac{\mathrm{d}H_n}{\mathrm{d}t} = L_{K_m + \epsilon R} H_n = \{H_m, H_n\}_{\theta_0} + \epsilon \langle \nabla H_n, R \rangle = \epsilon \langle \nabla H_n, R \rangle, \tag{5.113}$$

$$\frac{\mathrm{d}T_n}{\mathrm{d}t} = L_{K_m + \epsilon R} T_n = \{H_m, T_n\}_{\theta_0} + \epsilon \langle \nabla T_n, R \rangle$$
$$= -((n + m + \rho)H_{n+m} + \epsilon \langle \nabla T_n, R \rangle. \tag{5.114}$$

From the explicit form (5.108) of the conserved quantity H_n we find

$$\frac{\mathrm{d}H_n}{\mathrm{d}t} = \frac{\beta}{n + \rho} \sum_{i=1}^{N} (c_i^{n+\rho})_t = \beta \sum_{i=1}^{N} c_i^{n-1+\rho} (c_i)_t. \tag{5.115}$$

Hence, substituting (5.115) and the explicit form (5.109) of ∇H_n into equation (5.113) we obtain

$$\sum_{i=1}^{N} c_i^{n-1} \left(\beta c_i^{\rho}(c_i)_t - \epsilon \int_R A_i^* R(u_N) \mathrm{d}x \right) = 0. \tag{5.116}$$

Finally, from the arbitrariness of n, we find the time dependence of soliton velocities

$$(c_i)_t = \frac{\epsilon}{\beta} c_i^{-\rho} \int_R A_i^* R(u_N) \mathrm{d}x. \tag{5.117}$$

Analogously, from the explicit form (5.108) of the conserved quantities T_n, we have

$$\frac{dT_n}{dt} = -\beta \sum_{i=1}^{N}[(n+\rho)q_i c_i^{n-1+\rho}(c_i)_t + c_i^{n+\rho}(q_i)_t].\tag{5.118}$$

Substituting (5.118), the explicit form (5.110) of ∇T_n and the time dependence (5.117) of c_i into (5.114), we obtain

$$\sum_{i=1}^{N} c_i^{n+\rho}\left(-\beta(q_i)_t + \beta c_i^m - \epsilon c_i^{-\rho}\int_R B_i^* R(u_N)\,dx\right) = 0\tag{5.119}$$

and, from the arbitrariness of n, we have the second final formula

$$(q_i)_t = c_i^m - \frac{\epsilon}{\beta}c_i^{-\rho}\int_R B_i^* R(u_N)\,dx,\tag{5.120}$$

representing the time evolution of the N-soliton phases.

Example 5.5 The perturbed KdV hierarchy.
As $\theta_0 = D^3 + \frac{1}{3}a(Du + uD)$ and $J = D^{-1}$, then

$$A_i^* = D^{-1}\frac{\partial u_N}{\partial q_i}, \qquad B_i^* = D^{-1}\frac{\partial u_N}{\partial c_i}.\tag{5.121}$$

In order to compare the results with the literature we express the i-th velocity c_i by the spectral parameter κ_i as $c_i = 4\kappa_i^2$. Then, the perturbed N-soliton takes the form $u_N = u_N(\kappa_1(t),...,\kappa_N(t),q_1(t),...,q_N(t))$, where

$$(\kappa_i)_t = \epsilon\frac{a^2}{18}\frac{1}{16\kappa_i^2}\int_R R(u_N)D^{-1}\frac{\partial u_N}{\partial q_i}\,dx,$$

$$(q_i)_t = (2\kappa_i)^{2m} - \epsilon\frac{a^2}{18}\frac{1}{16\kappa_i^2}\int_R R(u_N)D^{-1}\frac{\partial u_N}{\partial \kappa_i}\,dx.\tag{5.122}$$

In the particular case, when $N = 1$ and $m = 1$, we find

$$u_s = \frac{12}{a}\kappa^2\text{sech}\, z, \quad z = \kappa(x + q(t)),$$

$$\kappa_t = \frac{a}{6}\frac{\epsilon}{4\kappa}\int_R R(u_s)\text{sech}^2 z\,dz,$$

$$q_t = 4\kappa^2 - \frac{a}{6}\frac{\epsilon}{4\kappa^3}\int_R R(u_s)(z + \frac{1}{2}\sinh 2z)\text{sech}^2 z\,dz,\tag{5.123}$$

which is the well known result of 1-soliton KdV perturbation theory [114].

Example 5.6 The perturbed MKdV equation.
In this case we have $J = D^{-1}, \rho = \frac{1}{2}, \beta = \frac{3}{a}$ and $c_i = 4\kappa_i^2$, where κ_i is a spectral parameter. We find the adiabatically perturbed N-soliton solution in the form $u_N = u_N(\kappa_1(t),...,\kappa_N(t),q_1(t),...,q_N(t))$, where

$$(\kappa_i)_t = \epsilon \frac{a}{12} \int_R R(u_N)D^{-1}\frac{\partial u_N}{\partial q_i}\, dx,$$

$$(q_i)_t = 4\kappa_i^2 - \epsilon \frac{a}{12} \int_R R(u_N)D^{-1}\frac{\partial u_N}{\partial \kappa_i}\, dx. \qquad (5.124)$$

Again, in the particular case of a 1-soliton, we find for $a = 6$

$$u_s = 2\kappa \sec h\, z, \qquad z = 2\kappa(x + q(t)),$$

$$\kappa_t = \frac{1}{2}\epsilon \int_R R(u_s)\text{sech}\, z\, dz,$$

$$q_t = 4\kappa^2 - \frac{\epsilon}{4\kappa^2} \int_R R(u_s)z \sec h\, z\, dz, \qquad (5.125)$$

which is the well known result of the inverse-scattering approach in the adiabatic 1-soliton perturbation [101],[102].

Example 5.7 The perturbed BO hierarchy.
According to our general formulae (5.12) and (5.120), the time evolution of the Benjamin–Ono N-soliton parameters is as follows

$$(c_i)_t = \frac{4a^2}{\pi}\epsilon \int_R R(u_N)D^{-1}\frac{\partial u_N}{\partial q_i}\, dx,$$

$$(q_i)_t = c_i^m - \frac{4a^2}{\pi}\epsilon \int_R R(u_N)D^{-1}\frac{\partial u_N}{\partial c_i}\, dx. \qquad (5.126)$$

In the particular case of a 1-soliton we find

$$u_s = \frac{1}{a}\frac{c}{z^2 + 1}, \qquad z = c(x + q(t)),$$

$$c_t = \frac{4a}{\pi}\epsilon \int_R \frac{R(u_N)}{1 + z^2}\, dz,$$

$$q_t = c^m - \frac{4a}{\pi}\frac{\epsilon}{c^2} \int_R \frac{zR(u_N)}{1 + z^2}\, dz. \qquad (5.127)$$

Example 5.8 The perturbed SK equation.
We consider here the perturbed Sawada–Kotera equation (3.65)

$$u_t = u_{5x} + \frac{5}{2}auu_{3x} + \frac{5}{2}au_xu_{2x} + \frac{5}{4}a^2u^2u_x + \epsilon R(u). \qquad (5.128)$$

For the above system $J = D^{-1}(D^2 + \frac{1}{2}au)D(D^2 + \frac{1}{2}au)D^{-1}, \rho = \frac{1}{2}, \beta = \frac{2}{a^2}$ and $c_i = (2\kappa_i)^6$ hold, so the time evolution of the N-soliton parameters are of the following form

$$(\kappa_i)_t = \epsilon \frac{a^2}{24} \int_R R(u_N) J \frac{\partial u_N}{\partial q_i}\, dx,$$

$$(q_i)_t = 4\kappa_i^2 - \epsilon \frac{a^2}{24} \int_R R(u_N) J \frac{\partial u_N}{\partial \kappa_i}\, dx. \tag{5.129}$$

In the particularly interesting 1-soliton case

$$u_s = \frac{12}{a}\kappa^2 \mathrm{sech}\, z, \quad z = \kappa(x + q(t)), \quad q(t) = (2\kappa)^4 t + q_0$$

the formulae (5.129) are confined to the form

$$\kappa_t = \epsilon \frac{a}{8\kappa} \int_R \left(\mathrm{sech}^2 z - \frac{3}{4}\mathrm{sech}^4 z \right) R(u_s)\, dz,$$

$$q_t = (2\kappa)^4 - \epsilon \frac{a}{8\kappa^3} \int_R \mathrm{sech}^2 z \left(z + \frac{1}{4}\sinh 2z - \frac{3}{4}\tanh z - \frac{3}{4}z\,\mathrm{sech}^2 z \right) R(u_s)\, dz. \tag{5.130}$$

The reader can find the case of perturbed complex multi-solitons in [26].

6. Multi-Hamiltonian Finite Dimensional Systems

In the following chapter we apply the results of the previous chapters to implement the bi-Hamiltonian formalism to finite-dimensional nonlinear systems. Obviously, we generalize the results on the bi-Hamiltonian formalism derived in Chap. 5 for the linear case.

The whole story of bi-Hamiltonian finite dimensional systems started with the nonperiodic Toda lattice and its generalizations. To find its Lax representation Flaschka [68] changed the original $(q, p) \in \mathbb{R}^{2N}$ variables into the new reduced variables $(a, b) \in \mathbb{R}^{2N-1}$. In the new coordinates the canonical bracket transformed into a degenerate Lie–Poisson bracket. The second Poisson bracket (quadratic in a, b coordinates) was constructed by Adler [4] in 1979 and the third one (cubic in the a, b coordinates) by Kupershmidt [112] in 1985. As all three structures are degenerate, the dynamics take place on a bi-Poisson manifold but not on a Poisson–Nijenhuis manifold, and consequently, there is no recursion operator with implectic-symplectic factorization. The second Poisson structure in the original coordinates $(q, p) \in \mathbb{R}^{2N}$, and hence the recursion operator were found a decade later by Das and Okubo [55] in 1989.

The crucial year in the theory of bi-Hamiltonian finite dimensional systems was 1987 when Antonowicz, Fordy and Wojciechowski [5] constructed two Poisson structures for KdV stationary flows. Since their pioneering work a great number of new integrable bi-Hamiltonian finite dimensional systems have been constructed by restricting infinite dimensional integrable systems to finite dimensional invariant submanifolds. What is important is that for the majority of such systems both Poisson structures are degenerate and the related bi-Hamiltonian chain starts with a Casimir of the first Poisson structure and terminates with a Casimir of the second Poisson structure. Moreover, such bi-Hamiltonian systems can be projected from a bi-Poisson manifold onto the underlying Poisson–Nijenhuis manifold losing their bi-Hamiltonian property and turning into so called quasi-bi-Hamiltonian systems. Most of the following chapter is devoted to the construction of such bi-Hamiltonian chains and then to proving their separability.

Let the equation

$$u_t = K(u) \tag{6.1}$$

be a member of a bi-Hamiltonian hierarchy of field dynamical systems

$$u_{t_m} = K_m(u) = \theta_0(u)\nabla H_m(u) = \theta_1(u)\nabla H_{m-1}(u), \qquad (6.2)$$

generated by a recursion hereditary operator $\Phi = \theta_1 \circ \theta_0^{-1}$. Then, let $\psi_i, i = 1, ..., N$, be N arbitrary eigenfunctions of Φ with eigenvalues α_i. In this chapter we are going to consider the following *constrained systems*

$$\Phi\psi_i = \alpha_i\psi_i, \quad i = 1, ..., N$$

$$K_m(u) = \sum_{i=1}^{N} \psi_i. \qquad (6.3)$$

The set of equations (6.3) is invariant with respect to the action of the hierarchy (6.2) since ψ_i are symmetries of (6.1). By the $N-1$ times repeated action of the recursion operator on $K_m(u) = \sum_{i=1}^{N} \psi_i$ we get higher vector fields on the left hand side and linear combinations of ψ_i on the right hand side. Expressing ψ_i as $K_{m+j}, j = 0, ..., N-1$, then substituting them into the first N equations of (6.3) and finally summing up these equations over all i we arrive at the form

$$\sum_{j=0}^{N} a_j K_{m+j}(u) = \Phi^m \sum_{j=0}^{N} a_j K_j(u) \Rightarrow \sum_{j=0}^{N} a_j K_j(u) = 0, \qquad (6.4)$$

with $a_j = a_j(\alpha_1, ..., \alpha_N)$, which is an alternative definition of the N-dimensional invariant submanifold of the dynamical system (6.1).

In the special case, when all α_i are different, non-zero eigenvalues, belonging to the discrete part of the spectrum of the recursion operator Φ, the expression

$$M_N = \left\{ u : \sum_{j=0}^{N} a_j K_j(u) = 0 \right\} \qquad (6.5)$$

defines the N-soliton manifold [88]. In the previous chapter we have considered the linear parametrization of the system (6.3). As we demonstrated, the linear parametrization leads to an equivalent noninteracting, finite dynamical system described in action/angle variables, whose bi-Hamiltonian algebraic theory was developed in Sect. 5.2. Now we shall concentrate on various nonlinear parametrizations of the system (6.3). They lead to a number of well known integrable systems of classical mechanics, as well as to a variety of new integrable finite dimensional systems. We present the way the bi- (or multi-) Hamiltonian structure of infinite dimensional flows (6.2) can be reconstructed in the finite dimensional constrained systems (6.3), give a proof of their Liouville integrability and solve them by quadratures.

6.1 Stationary Flows of Infinite Systems. Ostrogradsky Parametrizations

Let us consider the special case of the constrained system (6.3) when the ψ_i functions belong to $\ker \Phi$, i.e. where $\alpha_i = 0$. Then (6.3) reduce to

$$K_m(u) = \ker \Phi \tag{6.6}$$

and we refer to this equation as to the m-th *stationary flow* of system (6.1). A special case of (6.6) takes the form

$$0 = K(u) = \theta(u)\delta H(u), \tag{6.7}$$

and is also called the m-th stationary flow. In general $u = (u_1, ..., u_m)^{\mathrm{T}}$ and we have used the symbol of variation δ instead of the gradient ∇ to differentiate functionals from functions.

First we shall concentrate on the so called *Ostrogradsky representation* of (6.6),(6.7). Consider a Lagrangian density $L = L(u_j, u_{j,x}, ..., u_{j,nx})$ where $u_{j,kx} = \mathrm{d}^k u_j / \mathrm{d}x^k$. We assume that L is nondegenerate, i.e. that $\det[\partial^2 L / \partial u_{j,nx} \partial u_{i,nx}] \neq 0$. Then we apply the generalized Legendre transformation and construct the canonical Hamiltonian representation in terms of the so called Ostrogradsky variables [187]. The generalized coordinates $q_{j,k}$ and their conjugate momenta $p_{j,k}$ are defined by the formula

$$q_{j,k} = u_{j,(k-1)x}, \qquad k = 1, ..., n, \ j = 1, ..., m.$$

$$p_{j,k} = \frac{\delta L}{\delta u_{j,kx}} = \sum_{i=0}^{n-k} (-D)_x^i \frac{\partial L}{\partial u_{j,(k+i)x}}. \tag{6.8}$$

The Lagrangian equation $\delta L = 0$, $\delta = (\delta/\delta u_1, ..., \delta/\delta u_m)^{\mathrm{T}}$ is then equivalent to the Hamiltonian canonical equations with the Hamiltonian

$$h(q,p) = \sum_{j=1}^{m} \left[p_{j,n}(q_{j,n})_x + \sum_{k=1}^{n-1} p_{j,k} q_{j,k+1} \right] - L. \tag{6.9}$$

Thus, our fundamental task is to transform (6.6),(6.7) into a Lagrangian form. As we shall see, the m parameters c_i, $i = 1, ..., m$, always appear in this process. The Legendre transformation is to be performed for any set of fixed values of c_i independently. Thus, we end up with an m-parameter family of Hamiltonian flows on an m-parameter family of $2nm$-dimensional symplectic manifolds with canonical Poisson brackets. From the very construction it follows that we can glue all these manifolds together and consider the $(2nm + m)$-dimensional extended phase space M of a given stationary flow. This is no longer a symplectic manifold but a Poisson one with a canonical degenerate implectic operator θ_0 which, in local coordinates $(q, p, c) = (q_1, ..., q_n, p_1, ..., p_n, c_1, ..., c_m)$, is given by

$$\theta_0 = \left(\begin{array}{cc|c} 0 & I & 0 \\ -I & 0 & 0 \\ \hline 0 & 0 & 0 \end{array} \right), \tag{6.10}$$

where I is a $(nm) \times (nm)$ identity matrix. Now, c_i $i = 1, ..., m$, are Casimirs of this Poisson structure and the Hamiltonian form of the stationary flow reads

$$w_x = \theta_0 \circ \nabla h, \tag{6.11}$$

where $w = (q, p, c)^{\mathrm{T}}, \nabla = (\partial/\partial q, \partial/\partial p, \partial/\partial c)^{\mathrm{T}}$. Notice that now the x-variable plays the role of an evolution parameter. The original $2nm$-dimensional symplectic manifolds M_c are leaves of a symplectic foliation of the Poisson manifold (M, θ_0) given by

$$M_c = \{(q, p, c) \in M : c = const\}. \tag{6.12}$$

The above extended phase space picture seems to be of rather marginal importance for studying the intrinsic nature of a given stationary flow, but in fact it is not the case, when we can construct a second Hamiltonian representation. The second implectic operator θ_1 is noncanonical and degenerate as well. But the symplectic foliation of the Poisson manifold (M, θ_1) is not compatible with the symplectic foliation of the manifold (M, θ_0), i.e. θ_1 cannot be restricted to a particular leaf M_c. Hence, the extended phase space plays a crucial role in the bi-Hamiltonian description of stationary flows (6.6),(6.7), and turns to be a bi-Poisson manifold.

To find a variational problem equivalent to (6.6),(6.7) is far from trivial. We will do it for some subclasses of Poisson operators θ. Let $\mathcal{T}_m^{(2,0)}$ denote all tensor fields $\mathcal{T}^{(2,0)}$ which are $m \times m$ matrix differential operators.

Definition 6.1 *A Poisson operator $\theta \in \mathcal{T}_m^{(2,0)}$ is said to be regular if it is possible to bring it, through a suitable invertible change of variables $v_i = F_i(u), i = 1, ..., m$, into a form $\bar{\theta}$ such that $\ker \bar{\theta} = \mathbb{R}^m$. We refer to this as a normal form of the operator θ [9].*

From the above definition it follows that $\bar{\theta} = b[D]$, i.e. $\bar{\theta}$ is a constant coefficient differential operator. Now we shall confine ourselves to the case when $b[D] = bD$, b is a constant invertible matrix, i.e. $\bar{\theta}$ is a first order constant coefficient nondegenerate differential operator. We note that the kernel $\ker \theta$ of any regular Poisson operator $\theta \in \mathcal{T}_m^{(2,0)}$ can be represented as

$$\ker \theta = c_1 \delta F_1 + ... + c_m \delta F_m, \tag{6.13}$$

where $c_i \in \mathbb{R}$ and F_i are differential functions of u_i variables. This follows from the fact that $\ker \bar{\theta} = c_1 \delta v_1 + ... + c_m \delta v_m$ and since $v_i = F_i(u)$ we have the relation (6.13). The stationary equation (6.7) with a regular Poisson operator θ is then equivalent to

$$\delta H = \ker \theta = \delta R, \tag{6.14}$$

where $R = \int_R (c_1 \delta F_1 + ... + c_m \delta F_m) dx$. Hence, (6.14) is the Lagrangian equation $\delta L = 0$, with the Lagrangian $L = H - R$ depending on m parameters $c_1, ..., c_m$. If L is nondegenerate, we can construct the canonical Hamiltonian representation described at the beginning of this section.

Example 6.1 The first Ostrogradsky representation of the stationary KdV hierarchy [5].

Let us consider the first Hamiltonian representation of the KdV hierarchy

$$0 = K_n(u) = D \, \delta H_n[u]. \tag{6.15}$$

Obviously, the Poisson operator D is a regular operator in its normal form, hence, the Lagrangian representation of (6.15) is

$$\delta L_n[u] = 0, \qquad L_n[u] = H_n[u] - \int_R cu \, dx. \tag{6.16}$$

As for the KdV hierarchy $H_n[u] = H_n(u, u_x, ..., u_{nx})$ and $\partial^2 H_n / \partial u_{nx}^2 \neq 0$ we can always construct a canonical Hamiltonian representation (6.8)–(6.11). We illustrate it for $n = 2$ of the hierarchy (3.84) $(a = 6)$

$$0 = (D^2 + 4u + 2u_x D^{-1})^2 u_x$$

$$= (u_{4x} + 10uu_{xx} + 5u_x^2 + 10u^3)_x = D \, \delta \int_R \left(\frac{1}{2} u_{xx}^2 - 10uu_x^2 + \frac{5}{2} u^4 \right) dx. \tag{6.17}$$

Then, we have

$$q_1 = u, \; q_2 = u_x, \; p_1 = -10uu_x - u_{3x}, \; p_2 = u_{xx},$$

$$c = u_{4x} + 10uu_{xx} + 5u_x^2 + 10u^3$$

$$h = \frac{1}{2} p_2^2 + q_2 p_1 + 5q_1 q_2^2 - \frac{5}{2} q_1^4 + q_1 c,$$

$$\begin{pmatrix} q_1 \\ q_2 \\ p_1 \\ p_2 \\ c \end{pmatrix}_x = \begin{pmatrix} q_2 \\ p_2 \\ 10q_1^3 - 5q_2^2 - c \\ -10q_1 q_2 - p_1 \\ 0 \end{pmatrix} = \theta_0 \circ \nabla h(q, p, c). \tag{6.18}$$

Example 6.2 The first Ostrogradsky representation of the Harry–Dym hierarchy.

We consider the first Hamiltonian representation of the HD hierarchy (3.32) in the form

$$u_{t_n} = K_n(u) = \left(-\frac{1}{2}uD - \frac{1}{2}Du\right)\delta H_n[u], \quad n = 0, 1, \dots \qquad (6.19)$$

In this case we have

$$\ker \Phi = \theta_0 \ker(\theta_1) = -uw_x - \frac{1}{2}u_x w, \quad w = c_2 x^2 + c_1 x + c_0 \qquad (6.20)$$

and c_0, c_1, c_2 are constants. Here, we consider a special case of (6.20) with $w =$const, which gives the m-th stationary flow (6.6) of hierarchy (6.19), which in this case we consider in the form

$$K_n(u) = \ker \Phi(w = 2\beta) \Leftrightarrow K_n(u) + \beta u_x = 0, \quad \beta = \text{const}. \qquad (6.21)$$

As θ_0 is a regular operator with $\ker \theta_0 = cu^{-1/2}$, then (6.21) is equivalent to

$$\gamma_n - 2\beta - cu^{-1/2}, \quad \gamma_n = \delta H_n[u], \qquad (6.22)$$

which comes from the Euler-Lagrange equation $\delta L_n = 0$, where

$$L_n[u] = H_n[u] - \int_R \left(2\beta u + 2cu^{1/2}\right) dx. \qquad (6.23)$$

Let us consider the case $n = 2$. The appropriate Hamiltonian reads

$$H_2[u] = \int_R \left(-\frac{1}{64}u^{-7/2}u_{xx}^2 + \frac{3}{64}u^{-9/2}u_x^2 u_{xx} - \frac{37}{16\cdot 64}u^{-11/2}u_x^4\right) dx, \qquad (6.24)$$

so we get

$$q_1 = u, \ q_2 = u_x, \ p_1 = \frac{17}{4\cdot 64}u^{-11/2}u_x^3 - \frac{7}{64}u^{-9/2}u_x u_{xx} + \frac{1}{32}u^{-7/2}u_{3x},$$

$$p_2 = -\frac{1}{32}u^{-7/2}u_{xx} + \frac{3}{64}u^{-9/2}u_x^2,$$

$$c = \frac{1155}{32\cdot 64}u^{-6}u_x^4 - \frac{231}{4\cdot 64}u^{-5}u_x^2 u_{xx} + \frac{21}{2\cdot 64}u^{-4}u_{xx}^2 + \frac{7}{32}u^{-4}u_x u_{3x}$$

$$-\frac{1}{32}u^{-3}u_{4x} - 2\beta u^{1/2},$$

$$h = p_1 q_2 - 16p_2^2 q_1^{7/2} + \frac{3}{2}p_2 q_1^{-1}q_2^2 + \frac{1}{16\cdot 64}q_1^{-11/2}q_2^4 + 2cq_1^{1/2} + 2\beta q_1,$$

$$\begin{pmatrix} q_1 \\ q_2 \\ p_1 \\ p_2 \\ c \end{pmatrix}_x$$

$$= \begin{pmatrix} q_2 \\ -32 p_2 q_1^{7/2} + \frac{3}{2} p_2 q_1^{-1} q_2^2 \\ 56 p_2^2 q_1^{5/2} + \frac{3}{2} p_2 q_1^{-2} q_2^2 + \frac{11}{32\cdot 64} q_1^{-13/2} q_2^4 - c q_1^{-1/2} - 2\beta \\ -p_1 - 3 p_2 q_1^{-1} q_2 - \frac{1}{4\cdot 64} q_1^{-11/2} q_2^3 \\ 0 \end{pmatrix} = \theta_0 \circ \nabla h.$$

$$(6.25)$$

Example 6.3 The first Ostrogradsky representation of the DWW hierarchy [28].

Let us consider the first Hamiltonian representation of the DWW (3.35) stationary hierarchy

$$0 = K_n(u, v) = \begin{pmatrix} -\frac{1}{2} u D - \frac{1}{2} D u & D \\ D & 0 \end{pmatrix} \delta H_n[u, v]. \qquad (6.26)$$

The Poisson operator θ_0 from (6.26) is regular and

$$\ker \theta_0 = c_1 \delta \int_R \frac{1}{2} v dx + c_2 \delta \int_R \left(\frac{1}{2} u + \frac{1}{8} v^2 \right) dx, \qquad (6.27)$$

hence, the hierarchy (6.26) of stationary flows of the DWW is equivalent to the hierarchy of the Euler-Lagrange equations $\delta L_n = 0$, where

$$L_n[u, v] = H_n[u, v] - \int_R \left[\frac{1}{2} c_1 v + c_2 \left(\frac{1}{2} u + \frac{1}{8} v^2 \right) \right] dx. \qquad (6.28)$$

We illustrate the result by the flow $n = 2$ with the respective Hamiltonian

$$H_2[u, v] = \int_R \left(-\frac{1}{4} u_x v_x - \frac{5}{16} v v_x^2 + \frac{5}{8} u v^3 + \frac{3}{4} u^2 v + \frac{7}{64} v^5 \right) dx. \qquad (6.29)$$

Then we have

$$q_1 = u, \quad q_2 = v, \quad p_1 = -\frac{1}{4} v_x, \quad p_2 = -\frac{1}{4} u_x - \frac{5}{8} v v_x,$$

$$c_1 = 2 \frac{\delta H_2}{\delta v} - v \frac{\delta H_2}{\delta u} = \frac{1}{2} u_{xx} + \frac{5}{8} v_x^2 + v v_{xx} + \frac{9}{4} u v^2 + \frac{3}{2} u^2 + \frac{15}{32} v^4,$$

$$c_2 = 2 \frac{\delta H_2}{\delta u} = \frac{1}{2} v_{xx} + \frac{5}{4} v^3 + 3 u v,$$

$$h = -4p_1p_2 + 5q_2p_1^2 - \frac{5}{8}q_1q_2^3 - \frac{3}{4}q_1^2q_2 - \frac{7}{64}q_2^5 + \frac{1}{2}q_2c_1 + \frac{1}{2}q_1c_2 + \frac{1}{8}q_2^2c_2,$$

$$\begin{pmatrix} q_1 \\ q_2 \\ p_1 \\ p_2 \\ c_1 \\ c_2 \end{pmatrix}_x = \begin{pmatrix} 10q_2p_1 - 4p_2 \\ -4p_1 \\ \frac{5}{8}q_2^3 + \frac{3}{2}q_1q_2 - \frac{1}{2}c_2 \\ \frac{15}{8}q_1q_2^2 + \frac{3}{4}q_1^2 + \frac{35}{64}q_2^4 - 5p_1^2 - \frac{1}{2}c_1 - \frac{1}{4}q_2c_2 \\ 0 \\ 0 \end{pmatrix} = \theta_0 \circ \nabla h.$$

$$(6.30)$$

Now we demonstrate that the transformation of the stationary equation (6.7) with a regular Poisson operator θ into a variational Euler–Lagrange problem, described in this section, is closely related to the integration of (6.7). First we notice that from the definitions of the dual derivative (2.49) and the variational operator (2.56) we have

$$\ker \theta = c_1\delta F_1 + \ldots + c_m\delta F_m = F'^\dagger \circ c, \qquad (6.31)$$

where $F = (F_1, \ldots, F_m)^\mathrm{T}$ and $c = (c_1, \ldots, c_m)^\mathrm{T}$. Now, keeping in mind that $F' \circ \theta \circ F'^\dagger = bD$ we have

$$\begin{aligned} \theta \circ \delta H = 0 \quad &\Rightarrow F' \circ \theta \circ \delta H = 0 \\ &\Rightarrow D^{-1} \circ F' \circ \theta \circ \delta H = \bar{c} \\ &\Rightarrow b^{-1}D^{-1} \circ F' \circ \theta \circ \delta H = b^{-1}\bar{c} := c \\ &\Rightarrow F'^\dagger \circ b^{-1}D^{-1} \circ F' \circ \theta \circ \delta H = F'^\dagger \circ c \\ &\Rightarrow \theta^{-1} \circ \theta \circ \delta H = F'^\dagger \circ c \\ &\Rightarrow \delta H = F'^\dagger \circ c = \ker \theta. \end{aligned} \qquad (6.32)$$

Thus, the parameters c_i come from the integration of (6.7). The method of integration can be applied to stationary equations with a Poisson operator θ other than the regular one.

Let us consider the $(2n+3)$th order Hamiltonian evolution equation

$$u_t = \theta(u)\delta H[u], \qquad (6.33)$$

where

$$\theta = D^3 + auD + aDu, \qquad a = \text{const} \qquad (6.34)$$

is a Poisson operator, $H[u] = H(u, u_x, \ldots, u_{nx})$ is a Hamiltonian functional whose density is a differential function of order n and $\delta = \delta/\delta u$. Its stationary equation has the form

$$\left(D^3 + auD + aDu\right)\frac{\delta H[u]}{\delta u} = 0 \qquad (6.35)$$

and we are looking for an Euler–Lagrange representation of (6.35) [14],[29]. Let $\gamma = \delta H/\delta u$, so (6.35) reads

$$\gamma_{3x} + 2au\gamma_x + au_x\gamma = 0. \tag{6.36}$$

Multiplying (6.36) by γ and integrating once over the x variable we obtain

$$\gamma\gamma_{xx} - \frac{1}{2}\gamma_x^2 + au\gamma^2 = \frac{1}{8}a^2\alpha, \tag{6.37}$$

where α is the integration constant. Introducing a new variable v

$$\gamma = -\frac{1}{4}av^2, \tag{6.38}$$

we transform (6.37) to the form

$$2v^3v_{xx} + auv^4 = \alpha. \tag{6.39}$$

Thus, the stationary equation (6.35) is equivalent to the pair of equations (6.37) and (6.38), that is

$$\frac{\delta H[u]}{\delta u} + \frac{1}{4}av^2 = 0, \quad v_{xx} + \frac{1}{2}auv - \frac{\alpha}{2v^3} = 0. \tag{6.40}$$

But (6.40) come from the Euler-Lagrange equation $\delta L = 0$, where $\delta = (\delta/\delta u, \delta/\delta v)^{\mathrm{T}}$ and

$$L[u, v] = H[u] + \int_R \left(-\frac{1}{2}v_x^2 + \frac{1}{4}auv^2 + \frac{\alpha}{4v^2} \right) dx. \tag{6.41}$$

Again in the case of a nondegenerate $H[u]$ we can construct the canonical Hamiltonian representation in the terms of the Ostrogradsky variables.

Example 6.4 The second Ostrogradsky representation of the stationary KdV hierarchy.
Let us again consider the fifth KdV stationary flow (6.17), now in the second Hamiltonian representation

$$0 = (D^3 + 2uD + 2Du)\,\delta \int_R \left(-\frac{1}{2}u_x^2 + u^3 \right) dx, \tag{6.42}$$

which is a special case of (6.35). If we define $\bar{q}_1 = u$ and $\bar{q}_2 = v$ then (6.40) and the Lagrangian (6.41) read

$$\bar{q}_{1_{xx}} = -3\bar{q}_1^2 - \frac{1}{2}\bar{q}_2^2,$$

$$\bar{q}_{2_{xx}} = -\bar{q}_1\bar{q}_2 + \frac{\alpha}{\bar{q}_2^3},$$

$$L = \frac{1}{2}\bar{q}_1^2 - \bar{q}_1^3 - \frac{1}{2}\bar{q}_1\bar{q}_2^2 - \frac{1}{2}\bar{q}_2^2 - \frac{1}{2}\frac{\alpha}{\bar{q}_2^2}. \tag{6.43}$$

We have obtained the Newton equations of motion of the generalized (by the term $\alpha/\bar{q}_2^3$), well known Henon–Heiles system [13],[29], actually one of its three integrable cases. Introducing the conjugate momenta $\bar{p}_1 = (\bar{q}_1)_x$ and $\bar{p}_2 = (\bar{q}_2)_x$, (6.43) can be put into the canonical Hamiltonian form

$$
\begin{pmatrix} \bar{q}_1 \\ \bar{q}_2 \\ \bar{p}_1 \\ \bar{p}_2 \\ \alpha \end{pmatrix}_x = \begin{pmatrix} \bar{p}_1 \\ \bar{p}_2 \\ -3\bar{q}_1^2 - \frac{1}{2}\bar{q}_2^2 \\ -\bar{q}_1\bar{q}_2 \\ 0 \end{pmatrix} = \theta_0 \circ \nabla\bar{h},
\tag{6.44}
$$

where $\bar{h} = \frac{1}{2}\bar{p}_1^2 + \frac{1}{2}\bar{p}_2^2 + \bar{q}_1^3 + \frac{1}{2}\,\bar{q}_1\,\bar{q}_2^2 + \frac{1}{2}\alpha/\bar{q}_2^2$.

Example 6.5 The first Ostrogradsky representation of the stationary SK hierarchy [29].
As the first Poisson operator of the Sawada–Kotera hierarchy (see Example 3.12) has the form (6.34), it also fits into our scheme. Let us consider the example of a stationary flow of the seventh order SK equation

$$
0 = \Phi u_x = \theta \delta H[u] = (D^3 + auD + aDu)\,\delta \int_R \left(\frac{1}{2}u_{xx}^2 - \frac{3}{4}auu_x^2 + \frac{1}{24}a^2u^4 \right)\,\mathrm{d}x.
\tag{6.45}
$$

The system (6.45) is equivalent to the following:

$$
u_{4x} + \frac{3}{2}auu_{xx} + \frac{3}{4}au_x^2 + \frac{1}{6}a^2u^3 + \frac{1}{4}av = 0,
$$

$$
v_{xx} + \frac{1}{2}auv - \frac{\alpha}{2v^3} = 0,
\tag{6.46}
$$

with the Lagrangian

$$
L = \int_R \left(\frac{1}{2}u_{xx}^2 - \frac{3}{4}auu_x^2 - \frac{1}{2}v_x^2 + \frac{1}{24}a^2u^4 + \frac{1}{4}auv^2 + \frac{\alpha}{4v^2} \right)\,\mathrm{d}x.
\tag{6.47}
$$

In the Ostrogradsky variables

$$
q_1 = u, \quad q_2 = u_x, \quad q_3 = v, \quad p_1 = -\frac{3}{2}auu_x - u_{3x}, \quad p_2 = u_{2x}, \quad p_3 = -v_x,
$$

it is a canonical Hamiltonian system

$$
q_{i_x} = \frac{\partial h}{\partial p_i}, \quad p_{i_x} = -\frac{\partial h}{\partial q_i}, \quad \alpha_t = 0, \qquad i = 1, 2, 3,
$$

$$
h = -\frac{1}{2}p_3^2 - \frac{1}{2}p_2^2 + p_1q_2 - p_3q_3 + \frac{3}{4}aq_1q_2^2 - \frac{1}{24}a^2q_1^4 - \frac{1}{4}aq_1q_3^2 - \frac{\alpha}{4q_3^2}.
\tag{6.48}
$$

In the special case of the fifth order SK stationary flow, the Ostrogradsky representation turns out to be the second integrable case of the generalized Henon–Heiles (gHH) representation [13],[33].

Example 6.6 Second Ostrogradsky representation of the stationary Harry–Dym hierarchy.

With respect to the second Poisson operator (3.32) the HD stationary equation (6.21) can be presented in the form

$$0 = \theta_1 \gamma_n + \beta u_x = \frac{1}{4} D^3 \gamma_n + \beta u_x. \tag{6.49}$$

After one integration over the x variable we get

$$-\frac{1}{4}(\gamma_n)_{xx} = \beta u + \bar{c}. \tag{6.50}$$

Introducing a new variable v by $\gamma_n = \beta v$ we have

$$\gamma_n = \beta v, \quad -\frac{1}{4}\beta v_{xx} = \beta u + \bar{c}. \tag{6.51}$$

Equations (6.51) come from the Euler–Lagrange equation $\delta \bar{L} = 0$, where

$$\bar{L}[u,v] = H_{n-1}[u] + \int_R \left(\frac{1}{8}\beta v_x^2 - \beta uv - \bar{c}v \right) dx. \tag{6.52}$$

Hence, for the case $n = 2$, the second Ostrogradsky representation reads

$$\cdot \quad H_1[u] = \int_R \left(-\frac{1}{16}u^{-5/2}u_x^2 \right) dx,$$

$$\bar{q}_1 = u, \quad \bar{q}_2 = v = \frac{1}{\beta}\left(-\frac{5}{32}u^{-7/2}u_x^2 + \frac{1}{8}u^{-5/2}u_{xx} \right), \quad \bar{p}_1 = -\frac{1}{8}u^{-5/2}u_x,$$

$$\bar{p}_2 = \frac{1}{4}\beta v_x = \frac{35}{64}u^{-9/2}u_x^3 - \frac{5}{8}u^{-7/2}u_x u_{xx} + \frac{1}{8}u^{-5/2}u_{3x},$$

$$\bar{c} = \frac{9 \cdot 35}{8 \cdot 64}u^{-11/2}u_x^4 - \frac{7 \cdot 35}{4 \cdot 64}u^{-9/2}u_x^2 u_{xx} + \frac{5}{32}u^{-7/2}u_{xx}^2 + \frac{15}{64}u^{-7/2}u_x u_{3x}$$

$$-\frac{1}{32}u^{-5/2}u_{4x} - \beta u,$$

$$\bar{h} = -4\bar{p}_1^2 \bar{q}_1^{-5/2} + \frac{2}{\beta}\bar{p}_2^2 + \beta \bar{q}_1 \bar{q}_2 + \bar{c}\bar{q}_2,$$

$$\begin{pmatrix} \bar{q}_1 \\ \bar{q}_2 \\ \bar{p}_1 \\ \bar{p}_2 \\ \bar{c} \end{pmatrix}_x = \begin{pmatrix} -8\bar{p}_1\bar{q}_1^{-5/2} \\ 4\bar{p}_2/\beta \\ 10\bar{p}_1^2\bar{q}_1^{-3/2} - \beta\bar{q}_2 \\ -\beta\bar{q}_1 - \bar{c} \\ 0 \end{pmatrix} = \bar{\theta}_0 \circ \nabla \bar{h}. \tag{6.53}$$

Example 6.7 The second Ostrogradsky representation of the DWW hierarchy. The second Hamiltonian representation of the DWW (3.35) stationary hierarchy reads

$$0 = K_n(u, v) = \begin{pmatrix} \frac{1}{4}D^3 + \frac{1}{2}uD + \frac{1}{2}Du & 0 \\ 0 & D \end{pmatrix} \delta H_{n-1}[u, v]. \qquad (6.54)$$

Hence, introducing $\gamma = \delta H_{n-1}/\delta u$ and $\zeta = \delta H_{n-1}/\delta v$, (6.54) takes the form

$$\gamma_{3x} + 4u\gamma_x + 2u_x\gamma = 0, \qquad \zeta_x = 0, \qquad (6.55)$$

and we can again apply the scheme (6.36)–(6.40). Introducing the w variable by $\gamma = -\frac{1}{2}w^2$ and $\alpha := \bar{c}_1$, the system (6.55) is equivalent to the following:

$$\frac{\delta H_{n-1}}{\delta u} + \frac{1}{2}w^2 = 0, \quad w_{xx} + uw - \frac{\bar{c}_1}{2w^3} = 0, \quad \frac{\delta H_{n-1}}{\delta v} - \bar{c}_2 = 0. \qquad (6.56)$$

Equations (6.56) come from the Euler–Lagrange equation $\delta \bar{L} = 0$, where

$$\bar{L}[u, w, v] = H_{n-1}[u, v] + \int_R \left(-\frac{1}{2}w_x^2 + \frac{1}{2}uw^2 + \frac{\bar{c}_1}{4w^2} - \bar{c}_2 v \right) dx. \qquad (6.57)$$

In the special case when $n = 2$ as $-\frac{1}{2}w^2 = u + \frac{3}{4}v^2$ we can eliminate the u variable and then

$$\bar{L}[w, v] = \int_R \left(-\frac{1}{2}w_x^2 - \frac{1}{8}v_x^2 - \frac{1}{8}w^4 - \frac{3}{8}v^2w^2 - \frac{1}{8}v^4 + \frac{\bar{c}_1}{4w^2} - \bar{c}_2 v \right) dx,$$

and the Ostrogradsky representation takes the form

$$\bar{q}_1 = w, \ \bar{q}_2 = v, \ \bar{p}_1 = -w_x, \ \bar{p}_2 = -\frac{1}{4}v_x,$$

$$\bar{c}_1 = 2w^3 w_{xx} + 2uw^4 = 2w^3 w_{xx} - \frac{3}{2}v^2w^4 - w^6,$$

$$\bar{c}_2 = \frac{1}{4}v_{xx} + \frac{3}{2}uv + \frac{5}{8}v^3 = \frac{1}{2}c_2,$$

$$\bar{h} = -\frac{1}{2}\bar{p}_1^2 - 2\bar{p}_2^2 + \frac{1}{8}\bar{q}_1^4 + \frac{3}{8}\bar{q}_1^2\bar{q}_2^2 + \frac{1}{8}\bar{q}_2^4 - \frac{\bar{c}_1}{4\bar{q}_1^2} - \bar{c}_2\bar{q}_2,$$

$$\begin{pmatrix} \bar{q}_1 \\ \bar{q}_2 \\ \bar{p}_1 \\ \bar{p}_2 \\ \bar{c}_1 \\ \bar{c}_2 \end{pmatrix}_x = \begin{pmatrix} -\bar{p}_1 \\ -4\bar{p}_2 \\ -\frac{1}{2}\bar{q}_1^3 - \frac{3}{4}\bar{q}_1\bar{q}_2^2 - \frac{1}{2}\bar{c}_1/\bar{q}_2 \\ -\frac{1}{2}\bar{q}_2^3 - \frac{3}{4}\bar{q}_1^2\bar{q}_2 - \bar{c}_2 \\ 0 \\ 0 \end{pmatrix} = \bar{\theta}_0 \circ \nabla \bar{h}. \qquad (6.58)$$

As in all our examples with two Ostrogradsky representations, one can find an appropriate Miura map $\phi : (q, p, c) \to (\bar{q}, \bar{p}, \bar{c})$ which is a noncanonical transformation. Hence, for both representations we can find the second Hamiltonian structure

$$\theta_1 = \phi^* \circ \bar{\theta}_0 = (\phi')^{-1} \circ \bar{\theta}_0 \circ (\phi'^\dagger)^{-1},$$

$$\bar{\theta}_{-1} = \phi_* \circ \theta_0 = \phi' \circ \theta_0 \circ \phi'^T. \tag{6.59}$$

Example 6.8 The 5th KdV stationary flow.

The Miura map relating both Ostrogradsky representations of the 5th KdV stationary flow and its inverse are the following

$$q_1 = \bar{q}_1, \ q_2 = \bar{p}_1, \ p_1 = -4\bar{q}_1\bar{p}_1 - \bar{q}_2\bar{p}_2,$$

$$c = -\bar{p}_1^2 - \bar{p}_2^2 - 2\bar{q}_1^3 - \bar{q}_1\bar{q}_2 - \frac{\alpha}{\bar{q}_2^2} = -2\bar{h}, \tag{6.60}$$

and

$$\bar{q}_1 = q_1, \ \bar{q}_2^2 = 2p_2 + 6q_1^2, \ \bar{p}_1 = q_2, \ \bar{q}_2\bar{p}_2 = -p_1 - 4q_1q_2,$$

$$\alpha = p_1^2 + 4q_1p_2^2 - 2q_2^2p_2 + 20q_1^3p_2 + 8q_1q_2p_1$$
$$+ 10q_1^2q_2^2 + 24q_1^5 - 2p_2c - 6q_1^2c. \tag{6.61}$$

Hence, we find the second Hamiltonian structures in the form

$$\theta_1 = \begin{pmatrix} 0 & -1 & 4q_1 & 0 & 2q_2 \\ 1 & 0 & -4q_2 & -6q_1 & 2p_2 \\ -4q_1 & 4q_2 & 0 & 2p_2 + 30q_1^2 & A \\ 0 & 6q_1 & -2p_2 - 30q_1^2 & 0 & B \\ -2q_2 & -2p_2 & -A & -B & 0 \end{pmatrix},$$

$$A = 20q_1^3 - 10q_2^2 - 2c, \ B = -2p_1 - 20q_1q_2 \tag{6.62}$$

$$\bar{\theta}_{-1} = \begin{pmatrix} 0 & 0 & 0 & -1/\bar{q}_2 & 2\bar{q}_2\bar{p}_2 \\ 0 & 0 & -1/\bar{q}_2 & 2\bar{q}_1/\bar{q}_2^2 & 2\bar{q}_2\bar{p}_1 - 4\bar{q}_1\bar{p}_2 \\ 0 & 1/\bar{q}_2 & 0 & -\bar{p}_2/\bar{q}_2^2 & \bar{A} \\ 1/\bar{q}_2 & -2\bar{q}_1/\bar{q}_2^2 & \bar{p}_2/\bar{q}_2^2 & 0 & \bar{B} \\ -2\bar{q}_2\bar{p}_2 & -2\bar{q}_2\bar{p}_1 + 4\bar{q}_1\bar{p}_2 & -\bar{A} & -\bar{B} & 0 \end{pmatrix},$$

$$\bar{A} = -2\bar{q}_1\,\bar{q}_2^2 + 2\bar{p}_2^2 + \frac{2\alpha}{\bar{q}_2^2}, \ \bar{B} = -2\bar{p}_1\bar{p}_2 - 2\bar{q}_1^2\,\bar{q}_2 - \bar{q}_2^3 - \frac{4\alpha\bar{q}_1}{\bar{q}_2^2}. \tag{6.63}$$

As the pairs (θ_0, θ_1) and $(\bar{\theta}_{-1}, \bar{\theta}_0)$ are compatible, we have the following bi-Hamiltonian chains

$$0 = \theta_0 \circ \nabla h_0, \quad \theta_1 \circ \nabla h_0 = \theta_0 \circ \nabla h_1, \quad \theta_1 \circ \nabla h_1 = \theta_0 \circ \nabla h_2, \quad \theta_1 \circ \nabla h_2 = 0,$$

$$h_0 = \frac{1}{2}c, \ h_1 = h = \frac{1}{2}p_2^2 + q_2p_1 + 5q_1q_2^2 - \frac{5}{2}q_1^4 + q_1c,$$

$$h_2 = -\frac{1}{2}p_1^2 - 2q_1p_2^2 + q_2^2p_2 - 10q_1^3p_2 - 4q_1q_2p_1 + -5q_1^2q_2^2 - 12q_1^5$$

$$+ p_2c + 3q_1^2c, \tag{6.64}$$

and

$$0 = \bar{\theta}_{-1} \circ \nabla \bar{h}_0, \quad \bar{\theta}_0 \circ \nabla \bar{h}_0 = \bar{\theta}_{-1} \circ \nabla \bar{h}_1, \quad \bar{\theta}_0 \circ \nabla \bar{h}_1 = \bar{\theta}_{-1} \circ \nabla \bar{h}_2, \quad \bar{\theta}_0 \circ \nabla \bar{h}_2 = 0,$$

$$\bar{h}_0 = \frac{1}{2}\bar{p}_1^2 + \frac{1}{2}\bar{p}_2^2 + \bar{q}_1^3 + \frac{1}{2}\bar{q}_1\bar{q}_2^2 + \frac{1}{2}\frac{\alpha}{\bar{q}_2^2} = \bar{h},$$

$$\bar{h}_1 = -\bar{q}_2\bar{p}_1\bar{p}_2 + \bar{q}_1\bar{p}_2^2 - \frac{1}{8}\bar{q}_2^4 - \frac{1}{2}\bar{q}_1^2\bar{q}_2^2 + \frac{\bar{q}_1\alpha}{\bar{q}_2^2}, \quad \bar{h}_2 = -\frac{1}{2}\alpha. \qquad (6.65)$$

Note that each chain starts with the Casimir of the first Poisson structure and terminates with the Casimir of the second Poisson structure.

Example 6.9 The Harry–Dym stationary flows.
The invertible Miura map relating the the Ostrogradsky representations (6.25) and (6.53) is the following

$$q_1 = \bar{q}_1, \quad q_2 = -8\bar{p}_1\bar{q}_1^{5/2}, \quad p_1 = \bar{q}_1^{-1}\bar{p}_2 - 3\beta\bar{q}_1^{1/2}\bar{q}_2\bar{p}_1 + 6\bar{q}_1^2\bar{p}_1^3,$$

$$p_2 = -\frac{1}{4}\beta\bar{q}_1^{-1}\bar{q}_2 + \frac{1}{2}\bar{q}_1^{1/2}\bar{p}_1^2,$$

$$c = \frac{1}{2}\beta^2\bar{q}_1\bar{q}_2^2 - 2\beta\bar{q}_1^{5/2}\bar{q}_2\bar{p}_1^2 + 4\bar{q}_1\bar{p}_1\bar{p}_2 - \beta\bar{q}_1^{1/2} + \bar{q}_1^{-1/2}\bar{c}, \qquad (6.66)$$

and

$$\bar{q}_1 = q_1, \quad \bar{q}_2 = \frac{1}{\beta}\left(\frac{1}{32}q_1^{-7/2}q_2^2 - 4q_1p_2\right), \quad \bar{p}_1 = -\frac{1}{8}q_1^{-5/2}q_2,$$

$$\bar{p}_2 = q_1p_1 + \frac{3}{2}q_2p_2, \quad \bar{c} = \frac{1}{32\cdot 64}q_1^{-11/2}q_2^4 - 8q_1^{7/2}p_2^2 + \frac{3}{4}q_1^{-1}q_2^2p_2 + \frac{1}{2}p_1q_2 \quad (6.67)$$

and again is noncanonical transformation. Thus we get the second Hamiltonian structure

$$\bar{\theta}_1 =$$

$$\begin{pmatrix}
0 & 0 & 0 & \bar{q}_1 & -4\bar{p}_1\bar{q}_1^{5/2} \\
0 & 0 & -\frac{1}{\beta}\frac{1}{2}\bar{q}_1^{-3/2} & -\frac{1}{2}\bar{q}_2 & \frac{2}{\beta}\bar{p}_2 \\
0 & \frac{1}{\beta}\frac{1}{2}\bar{q}_1^{-3/2} & 0 & -\bar{p}_1 & -\frac{2}{\beta}\bar{q}_2 + 5\bar{q}_1^{3/2}\bar{p}_1 \\
-\bar{q}_1 & \frac{1}{2}\bar{q}_2 & \bar{p}_1 & 0 & -\frac{1}{2}\bar{c} - \frac{1}{2}\beta\bar{q}_1 \\
4\bar{p}_1\bar{q}_1^{5/2} & -\frac{2}{\beta}\bar{p}_2 & \frac{2}{\beta}\bar{q}_2 - 5\bar{q}_1^{3/2}\bar{p}_1 & \frac{1}{2}\bar{c} + \frac{1}{2}\beta\bar{q}_1 & 0
\end{pmatrix}$$

$$(6.68)$$

and the following bi-Hamiltonian chain

$$\bar{\theta}_0 \circ \nabla \bar{h}_0 = 0, \quad \bar{\theta}_0 \circ \nabla \bar{h}_1 = \bar{\theta}_1 \circ \nabla \bar{h}_0, \quad \bar{\theta}_0 \circ \nabla \bar{h}_2 = \bar{\theta}_1 \circ \nabla \bar{h}_1, \quad 0 = \bar{\theta}_1 \circ \nabla \bar{h}_2,$$

$$(6.69)$$

where $\bar{h}_0 = 2\bar{c}$, $\bar{h}_1 = \bar{h}$ and

$$\bar{h}_2 = \frac{4}{\beta}\bar{q}_1\bar{p}_1\bar{p}_2 - 2\bar{q}_1^{5/2}\bar{q}_2\bar{p}_1^2 - \bar{q}_1^{1/2} + \frac{1}{2}\beta\bar{q}_1\bar{q}_2^2 + \frac{1}{\beta}\bar{q}_1^{-1/2}\bar{c}. \qquad (6.70)$$

In a similar way one can find the Hamiltonian structure θ_{-1} which we do not present here in explicit form as the formulas are long.

Example 6.10 The DWW stationary flows.
The Miura map relating the Ostrogradsky representations (6.30) and (6.58) is the following

$$q_1 = -\frac{1}{2}\bar{q}_1^2 - \frac{3}{4}\bar{q}_2^2, \quad q_2 = \bar{q}_2, \quad p_1 = \bar{p}_2, \quad p_2 = -\frac{1}{4}\bar{q}_1\bar{p}_1 + \bar{q}_2\bar{p}_2,$$

$$c_1 = \bar{h} = -\frac{1}{2}\bar{p}_1^2 - 2\bar{p}_2^2 + \frac{1}{8}\bar{q}_1^4 + \frac{3}{8}\bar{q}_1^2\bar{q}_2^2 + \frac{1}{8}\bar{q}_2^4 - \frac{\bar{c}_1}{4\bar{q}_1^2} - \bar{c}_2\bar{q}_2, \quad c_2 = 2\bar{c}_2, \quad (6.71)$$

and the respective second Hamiltonian structure reads

$$\theta_1 = \begin{pmatrix} 0 & 0 & -\frac{3}{2}q_2 & -\frac{1}{2}q_1 - \frac{15}{8}q_2^2 & q_{1_x} & 0 \\ 0 & 0 & 1 & q_2 & q_{2_x} & 0 \\ \frac{3}{2}q_2 & -1 & 0 & -p_1 & p_{1_x} & 0 \\ \frac{1}{2}q_1 + \frac{15}{8}q_2^2 & -q_2 & p_1 & 0 & p_{2_x} & 0 \\ -q_{1_x} & -q_{2_x} & -p_{1_x} & -p_{2_x} & 0 & 0 \\ 0 & 0 & 0 & 0 & 0 & 0 \end{pmatrix}, \quad (6.72)$$

where $q_{1_x}, q_{2_x}, p_{1_x}$ and p_{2_x} are given by (6.30). The respective bi-Hamiltonian chain takes the form

$$0 = \theta_0 \circ \nabla h_0, \quad \theta_1 \circ \nabla h_0 = \theta_0 \circ \nabla h_1, \quad \theta_1 \circ \nabla h_1 = \theta_0 \circ \nabla h_2, \quad \theta_1 \circ \nabla h_2 = 0,$$

where $h_0 = c_1, h_1 = h$ and

$$h_2 = -20q_2^2p_1^2 + 16q_1p_1^2 - 32p_2^2 + 64q_2p_1p_2 - 3q_1^2q_2^2$$

$$-4q_1^3 + c_1(6q_2^2 + 8q_1) - c_2(3q_2^3 + 4q_1q_2).$$

All pairs of compatible Poisson operators constructed in the way presented in the examples have a common feature. We shall analyse it on the pair from Example 6.9. More detailed inspection of the second Poisson tensor (6.63) reveals that its last column is equal to the vector field $\bar{\theta}_0\nabla(-2\bar{h}_1)$, where $\bar{h}_1$ is the second integral of motion of the generalized Henon-Heiles system (6.43). So, the possibility of a natural extension of the bi-Hamiltonian chain (6.65) arises [33]. Actually we are looking for functions

$$\bar{h}_0 = \frac{1}{2}\bar{p}_1^2 + \frac{1}{2}\bar{p}_2^2 + V(\bar{q}_1, \bar{q}_2) + \frac{1}{2}\frac{\alpha}{\bar{q}_2^2},$$

$$\bar{h}_1 = -\bar{q}_2\bar{p}_1\bar{p}_2 + \bar{q}_1\bar{p}_2^2 + M(\bar{q}_1, \bar{q}_2) + \frac{\bar{q}_1\alpha}{\bar{q}_2^2}, \quad \bar{h}_2 = -\frac{1}{2}\alpha, \quad (6.73)$$

which form the bi-Hamiltonian chain (6.65) with respect to the canonical Poisson operator $\bar{\theta}_0$ and the noncanonical one

$$\bar{\theta}_{-1} =$$

$$
\begin{pmatrix}
0 & 0 & 0 & -1/\bar{q}_2 & 2\bar{q}_2\bar{p}_2 \\
0 & 0 & -1/\bar{q}_2 & 2\bar{q}_1/\bar{q}_2^2 & 2\bar{q}_2\bar{p}_1 - 4\bar{q}_1\bar{p}_2 \\
0 & 1/\bar{q}_2 & 0 & -\bar{p}_2/\bar{q}_2^2 & 2\bar{p}_2^2 + \partial M/\partial\bar{q}_1 + 2\alpha/\bar{q}_2^2 \\
1/\bar{q}_2 & -2\bar{q}_1/\bar{q}_2^2 & \bar{p}_2/\bar{q}_2^2 & 0 & -2\bar{p}_1\bar{p}_2 + \partial M/\partial\bar{q}_2 - 4\alpha\bar{q}_1/\bar{q}_2^3 \\
* & * & * & & 0
\end{pmatrix},
$$

$$(6.74)$$

where (*) stands for matrix elements which make (6.74) skew-symmetric and again the last column is equal to $\bar{\theta}_0 \nabla(-2\bar{h}_1)$. Of course the functions V and M are not independent. The appropriate relation follows from the equality $\bar{\theta}_0 \circ \nabla\bar{h}_0 = \bar{\theta}_{-1} \circ \nabla\bar{h}_1$ and reads

$$\frac{\partial M}{\partial\bar{q}_2} = 2\bar{q}_1\frac{\partial V}{\partial\bar{q}_2} - \bar{q}_2\frac{\partial V}{\partial\bar{q}_1}, \qquad \frac{\partial M}{\partial\bar{q}_1} = -\bar{q}_2\frac{\partial V}{\partial\bar{q}_2}. \qquad (6.75)$$

One additional restriction on the potentials V is required and it comes from the demand of compatibility between the $\bar{\theta}_0$ and $\bar{\theta}_{-1}$ Poisson tensors. The linear combination of $\bar{\theta}_0$ and $\bar{\theta}_{-1}$ is also a Poisson tensor if V satisfies

$$3\frac{\partial V}{\partial\bar{q}_2} + 2\bar{q}_1\frac{\partial^2 V}{\partial\bar{q}_1\partial\bar{q}_2} - \bar{q}_2\left(\frac{\partial^2 V}{\partial\bar{q}_1^2} - \frac{\partial^2 V}{\partial\bar{q}_2^2}\right). \qquad (6.76)$$

Note that the equality of mixed derivatives of M in (6.75) provides the same condition. Moreover, for the special case of $M(\bar{q}_1,\bar{q}_2) = -\frac{1}{8}\bar{q}_2^4 - \frac{1}{2}\bar{q}_1^2\bar{q}_2^2$ and $V(\bar{q}_1,\bar{q}_2) = \bar{q}_1^3 + \frac{1}{2}\bar{q}_1\bar{q}_2^2$ we generate our gHH system of the KdV type.

Equation (6.76) is the necessary and sufficient condition for separability of a potential $V(\bar{q}_1,\bar{q}_2)$ in parabolic coordinates [131]. An infinite family of solutions of (6.76) can be constructed recursively. Let $\bar{h}_0^{(m)} = \frac{1}{2}\bar{p}_1^2 + \frac{1}{2}\bar{p}_2^2 + V^{(m)}$ denote a Hamiltonian with $V^{(m)}$ satisfying (6.76), and $\bar{h}_1^{(m)} = \bar{q}_2\bar{p}_1\bar{p}_2 - \bar{q}_1\bar{p}_2^2 + M^{(m)}$ be its integral of motion $\{\bar{h}_0^{(m)}, \bar{h}_1^{(m)}\} = 0$. Then, the next Hamiltonian $\bar{h}_0^{(m+1)}$ and integral of motion $\bar{h}_1^{(m+1)}$ are defined by the recursive formula

$$W^{(\bar{m}1)} = \bar{q}_2^2 V^{(m)}, \qquad V^{(m+1)} = 2\bar{q}_1 V^{(m)} + W^{(m)}. \qquad (6.77)$$

By starting with $V^{(0)} = 1$ and $W^{(0)} = 0$ we obtain $V^{(1)} = 2\bar{q}_1$, $V^{(2)} = 4\bar{q}_1^2 + \bar{q}_2^2$, $V^{(3)} = \bar{q}_1^3 + 4\bar{q}_1\bar{q}_2^2$ — the HH potential, $V^{(4)} = 16\bar{q}_1^4 + 12\bar{q}_1^2\bar{q}_2^2 + \bar{q}_2^4$, ..., and so on. The recursion (6.77) is invertible and starting with $V^{(0)} = 1$ and $W^{(0)} = \kappa = $ const we obtain $V^{(-1)} = \kappa/\bar{q}_2^2$, $V^{(-2)} = 1/\bar{q}_2^2 - 2\kappa\,\bar{q}_1/\bar{q}_2^4$, ..., and so on. Such an extension procedure can be applied for an arbitrary bi-Hamiltonian chain in order to find the whole family of new bi-Hamiltonian chains.

What we have presented above is a more or less systematic method of constructing a bi-Hamiltonian finite dimensional system. More examples the reader can find in literature [17],[80],[81]. The complete integrability of stationary flows relies on the completeness of the set of integrals h_i generated by a pair of Poisson matrices (θ_0, θ_1). This is often easy to check for specific examples but difficult to determine in general. Moreover a more general theory which could answer the question of whether stationary flows of bi-Hamiltonian field systems are always bi-Hamiltonian finite dimensional systems and further how to construct them in an arbitrary case is still lacking.

6.2 Stationary Flows of Infinite Systems. Newton Parametrization

The main disadvantage of the Ostrogradsky representation is that it is not 'physical'. This means that the Hamiltonian function of the canonical representation, in general, does not separate into the kinetic part which is a quadratic form of momenta and the potential part which depends only on coordinates. So, it would be convenient to find another more physical representation. Recently some progress in this direction has been made. Actually, the so called Newton representation of ordinary differential equations (ODEs) it has been found [30],[33],[35],[130],[164],[166].

Consider a fourth order ODE of the form

$$u_{4x} + f[u] = c, \qquad c = \text{const}, \tag{6.78}$$

where $f[u] = f(u, u_x, u_{2x}, u_{3x})$ is a polynomial depending on u and its x-derivatives. Suppose that (6.78) is homogeneous with respect to the scaling

$$u \rightarrow \epsilon u, \quad x \rightarrow \epsilon^{-1/2} x, \qquad c \rightarrow \epsilon^3 c. \tag{6.79}$$

The general form of this equation is

$$u_{4x} + auu_{xx} + bu_x^2 + du^3 = c \tag{6.80}$$

and its homogeneity order with respect to (6.79) is ϵ^3. A simple substitution

$$r_1 = u, \; r_2 = u_{xx} + \frac{1}{2}bu^2 \tag{6.81}$$

turns (6.80) into the system of equations

$$r_{1_{xx}} = r_2 - \frac{1}{2}br_1^2,$$

$$r_{2_{xx}} = c + (b-a)r_1r_2 + (\frac{1}{2}ab - \frac{1}{2}b^2 - d)r_1^3. \tag{6.82}$$

The equations (6.82) have the form of Newton's equations of motion for a particle with two degrees of freedom, accelerated by an external force given by the right hand side of (6.82). The special feature of (6.82) is that these equations do not contain 'velocities' r_{1_x}, r_{2_x} as is the case for a large variety of dynamical systems in classical mechanics. This form of (6.80) is called its *Newton representation*.

Definition 6.2 *A set of second order ODEs*

$$r_{k_{xx}} = F_k(r_1, ..., r_p), \qquad k = 1, ..., p \tag{6.83}$$

is called a Newton representation of a 2p-th order ODE

$$0 = u_{2px} + f(u, u_x, ..., u_{2(p-1)x}) \tag{6.84}$$

if for every solution $r = (r_1, ..., r_p)$ of (6.83) the variable $u = r_1$ satisfies (6.84).

For a high order ODE, admitting a Lagrangian formulation $\delta L/\delta u = 0$, we have presented an elegant method of turning it into a set of Hamilton equations by introducing the so called the Ostrogradsky variables. However, not every high order ODE admits a Lagrangian representation. For instance (6.80) can be written as $\delta L/\delta u = 0$ if and only if $a = 2b$, and in this case the Ostrogradsky variables $(q_1 = u, q_2 = u_x, p_1 = -auu_x - u_{3x}, p_2 = u_{xx})$ differ from canonical variables for the Newton representation $(r_1 = u, r_2 = u_{xx} + \frac{1}{4}au^2, p_1 = r_{2_x} = u_{3x} + \frac{1}{2}auu_x, p_2 = r_{1_x} = u_x)$. In general (6.80) has no the Ostrogradsky representation. In spite of this it still admits the Newton representation (6.82).

It is a nontrivial property of an ODE to have a Newton representation, and the study of algebraic methods for finding Newton representations of high order ODEs is a separate problem. The existence of such Newton representations has been proved [30] for an arbitrary sixth order ODE invariant with respect to the scaling (6.79) and has been conjectured for a general 2p-th order ODE of the form

$$u_{2px} + f_{p+1}[u] = c, \tag{6.85}$$

where $f_{p+1}[u] = f_{p+1}(u, u_x, ..., u_{2(p-1)x})$. Newton representations, whenever they exist, are an important tool for studying high order ODEs. They enable us to use the well developed theory of qualitative analysis of trajectories of second order mechanical systems. If the Newton representation of a high order ODE is integrable in the Liouville sense we can apply the well known techniques for solving it. Hence, these Newton representations for which we can find an appropriate set of integrals of motion via the multi-Hamiltonian formalism are of special interest.

In this section we analyse in detail the case of Newton representations for the KdV hierarchy of stationary flows, but such representations can be found for many other ODEs including stationary flows of soliton hierarchies such as Harry–Dym, DWW and others [129],[130],[164]. We show that for each stationary flow of the KdV hierarchy, there are two distinct Newton representations (in analogy to two distinct Ostrogradsky representations) with polynomial forces. These forces are potential and the Newton equations follow from a Lagrangian with an indefinite kinetic energy term. These Newton representations are canonically inequivalent. This leads to a bi-Hamiltonian formulation for each set of Newton equations. The majority of these Newton equations represent new integrable mechanical systems.

Let us consider the KdV spectral problem in a slightly modified form [9], which will be suitable for our further considerations. The KdV from Example 4.4 follows from the Schrödinger spectral problem

$$0 = [D^2 + (u - \lambda)]\varphi := (L\varphi)_0, \tag{6.86}$$

(λ is the spectral parameter) when we required that the time evolution of the eigenfunction φ is governed by the linear equation

$$\varphi_t = \left(\frac{1}{2}P[\lambda, u] + Q[\lambda, u]\right)\varphi = (A\varphi)_0. \tag{6.87}$$

The $P[\lambda, u]$ and $Q[\lambda, u]$ are differential functions of u and they also depend on the spectral parameter λ. Equations (6.86) and (6.87) have a common solution if the compatibility condition

$$0 = (L_t\varphi + L\varphi_t)_0 = (L_t\varphi + LA\varphi)_0 = \left[u_t + \left(-P_x u - \frac{1}{2}Pu_x + Q_{xx}\right)\right]\varphi$$

$$+ \left(\frac{1}{2}P_{xx} + 2Q_x\right)\varphi_x \tag{6.88}$$

(where φ_{xx} has been replaced by $-u\varphi$) is satisfied. Equation (6.88) yields $P_{xx} + 4Q_x = 0$ at φ_x and therefore the compatibility condition reads

$$u_t = \left(\frac{1}{4}D^3 + (u - \lambda)D + \frac{1}{2}u_x\right)P = (\theta_1 - \lambda\theta_0)P, \tag{6.89}$$

where $\theta_1 = \frac{1}{4}D^3 + uD + \frac{1}{2}u_x$ and $\theta_0 = D$ are the two well known Poisson operators of the KdV. Notice that condition (6.88) acquires the standard Lax form (1.2)

$$0 = (L_t\varphi + LA\varphi)_0 = (L_t\varphi + LA\varphi - AL\varphi)_0 = ((L_t + [L, A])\varphi)_0 \tag{6.90}$$

if we remember that $(L\varphi)_0 = 0$.

The explicit form of (6.89) depends on the particular assumption about the λ–dependence of $P[\lambda, u]$. A polynomial in λ ansatz $P = \sum_{k=0}^{m} \gamma_k \lambda^{m-k}$ leads to the KdV hierarchy. Most conveniently the differential functions γ_k can be calculated from the recursion relation

$$0 = (\theta_1 - \lambda\theta_0)\mathcal{P} = \left(\frac{1}{4}D^3 + (u - \lambda)D + \frac{1}{2}u_x\right)\mathcal{P}, \tag{6.91}$$

where $\mathcal{P} = \sum_{r=0}^{\infty} \gamma_r \lambda^{-r}$ is a formal power series. The first Lax operators A, obtained in this way, from (6.87) are

$$A_0 = \partial, \quad A_1 = \left(\lambda + \frac{1}{2}u\right)\partial - \frac{1}{4}u_x,$$

$$A_2 = \left(\lambda^2 + \frac{1}{2}\lambda u + \frac{3}{8}u^2 + \frac{1}{8}u_{xx}\right)\partial - \frac{1}{4}\lambda u_x - \frac{3}{8}uu_x - \frac{1}{16}u_{3x}, \quad \cdots \quad (6.92)$$

Note that A_i from (6.92) are equal to those given by the formula (4.45) after the substitution $L = \partial^2 + u$ by λ in (4.45). The dynamical equations $u_{t_{m+1}} = \theta_0 \circ \gamma_{m+1} = \theta_1 \circ \gamma_m = K_{m+1}(u)$ arise at the lowest powers of λ in (6.91).

Stationary equations are determined by the condition $0 = \theta_0 \circ \gamma_{m+1} = \theta_1 \circ \gamma_m$. The γ_m are determined by the recursion (6.91) which can be integrated once. Thus, each stationary equation is described by the conditions

$$\left(\frac{1}{2}\mathcal{P}\mathcal{P}_{xx} - \frac{1}{4}\mathcal{P}_x^2\right) + (u - \lambda)\mathcal{P}^2 = C(\lambda) \tag{6.93}$$

and $\theta_0 \circ \gamma_{m+1} = 0$ or $\theta_1 \circ \gamma_m = 0$. The equation $\theta_0 \circ \gamma_{m+1} = 0$ integrates to

$$\gamma_{m+1} = \ker\theta_0 = c = \text{const.} \tag{6.94}$$

The symbol $C(\lambda)$ in (6.93) denotes constant of integration which may depend on the spectral parameter λ. It comes up when we integrate (6.91) by multiplying it by $\mathcal{P}$. For the sake of convenience we choose $C(\lambda) = -4\lambda$. At the highest power of λ this choice yields $\gamma_0 = 2$ and $\gamma_1 = u$. Equation (6.93) allows successive determination of all $\gamma_r (r = 0, 1, ..., m)$ through u and its derivatives, and then the stationary equation (6.94) becomes a $(2m + 1)st$ order ODE for u.

Alternatively (after dividing by $\mathcal{P}$) we may look at (6.93) as a set of Newton equations for $\gamma_r, r = 0, 1, ..., m$, where γ_{m+1} in the last equation is substituted by $\ker\theta_0$. Forces in these Newton equations depend, however, on velocities. A more interesting set of Newton equations, with velocity independent forces, are obtained through the simple formal substitution $\mathcal{P} = 2\mathcal{R}^2$, $\mathcal{R} = \sum_{m=0}^{\infty} r_m \lambda^{-m}$. Then (6.93) becomes

$$0 = \mathcal{R}_{xx} + u - \lambda(\mathcal{R} - \mathcal{R}^{-3}) \tag{6.95}$$

and constitutes an infinite recursion defining higher r_m through the lower ones and their derivatives. The stationary condition (6.94) expressed in terms of r-variables reads

$$\text{res}(2\lambda^m\mathcal{R}^2) = \text{res}(\lambda^m\mathcal{P}) = \gamma_{m+1} = \ker\theta_0, \tag{6.96}$$

where $rmres$ means residuum, the coefficient at λ^{-1} in the formal power series of λ. If we use it in order to eliminate the r_{m+1} variable from the $(m + 1)st$ equation in the sequence (6.95) (the equation which starts with the term $r_{m_{xx}} + ...$) then we obtain an autonomous set of Newton equations for $r_0, r_1, ..., r_m$. This set is equivalent to (6.93),(6.94). The first term in the recursion (6.95) yields $r_0 = 1, u = 4r_1$.

In analogy to the Ostrogradsky representation, the second form of the stationary condition

$$\theta_1 \text{res}(2\lambda^{m-1}\mathcal{R}^2) = \theta_1\gamma_m = 0 \qquad (6.97)$$

can be integrated. It leads to a new sequence of Newton equations which look quite similar to (6.95),(6.96) but are equipped with different (canonically inequivalent) canonical formulation. New Newton variables introduced after integrating θ_1 are denoted as $\bar{r}_k, k = 1, ..., m$. The 2nd form of the stationary condition $0 = \theta_1 \circ \gamma_m = (\frac{1}{4}D^3uD + \frac{1}{2}u_x)\gamma_m$ multiplied by γ_m yields

$$\frac{1}{4}\gamma_m\gamma_{m_{xx}} - \frac{1}{8}\gamma_{m_x}^2 + \frac{1}{2}u\gamma_m^2 = \text{const.} \qquad (6.98)$$

By introducing a new variable $\bar{p}_m$ defined as $\alpha\bar{p}_m = \gamma_m$ we eliminate the velocity dependent term $\frac{1}{8}\gamma_{m_x}^2$ and obtain a complete set of equations

$$\gamma_m = \alpha\bar{p}_m^2, \qquad 0 = \bar{p}_{m_{xx}} + u\bar{p}_m - \bar{c}\bar{p}_m^{-3}, \qquad \bar{c} = \text{const.} \qquad (6.99)$$

Since γ_m can be expressed in terms of the r-variables (which we shall now call $\bar{r}_1, ..., \bar{r}_{m-1}$) we obtain the second Newton representation.

Theorem 6.1 *Each stationary KdV equation* $0 = \theta_0 \circ \gamma_{m+1} = \theta_1 \circ \gamma_m$ *has two parametrizations in the form of Newton equations, which are potential and have a Lagrangian functional. The integrated equation* $\gamma_{m+1}(u) = c$ *is, under the substitution* $u = 4r_1$, *equivalent to the following set of m Newton equations*

$$0 = \text{res}[\mathcal{R}_{xx} + 4r_1\mathcal{R} - \lambda(\mathcal{R} - \mathcal{R}^{-3})] = \delta L_{m+1}/\delta r_m,$$

$$\vdots$$

$$0 = \text{res}\lambda^{m-2}[\mathcal{R}_{xx} + 4r_1\mathcal{R} - \lambda(\mathcal{R} - \mathcal{R}^{-3})] = \delta L_{m+1}/\delta r_2, \qquad (6.100)$$

$$0 = \text{res}\lambda^{m-1}[\mathcal{R}_{xx} + 4r_1\mathcal{R} - \lambda(\mathcal{R} - \mathcal{R}^{-3})] + 2\text{res}(\lambda^m\mathcal{R}^2) - c$$

$$= \delta L_{m+1}/\delta r_1,$$

which follows from the Lagrangian density

$$L_{m+1} = \text{res}\left[-\frac{1}{2}\lambda^m\mathcal{R}_x^2 + 2\lambda^m r_1\mathcal{R}^2 - \lambda^{m+1}\frac{1}{2}(\mathcal{R}^2 - \mathcal{R}^{-2})\right] + cr_1. \qquad (6.101)$$

The integrated equation $\frac{1}{4}\gamma_m\gamma_{m_{xx}} - \frac{1}{8}\gamma_{m_x}^2 + \frac{1}{2}u\gamma_m^2 = -\frac{1}{4}\alpha$ *is, under the substitution* $\gamma_m = \alpha\bar{p}_m^2$ *and* $u = 4\bar{r}_1$, *equivalent to the following set of m Newton equations*

$$0 = \text{res}[\overline{\mathcal{R}}_{xx} + 4\overline{r}_1\overline{\mathcal{R}} - \lambda(\overline{\mathcal{R}} - \overline{\mathcal{R}}^{-3})] = \delta\overline{L}_{m+1}/\delta\overline{r}_{m-1},$$

$$\vdots$$

$$0 = \text{res}\lambda^{m-3}[\overline{\mathcal{R}}_{xx} + 4\overline{r}_1\overline{\mathcal{R}} - \lambda(\overline{\mathcal{R}} - \overline{\mathcal{R}}^{-3})] = \delta\overline{L}_{m+1}/\delta\overline{r}_2,$$

$$0 = \text{res}\lambda^{m-2}[\overline{\mathcal{R}}_{xx} + 4\overline{r}_1\overline{\mathcal{R}} - \lambda(\overline{\mathcal{R}} - \overline{\mathcal{R}}^{-3})] + 2\text{res}(\lambda^{m-1}\overline{\mathcal{R}}^2) - \alpha\overline{\rho}_m^2$$

$$= \delta\overline{L}_{m+1}/\delta\overline{r}_1,$$

$$0 = \overline{\rho}_{m_{xx}} + 4\overline{r}_1\overline{\rho}_m - \overline{c}\overline{\rho}_m^{-3} = \delta\overline{L}_{m+1}/\delta\overline{\rho}_m,$$

$$(6.102)$$

where $\overline{\mathcal{R}} = \sum_{k=0}^{\infty} \overline{r}_k \lambda^{-k}$. *They are generated by the Lagrangian density*

$$\overline{L}_{m+1} = \text{res}\left[-\frac{1}{2}\lambda^{m-1}\overline{\mathcal{R}}_x^2 + 2\lambda^{m-1}\overline{r}_1\overline{\mathcal{R}}^2 - \lambda^m\frac{1}{2}(\overline{\mathcal{R}}^2 - \overline{\mathcal{R}}^{-2})\right] + 2\overline{\rho}_m^2\overline{r}_1$$

$$-\frac{1}{2}\overline{\rho}_{m_x} + \frac{1}{2}\overline{c}\overline{\rho}_m^{-2} \qquad (6.103)$$

if we choose $\alpha = -2$ *and* $\delta L/\delta r = \partial L/\partial r - (d/dx)\partial L/\partial r_x$.

Proof. Equation (6.100) follows directly from (6.95) which is an infinite hierarchy of equalities defining the higher variables γ_m in terms of the lower ones and their derivatives. The condition $c = \gamma_{m+1}(u) = 2\text{res}(\lambda^m\mathcal{R}^2)$ terminates this sequence at the m-th equation by expressing the variable r_{m+1} through the lower ones. The additional term $c - 2\text{res}(\lambda^m\mathcal{R}^2)$ in the last equation of (6.100), which contains the information that we consider the $(m+1)$-st stationary KdV equation, produces the term $4r_{m+1}$ cancelling the corresponding term $-(r_{m+1} + 3r_{m+1})$ in $\lambda^{m+1}\frac{1}{2}(\mathcal{R}^2 - \mathcal{R}^{-2})$. Thus we obtain an autonomous set of Newton equations for $r_1, ..., r_m$. We recall that the substitution $u = 4r_1$ follows from the λ^0 term in (6.95).

To prove that (6.101) is a Lagrangian for (6.100) is a simple exercise in deriving a formal series for $\mathcal{R}$. For instance for the last equation (6.100) we obtain

$$-\frac{d}{dx}\frac{\partial L_{m+1}}{\partial r_{1_x}} = \frac{d}{dx}\{\text{res}\lambda^m\mathcal{R}_x(\partial\mathcal{R}_x/\partial r_{1_x})\} = \text{res}\lambda^{m-1}\{\mathcal{R}_{xx}\} = r_{m_{xx}}$$

and

$$\frac{\partial L_{m+1}}{\partial r_1} = \text{res}\{4\lambda^m r_1\mathcal{R}(\partial\mathcal{R}/\partial r_1) + 2\lambda^m\mathcal{R}^2$$

$$-\lambda^{m+1}2\mathcal{R}(\partial\mathcal{R}/\partial r_1) - \mathcal{R}^{-3}(\partial\mathcal{R}/\partial r_1)]\} - c$$

$$= \text{res}\lambda^{m-1}[4r_1\mathcal{R} - \lambda(\mathcal{R} - \mathcal{R}^{-3})] + 2\text{res}(\lambda^m\mathcal{R}^2) - c.$$

Note that terms containing the variable r_{m+1} always cancel in the Lagrangian L_{m+1}. The first $(m-2)$ equations (6.102) are identical to (6.100) since $4\overline{r}_1 =$

$u = 4r_1$. Only the two last equations are different due to the new coupling term $\alpha \bar{\rho}_m^2$ in place of the constant c. However, the variables $\bar{r}_1, ..., \bar{r}_{m-1}$ are now prescribed to equations (6.102) in a shifted order since the last variable $\bar{r}_m$ is attached to the m-th equation. The remaining calculations are very similar to those of (6.100). We notice only that in the Lagrangian density (6.103) the term under the residuum symbol does not depend on $\bar{r}_m$ so it is easy to calculate the last equation (6.102). The constant α has to be chosen as -2 to get a Lagrangian for (6.102). $\square$

Lemma 6.1 *Newton equations (6.100),(6.102) are equivalent with respect to the map*

$$r_k = \bar{r}_k, \quad k = 1, ..., m-1,$$

$$r_m = -\bar{r}_1 \bar{r}_{m-1} - \bar{r}_2 \bar{r}_{m-2} - ... + \left\{ \begin{array}{ll} -\bar{r}_{(m-1)/2} \bar{r}_{(m+1)/2} - \frac{1}{2}\bar{\rho}_m^2 & m - odd \\[2mm] -\frac{1}{2}\bar{r}_{m/2}^2 - \frac{1}{2}\bar{\rho}_m^2 & m - even \end{array} \right\}$$
(6.104)

$$c = 2\mathrm{res}\left[\frac{1}{2}\lambda^{m-1}\mathcal{R}_x^2(\bar{r}) \right] + 2\mathrm{res}\left[\lambda^{m-1}\bar{r}_1\mathcal{R}^2 - \lambda^m \frac{1}{2}\left(\mathcal{R}^2 - \mathcal{R}^{-2} \right) \right]$$

$$+ 2\left[2\bar{\rho}_m^2 \bar{r}_1 + \frac{1}{2}\bar{\rho}_{m_x}^2 + \frac{1}{2}\overline{c\rho}_m^2 \right] = -2\bar{h}(\bar{r}, \bar{r}_x, \bar{c}).$$
(6.105)

This map extends for the canonical momenta as

$$s_1 = -\bar{r}_1 \bar{s}_1 - ... - \bar{r}_{m-1}\bar{s}_{m-1} - \bar{\rho}_m \bar{s}_m, \quad s_2 = \bar{s}_1, \quad ... \quad s_m = \bar{s}_{m-1}, \quad (6.106)$$

where $s_k = \partial L_{m+1}/\partial r_{k_x}$ $k = 1, ..., m$, and $\bar{s}_k = \partial \bar{L}_{m+1}/\partial \bar{r}_{k_x}$, $k = 1, ..., m-1$, $\bar{s}_m = \partial \bar{L}_{m+1}/\partial \bar{\rho}_{m_x}$.

Proof. We see that $r_1 = \frac{1}{4}u = \bar{r}_1$ and the first $(m-2)$ equations of (6.100) and (6.102) yield $r_k = \bar{r}_k, k = 1, ..., m-1$. The $(m-1)$-st equation gives (6.104). The corresponding map for the momenta follows from their definitions. The formula connecting the parameters c and $\bar{c}$ can be calculated from the last equations of (6.100) and (6.102). We can write the last of the equations (6.104) as $r_m = -\frac{1}{2}\mathrm{res}(\lambda^{m-1}\overline{\mathcal{R}}^2) + \bar{r}_m - \frac{1}{2}\bar{\rho}_m^2$, ($\bar{r}_m$ cancels with $-\bar{r}_m$ in the residual term). After differentiating twice we eliminate the second order derivatives by using the last of the equations in (6.102) and by writing $\mathcal{R}_{xx}$ as

$$\mathcal{R}_{xx} = \left[(\lambda - 4\bar{r}_1)\overline{\mathcal{R}} - \lambda \overline{\mathcal{R}}^{-3} - \alpha \right]_{\geq -m+1} + (\bar{r}_m)_{xx}\lambda^{-m} + (\bar{r}_{m+1})_{xx}\lambda^{-m-1} + ...$$

(this full form of stationary condition implies cancelling of $(\bar{r}_m)_{xx}, \bar{r}_m$ and $\bar{r}_{m+1}$), where the symbol $[...]_{\geq k}$ denotes that part of a power series which contains powers of λ greater than or equal to k. Inserting this all into the last of equations (6.100) we arrive at (6.104). $\square$

Example 6.11 Two Newton representations for the fifth KdV stationary flow: $m = 2$.

In terms of the r-variables we get

$$0 = r_{1_{xx}} + 10r_1^2 - 4r_2 = \frac{\delta L_3}{\delta r_2},$$

$$0 = r_{2_{xx}} + 20r_1 r_2 - 10r_1^3 - c = \frac{\delta L_3}{\delta r_1}, \tag{6.107}$$

with the Lagrangian

$$L_3 = -r_{1_x} r_{2_x} + 10r_1^2 r_2 - 2r_2^2 - \frac{5}{2}r_1^4 - cr_1. \tag{6.108}$$

The second set of Newton equations (6.102)

$$0 = \bar{r}_{1_{xx}} + 12\bar{r}_1^2 + 2\bar{\rho}_2^2 = \frac{\delta \bar{L}_3}{\delta \bar{r}_1}$$

$$0 = \bar{\rho}_{2_{xx}} + 4\bar{r}_1 \bar{\rho}_2 - \overline{c\rho}_2^{-3} = \frac{\delta \bar{L}_3}{\delta \bar{\rho}_2}, \tag{6.109}$$

with

$$\bar{L}_3 = -\frac{1}{2}\bar{r}_{1_x}^2 - \frac{1}{2}\bar{\rho}_{2_x}^2 + 4\bar{r}_1^3 + 2\bar{r}_1\bar{\rho}_2^2 + \frac{1}{2}\overline{c\rho}^{-2} \tag{6.110}$$

becomes, after the rescaling of variables $4\bar{r}_1 = \bar{q}_1, 4\bar{\rho}_2 = \bar{q}_2, 16\bar{c} = \alpha$, an integrable case of the celebrated Henon-Heiles system (6.43).

Example 6.12 Two Newton representations of the seventh order KdV stationary flow: $m = 3$.

For $m = 3$ (6.100) read

$$0 = r_{1_{xx}} + 10r_1^2 - 4r_2,$$

$$0 = r_{2_{xx}} + 16r_1 r_2 - 10r_1^3 - 4r_3,$$

$$0 = r_{3_{xx}} + 20r_1 r_3 + 8r_2^2 - 30r_1^2 r_2 + 15r_1^4 - c, \tag{6.111}$$

with

$$L_4 = -r_{1_x} r_{3_x} - \frac{1}{2}r_{2_x}^2 - 4r_2 r_3 + 10r_1^2 r_3 + 8r_1 r_2^2 - 10r_1^3 r_2 + 5r_1^5 - cr_1. \tag{6.112}$$

From (6.102) we obtain a new set of Newton equations

$$0 = \bar{r}_{1_{xx}} + 10\bar{r}_1^2 - 4\bar{r}_2,$$

$$0 = \bar{r}_{2_{xx}} + 20\bar{r}_1 \bar{r}_2 - 10\bar{r}_1^3 + 2\bar{\rho}_3,$$

$$0 = \bar{\rho}_{3_{xx}} + 4\bar{r}_1 \bar{\rho}_3 - \overline{c\rho}_3^{-3}, \tag{6.113}$$

with

$$\bar{L}_4 = -\bar{r}_{1_x}\bar{r}_{2_x} - \frac{1}{2}\bar{p}_{3_x}^2 + 10\bar{r}_1^2\bar{r}_2 - 2\bar{r}_2^2 - \frac{5}{2}\bar{r}_1^4 + 2\bar{r}_1\bar{p}_3^2 + \frac{1}{2}\bar{c}\bar{p}_3^{-2}. \quad (6.114)$$

Remark 6.1 In general, the Lagrangians L_{m+1} in (6.101) take the form

$$L_{m+1}[r, r_x, c] = -\frac{1}{2}\sum_{i,j=1}^{m}(r_i)_x\mu_{ij}(r_j)_x + V(r) - cr_1, \quad (6.115)$$

where $r = (r_1, ..., r_m)$ and the metric μ reads

$$\mu = \begin{pmatrix} & & & 1 \\ & 0 & 1 & \\ & \cdots & & \\ & 1 & 0 & \\ 1 & & & \end{pmatrix}. \quad (6.116)$$

On the other hand, the Lagrangians $\bar{L}_{m+1}$ in (6.103) take the form

$$\bar{L}_{m+1}[\bar{r}, \bar{r}_x, \bar{c}] = -\frac{1}{2}\sum_{i,j=1}^{m}(\bar{r}_i)_x\bar{\mu}_{ij}(\bar{r}_j)_x + \bar{V}(\bar{r}) + 2\bar{p}_m^2\bar{r}_1 + \frac{1}{2}\bar{c}\bar{p}_m^{-2}, \quad (6.117)$$

where $\bar{r} = (\bar{r}_1, ..., \bar{r}_{m-1}, \bar{p}_m)$ and the metric $\bar{\mu}$ takes the form

$$\bar{\mu} = \left(\begin{array}{ccc|c} & & 1 & \\ & \cdots & & 0 \\ & 1 & & \\ 1 & & & \\ \hline & 0 & & 1 \end{array} \right). \quad (6.118)$$

$V(r)$ and $\bar{V}(\bar{r})$ are appropriate potentials.

The bi-Hamiltonian form of Newton representations is illustrated for the case $m = 3$, as from this value of m both Newton representations differ from the Ostrogradsky ones (the reader ca find the bi-Hamiltonian representation for the case $m = 2$ (6.107) in [166]). The map (6.104)–(6.105) is noncanonical and can be used to derive a bi-Hamiltonian formulation for both systems of Newton equations (6.111) and (6.113). In the extended phase spaces of variables (r, s, c) and $(\bar{r}, \bar{s}, \bar{c})$ each system has a canonical Hamiltonian structure following from the Legendre transformations and Lagrangians (6.112) and (6.114), respectively. Each canonical Poisson matrix (6.10) induces a new noncanonical one through the Miura map $(\bar{r}, \bar{s}, \bar{c}) \rightarrow (r, s, c)$ given by the formulas (6.104)–(6.105)

$$r_1 = \bar{r}_1, \ r_2 = \bar{r}_2, \ r_3 = -\bar{r}_1\bar{r}_2 - \frac{1}{2}\bar{p}_3^2,$$

$$s_1 = -\bar{r}_1\bar{s}_1 - \bar{r}_2\bar{s}_2 - \bar{p}_3\bar{s}_3, \ s_2 = \bar{s}_1, \ s_3 = \bar{s}_2,$$

$$c = -2\left(\bar{s}_1\bar{s}_2 + \frac{1}{2}\bar{s}_3 + 10\bar{r}_1^2\bar{r}_2 - 2\bar{r}_2^2 - \frac{5}{2}\bar{r}_1^4 + 2\bar{r}_1\bar{p}_3^2 + \frac{1}{2}\overline{c}\overline{p}_3^{-2}\right)$$

$$= -2\bar{h}(\bar{r},\bar{s},\bar{c}). \tag{6.119}$$

Thus, we obtain the following bi-Hamiltonian formulation for the Newton equations (6.111)

$$\begin{pmatrix} r \\ s \\ c \end{pmatrix}_x = \theta_0 \circ \nabla h_1 = \theta_1 \circ \nabla h_0 \tag{6.120}$$

with the 7×7 Poisson matrices θ_0 given by the form (6.10) and

$$\theta_1 =$$

$$\begin{pmatrix}
0 & 0 & 0 & r_1/2 & -1/2 & 0 & s_3 \\
0 & 0 & 0 & r_2/2 & 0 & -1/2 & s_2 \\
0 & 0 & 0 & r_3 & r_2/2 & r_1/2 & s_1 \\
-r_1/2 & -r_2/2 & -r_3 & 0 & s_2/2 & s_3/2 & c - \partial V/\partial r_1 \\
1/2 & 0 & -r_2/2 & -s_2/2 & 0 & 0 & -\partial V/\partial r_2 \\
0 & 1/2 & -r_1/2 & -s_3/2 & 0 & 0 & -\partial V/\partial r_3 \\
-s_3 & -s_2 & -s_1 & \partial V/\partial r_1 - c & \partial V/\partial r_2 & \partial V/\partial r_3 & 0
\end{pmatrix}$$

$$\tag{6.121}$$

where

$$h_0 = c, \quad h_1 = s_1 s_3 + \frac{1}{2}s_2^2 + V(r) - cr_1 \tag{6.122}$$

and $V(r) = 10r_1^2 r_3 - 4r_2 r_3 + 8r_1 r_2^2 - 10r_1^3 r_2 + 3r_1^5$ is the potential of the Hamiltonian h_1. The last column of θ_1 is equal to the Hamiltonian vector field generated by h_1. The bi-Hamiltonian vector field (6.120) extends to the following bi-Hamiltonian chain

$$\theta_0 \circ \nabla h_0 = 0$$

$$\theta_0 \circ \nabla h_1 = \theta_1 \circ \nabla h_0$$

$$\theta_0 \circ \nabla h_2 = \theta_1 \circ \nabla h_1 \tag{6.123}$$

$$\theta_0 \circ \nabla h_3 = \theta_1 \circ \nabla h_2$$

$$0 = \theta_1 \circ \nabla h_3$$

where

$$h_2 = \frac{1}{2}r_3 s_3^2 - \frac{1}{2}r_1 s_2^2 + \frac{1}{2}r_2 s_2 s_3 - \frac{1}{2}s_1 s_2 - \frac{1}{2}r_1 s_1 s_3$$

$$+ W(r) + \frac{1}{4}cr_1^2 + \frac{1}{2}cr_2, \tag{6.124}$$

with $W(r) = 2r_1^2 r_2^2 + \frac{5}{2}r_1^4 r_2 - \frac{5}{4}r_1^6 - 2r_2^3 + r_3^2 - 6r_1 r_2 r_3$ and

$$h_3 = \frac{1}{8}r_2^2 s_3^2 + \frac{1}{8}r_1^2 s_2^2 + \frac{1}{8}s_1^2 + \frac{1}{4}r_1 s_1 s_2 + \frac{1}{4}r_2 s_1 s_3 - \frac{1}{4}r_1 r_2 s_2 s_3 - \frac{1}{2}r_3 s_2 s_3$$

$$+ U(r) - \frac{1}{4}c r_1 r_2 - \frac{1}{4}c r_3 \qquad (6.125)$$

with $U(r) = -3r_1^3 r_2^2 + r_1 r_2^3 + \frac{5}{4}r_1^5 r_2 + 2r_3^2 r_1 + \frac{5}{4}r_1^4 r_3 - r_1^2 r_2 r_3 + r_2^2 r_3$. The function h_3 is the Casimir function of θ_1. This bi-Hamiltonian chain contains three nontrivial vector fields which commute in pairs. Integrals of motion also commute and are functionally independent, since the Jacobian $\det[\partial(h_1, h_2, h_3)/\partial(s_1, s_2, s_3)]$ does not vanish. The Poisson operator θ_1 characterizes a new family of integrable three-dimensional potentials specified in Lemma 6.2 below.

Lemma 6.2 *If $V(r)$ satisfies the equations*

$$\frac{\partial^2 V}{\partial r_1^2} + r_2 \frac{\partial^2 V}{\partial r_2^2} + r_1 \frac{\partial^2 V}{\partial r_1 \partial r_2} - r_2 \frac{\partial^2 V}{\partial r_1 \partial r_3} + 2r_3 \frac{\partial^2 V}{\partial r_2 \partial r_3} + 3\frac{\partial V}{\partial r_2} = 0,$$

$$2r_3 \frac{\partial^2 V}{\partial r_3^2} + r_2 \frac{\partial^2 V}{\partial r_2 \partial r_3} + \frac{\partial^2 V}{\partial r_1 \partial r_2} + 3\frac{\partial V}{\partial r_3} = 0,$$

$$\frac{\partial^2 V}{\partial r_1 \partial r_3} - r_2 \frac{\partial^2 V}{\partial r_3^2} + r_1 \frac{\partial^2 V}{\partial r_2 \partial r_3} - \frac{\partial^2 V}{\partial r_2^2} = 0, \qquad (6.126)$$

then the Hamiltonians h_1 (6.122) are completely integrable; they admit the bi-Hamiltonian formulation (6.120) and belong to the bi-Hamiltonian chain (6.123). The higher Hamiltonians h_2, h_3 have the form (6.124) and (6.125) respectively, where $W(r)$ and $U(r)$ are given by the conditions

$$\frac{\partial W}{\partial r_1} = -\frac{1}{2}r_1 \frac{\partial V}{\partial r_1} + \frac{1}{2}r_2 \frac{\partial V}{\partial r_2} + r_3 \frac{\partial V}{\partial r_3},$$

$$\frac{\partial W}{\partial r_2} = -\frac{1}{2}\frac{\partial V}{\partial r_1} - r_1 \frac{\partial V}{\partial r_2} + \frac{1}{2}r_2 \frac{\partial V}{\partial r_3},$$

$$\frac{\partial W}{\partial r_1} = -\frac{1}{2}\frac{\partial V}{\partial r_2} - \frac{1}{2}r_1 \frac{\partial V}{\partial r_3} \qquad (6.127)$$

and

$$\frac{\partial U}{\partial r_1} = \frac{1}{4}r_2 \frac{\partial V}{\partial r_1} - \left(\frac{1}{4}r_1 r_2 + \frac{1}{2}r_3\right)\frac{\partial V}{\partial r_2} + \frac{1}{4}r_2^2 \frac{\partial V}{\partial r_3},$$

$$\frac{\partial U}{\partial r_2} = \frac{1}{4}r_1 \frac{\partial V}{\partial r_1} + \frac{1}{4}r_1^2 \frac{\partial V}{\partial r_2} - \left(\frac{1}{4}r_1 r_2 + \frac{1}{2}r_3\right)\frac{\partial V}{\partial r_3},$$

$$\frac{\partial U}{\partial r_3} = \frac{1}{4}\frac{\partial V}{\partial r_1} + \frac{1}{4}r_1 \frac{\partial V}{\partial r_2} + \frac{1}{4}r_2 \frac{\partial V}{\partial r_3}. \qquad (6.128)$$

Proof. The Jacobi identity for the Poisson operator θ_1 leads to (6.126). The term $W(r)$ is a primitive function of the partial differential equations (6.127) since the mixed derivatives are equal as can be verified by using the equation (6.126). The same applies to $U(r)$. Complete integrability of the Hamiltonian h_1 follows since h_1, h_2 and h_3 are functionally independent commuting integrals of motion. $\square$

The same map (6.119) used in the other direction provides a bi-Hamiltonian formulation for the system (6.113). The appropriate formulas are given in Sect. 6.5.

Now we will show that both Newton representations (6.100) and (6.102) have a matrix Lax representation. First, let us write the compatibility condition for (6.86) and (6.87) in the form of a zero-curvature equation. Let

$$V = \begin{pmatrix} -\frac{1}{4}P_x & \frac{1}{2}P \\ \frac{1}{2}P(u-\lambda) - \frac{1}{4}P_{xx} & \frac{1}{4}P_x \end{pmatrix}, \quad U = \begin{pmatrix} 0 & 1 \\ \lambda - u & 0 \end{pmatrix}. \tag{6.129}$$

Then

$$V_x - U_t + [V,U] = 0 \Leftrightarrow u_t = (\theta_1 - \lambda\theta_0)P. \tag{6.130}$$

Notice that as $P = \sum_{k=0}^m \gamma_k \lambda^{m-k}$, then representation (6.129) is equal to the one from Example 3.6.

The stationary flows are given by the Lax equation $V_x + [V,U] = 0$. Observe that now U plays the role of the Lax operator while V becomes a spectral operator. This is due to the fact, that in the case of stationary flows the variable x plays now the role of an evolution parameter.

Theorem 6.2

(i)   The first Newton representation of the m-th stationary KdV flow (6.100) has the Lax representation $V_x + [V,U] = 0$ with

$$V = \begin{pmatrix} -[\lambda^m \mathcal{R}\mathcal{R}_x]_{\geq 0} & [\lambda^m \mathcal{R}^2]_{\geq 0} \\ -\mathrm{res}[\lambda^m \mathcal{R}^2] + [\lambda^m(-\mathcal{R}_x^2 + \lambda\mathcal{R}^{-2} + \alpha\mathcal{R})]_{\geq 0} & [\lambda^m \mathcal{R}\mathcal{R}_x]_{\geq 0} \end{pmatrix}$$

$$U = \begin{pmatrix} 0 & 1 \\ \lambda - 4r_1 & 0 \end{pmatrix}, \quad \alpha = \lambda^{-m}[2\mathrm{res}(\lambda^m \mathcal{R}^2) - c]. \tag{6.131}$$

(ii)   The second Newton representation for the m-th stationary KdV flow (6.102) has the Lax representation $V_x + [V,U] = 0$ with

$$V = \begin{pmatrix} -[\lambda^m \overline{\mathcal{R}}\overline{\mathcal{R}}_x]_{\geq 1} + \bar{p}_m\bar{p}_{m_x} & [\lambda^m \overline{\mathcal{R}}^2]_{\geq 1} - \bar{p}_m^2 \\ \begin{array}{l} -\lambda\mathrm{res}[[\lambda^{m-1}\overline{\mathcal{R}}^2] + [\lambda^m(-\overline{\mathcal{R}}_x^2 + \lambda\overline{\mathcal{R}}^{-2} \\ +\alpha\overline{\mathcal{R}})]_{\geq 1} - \lambda\bar{p}_m^2 + \bar{p}_{m_x}^2 + \alpha\bar{p}_m^{-2} \end{array} & [\lambda^m \overline{\mathcal{R}}\overline{\mathcal{R}}_x]_{\geq 1} - \bar{p}_m\bar{p}_{m_x} \end{pmatrix},$$

$$U = \begin{pmatrix} 0 & 1 \\ \lambda - 4\bar{r}_1 & 0 \end{pmatrix}, \quad \alpha = \lambda^{-m+1}[2\mathrm{res}(\lambda^{m-1}\overline{\mathcal{R}}^2) + 2\bar{p}_m^2]. \tag{6.132}$$

where the symbol $[...]_{\geq k}$ denotes that part of the power series of elements in powers of λ greater than or equal to k.

Proof.

(i) The U matrix follows from (6.129) since now $u = 4r_1$. We remind the reader that in the case of the $\mathcal{R}$−representation we have $P = 2[\lambda^m \mathcal{R}^2]_{\geq 0}$, $\mathcal{R} = \sum_{m=0}^{\infty} r_m \lambda^{-m}, r_0 = 1$. After inserting this into (6.129) and taking into account the fact that the projection operation $[...]_{\geq k}$ and the differentiation commute, we find that

$$V = \begin{pmatrix} -[\lambda^m \mathcal{R}\mathcal{R}_x]_{\geq 0} & [\lambda^m \mathcal{R}^2]_{\geq 0} \\ -[\lambda^m \mathcal{R}^2]_{\geq 0} u - [\lambda^m (\mathcal{R}_x^2 + \mathcal{R}\mathcal{R}_{xx})]_{\geq 0} & [\lambda^m \mathcal{R}\mathcal{R}_x]_{\geq 0} \end{pmatrix}. \quad (6.133)$$

However, due to relations (6.100), the stationary condition for the KdV flow can be written as $\mathcal{R}_{xx} + u\mathcal{R} + \lambda \mathcal{R}^{-3} + \alpha = 0$. After inserting this into (6.133) and using the fact that $[\lambda^m u\mathcal{R}^2]_{\geq 0} - [\lambda^m \mathcal{R}^2]_{\geq 0} = -\text{res}[\lambda^m \mathcal{R}^2]$ we obtain the formula (6.131). □

(ii) Again, the form of V in (6.132) follows from the substitution of $u = 4\bar{r}_1$ into (6.129). However, this time P can be expressed through the $\overline{\mathcal{R}}$−variable as $P = 2[\lambda^m \overline{\mathcal{R}}^2]_{\geq 1} - 2\bar{\rho}_m^2$ (since $P_m = -2\bar{\rho}_m^2$), where $\overline{\mathcal{R}} = \sum_{m=0}^{\infty} \bar{r}_m \lambda^{-m}, \bar{r}_0 = 1$. The stationary equations (6.102) expressed in terms of a formal series $\overline{\mathcal{R}}$ are $\overline{\mathcal{R}}_{xx} + u\overline{\mathcal{R}} + \lambda \overline{\mathcal{R}}^{-3} + \alpha = 0$ and $\bar{\rho}_{m_{xx}} + 4\bar{r}_1 \bar{\rho}_m - \bar{c} \, \bar{\rho}_m^{-3} = 0$. Inserting all this into (6.129) we arrive at formula (6.132). □

Example 6.13 Lax representation of the fifth order KdV stationary flows (Newton variables).
In the case of $m = 2$ the formula (6.131) reads

$$V = \begin{pmatrix} -r_{1_x}\lambda - r_{2_x} - r_1 r_{1_x} & \lambda^2 + 2r_1\lambda + r_1^2 \\ \lambda^3 - 2r_1\lambda^2 + (-2r_2 + 3r_1^2)\lambda & r_{1_x}\lambda + r_{2_x} + r_1 r_{1_x} \\ +8r_1 r_2 - 4r_1^3 - r_{1_x}^2 - c & \end{pmatrix},$$

$$U = \begin{pmatrix} 0 & 1 \\ \lambda - 4r_1 & 0 \end{pmatrix}, \quad (6.134)$$

and the Lax equation $V_x + [V, U] = 0$ produces (6.107). On the other hand, formulas (6.132) specify

$$V = \begin{pmatrix} -\bar{r}_{1_x}\lambda + \bar{\rho}_2 \bar{\rho}_{2_x} & \lambda^2 + 2\bar{r}_1\lambda - \bar{\rho}_2^2 \\ \lambda^3 - 2\bar{r}_1\lambda^2 + (4\bar{r}_1^2 + \bar{\rho}_2^2)\lambda + \bar{\rho}_{2_x}^2 + \bar{c}\bar{\rho}_2^{-2} & \bar{r}_{1_x}\lambda - \bar{\rho}_2 \bar{\rho}_{2_x} \end{pmatrix}$$

$$U = \begin{pmatrix} 0 & 1 \\ \lambda - 4\bar{r}_1 & 0 \end{pmatrix}, \quad (6.135)$$

and the Lax equation $V_x + [V, U] = 0$ yields (6.109).

All examples of Ostrogradsky representations, presented in the previous section, have their Newton counterparts. The reader can find the details on the systematic construction of Newton representations and their bi-Hamiltonian forms for the Harry–Dym hierarchy and coupling KdV hierarchy (including the DWW hierarchy) in [165] and [129]. Here, we end this section with the example of the Harry–Dym stationary equation. Apart from two the Ostrogradsky representations (see Examples 6.2 and 6.6), the stationary HD also has two Newton representations following from the appropriate Lagrangians. The first representation related to $h_2[u]$ has the form

$$r_{1_{xx}} = -4r_1^{-4}r_2,$$

$$r_{2_{xx}} = 8r_1^{-5}r_2^2 + 2\gamma^2 r_1^{-5} + dr_1^{-3}, \tag{6.136}$$

with

$$L = (r_1)_x(r_2)_x - 2r_1^{-4}r_2^2 - \frac{1}{2}\gamma^2 r_1^{-4} - \frac{1}{2}dr_1^{-2}, \tag{6.137}$$

and the second one is

$$\bar{r}_{1_{xx}} = -4\gamma\bar{r}_1^{-5}\bar{p}_2,$$

$$\bar{p}_{2_{xx}} = \gamma\bar{r}_1^{-4} - \bar{d}, \tag{6.138}$$

where

$$\bar{L} = \frac{1}{2}(\bar{r}_{1_x})^2 + \frac{1}{2}(\bar{p}_{2_x})^2 + \gamma\bar{p}_2\bar{r}_1^{-4} - \bar{d}\bar{p}_2. \tag{6.139}$$

The constant γ is related to the HD parameters w and β as follows: $\gamma = (\frac{1}{2}\beta w)^{1/2}$. Equations (6.136), (6.137) and (6.138), (6.139) are equivalent through the map which follows from $\bar{r} = u^{-1/4} = r_1$. The first equations yield $\bar{p}_2 = \gamma^{-1}r_1r_2$ and the second ones give

$$\bar{d} = -2\gamma^{-1}\left(r_{1_x}r_{2_x} + 2r_1^{-4}r_2^2 + \frac{1}{2}\gamma^2 r_1^{-4} + \frac{1}{2}dr_1^{-2}\right) = -2\gamma^{-1}h(r, r_x, d),$$

where $h(r, r_x, d)$ is the energy integral for the Lagrangian equations generated by L. This map between configurational variables $(r, r_x, d), (\bar{r}, \bar{r}_x, \bar{d})$ extends to a map between canonically conjugate momenta $(s_1, s_2) = s = \delta L/\delta r_x = (r_{2_x}, r_{1_x})$ and $(\bar{s}_1, \bar{s}_2) = \bar{s} = \delta\bar{L}/\delta\bar{r}_x = (\bar{r}_{1_x}, \bar{p}_{2_x})$ and has the form

$$\bar{r}_1 = r_1, \quad \bar{p}_2 = \gamma^{-1}r_1r_2, \quad \bar{s}_1 = s_2, \quad \bar{s}_2 = \gamma^{-1}(r_1s_1 + r_2s_2), \quad \bar{d} = -2\gamma^{-1}h. \tag{6.140}$$

The bi-Hamiltonian structure of (6.138) is derived from the map (6.140), as it is noncanonical. A bi-Hamiltonian formulation for the Newton equations (6.138) is

$$\begin{pmatrix} \bar{r} \\ \bar{s} \\ \bar{d} \end{pmatrix}_x = \bar{\theta}_0 \circ \nabla\bar{h}_1 = \bar{\theta}_1 \circ \nabla\bar{h}_0, \tag{6.141}$$

with the standard 5×5 Poisson matrix (6.10) and

$$\bar{\theta}_1 = \gamma^{-1} \begin{pmatrix} 0 & 0 & 0 & \bar{r}_1 & -2\bar{s}_1 \\ 0 & 0 & \bar{r}_1 & 2\bar{p}_2 & -2\bar{s}_2 \\ 0 & -\bar{r}_1 & 0 & -\bar{r}_1 & 2\partial V/\partial \bar{r}_1 \\ -\bar{r}_1 & -2\bar{p}_2 & \bar{r}_1 & 0 & 2(\partial V/\partial \bar{p}_2 + \bar{d}) \\ 2\bar{s}_1 & 2\bar{s}_2 & -2\partial V/\partial \bar{r}_1 & -2(\partial V/\partial \bar{p}_2 + \bar{d}) & \end{pmatrix}$$

$$(6.142)$$

where $V(\bar{r}_1, \bar{p}_2) = -\gamma \bar{p}_2 \bar{r}_1^{-4}$ is the potential in $\bar{h}_1 = \frac{1}{2}\bar{s}_1^2 + \frac{1}{2}\bar{s}_2^2 + V(\bar{r}_1, \bar{p}_2) + \bar{d}\bar{p}_2$ and $\bar{h}_0 = -\frac{1}{2}\gamma\bar{d}$. The bi-Hamiltonian vector field (6.141) extends to the following bi-Hamiltonian chain

$$\bar{\theta}_0 \circ \nabla \bar{h}_0 = 0, \; \bar{\theta}_0 \circ \nabla \bar{h}_1 = \bar{\theta}_1 \circ \nabla \bar{h}_0, \; \bar{\theta}_0 \circ \nabla \bar{h}_2 = \bar{\theta}_1 \circ \nabla \bar{h}_1, \; 0 = \bar{\theta}_1 \circ \nabla \bar{h}_2, \quad (6.143)$$

where $\bar{h}_2 = -\gamma^{-1}\bar{p}_2\bar{s}_1^2 + \gamma^{-1}\bar{r}_1\bar{s}_2\bar{s}_2 + \frac{1}{2}\bar{r}_1^{-2} + 2\bar{r}_1^{-4}\bar{p}_2^2 + \frac{1}{2}\gamma^{-1}\bar{r}_1^2\bar{d}$ is a Casimir function of $\bar{\theta}_1$. The operator $\bar{\theta}_1$ leads to the following family of integrable two-dimensional potentials. All Hamiltonians $\bar{h}_1 = \frac{1}{2}\bar{s}_1^2 + \frac{1}{2}\bar{s}_2^2 + V(\bar{r}_1, \bar{p}_2) + \bar{d}\bar{p}_2$ for which $V(\bar{r}_1, \bar{p}_2)$ satisfies the equation

$$3\frac{\partial V}{\partial \bar{r}_1} + \bar{r}_1 \left(\frac{\partial^2 V}{\partial \bar{r}_1^2} - \frac{\partial^2 V}{\partial \bar{p}_2^2} \right) + 2\bar{p}_2\frac{\partial^2 V}{\partial \bar{r}_1 \partial \bar{p}_2} = 0 \qquad (6.144)$$

are completely integrable. They admit the bi-Hamiltonian formulation (6.141) and belong to the bi-Hamiltonian chain (6.143). The Casimir function of $\bar{\theta}_1$

$$\bar{h}_2 = -\gamma^{-1}\bar{p}_2\bar{s}_1^2 + \gamma^{-1}\bar{r}_1\bar{s}_2\bar{s}_2 + \gamma^{-1}W(\bar{r}_1, \bar{p}_2) + \frac{1}{2}\gamma^{-1}\bar{r}_1^2\bar{d}, \qquad (6.145)$$

where $W(\bar{r})$ is given by the conditions

$$\frac{\partial W}{\partial \bar{r}_1} = \bar{r}_1\frac{\partial V}{\partial \bar{p}_2} - 2\bar{p}_2\frac{\partial V}{\partial \bar{r}_1}, \quad \frac{\partial W}{\partial \bar{p}_2} = \bar{r}_1\frac{\partial V}{\partial \bar{r}_1} \qquad (6.146)$$

is the second functionally independent integral of motion. The proof is analogous to the one from the fifth KdV example. The same map (6.140) used in the opposite direction provides a bi-Hamiltonian formulation for the system (6.136). The appropriate formulas are reported in [164].

The corresponding Newton and the Ostrogradsky representations are related to each other through canonical transformations. For example

$$\bar{r}_1 = \bar{q}_1^{-1/4}, \; \bar{p}_2 = \frac{1}{2}\frac{\beta}{\gamma}\bar{q}_2, \; \bar{s}_1 = 2\bar{q}_1^{5/4}\bar{p}_1, \; \bar{s}_2 = \frac{2}{\gamma}\bar{p}_2, \; \bar{d} = -\frac{\gamma}{\beta}\bar{c}, \qquad (6.147)$$

where $(\gamma/\beta)^2 = -1/2$.

6.3 Constrained Flows of Lax Equations

Another important special case of the constraint system (6.3) is the one when the u variable can be eliminated. One can do it by the choice of $m \leq 0$, where $K_0 = u_x$ is the symmetry, being the generator of space translations. Moreover, for convenience (to get integrable systems more interesting from the physical point of view), we replace the eigenvalue problem for the recursion operator $\Phi \psi_i = \alpha_i \psi_i, i = 1, ..., N$, by the equivalent (see Sect. 4.1) spectral equation $L(u; \alpha_i)\varphi_i = 0, i = 1, ..., N$. Hence, now, the constraint system (6.3) reads

$$L(u; \alpha_i)\varphi_i = 0, \quad i = 1, ..., N$$

$$K_{m \leq 0}(u) = \sum_{i=1}^{N} \psi_i(\varphi_i). \tag{6.148}$$

The elimination of the u variable from system (6.148) leads to the *nonlinearly constraint spectral equations*

$$L(\varphi; \alpha_i)\varphi_i = 0, \quad i = 1, ..., N. \tag{6.149}$$

In a series of papers [45],[46],[47],[48],[95],[96],[97],[197],[198],[199],[201], a number of classical Hamiltonian integrable systems were identified as resulting from the constraint spectral equations (6.149).

Example 6.14 Constrained Schrödinger spectral problem.
As for the KdV hierarchy, the eigenfunctions of the recursion operator are related to the eigenfunctions of the Schrödinger spectral problem through the relation $\psi_i = (\varphi_i^2)_x$ (see Example 4.1), and, moreover, $K_m(u) = \theta_0 \gamma_m(u) = (\gamma_m)_x$, hence, system (6.148) takes the form

$$\varphi_{k_{xx}} + u\varphi_k = \alpha_k \varphi_k, \quad k = 1, ..., N$$

$$K_{m \leq 0}(u) = \left(\sum_{i=1}^{N} \varphi_i^2\right)_x \Rightarrow \gamma_{m \leq 0} = \sum_{i=1}^{N} \varphi_i^2 + c. \tag{6.150}$$

For the case $m = 0, \gamma_0 = u$ and after eliminating the variable u, we obtain the Garnier system, well known in classical mechanics:

$$\varphi_{k_{xx}} + \varphi_k \sum_{i=1}^{N} \varphi_i^2 + c\varphi_k = \alpha_k \varphi_k, \quad k = 1, ..., N. \tag{6.151}$$

We shall use the notation $\varphi = (\varphi_1, ..., \varphi_N)^T$, $A = \text{diag}(\alpha_1, ..., \alpha_N)$, $(\varphi, \varphi) = \sum_{i=1}^{N} \varphi_i^2$, $(\varphi, A\varphi) = \sum_{i=1}^{N} \alpha_i \varphi_i^2$, etc. Hence, we can rewrite (6.151) in the compact form

$$0 = \varphi_{xx} + [(\varphi, \varphi) + c - A]\varphi = \varphi_{xx} + \frac{\partial V(\varphi)}{\partial \varphi},$$

$$V(\varphi) = \frac{1}{4}(\varphi, \varphi)^2 - \frac{1}{2}(\varphi, A\varphi) + \frac{1}{2}c(\varphi, \varphi), \qquad (6.152)$$

whose Lagrangian is $L = \frac{1}{2}(\varphi_x, \varphi_x) - V(\varphi)$, Hamiltonian $h = \frac{1}{2}(p, p) + V(q)$, where $q = \varphi$ and $p = \frac{\partial L}{\partial \varphi_x} = \varphi_x$, and which has been shown to be a completely integrable Hamiltonian system [45].

As we know from the previous considerations, for the KdV hierarchy there exists an additional cosymmetry $\gamma_{-1} = c = $ const. and hence (6.150) reads

$$\varphi_{xx} + (u - A)\varphi = 0, \quad (\varphi, \varphi) = c. \qquad (6.153)$$

The elimination of u from the system (6.153) $(u = (\varphi_x, \varphi_x) + (\varphi, A\varphi))$ yields the famous Neumann system

$$\varphi_{xx} + c^{-1}[(\varphi_x, \varphi_x) + (\varphi, A\varphi)]\varphi - A\varphi, \quad (\varphi, \varphi) = c, \qquad (6.154)$$

which is completely integrable on the tangent bundle of the sphere TS^N, whose integrability is given by the Moser constraint [142] of an integrable system in $\mathbb{R}^{2N+2}$ restricted on TS^N.

Remark 6.2 The constraint $u_x = \sum_i \psi_i$ is known in the literature as the Bargmann constraint while the constraint $0 = \sum_i \psi_i$ is known as the Neumann constraint.

Example 6.15 Constrained DWW isospectral problem.
Consider the isospectral problem

$$\begin{pmatrix} \varphi_1 \\ \varphi_2 \end{pmatrix}_x = \begin{pmatrix} -\frac{1}{2}\alpha + \frac{1}{2}r & -q \\ 1 & \frac{1}{2}\alpha - \frac{1}{2}r \end{pmatrix} \begin{pmatrix} \varphi_1 \\ \varphi_2 \end{pmatrix} \qquad (6.155)$$

of the DWW hierarchy (Example 3.5). The eigenfunctions $(\psi_1, \psi_2)^T$ of the recursion operator

$$\Phi = \theta_1 \circ \theta_0^{-1} = \frac{1}{2}\begin{pmatrix} -D + r & q + DqD^{-1} \\ 2 & D + DrD^{-1} \end{pmatrix} \qquad (6.156)$$

are related to those of the spectral problem (6.155) in the following way

$$\psi_1 = -(\varphi_1\varphi_2)_x, \quad \psi_2 = (\varphi_2^2)_x. \qquad (6.157)$$

Hence, the Bargmann constraint

$$\begin{pmatrix} q \\ r \end{pmatrix}_x = \sum_{i=1}^{N}\begin{pmatrix} \psi_{1i} \\ \psi_{2i} \end{pmatrix} = \sum_{i=1}^{N}\begin{pmatrix} -(\varphi_{1i}\varphi_{2i}) \\ (\varphi_2^2) \end{pmatrix}_x \qquad (6.158)$$

apply to N copies of the isospectral problem (6.155) for different α_i yields

$$Q_x = -\frac{1}{2}AQ + \frac{1}{2}(P, P)Q + (Q, P)P,$$

$$P_x = Q + \frac{1}{2}AP - \frac{1}{2}(P,P)P, \qquad (6.159)$$

where $Q = (Q_1, ..., Q_N)^{\mathrm{T}}$, $P = (P_1, ..., P_N)^{\mathrm{T}}$ and $(Q_i, P_i)^{\mathrm{T}}$ is the eigenfunction $(\varphi_1, \varphi_2)^{\mathrm{T}}$ related to the eigenvalue α_i. System (6.159) has been shown to be a completely integrable Hamiltonian system [47], where the Hamiltonian function is

$$h = -\frac{1}{2}(AQ, P) + \frac{1}{2}(P,P)(Q,P) - \frac{1}{2}(Q,Q). \qquad (6.160)$$

The Neumann constraint also leads to a completely integrable DWW-Neumann system [47].

Example 6.16 Constrained AKNS spectral problem.
Our last example consists of N copies of the AKNS spectral problem

$$\begin{pmatrix} \varphi_{1k} \\ \varphi_{2k} \end{pmatrix}_x = \begin{pmatrix} -\mathrm{i}\alpha_k & q \\ r & \mathrm{i}\alpha_k \end{pmatrix} \begin{pmatrix} \varphi_{1k} \\ \varphi_{2k} \end{pmatrix}, \quad k = 1, ..., N \qquad (6.161)$$

together with the constraint

$$K_{-1} = \begin{pmatrix} -\mathrm{i}q \\ \mathrm{i}r \end{pmatrix} = \sum_{k=0}^{N} \begin{pmatrix} \psi_{1k} \\ \psi_{2k} \end{pmatrix}. \qquad (6.162)$$

From Example 3.3 we know that $\psi_{1k} = \mathrm{i}\varphi_{1k}^2$, and $\psi_{2k} = \mathrm{i}\varphi_{2k}^2$, hence elimination of the q and r variables from (6.161) leads to the Hamiltonian system

$$\varphi_{1_x} = -\mathrm{i}A\varphi_1 - (\varphi_1, \varphi_1)\varphi_2 = -\frac{\partial h}{\partial \varphi_2},$$

$$\varphi_{2_x} = \mathrm{i}A\varphi_2 - (\varphi_2, \varphi_2)\varphi_1 = \frac{\partial h}{\partial \varphi_1}, \qquad (6.163)$$

where

$$h = (\mathrm{i}A\varphi_1, \varphi_2) + \frac{1}{2}(\varphi_1, \varphi_1)(\varphi_2, \varphi_2), \qquad (6.164)$$

which again happen to be completely integrable [46],[96].

The reader can find other examples of nonlinearly constrained spectral problems in the literature quoted at the beginning of this section. In all these cases, the equations considered happened to be integrable Hamiltonian systems.

The purpose of this section is to outline a construction of the bi-Hamiltonian structure of systems (6.149) and hence the simplest and immediate proof of their integrability in the Liouville sense. First, one should notice that the constraint $K_m(u) = \sum_i \psi_i$ can be presented in two different (but equivalent) ways

$$K_m(u) = \sum_{i=1}^{N} \theta_0(u)\psi_i^{\dagger} = \sum_{i=1}^{N} \theta_1(u)\alpha_i^{-1}\psi_i^{\dagger}, \qquad (6.165)$$

where $\psi_i^\dagger$ are eigenfunctions of the recursion operator $\Phi^\dagger$ for cosymmetries. As $\psi_i^\dagger = \psi_i^\dagger(\varphi_i)$, where φ_i are eigenfunctions of the related isospectral problem, for the case $m \leq 0$, one can resolve (6.165) with respect to the u variable in two different ways and obtain two different (but equivalent) representations of the constrained spectral equations (6.149). They are either directly canonical Hamiltonian systems or are Lagrangian ones. In the latter case, from two different Lagrangian functions, via the Legendre transformation, two canonical Hamiltonian representations can be constructed as well. In both cases, canonical representations are related to each other in the extended phase space by an appropriate Miura map. Finally, the Miura map yields the bi-Hamiltonian structure of each representation, and hence, the hierarchy of constants of motion, necessary to prove the Liouville integrability of the considered system. The method is illustrated with the representative example of the Garnier system. For other examples we refer the reader to the literature.

Construction of the bi-Hamiltonian structure for the Garnier system relies essentially on the existence of two equivalent constraint equations (6.165), corresponding to two local Hamiltonian structures of the KdV equation: $\theta_0 = D$ and $\theta_1 = \frac{1}{4}D^3 + \frac{1}{2}uD + \frac{1}{2}Du$. As was demonstrated in Example 6.13, the first constraint $u_x = \theta_0(\varphi, \varphi)$ leads through the Legendre transformation

$$q_k = \varphi_k, \quad p_k = \frac{\partial L}{\partial \varphi_{k_x}} = \varphi_{k_x}, \quad k = 1, ..., N,$$

$$h_1 = pq_x - L = \frac{1}{2}(p, p) + \frac{1}{4}(q, q)^2 - \frac{1}{2}(q, Aq) + \frac{1}{2}c(q, q)$$

to the canonical Hamiltonian description of the Garnier system (6.152) in the form

$$\begin{pmatrix} q \\ p \end{pmatrix}_x = \begin{pmatrix} 0 & I \\ -I & 0 \end{pmatrix} \nabla h_1, \tag{6.166}$$

where $q = (q_1, ..., q_N)^T$ and $p = (p_1, ..., p_N)^T$. The second constraint condition $u_x = \theta_1(\varphi, A^{-1}\varphi)$ can also be resolved

$$u = \frac{(\varphi, A^{-1}\varphi_{xx}) + (\varphi_x, A^{-1}\varphi_x)}{2W} + \frac{\bar{c} + \frac{1}{4}(\varphi, A^{-1}\varphi_x)^2}{W^2}, \tag{6.167}$$

where $W = 1 - \frac{1}{2}(\varphi, A^{-1}\varphi)$ and $\bar{c}$ is the new constant of integration. Elimination of u from the linear problem (6.152) gives

$$0 = \varphi_{xx}$$

$$+ \left\{ \frac{1}{2W}[(\varphi, A^{-1}\varphi_{xx}) + (\varphi_x, A^{-1}\varphi_x)] + \frac{1}{W^2}[\bar{c} + \frac{1}{4}(\varphi, A^{-1}\varphi_x)^2] - A \right\}\varphi. \tag{6.168}$$

The nonlinear second order term $(\varphi, A^{-1}\varphi_{xx})$ can be eliminated from the equation (6.168) by multiplying it on the left by φA^{-1}, and we get

$$0 = \varphi_{xx} + \left[\frac{1}{2}(\varphi_x, A^{-1}\varphi_x) + \frac{1}{2}(\varphi, \varphi) + \frac{\bar{c} + \frac{1}{4}(\varphi, A^{-1}\varphi_x)^2}{W} - A\right]\varphi. \quad (6.169)$$

This can be written as

$$0 = \varphi_{xx} + [(\varphi, \varphi) + c - A]\varphi + (\bar{E} - c)\varphi, \quad (6.170)$$

where

$$\bar{E} = \frac{1}{2}(\varphi_x, A^{-1}\varphi_x) - \frac{1}{2}(\varphi, \varphi) + \frac{\bar{c} + \frac{1}{4}(\varphi, A^{-1}\varphi_x)^2}{W} \quad (6.171)$$

is the first integral of (6.169). Taking $c = \bar{E}$ we reduce (6.169) to (6.152), that is to the Garnier system. More precisely, the integral curves of (6.169) with a given energy $\bar{E}$ coincide with the integral curves of (6.152) corresponding to the constant c equal to $\bar{E} : c = \bar{E}$.

To interpret this in terms of Hamiltonian structures we observe that (6.168) is a Lagrangian system with the Lagrangian

$$\bar{L} = \frac{1}{2}(\varphi_x, A^{-1}\varphi_x) + \frac{1}{2}(\varphi, \varphi) + \frac{1}{W}\left[\frac{1}{4}(\varphi, A^{-1}\varphi_x)^2 - \bar{c}\right]. \quad (6.172)$$

The standard Legendre transformation

$$\bar{q}_k = \varphi_k, \quad \bar{p}_k = \frac{\partial \bar{L}}{\partial \bar{q}_{k_x}} = \alpha_k^{-1}\varphi_{k_x} + \frac{1}{W}\frac{1}{2}(\varphi, A^{-1}\varphi_x)\alpha_k^{-1}\varphi_k, \quad k = 1, ..., N,$$

$$\bar{h}_1 = (\bar{p}, \varphi_x) - \bar{L} = \frac{1}{2}(\bar{p}, A\bar{p}) - \frac{1}{2}(\bar{p}, \bar{q}) - \frac{1}{2}(\bar{q}, \bar{q}) + \frac{\bar{c}}{W}, \quad (6.173)$$

gives the Hamiltonian description of equation (6.168)

$$\begin{pmatrix} \bar{q} \\ \bar{p} \end{pmatrix}_x = \begin{pmatrix} 0 & I \\ -I & 0 \end{pmatrix} \nabla \bar{h}_1, \quad (6.174)$$

where $\bar{q} = (\bar{q}_1, ..., \bar{q}_N)^{\mathrm{T}}$ and $\bar{p} = (\bar{p}_1, ..., \bar{p}_N)^{\mathrm{T}}$. The equality $c = \bar{E}(\varphi, \varphi_x, \bar{c})$ together with (6.173) is interpreted as the Miura map $(q, p, c) = \phi(\bar{q}, \bar{p}, \bar{c})$ between extended phase spaces, given by

$$q = \varphi = \bar{q}, \quad p = \varphi_x = A\bar{p} - \frac{1}{2}(\bar{p}, \bar{q})\bar{q},$$

$$c = \bar{E}(\varphi, \varphi_x, \bar{c}) = \frac{1}{2}(\bar{p}, A\bar{p}) - \frac{1}{2}(\bar{p}, \bar{q}) - \frac{1}{2}(\bar{q}, \bar{q}) + \frac{\bar{c}}{W} = \bar{h}_1, \quad (6.175)$$

which transforms integral curves of the Hamiltonian system

$$\begin{pmatrix} \bar{q} \\ \bar{p} \\ \bar{c} \end{pmatrix}_x = \begin{pmatrix} 0 & I & 0 \\ -I & 0 & 0 \\ 0 & 0 & 0 \end{pmatrix} \nabla \bar{h}_1 = \bar{\theta}_0 \circ \nabla \bar{h}_1 \quad (6.176)$$

into those of the Hamiltonian system

$$\begin{pmatrix} q \\ p \\ c \end{pmatrix}_x = \begin{pmatrix} 0 & I & 0 \\ -I & 0 & 0 \\ 0 & 0 & 0 \end{pmatrix} \nabla h_1 = \bar{\theta}_0 \circ \nabla h_1. \tag{6.177}$$

Map (6.175) transports the canonical Poisson structure $\bar{\theta}_0$ of system (6.176) into a noncanonical one for system (6.177) [10]

$$\theta_1 = \phi' \circ \bar{\theta}_0 \circ \phi'^T = \begin{pmatrix} 0 & A - \frac{1}{2}q \otimes q & p \\ -A + \frac{1}{2}q \otimes q & \frac{1}{2}p \otimes q - \frac{1}{2}q \otimes p & [A - c - (q,q)]q \\ -p^T & [-A + c + (q,q)]q^T & 0 \end{pmatrix} \tag{6.178}$$

where $\otimes$ denotes the tensor product. The Garnier system becomes bi-Hamiltonian

$$\begin{pmatrix} q \\ p \\ c \end{pmatrix}_x = \begin{pmatrix} p \\ [A - c - (q,q)]\, q \\ 0 \end{pmatrix} = \theta_0 \circ \nabla h_1 = \theta_1 \circ \nabla h_0, \tag{6.179}$$

with $h_0 = \bar{h}_1 \cdot \phi^{-1} = c$. Starting with Garnier Hamiltonians h_0 and h_1, the chain equation

$$\theta_0 \circ \nabla h_{k+1} = \theta_1 \circ \nabla h_k \tag{6.180}$$

generates an infinite sequence of Hamiltonian functions

$$h_{k+1} = \frac{1}{2}(p, A^k p) - \frac{1}{2}(q, A^{k+1}q) + \frac{1}{4}(q,q)(q, A^k q) + \frac{1}{2}c(q, A^k q)$$

$$+ \frac{1}{4} \sum_{j=1}^{k} \left[(q, A^{j-1}q)(p, A^{k-j}p) - (p, A^{j-1}q)(p, A^{k-j}q) \right], \quad k = 0, 1, ... \tag{6.181}$$

Only the first $N + 1$ of these Hamiltonians are functionally independent but it is, of course, enough to prove Liouville integrability of any of the corresponding flows. Having the basic sequence (6.181) it is possible to construct a finite bi-Hamiltonian chain starting and terminating with Casimirs of θ_0 and θ_1, respectively. The details are given in Sect. 6.5.

The reader can find other examples of the bi-Hamiltonian structure of nonlinearly constrained isospectral problems in [13],[200].

Natural extensions of the constrained system (6.148) have the following form

$$(\varphi_i)_{t_n} = B_n(u; \lambda_i)\varphi_i, \qquad i = 1, ..., N$$

$$K_{m \leq 0}(u) = \sum_{i=1}^{N} \psi_i(\varphi_i), \tag{6.182}$$

where the first N equations of (6.182) represent the time evolution part of the eigenfunctions of the underlying isospectral problem. A few questions appear concerning the constraint system (6.182). The main one deals with the possible existence of bi-Hamiltonian structure. In the case of a positive answer, the next question is whether the bi-Hamiltonian hierarchy of (6.182) is something essentially new when compared with the bi-Hamiltonian hierarchy of (6.148). To get some insight into this problem we shall reconsider the KdV case with the Bargmann constraint, but now in a slightly different way [27].

Let us rewrite the Schrödinger spectral problem together with the time evolution of its eigenfunctions in the matrix form (see Example 4.6)

$$\begin{pmatrix} \varphi_1 \\ \varphi_2 \end{pmatrix}_{t_n} = L_n \begin{pmatrix} \varphi_1 \\ \varphi_2 \end{pmatrix}, \qquad \varphi_1 = \varphi, \ \varphi_2 = \varphi_x, \tag{6.183}$$

$$L_n = \begin{pmatrix} -\frac{1}{2}\alpha_x^{(n)} & \alpha^{(n)} \\ -\frac{1}{2}\alpha_{xx}^{(n)} + \alpha^{(n)}(\lambda - u) & \frac{1}{2}\alpha_x^{(n)} \end{pmatrix}, \qquad \alpha^{(n)} = \sum_{k=0}^{n} \frac{1}{2}\gamma_{k-1}\alpha^{n-k},$$

$$\gamma_{-1} = 2, \quad \gamma_{k+1} = \Phi^\dagger\gamma_k, \quad k \geq 0, \quad \Phi^\dagger = \frac{1}{4}D^2 + u - \frac{1}{2}D^{-1}u_x, \tag{6.184}$$

where γ_k are cosymmetries of the KdV, related to the KdV hierarchy by $K_n = (\gamma_n)_x$, and $\Phi^\dagger$ is the recursion operator for cosymmetries. Note that for $t_0 = x$, (6.183) is equivalent to the Schrödinger spectral problem. In the representation (6.183), the KdV hierarchy appears as the zero-curvature condition

$$(L_0)_{t_n} = (L_n)_x + [L_n, L_0]. \tag{6.185}$$

Next let us consider the MKdV hierarchy $v_{t_n} = \overline{K}_n(v)$ and the corresponding Lax hierarchy [97]

$$\begin{pmatrix} \psi_1 \\ \psi_2 \end{pmatrix}_{t_n} = \overline{L}_n \begin{pmatrix} \psi_1 \\ \psi_2 \end{pmatrix}, \tag{6.186}$$

where

$$\overline{L}_n = \sum_{k=0}^{n} \begin{pmatrix} c_k & \frac{1}{2}\alpha(b_k + a_k) \\ \frac{1}{2}(b_k - a_k) & -c_k \end{pmatrix} \alpha^{n-k}, \tag{6.187}$$

with

$$a_0 = 0, \ b_0 = 2, \ c_k = -\overline{\gamma}_k(v), \ a_k = -\overline{K}_{k-1}(v), \ b_k = 2D^{-1}v\overline{K}_{k-1}(v),$$

$$\overline{\gamma}_{k+1}(v) = \overline{\Phi}^\dagger\overline{\gamma}_k(v), \quad \overline{\Phi}^\dagger = \frac{1}{4}D^2 - vD^{-1}vD,$$

where again $\overline{\gamma}_k(v)$ are cosymmetries of the MKdV equation related to the hierarchy $\overline{K}_n(v)$ by $\overline{K}_n(v) = -(\overline{\gamma}_k(v))_x$ and $\overline{\Phi}^\dagger$ is the MKdV recursion operator for cosymmetries. Moreover, for $t_0 = x$, (6.186) is the MKdV spectral problem

$$\begin{pmatrix} \psi_1 \\ \psi_2 \end{pmatrix}_x = \begin{pmatrix} v & \alpha \\ 1 & -v \end{pmatrix} \begin{pmatrix} \psi_1 \\ \psi_2 \end{pmatrix}, \tag{6.188}$$

and for $t_n \neq t_0$, (6.186) gives the time evolutions for eigenfunctions of the spectral problem (6.188), which generate the MKdV hierarchy through the zero-curvature condition

$$(\overline{L}_0)_{t_n} = (\overline{L}_n)_x + [\overline{L}_n, \overline{L}_0]. \tag{6.189}$$

Lemma 6.3 *The gauge transformation*

$$\begin{pmatrix} \varphi_1 \\ \varphi_2 \end{pmatrix} = \begin{pmatrix} 1 & 0 \\ v & \alpha \end{pmatrix} \begin{pmatrix} \psi_1 \\ \psi_2 \end{pmatrix} \tag{6.190}$$

and the Miura map $M : u = -v^2 - v_x$ transform the MKdV Lax hierarchy (6.186) into the KdV Lax hierarchy (6.183).

Proof. This is done by direct calculation, using the relations

$$K_n(u)_{|u=M(v)} = -(D + 2v)\overline{K}_n(v)$$

and

$$\gamma_n(u)_{|u=M(v)} = (D + 2D^{-1}vD)\overline{\gamma}_n(v),$$

which follow easily from the Miura map and the connection between the cosymmetries $\gamma_n(u)$ and the vector fields $K_n(n)$. □

Again, let $(.,.)$ denote the standard inner product in $\mathbb{R}^N$ and put $\varphi_{1k} = q_k, \varphi_{2k} = p_k, q = (q_1, ..., q_N)^{\mathrm{T}}, p = (p_1, ..., p_N)^{\mathrm{T}}, \; q_k, p_k \in \mathbb{R}, A = \operatorname{diag}(\alpha_1, ..., \alpha_N)$. Then the Lax hierarchy (6.183) with $\alpha = \alpha_k, k = 1, ..., N$, is given by the following vector equations

$$q_{t_n} = A^n p + \sum_{k=1}^{n} \left[-\frac{1}{4}(\gamma_{k-1})_x A^{n-k} q + \frac{1}{2}\gamma_{k-1} A^{n-k} p \right],$$

$$p_{t_n} = A^{n+1}p - uA^n q + \frac{1}{4}\sum_{k=1}^{n} \{[-(\gamma_{k-1})_{xx} - 2u\gamma_{k-1}]A^{n-k}q + 2\gamma_{k-1}A^{n-k+1}q$$

$$+ (\gamma_{k-1})_x A^{n-k} p\}. \tag{6.191}$$

In this way we obtain a hierarchy of systems. The same procedure with the following substitutions, $\psi_{1k} = \overline{q}_k, \psi_{2k} = \overline{p}_k, \; \overline{q} = (\overline{q}_1, ..., \overline{q}_N)^{\mathrm{T}}, \overline{p} = (\overline{p}_1, ..., \overline{p}_N)^{\mathrm{T}}, \; \overline{q}_k, \overline{p}_k \in \mathbb{R}, A = \operatorname{diag}(\alpha_1, ..., \alpha_N)$, gives us the MKdV Lax hierarchy (6.187) in the following vector equations

$$\overline{q}_{t_n} = vA^n\overline{q} + A^{n+1}\overline{p}$$

$$+ \sum_{k=1}^{n} \left\{ -\overline{\gamma}_k A^{n-k}\overline{q} + \frac{1}{2}[2D^{-1}v(\overline{\gamma}_{k-1})_x + (\overline{\gamma}_{k-1})_x]A^{n-k+1}\overline{p} \right\},$$

$$\bar{p}_{t_n} = -vA^n\bar{p} + A^n\bar{q}$$

$$+ \sum_{k=1}^{n} \left\{ \frac{1}{2} [2D^{-1}v(\bar{\gamma}_{k-1})_x - (\bar{\gamma}_{k-1})_x] A^{n-k}\bar{q} + \bar{\gamma}_k A^{n-k}\bar{p} \right\}. \tag{6.192}$$

Now, let us impose the following constraint on the KdV Lax hierarchy (6.191)

$$u = (q,q) + c, \quad \gamma_k = (A^k p, q), \quad k = 1, ..., N, \quad c = \text{const.} \tag{6.193}$$

The first system of the hierarchy (6.191), namely $t_0 = x$, is

$$q_x = p, \quad p_x = Aq - (q,q)q - cq \tag{6.194}$$

and is just the Garnier system (6.152) in our new variables. Equations (6.193), (6.194) give us the following relations

$$(\gamma_{k-1})_x = 2(A^{k-1}q, p),$$

$$(\gamma_{k-1})_{xx} = 2(A^{k-1}p, p) + 2(A^k q, q) - 2(q,q)(A^{k-1}q, q) - 2c(A^{k-1}q, q),$$

$$-\frac{1}{4}(\gamma_{k-1})_{xx} - \frac{1}{2}u\gamma_{k-1} = -\frac{1}{2}(A^{k-1}p, p) - \frac{1}{2}(A^k q, q). \tag{6.195}$$

In this way, the constraint (6.193) turns the KdV Lax hierarchy (6.191) into the following hierarchy of finite dimensional dynamical systems

$$q_{t_n} = A^n p + \frac{1}{2}\sum_{k=1}^{n} \left[-(A^{k-1}q, p)A^{n-k}q + (A^{k-1}q, q)A^{n-k}p \right],$$

$$p_{t_n} = A^{n+1}q - \frac{1}{2}(q,q)A^n q - \frac{1}{2}(A^n q, q)q - cA^n q$$

$$+ \frac{1}{2}\sum_{k=1}^{n} \left[-(A^{k-1}p, p)A^{n-k}q + (A^{k-1}q, p)A^{n-k}p \right], \tag{6.196}$$

which we can write as the Hamiltonian system

$$\begin{pmatrix} q \\ p \\ c \end{pmatrix}_{t_n} = \begin{pmatrix} 0 & I & 0 \\ -I & 0 & 0 \\ 0 & 0 & 0 \end{pmatrix} \nabla h_n, \tag{6.197}$$

with

$$h_{k+1} = \frac{1}{2}(p, A^k p) - \frac{1}{2}(q, A^{k+1}q) + \frac{1}{4}(q,q)(q, A^k q) + \frac{1}{2}c(q, A^k q)$$

$$+ \frac{1}{4}\sum_{j=1}^{k} \left[(q, A^{j-1}q)(p, A^{k-j}p) - (p, A^{j-1}q)(p, A^{k-j}q) \right], \quad k = 0, 1, ...;$$

$$\tag{6.198}$$

but this is precisely the Garnier hierarchy (6.181).

Next we impose the following constraints on the MKdV Lax hierarchy 6.192)

$$v = -\frac{1}{2}(\bar{q},\bar{p}) + \bar{c}, \quad \bar{\gamma}_k = \frac{1}{2}(A^k\bar{q},\bar{p}), \quad c = \text{const}, \quad k = 1, ..., N, \quad (6.199)$$

and with these constraints the first Lax equation in (6.192) becomes the modified Garnier system

$$\bar{q}_x = -\frac{1}{2}(\bar{q},\bar{p})\bar{q} + A\bar{p} + \bar{c}\bar{q}, \quad \bar{p}_x = \frac{1}{2}(\bar{q},\bar{p}) + \bar{q} - \overline{cp}. \quad (6.200)$$

Then (6.199),(6.200) give the following relations

$$(\bar{\gamma}_{k-1})_x = \frac{1}{2}(A^k\bar{p},\bar{p}) + \frac{1}{2}(A^{k-1}\bar{q},\bar{q}), \quad 2D^{-1}v(\bar{\gamma}_{k-1})_x + (\bar{\gamma}_{k-1})_x = (A^{k-1}\bar{q},\bar{q}),$$

$$-2D^{-1}v(\bar{\gamma}_{k-1})_x + (\bar{\gamma}_{k-1})_x = (A^{k-1}\bar{p},\bar{p}). \quad (6.201)$$

Using these relations, the constraints (6.199) applied to the MKdV Lax hierarchy (6.192) give the following hierarchy of finite dimensional systems

$$\bar{q}_{t_n} = -\frac{1}{2}(\bar{q},\bar{p})A^n\bar{q} + A^{n+1}\bar{p} + \bar{c}A^n\bar{q}$$

$$+\frac{1}{2}\sum_{k=1}^{n}\left[(A^{k-1}\bar{q},\bar{q})A^{n-k+1}\bar{p} - (A^k\bar{q},\bar{p})A^{n-k}\bar{q}\right],$$

$$\bar{p}_{t_n} = \frac{1}{2}(\bar{q},\bar{p})A^n\bar{p} + A^n\bar{q} - \bar{c}A^n\bar{p} + \frac{1}{2}\sum_{k=1}^{n}\left[(A^k\bar{q},\bar{p})A^{n-k}\bar{p} - (A^k\bar{p},\bar{p})A^{n-k}\bar{q}\right],$$

$$(6.202)$$

which can also be written as

$$\bar{q}_{t_n} = -\frac{1}{2}(A^n\bar{q},\bar{q})\bar{p} + A^{n+1}\bar{p} + \bar{c}A^n\bar{q} + \frac{1}{2}\sum_{k=0}^{n}\left[(A^k\bar{q},\bar{q})A^{n-k}\bar{p} - (A^k\bar{q},\bar{p})A^{n-k}\bar{q}\right],$$

$$\bar{p}_{t_n} = \frac{1}{2}(\bar{p},\bar{p})A^n\bar{q} + A^n\bar{q} - \bar{c}A^n\bar{p} + \frac{1}{2}\sum_{k=0}^{n}\left[(A^k\bar{q},\bar{p})A^{n-k}\bar{p} - (A^k\bar{p},\bar{p})A^{n-k}\bar{q}\right].$$

$$(6.203)$$

This is also a Hamiltonian hierarchy

$$\begin{pmatrix} \bar{q} \\ \bar{p} \\ \bar{c} \end{pmatrix}_{t_n} = \begin{pmatrix} 0 & I & 0 \\ -I & 0 & 0 \\ 0 & 0 & 0 \end{pmatrix} \nabla \bar{h}_n, \quad (6.204)$$

where

$$\bar{h}_{n+1} = -\frac{1}{4}(p,p)(A^n q, q) - \frac{1}{2}(A^n q, q) + \frac{1}{2}(A^{n+1}p, p) + c(A^n q, p)$$

$$+ \frac{1}{4}\sum_{k=0}^{n}\left[(A^{n-k}q, q)(A^k p, p) - (A^{n-k}q, p)(A^k q, p)\right]. \tag{6.205}$$

Our last step is to construct the Miura map which relates the two hierarchies.

Lemma 6.4 *The Miura map relating the MKdV Lax hierarchy (6.192) under constraint (6.199) with the KdV Lax hierarchy (6.191) under constraint (6.193) is*

$$M: \quad q = \bar{q}, \quad p = -\frac{1}{2}(\bar{q},\bar{p})\bar{q} + \bar{c}\bar{q} + A\bar{p},$$

$$\bar{c} = \frac{1}{2}(A\bar{p},\bar{p}) - \frac{1}{2}(\bar{q},\bar{p}) - \frac{1}{4}(\bar{q},\bar{p})^2 + \bar{c}(\bar{q},\bar{p}) - \bar{c}^2, \tag{6.206}$$

and the Hamiltonians $h_n, \bar{h}_n$ of (6.198),(6.205) satisfy

$$\bar{h}_{n+1}(\bar{q},\bar{p},\bar{c}) = h_n(q,p,c)_{|(6.206)}, \quad n = 0, 1, ..., N. \tag{6.207}$$

Proof. Apply (6.193),(6.199) to the gauge (6.190) and to the Miura map $u = -v^2 - v_x$. The property of being a Miura map follows from Lemma 6.3, and the relationship between Hamiltonians is obtained by direct (if involved) calculations. □

Now, the second Hamiltonian structure for each equation of the constrained KdV Lax hierarchy has the same form, and is given by

$$M' \circ \bar{\theta}_0 \circ M'^T = \begin{pmatrix} 0 & A - \frac{1}{2}q \otimes q & p \\ -A + \frac{1}{2}q \otimes q & \frac{1}{2}p \otimes q - \frac{1}{2}q \otimes p & [A - c - (q,q)]q \\ -p^T & [-A + c + (q,q)]q^T & 0 \end{pmatrix} \tag{6.208}$$

and is equal to the second Hamiltonian structure of the Garnier system.

The above results show that all constrained higher order Lax equations of the KdV are integrable bi-Hamiltonian finite dimensional systems. Furthermore, we do not obtain any new systems from these restrictions, in the sense that they belong to the bi-Hamiltonian chain of the Garnier system. We have exploited the Miura map between the MKdV and KdV equations, but the method works for other pairs of equations related by Miura maps: if the first field equation is mapped onto the second by a Miura map, then the presented results suggest that in other cases the constraints of the higher order Lax equations (6.182) do not produce any finite dimensional dynamical systems which do not already belong to the bi-Hamiltonian family obtained from the constrained first Lax equation (6.148).

6.4 Restricted Flows of Infinite Systems

The natural generalization of the constrained systems (6.148), described in the previous section, is the set of equations

$$L(u; \alpha_i)\varphi_i = 0, \qquad i = 1, ..., N$$

$$K_{m>0}(u) = \theta_0 \left(\sum_{i=1}^{N} \frac{\delta \alpha_i}{\delta u} \right) = \theta_0 \sum_{i=1}^{N} \psi_i(\varphi) \qquad (6.209)$$

known as *restricted flows* [10] of the underlying field system. Contrary to constrained flows, in this case we cannot now eliminate the u variable.

Example 6.17 Restricted flows of the KdV hierarchy: Newton representation. All restricted flows (6.209) of the KdV hierarchy written as

$$0 = \varphi_{k_{xx}} + (u - \alpha_k)\varphi_k, \qquad k = 1, ..., N$$

$$\gamma_m(u) = \sum_{i=1}^{N} \varphi_i^2 + c, \qquad m > 0 \qquad (6.210)$$

admit the Newton representation (6.100) if c in the last equation of (6.100) is replaced by $\sum_{i=1}^{N} \varphi_i^2 + c$ and u in the spectral problem is expressed by $4r_1$. Equations (6.210) follow from the Lagrangian

$$L_{m+1}(r, r_x, \varphi, \varphi_x, c) = L_{m+1}(r, r_x, c) - \frac{1}{2} \left(\sum_{k=1}^{N} \varphi_{k_x}^2 \right) + 2r_1 \left(\sum_{k=1}^{N} \varphi_k^2 \right)$$

$$- \frac{1}{2} \left(\sum_{k=1}^{N} \alpha_k \varphi_k^2 \right). \qquad (6.211)$$

For $m = 2$ and $N = 2$ we get

$$\begin{aligned}
0 &= \varphi_{1_{xx}} + 4r_1\varphi_1 - \alpha_1\varphi_1, \\
0 &= \varphi_{2_{xx}} + 4r_1\varphi_2 - \alpha_2\varphi_2, \\
0 &= r_{1_{xx}} - 4r_2 + 10r_1^2, \\
0 &= r_{2_{xx}} + 20r_1r_2 - 10r_1^3 - c - \varphi_1^2 - \varphi_2^2,
\end{aligned} \qquad (6.212)$$

with the Lagrangian

$$L = -(r_1)_x(r_2)_x - \frac{1}{2}(\varphi_1)_x(\varphi_2)_x + 10r_1^2r_2 - 2r_2^2 - \frac{5}{2}r_1^4 - cr_1$$

$$+ 2r_1\varphi_1^2 + 2r_1\varphi_2^2 - \frac{1}{2}\alpha_1\varphi_1^2 - \frac{1}{2}\alpha_2\varphi_2^2.$$

Example 6.18 Restricted flows of the HD hierarchy: Newton representation. The spectral problem related to the hierarchy of the Harry–Dym equation (3.32) takes the form

$$\varphi_{xx} + \alpha u \varphi = 0, \tag{6.213}$$

with the following square eigenfunction relation

$$\theta_0(\alpha \varphi^2) = \theta_1(\varphi^2). \tag{6.214}$$

As $\ker \theta_0 = du^{-1/2}$, the restricted flows (6.210) take the form

$$0 = \varphi_{k_{xx}} + \alpha_k u \varphi_k, \qquad k = 1, ..., N$$

$$\gamma_m + \sum_{i=1}^{N} \varphi_i^2 = du^{-1/2}. \tag{6.215}$$

Presenting $\gamma_m = 0$ in the Newton form (where $u = r_1^{-4}$) [167] with Lagrangian $L_m(r, r_x, d)$, the restricted flow (6.215) follows from the Lagrangian

$$L_m(r, r_x, \varphi, \varphi_x, d) = L_m(r, r_x, d) - \frac{1}{2}\left(\sum_{k=1}^{N} \varphi_{k_x}^2 - r_1^{-4}\alpha_k \varphi_k^2\right). \tag{6.216}$$

The Newton representation of the stationary flow with $N = 1$, related to the stationary flow (6.136), takes the form

$$\varphi_{1_{xx}} = -\alpha_1 r_1^{-4}\varphi_1,$$

$$r_{1_{xx}} = -4r_1^{-4}r_2,$$

$$r_{2_{xx}} = 8r_1^{-5}r_2^2 + 2\gamma^2 r_1^{-5} + dr_1^{-3} - 2\alpha_1 r_1^{-5}\varphi_1^2, \tag{6.217}$$

with

$$L = (r_1)_x(r_2)_x - \frac{1}{2}(\varphi_1)_x^2 - 2r_1^{-4}r_2^2 - \frac{1}{2}\gamma^2 r_1^{-4} - \frac{1}{2}dr_1^{-2} + \frac{1}{2}\alpha_1 r_1^{-4}\varphi_1^2.$$

More examples of restricted flows related to the AKNS, coupled KdV and coupled HD hierarchies can be found in [163] and [12].

Again we are interested in constructing a bi-Hamiltonian representation of restricted flows. It is not yet clear whether one can construct them in the arbitrary case; nevertheless for some cases we know how to do it. We illustrate the method on the simplest example of restricted flows of the KdV [31], where once more the Miura map plays the crucial role in the construction.

Our first goal is to find a Hamiltonian formulation of the dynamics (6.210). To do this let us rewrite them in a more convenient form

$$\begin{pmatrix} \varphi_{1k} \\ \varphi_{2k} \end{pmatrix}_x = \begin{pmatrix} 0 & 1 \\ \alpha_k - u & 0 \end{pmatrix}\begin{pmatrix} \varphi_{1k} \\ \varphi_{2k} \end{pmatrix}, \qquad \varphi_1 = \varphi, \; \varphi_2 = \varphi_x, \; k = 1, ..., N,$$

$$\gamma_m(u) = \sum_{i=1}^{N} \varphi_{1i}^2 + c. \tag{6.218}$$

where γ_m are cosymmetries of the KdV, such that $K_m = \theta_0 \gamma_m = (\gamma_m)_x$ and c is a constant of integration. Before we treat this system let us consider for a moment an uncoupled version

$$\left(\begin{array}{c} \varphi_{1k} \\ \varphi_{2k} \end{array} \right)_x = \left(\begin{array}{cc} 0 & 1 \\ \alpha_k & 0 \end{array} \right) \left(\begin{array}{c} \varphi_{1k} \\ \varphi_{2k} \end{array} \right), \qquad k = 1, ..., N,$$

$$\gamma_m(u) = c. \tag{6.219}$$

The first system is the asymptotic $(u \to 0)$ KdV spectral problem and the second one is the m-th stationary flow of the KdV. Both of them can be put into a Hamiltonian form. Let $\varphi_{1k} = q_k, \varphi_{2k} = p_k, q = (q_1, ..., q_N)^{\mathrm{T}}, p = (p_1, ..., p_N)^{\mathrm{T}}$ and $A = \mathrm{diag}(\alpha_1, ..., \alpha_N)$. Then

$$\left(\begin{array}{c} q \\ p \end{array} \right)_x = \left(\begin{array}{cc} 0 & I_N \\ -I_N & 0 \end{array} \right) \nabla \left[\frac{1}{2}(p, p) - \frac{1}{2}(q, Aq) \right] = \theta_0 \circ \nabla h_L(q, p). \tag{6.220}$$

On the other hand the stationary flow $\gamma_m = c$ can be put into a Hamiltonian form according to the procedure presented in the first section. Actually

$$\left(\begin{array}{c} r \\ s \\ c \end{array} \right)_x = \left(\begin{array}{c} K_{rm} \\ K_{sm} \\ 0 \end{array} \right) = \left(\begin{array}{ccc} 0 & I_m & 0 \\ -I_m & 0 & 0 \\ 0 & 0 & 0 \end{array} \right) \nabla h_m(r, s, c), \tag{6.221}$$

where $r = (r_1, ..., r_m)^{\mathrm{T}}$, $s = (s_1, ..., s_m)^{\mathrm{T}}$ represent Ostrogradsky variables.

Now, let us introduce the coupling between both systems through the interacting term in the Hamiltonian

$$h_{\mathrm{int}} = \frac{1}{2} r_1(q, q). \tag{6.222}$$

This gives the following new Hamiltonian system

$$\left(\begin{array}{c} q \\ r \\ p \\ s \\ c \end{array} \right)_x = \left(\begin{array}{c} p \\ K_{rm} \\ (A - r_1 I_N)q \\ K_{sm} + K_{\mathrm{int}} \\ 0 \end{array} \right) = \left(\begin{array}{ccccc} 0 & 0 & I_N & 0 & 0 \\ 0 & 0 & 0 & I_m & 0 \\ -I_N & 0 & 0 & 0 & 0 \\ 0 & -I_m 0 & 0 & 0 & 0 \\ 0 & 0 & 0 & 0 & 0 \end{array} \right) \nabla h_{N,m} \tag{6.223}$$

where

$$K_{\mathrm{int}} = \left(-\frac{1}{2}(q, q), 0, ..., 0 \right)^{\mathrm{T}}, \quad h_{N,m} = h_L + h_m + h_{\mathrm{int}}, \tag{6.224}$$

which is just the first Hamiltonian formulation of the restricted flow (6.218). Note that in the case when $m = 1$, $\gamma_1(u) = \frac{1}{4} u_{xx} + \frac{3}{4} u^2$ the restricted flow (6.210) is equivalent to the stationary flow of the Melnikov system [135] also known as the KdV with sources.

The candidates for a modified system are of course restricted flows of the MKdV hierarchy

$$
\begin{pmatrix} \psi_{1k} \\ \psi_{2k} \end{pmatrix}_x = \begin{pmatrix} v & \alpha_k \\ 1 & -v \end{pmatrix} \begin{pmatrix} \psi_{1k} \\ \psi_{2k} \end{pmatrix}, \quad k = 1, ..., N
$$

$$
\bar{\theta}_0 \left(-\frac{1}{2} \sum_{i=1}^{N} \psi_{1i}\psi_{2i} \right) = \overline{K}_m(v), \quad \bar{\theta}_0 = -D_x. \tag{6.225}
$$

First, let us rewrite system (6.225) in a more convenient form, integrating the second equation once

$$
\begin{pmatrix} \psi_{1k} \\ \psi_{2k} \end{pmatrix}_x = \begin{pmatrix} v & \alpha_k \\ 1 & -v \end{pmatrix} \begin{pmatrix} \psi_{1k} \\ \psi_{2k} \end{pmatrix}, \quad k = 1, ..., N
$$

$$
\bar{\gamma}_m(v) = -\frac{1}{2} \sum_{i=1}^{N} \psi_{1i}\psi_{2i} + \bar{c}, \tag{6.226}
$$

where $\bar{\gamma}_m(v)$ are cosymmetries of the MKdV and $\bar{c}$ is a constant of integration. Again, let us write down the uncoupled case

$$
\begin{pmatrix} \psi_{1k} \\ \psi_{2k} \end{pmatrix}_x = \begin{pmatrix} 0 & \alpha_k \\ 1 & 0 \end{pmatrix} \begin{pmatrix} \psi_{1k} \\ \psi_{2k} \end{pmatrix}, \quad k = 1, ..., N
$$

$$
\bar{\gamma}_m(v) = \bar{c}, \tag{6.227}
$$

where the first equation represents the asymptotic MKdV spectral problem and the second one the MKdV stationary flows. Let $\psi_{1k} = \bar{q}_k, \psi_{2k} = \bar{p}_k$, $\bar{q} = (\bar{q}_1, ..., \bar{q}_N)^T, \bar{p} = (\bar{p}_1, ..., \bar{p}_N)^T$ and $A = \mathrm{diag}(\alpha_1, ..., \alpha_N)$. Then the Hamiltonian form of the asymptotic spectral problem is the following

$$
\begin{pmatrix} \bar{q} \\ \bar{p} \end{pmatrix}_x = \begin{pmatrix} A\bar{p} \\ \bar{q} \end{pmatrix} = \begin{pmatrix} 0 & I_N \\ -I_N & 0 \end{pmatrix} \nabla \left[\frac{1}{2}(\bar{p}, A\bar{p}) - \frac{1}{2}(\bar{q}, \bar{q}) \right]
$$

$$
= \bar{\theta}_0 \circ \nabla \bar{h}_L(\bar{q}, \bar{p}). \tag{6.228}
$$

We can do the same with the stationary flows. Applying the procedure from the first section we can transform $\bar{\gamma}_m(v) = \bar{c}$ to the Hamiltonian form

$$
\begin{pmatrix} \bar{r} \\ \bar{s} \\ \bar{c} \end{pmatrix}_x = \begin{pmatrix} \overline{K}_{rm} \\ \overline{K}_{sm} \\ 0 \end{pmatrix} = \begin{pmatrix} 0 & I_m & 0 \\ -I_m & 0 & 0 \\ 0 & 0 & 0 \end{pmatrix} \nabla \bar{h}_m(\bar{r}, \bar{s}, \bar{c}), \tag{6.229}
$$

where $\bar{r} = (\bar{r}_1, ..., \bar{r}_m)^T, \bar{s} = (\bar{s}_1, ..., \bar{s}_m)^T$ are the Ostrogradsky variables. Now, we shall again introduce the coupling between both systems through the interacting term in the Hamiltonian

$$
\bar{h}_{\mathrm{int}} = \bar{r}_1(\bar{q}, \bar{p}) \tag{6.230}
$$

which leads to the following Hamiltonian dynamics

$$
\begin{pmatrix} \bar{q} \\ \bar{r} \\ \bar{p} \\ \bar{s} \\ \bar{c} \end{pmatrix}_x = \begin{pmatrix} A\bar{p} + \bar{r}_1\bar{q} \\ \overline{K}_{rm} \\ \bar{q} - \bar{r}_1\bar{p} \\ \overline{K}_{sm} + \overline{K}_{int} \\ 0 \end{pmatrix} = \begin{pmatrix} 0 & 0 & I_N & 0 & 0 \\ 0 & 0 & 0 & I_m & 0 \\ -I_N & 0 & 0 & 0 & 0 \\ 0 & -I_m & 0 & 0 & 0 \\ 0 & 0 & 0 & 0 & 0 \end{pmatrix} \nabla \bar{h}_{N,m},
$$

$$(6.231)$$

where

$$
\overline{K}_{int} = (-(\bar{q}, \bar{p}), 0, ..., 0)^T, \quad \bar{h}_{N,m} = \bar{h}_L + \bar{h}_m + \bar{h}_{int},
$$

and which is equivalent to the modified restricted flow (6.226).

Now, to derive the second Hamiltonian formulation for the considered system we have to find an explicit form of the Miura map relating the modified restricted flow (6.226) to the given one (6.218). The Miura map between uncoupled systems has the well known form

$$
M: \quad \begin{aligned} q &= \bar{q}, \\ p &= A\bar{p}, \\ r &= M_{rm}(\bar{r}, \bar{s}, \bar{c}), \\ s &= M_{sm}(\bar{r}, \bar{s}, \bar{c}), \\ c &= M_{cm}(\bar{r}, \bar{s}, \bar{c}), \end{aligned} \quad\quad (6.232)
$$

where the first two equalities come from the gauge relation between the asymptotics of the spectral equations and the other equalities are a consequence of the Miura map $u = -v^2 - v_x$ between the fields v and u.

Lemma 6.5 *The Miura map relating the modified coupled system (6.231) to the given one (6.223) is of the following form*

$$
M: \quad \begin{aligned} q &= \bar{q}, \\ r &= M_{rm}(\bar{r}, \bar{s}, \bar{c}), \\ p &= A\bar{p} + \bar{r}_1\bar{q}, \\ s &= M_{sm}(\bar{r}, \bar{s}, \bar{c}) + M_{int}, \\ c &= M_{cm}(\bar{r}, \bar{s}, \bar{c}) + \tfrac{1}{2}(\bar{p}, A\bar{p}) - \tfrac{1}{2}(\bar{q}, \bar{q}) + \bar{r}_1(\bar{q}, \bar{p}), \end{aligned} \quad (6.233)
$$

where $M_{int} = \left(-\tfrac{1}{4}(\bar{q}, \bar{p}), 0, ..., 0\right)^T$.

The proof is through direct calculations.

Now, by applying the map M of (6.233) to the first Hamiltonian structure of the MKdV restricted flows we generate the second Hamiltonian structure of the KdV restricted flows.

Example 6.19 Bi-Hamiltonian structure of the KdV restricted flow: $N = 1$, $m = 2$.

In this case we have

$$\gamma_2(u) = \frac{1}{16}u_{4x} + \frac{5}{8}uu_{xx} + \frac{5}{16}u_x^2 + \frac{5}{8}u^3,$$

$$\overline{\gamma}_2(v) = \frac{1}{16}v_{4x} - \frac{5}{8}vv_x^2 - \frac{5}{8}v^2 v_{xx} + \frac{3}{8}v^5, \tag{6.234}$$

hence, the appropriate first Hamiltonian formulation of the KdV and the MKdV restricted flows are the following

$$
\begin{pmatrix} q \\ r_1 \\ r_2 \\ p \\ s_1 \\ s_2 \\ c \end{pmatrix}_x
=
\begin{pmatrix}
p \\
r_2 \\
32s_2 \\
(\alpha - r_1)q \\
\frac{5}{16}r_1^3 - \frac{5}{32}r_2^2 - \frac{1}{2}c - \frac{1}{2}q^2 \\
-\frac{5}{16}r_1 r_2 - s_1 \\
0
\end{pmatrix}
\equiv K = \theta_0 \circ \nabla h
$$

$$
=
\begin{pmatrix} 0 & I_3 & 0 \\ -I_3 & 0 & 0 \\ 0 & 0 & 0 \end{pmatrix}
\nabla\left(\frac{1}{2}p^2 + 16s_2^2 + r_2 s_1 - \frac{1}{2}\alpha q^2\right.
$$

$$
\left. + \frac{1}{2}r_1 q^2 + \frac{5}{32}r_1 r_2^2 - \frac{5}{64}r_1^4 + \frac{1}{2}cr_1\right) \tag{6.235}
$$

where $r_1 = u$, $r_2 = u_x$, $s_1 = -\frac{5}{16}uu_x - \frac{1}{32}u_{3x}$, $s_2 = \frac{1}{32}u_{xx}$, $c = \frac{1}{16}u_{4x} + \frac{5}{8}uu_{xx} + \frac{5}{16}u_x^2 + \frac{5}{8}u^3 - q^2$, and

$$
\begin{pmatrix} \overline{q} \\ \overline{r}_1 \\ \overline{r}_2 \\ \overline{p} \\ \overline{s}_1 \\ \overline{s}_2 \\ \overline{c} \end{pmatrix}_x
=
\begin{pmatrix}
\alpha\overline{p} + \overline{r}_1\overline{q} \\
\overline{r}_2 \\
-8\overline{s}_2 \\
\overline{q} - \overline{r}_1\overline{p} \\
-\frac{3}{4}\overline{r}_1^5 - \frac{5}{4}\overline{r}_1\overline{r}_2^2 - 2\overline{c} - \overline{q}\,\overline{p} \\
-\overline{s}_1 - \frac{5}{4}\overline{r}_1^2\overline{r}_2 \\
0
\end{pmatrix}
\equiv \overline{K} = \overline{\theta}_0 \circ \nabla\overline{h}
$$

$$
=
\begin{pmatrix} 0 & I_3 & 0 \\ -I_3 & 0 & 0 \\ 0 & 0 & 0 \end{pmatrix}
\nabla(\frac{1}{2}\alpha\overline{p}^2 + \overline{r}_1\overline{q}\,\overline{p} + \overline{r}_2\overline{s}_1 - 4\overline{s}_2^2
$$

$$
-\frac{1}{2}\overline{q}^2 + \frac{3}{24}\overline{r}_1^6 + \frac{5}{8}\overline{r}_1^2\overline{r}_2^2 + 2\overline{c}\,\overline{r}_1), \tag{6.236}
$$

where $\overline{r}_1 = v$, $\overline{r}_2 = v_x$, $\overline{s}_1 = -\frac{5}{4}v^2 v_x + \frac{1}{8}v_{3x}$, $\overline{s}_2 = -\frac{1}{8}v_{xx}$, $\overline{c} = \frac{1}{16}v_{4x} - \frac{5}{8}vv_x^2 - \frac{5}{8}v^2 v_{xx} + \frac{3}{8}v^5 - \frac{1}{2}\overline{q}\,\overline{p}$.

The Miura map linking both systems is

$$M: \quad \begin{aligned}
q &= \bar{q} \\
r_1 &= -\bar{r}_1^2 - \bar{r}_2 \\
r_2 &= 8\bar{s}_2 - 2\bar{r}_1\bar{r}_2 \\
p &= \bar{r}_1\bar{q} + \alpha\bar{p} \\
s_1 &= \tfrac{1}{2}\bar{r}_1\bar{s}_1 + \bar{r}_2\bar{s}_2 - \tfrac{5}{16}\bar{r}_1\bar{r}_2^2 - \tfrac{3}{16}\bar{r}_1^5 - \tfrac{1}{2}\bar{c} - \tfrac{1}{4}\bar{q}\bar{p} \\
s_2 &= -\tfrac{1}{4}\bar{s}_1 + \tfrac{1}{2}\bar{r}_1\bar{s}_2 - \tfrac{5}{16}\bar{r}_1^2\bar{r}_2 - \tfrac{1}{16}\bar{r}_2^2 \\
c &= \tfrac{1}{2}\alpha\bar{p}^2 + \bar{r}_1\bar{q}\bar{p} + \bar{r}_2\bar{s}_1 - 4\bar{s}_2^2 - \tfrac{1}{2}\bar{q}^2 + \tfrac{3}{24}\bar{r}_1^6 + \tfrac{5}{8}\bar{r}_1^2\bar{r}_2^2 + 2\bar{c}\bar{r}_1
\end{aligned}$$

$$(6.237)$$

and thus the second Hamiltonian structure of (6.235) is

$$\theta_1 = M' \circ \bar{\theta}_0 \circ M'^T$$

$$= \begin{pmatrix}
0 & 0 & 0 & \alpha & -\tfrac{1}{4}q & 0 & p \\
0 & 0 & -8 & 0 & r_1 & 0 & r_2 \\
0 & 8 & 0 & 0 & -r_2 & -\tfrac{3}{2}r_1 & 32s_2 \\
-\alpha & 0 & 0 & 0 & \tfrac{1}{4}p & -\tfrac{1}{4}q & (\alpha - r_1)q \\
\tfrac{1}{4}q & -r_1 & r_2 & -\tfrac{1}{4}p & 0 & \tfrac{1}{2}s_2 + \tfrac{15}{16}r_1^2 & A \\
0 & 0 & \tfrac{3}{2}r_1 & \tfrac{1}{4}q & -\tfrac{1}{2}s_2 - \tfrac{15}{16}r_1^2 & 0 & -\tfrac{5}{16}r_1r_2 - s_1 \\
* & * & * & * & * & * & 0
\end{pmatrix},$$

$$(6.238)$$

where $A = \tfrac{5}{16}r_1^3 - \tfrac{5}{32}r_2^2 - \tfrac{1}{2}c - \tfrac{1}{2}q^2$. So, we have the following bi-Hamiltonian formulation of dynamics (6.235)

$$K = \theta_0 \circ \nabla h_1 = \theta_1 \circ \nabla h_0, \qquad h_1 = h_{1,2}, \ h_0 = c.$$

The reader can find examples of bi-Hamiltonian structure for other KdV restricted flows in [31].

We will complete the considerations with some comments on possible application of the method to other systems. Let

$$u_t = K_m[u], \quad L(u)\varphi = \alpha\varphi, \ : \ u = (u_1, ..., u_n)^T \qquad (6.239)$$

and

$$v_t = \bar{K}_m[v], \quad \bar{L}(v)\psi = \alpha\psi, \quad v = (v_1, ..., v_n)^T \qquad (6.240)$$

be two soliton hierarchies, related by some Miura map, and their Lax equations, related by respective gauge transformations. A general form of restricted flows for (6.239) and (6.240) reads

$$L(u)\varphi_k = \alpha_k\varphi_k, \quad k = 1, ..., N, \quad \theta_0 \sum_{k=1}^{N} f_k(\varphi) = K_m[u], \quad (6.241)$$

$$\bar{L}(v)\chi_k = \alpha_k\psi_k, \quad k = 1, ..., N, \quad \bar{\theta}_0 \sum_{k=1}^{N} \bar{f}_k(\psi) = \bar{K}_m[v], \quad (6.242)$$

where $\theta_0, \bar{\theta}_0$ are first Hamiltonian structures of (6.239) and (6.240), respectively.

The extension of the method, here applied to the KdV restricted flows, to other restricted flows is possible under a few conditions. First, the uncoupled versions of (6.239) and (6.240)

$$L(u \to 0)\psi_k = \alpha_k \psi_k, \quad K_m[u] = 0, \quad k = 1, ..., N,$$

$$\bar{L}(v \to 0)\varphi_k = \alpha_k \varphi_k, \quad \bar{K}_m[v] = 0, \quad k = 1, ..., N. \tag{6.243}$$

have to be Hamiltonian systems. The pair of asymptotic Lax equations is related by a map obtained from the asymptotic gauge transformation and in many cases is Hamiltonian. The second pair of equations in (6.243) represents stationary flows of underlying systems. As was shown in Sect. 6.1 and 6.2, the projection onto stationary flows preserves the Hamiltonian structure and Miura map for a wide class of soliton equations. Secondly, the coupling between the asymptotic spectral problem and the stationary flow also has to be Hamiltonian.

These conditions are fulfilled at least for some classes of soliton systems, which makes the method more general. A possible candidate are restricted flows of dispersive water waves [9],[113] and their modifications, for example.

6.5 Separability of Bi-Hamiltonian Chains with Degenerate Poisson Structures

The results of the previous section allow us to re-examine the theory of classical Hamiltonian systems with separable potentials, in the framework of the bi-Hamiltonian theory. Here we prove a close relationship between separability and the bi-Hamiltonian formulation for a given dynamics [38].

First let us re-examine some facts about bi-Hamiltonian structures from the previous sections of this chapter. In the extended phase $\overline{M} \ni (q, p, c)$ canonical θ_0 and noncanonical θ_k, $k = 1, -1$ Poisson structures (both degenerate) take the general form

$$\theta_0 = \begin{pmatrix} \theta_c & 0 \\ 0 & 0 \end{pmatrix}, \qquad \theta_k = \begin{pmatrix} \theta_{nc} & K \\ -K^{\mathrm{T}} & 0 \end{pmatrix}, \tag{6.244}$$

where

$$\theta_c = \begin{pmatrix} 0 & I \\ -I & 0 \end{pmatrix}, \qquad \theta_{nc} = \begin{pmatrix} D & A \\ -A^{\mathrm{T}} & B \end{pmatrix} \tag{6.245}$$

are nondegenerate matrices from the space $M \ni (q, p)$ and $K = K(q, p, c)$ is an appropriate vector field.

In our further considerations we shall concentrate on the following form of θ_{nc}

$$\theta_{nc} = \begin{pmatrix} 0 & A(q) \\ -A^{\mathrm{T}}(q) & B(q,p) \end{pmatrix}. \tag{6.246}$$

We have to stress that for all known examples of constrained flows of Lax equations and stationary flows in the Newton representation, the matrix θ_{nc} takes the form (6.246), and for stationary flows in the Ostrogradsky representation θ_{nc} can be put into the form (6.246) by an appropriate canonical transformation of variables. Note that for the matrix θ_{nc} in the form (6.246) the Hamiltonians $h_r(q,p,c)$ take the general form

$$h_r(q,p,c) = h_r(q,p) + c\rho_r(q). \tag{6.247}$$

Before we continue our considerations of bi-Hamiltonian chains we shall introduce the concept of quasi-bi-Hamiltonian systems [40].

Definition 6.3 *An element $K \in \mathcal{L}$ is called a quasi-bi-Hamiltonian with respect to Poisson operators θ_0 and θ_1 if there exist elements $\rho, H, F \in \mathcal{F}$ such that*

$$K = \theta_0 \circ \nabla H = \frac{1}{\rho} \cdot \theta_1 \circ \nabla F, \tag{6.248}$$

where both Poisson matrices are compatible and nondegenerate (invertible). The function ρ is called the integrating factor.

To make this concept more familiar and relate it to the bi-Hamiltonian formalism let us review the well known example of the integrable case of the Henon–Heiles system [33],[40]

$$q_{1tt} = -3q_1^2 - \frac{1}{2}q_2^2,$$

$$q_{2tt} = -q_1 q_2, \tag{6.249}$$

which has two constants of motion of the form

$$H = \frac{1}{2}p_1^2 + \frac{1}{2}p_2^2 + q_1^3 + \frac{1}{2}q_1 q_2^2,$$

$$F = -2q_2 p_1 p_2 + 2q_1 p_2^2 - \frac{1}{4}q_2^4 - q_1^2 q_2^2, \tag{6.250}$$

where $p_1 = q_{1t}$ and $p_2 = q_{2t}$. In the phase space $M \ni (q_1, q_2, p_1, p_2)$ the related vector fields admit the following quasi-bi-Hamiltonian formulations

$$\begin{pmatrix} q_1 \\ q_2 \\ p_1 \\ p_2 \end{pmatrix}_{t_H} = \begin{pmatrix} p_1 \\ p_2 \\ -3q_1^2 - \frac{1}{2}q_2^2 \\ -q_1 q_2 \end{pmatrix} \equiv K_H = \theta_0 \circ \nabla H = -\frac{1}{q_2^2} \cdot \theta_1 \circ \nabla F,$$

$$\tag{6.251}$$

$$\begin{pmatrix} q_1 \\ q_2 \\ p_1 \\ p_2 \end{pmatrix}_{t_F} = \begin{pmatrix} -2q_2p_2 \\ -2q_2p_1 + 4q_1p_2 \\ 2q_1q_2^2 - 2p_2^2 \\ 2p_1p_2 + q_2^3 + 2q_1^2q_2 \end{pmatrix} \equiv K_F = \theta_0 \circ \nabla F = -q_2^2 \cdot \theta_{-1} \circ \nabla H,$$

$$(6.252)$$

where

$$\theta_0 = \begin{pmatrix} 0 & 0 & 1 & 0 \\ 0 & 0 & 0 & 1 \\ -1 & 0 & 0 & 0 \\ 0 & -1 & 0 & 0 \end{pmatrix}, \quad \theta_1 = \begin{pmatrix} 0 & 0 & q_1 & \frac{1}{2}q_2 \\ 0 & 0 & \frac{1}{2}q_2 & 0 \\ -q_1 & -\frac{1}{2}q_2 & 0 & \frac{1}{2}p_2 \\ -\frac{1}{2}q_2 & 0 & -\frac{1}{2}p_2 & 0 \end{pmatrix},$$

$$\theta_{-1} = \begin{pmatrix} 0 & 0 & 0 & 1/q_2 \\ 0 & 0 & 1/q_2 & -2q_1/q_2^2 \\ 0 & -1/q_2 & 0 & p_2/q_2^2 \\ -1/q_2 & 2q_1/q_2^2 & -p_2/q_2^2 & 0 \end{pmatrix} \qquad (6.253)$$

are appropriate compatible Poisson tensors. The Nijenhuis operator

$$\Phi = \theta_1 \circ \theta_0^{-1} \qquad (6.254)$$

relates both systems, i.e. applying Φ to (6.252) we get (6.251). Note that H, F and θ_0 are tensor invariants of $u_t = K_H$ but θ_1 and ρ are not. Hence, Φ is not a recursion operator.

Now, let us consider the first generalization of the Henon–Heiles system

$$q_{1tt} = -3q_1^2 - \frac{1}{2}q_2^2 + c,$$

$$q_{2tt} = -q_1q_2, \qquad (6.255)$$

with the following constants of the motion

$$\overline{H} = \frac{1}{2}p_1^2 + \frac{1}{2}p_2^2 + q_1^3 + \frac{1}{2}q_1q_2^2 - cq_1,$$

$$-\frac{1}{4}\overline{F} = \frac{1}{2}q_2p_1p_2 - \frac{1}{2}q_1p_2^2 + \frac{1}{16}q_2^4 + \frac{1}{4}q_1^2q_2^2 - \frac{1}{4}q_2^2c,$$

$$\overline{G} = c. \qquad (6.256)$$

In the extended phase space $\overline{M} \ni (q_1, q_2, p_1, p_2, c)$ the related dynamical systems admit the following bi-Hamiltonian formulations

$$\begin{pmatrix} q_1 \\ q_2 \\ p_1 \\ p_2 \\ c \end{pmatrix}_{t_H} = \begin{pmatrix} p_1 \\ p_2 \\ -3q_1^2 - \frac{1}{2}q_2^2 + c \\ -q_1q_2 \\ 0 \end{pmatrix} \equiv \overline{K_{\overline{H}}} = \overline{\theta}_0 \circ \overline{\nabla}\, \overline{H} = \overline{\theta}_1 \circ \overline{\nabla}\, \overline{G},$$

$$(6.257)$$

$$\begin{pmatrix} q_1 \\ q_2 \\ p_1 \\ p_2 \\ c \end{pmatrix}_{t_F} = \begin{pmatrix} \frac{1}{2}q_2p_2 \\ \frac{1}{2}q_2p_1 - q_1p_2 \\ -\frac{1}{2}q_1q_2^2 + \frac{1}{2}p_2^2 \\ -\frac{1}{2}p_1p_2 - \frac{1}{4}q_2^3 - \frac{1}{2}q_1^2q_2 + \frac{1}{2}q_2c \\ 0 \end{pmatrix} \equiv \overline{K}_{\overline{F}}$$

$$= \overline{\theta}_0 \circ \overline{\nabla}(-\frac{1}{4}\overline{F}) = \overline{\theta}_1 \circ \overline{\nabla}\,\overline{H}, \qquad (6.258)$$

where compatible Poisson tensors are of the following form

$$\overline{\theta}_0 = \begin{pmatrix} 0 & 0 & 1 & 0 & 0 \\ 0 & 0 & 0 & 1 & 0 \\ -1 & 0 & 0 & 0 & 0 \\ 0 & -1 & 0 & 0 & 0 \\ 0 & 0 & 0 & 0 & 0 \end{pmatrix},$$

$$\overline{\theta}_1 = \begin{pmatrix} 0 & 0 & q_1 & \frac{1}{2}q_2 & p_1 \\ 0 & 0 & \frac{1}{2}q_2 & 0 & p_2 \\ -q_1 & -\frac{1}{2}q_2 & 0 & \frac{1}{2}p_2 & -3q_1^2 - \frac{1}{2}q_2^2 + c \\ -\frac{1}{2}q_2 & 0 & -\frac{1}{2}p_2 & 0 & -q_1q_2 \\ -p_1 & -p_2 & 3q_1^2 + \frac{1}{2}q_2^2 - c & q_1q_2 & 0 \end{pmatrix}, \qquad (6.259)$$

each degenerate. Note that $\overline{G}$ is a Casimir of $\overline{\theta}_0$ and $\overline{F}$ is a Casimir of $\overline{\theta}_1$.

Finally, let us consider the second generalized Henon–Heiles system in the form

$$q_{1tt} = -3q_1^2 - \frac{1}{2}q_2^2,$$

$$q_{2tt} = -q_1q_2 + \frac{d}{q_2^3}, \qquad (6.260)$$

(see Example 6.4), with the following constants of the motion

$$-2\overline{H} = -p_1^2 - p_2^2 - 2q_1^3 - q_1q_2^2 - \frac{d}{q_2^2},$$

$$\overline{F} = -2q_2p_1p_2 + 2q_1p_2^2 - \frac{1}{4}q_2^4 - q_1^2q_2^2 + \frac{2dq_1}{q_2^2},$$

$$\overline{G} = d. \qquad (6.261)$$

In the extended phase space $\overline{M} \ni (q_1, q_2, p_1, p_2, d)$ the related dynamical systems admit the following bi-Hamiltonian formulations

$$\begin{pmatrix} q_1 \\ q_2 \\ p_1 \\ p_2 \\ c \end{pmatrix}_{t_H} = \begin{pmatrix} -2p_1 \\ -2p_2 \\ 6q_1^2 + q_2^2 \\ 2q_1q_2 - 2d/q_2^3 \\ 0 \end{pmatrix} \equiv \overline{K}_{\overline{H}} = \overline{\theta}_0 \circ \overline{\nabla}(-2\overline{H}) = \overline{\theta}_{-1} \circ \overline{\nabla}\,\overline{F},$$

$$(6.262)$$

$$
\begin{pmatrix} q_1 \\ q_2 \\ p_1 \\ p_2 \\ c \end{pmatrix}_{t_F} = \begin{pmatrix} -2q_2 p_2 \\ -2q_2 p_1 + 4q_1 p_2 \\ 2q_1 q_2^2 - 2p_2^2 - 2d/q_2^2 \\ 2p_1 p_2 + q_2^3 + 2q_1^2 q_2 + 4dq_1/q_2^3 \\ 0 \end{pmatrix} \equiv K_{\overline{F}}
$$

$$
= \overline{\theta}_0 \circ \overline{\nabla} F = \overline{\theta}_{-1} \circ \overline{\nabla} G, \tag{6.263}
$$

where compatible Poisson tensors are $\overline{\theta}_0$ and

$$
\overline{\theta}_{-1} =
$$

$$
\begin{pmatrix}
0 & 0 & 0 & 1/q_2 & -2q_2 p_2 \\
0 & 0 & 1/q_2 & -2q_1/q_2^2 & -2q_2 p_1 + 4q_1 p_2 \\
0 & -1/q_2 & 0 & p_2/q_2^2 & 2q_1 q_2^2 - 2p_2^2 - 2d/q_2^2 \\
-1/q_2 & 2q_1/q_2^2 & -p_2/q_2^2 & 0 & 2p_1 p_2 + q_2^3 + 2q_1^2 q_2 + 4dq_1/q_2^3 \\
* & * & * & * & 0
\end{pmatrix},
$$

$$
\tag{6.264}
$$

again both degenerate. The stars in the last row mean matrix elements which make (6.264) skew-symmetric. Note that $\overline{G}$ and $\overline{H}$ from (6.261) are respective Casimirs of $\overline{\theta}_0$ and $\overline{\theta}_{-1}$.

The generalized Henon–Heiles systems (6.257) and (6.263) additionally have a quasi-bi-Hamiltonian representation. Actually for the system (6.257) we find

$$
\begin{pmatrix} q_1 \\ q_2 \\ p_1 \\ p_2 \end{pmatrix}_{t_H} = \begin{pmatrix} p_1 \\ p_2 \\ -3q_1^2 - \frac{1}{2}q_2^2 + c \\ -q_1 q_2 \end{pmatrix} \equiv K_{\overline{H}} = \theta_0 \circ \nabla \overline{H} = -\frac{1}{q_2^2} \cdot \theta_1 \circ \nabla \overline{F},
$$

$$
\tag{6.265}
$$

where the Poisson tensors θ_0, θ_1 are given by (6.253) and the conserved functions $\overline{H}$, $\overline{F}$ take the form (6.256), while for the system (6.263) its quasi-bi-Hamiltonian representation reads

$$
\begin{pmatrix} q_1 \\ q_2 \\ p_1 \\ p_2 \end{pmatrix}_{t_F} = \begin{pmatrix} -2q_2 p_2 \\ -2q_2 p_1 + 4q_1 p_2 \\ 2q_1 q_2^2 - 2p_2^2 - 2d/q_2^2 \\ 2p_1 p_2 + q_2^3 + 2q_1^2 q_2 + 4dq_1/q_2^3 \end{pmatrix} \equiv K_{\overline{F}}
$$

$$
= \theta_0 \circ \nabla \overline{F} = -q_2^2 \cdot \theta_{-1} \circ \nabla \overline{H}, \tag{6.266}
$$

where the conserved functions $\overline{H}$, $\overline{F}$ take the form (6.261). Hence one concludes that the quasi-bi-Hamiltonian representation survives the reduction $c = d = 0$ (6.251), (6.252) the while bi-Hamiltonian representation does not.

The Henon–Heiles example gave us some insight into the mutual relations between bi-Hamiltonian and quasi-bi-Hamiltonian representations. Now we shall pass to a more systematic treatment of the problem. In the further considerations we shall only use the canonical coordinates (q, p) with respect to which θ_0 takes the canonical form $\theta_0 = \theta_c$ (6.245). As θ_0 and θ_1 are compatible and invertible, the tensor $\Phi = \theta_1 \circ \theta_0^{-1}$ is the Nijenhuis (hereditary) tensor. As Φ is implectic-symplectic factorizable all its eigenvalues have at least algebraic multiplicity 2 [182]. Here, we assume the existence of an open set $U \subset M$ such that in U the matrix Φ is maximal, i.e. has n distinct eigenvalues $\lambda = (\lambda_1, ..., \lambda_n)$. So, in a neighbourhood of a regular point, where the eigenvalues λ are independent, one can construct a canonical transformation $(q, p) \to (\lambda, \mu)$ $((\lambda, \mu)$ referred to as *Nijenhuis coordinates*) such that θ_1, θ_{-1} and Φ take the Darboux form

$$\theta_1 = \begin{pmatrix} 0 & \Lambda \\ -\Lambda & 0 \end{pmatrix}, \quad \theta_{-1} = \begin{pmatrix} 0 & \Lambda^{-1} \\ -\Lambda^{-1} & 0 \end{pmatrix} \quad \Phi = \begin{pmatrix} \Lambda & 0 \\ 0 & \Lambda \end{pmatrix}, \quad (6.267)$$

where $\Lambda = \mathrm{diag}(\lambda_1, ..., \lambda_n)$ and $\Lambda^{-1} = \mathrm{diag}(\frac{1}{\lambda_1}, ..., \frac{1}{\lambda_n})$.

Definition 6.4

(i) A quasi-bi-Hamiltonian vector field (6.248) is said to be in a Pfaffian form if in Nijehuis coordinates it takes the form

$$\begin{pmatrix} 0 & I \\ -I & 0 \end{pmatrix} \nabla H(\lambda, \mu) = \frac{(-1)^{n+1}}{\lambda_1 \cdot ... \cdot \lambda_n} \begin{pmatrix} 0 & \Lambda \\ -\Lambda & 0 \end{pmatrix} \nabla F(\lambda, \mu), \quad (6.268)$$

i.e. the integrating factor ρ is the product of the eigenvalues of Φ

$$\rho = (-1)^{n+1} \prod_{i=1}^{n} \lambda_i. \quad (6.269)$$

(ii) A quasi-bi-Hamiltonian vector field (6.248) is said to be in an inverse Pfaffian form if in Nijehuis coordinates it takes the form

$$\begin{pmatrix} 0 & I \\ -I & 0 \end{pmatrix} \nabla \overline{H}(\lambda, \mu) = (-1)^{n+1} \lambda_1 \cdot ... \cdot \lambda_n \begin{pmatrix} 0 & \Lambda^{-1} \\ -\Lambda^{-1} & 0 \end{pmatrix} \nabla \overline{F}(\lambda, \mu).$$

$$(6.270)$$

Notice that the forms (6.269) and (6.270) are related by the Nijenhuis operator Φ (6.267) and its inverse Φ^{-1}.

Lemma 6.6 *In the Pfaffian case (6.268) and in the inverse Pfaffian case (6.270), the general solution of the equation (6.248) is given by*

$$H = \sum_{i=1}^{n} \frac{1}{\Delta_i} f_i(\lambda_i, \mu_i) + g_H(\lambda),$$

$$F = \sum_{i=1}^{n} \frac{\rho}{\lambda_i} \frac{1}{\Delta_i} f_i(\lambda_i, \mu_i) + g_F(\lambda), \quad (6.271)$$

where

$$\frac{\partial g_H}{\partial \lambda_i} = \frac{\lambda_i}{\rho} \frac{\partial g_F}{\partial \lambda_i}, \quad \Delta_i = \prod_{i \neq j} \lambda_{ij}, \quad \lambda_{ij} := \lambda_i - \lambda_j, \tag{6.272}$$

and the n functions $f_i(\lambda_i, \mu_i)$ (each one depending on one pair of coordinates) are arbitrary smooth functions.

Proof. Equations (6.268) and (6.270) correspond to the two sets of equations

$$\frac{\partial H}{\partial \mu_i} = \frac{\lambda_i}{\rho} \frac{\partial F}{\partial \mu_i}, \quad i = 1, ..., n \tag{6.273}$$

$$\frac{\partial H}{\partial \lambda_i} = \frac{\lambda_i}{\rho} \frac{\partial F}{\partial \lambda_i}, \quad i = 1, ..., n. \tag{6.274}$$

Ignoring for a moment the arbitrary functions $g_H(\lambda)$ and $g_F(\lambda)$, we will follow the argument of [140].

Let the general solution of the first set (6.273) be

$$H = \frac{1}{\rho} \sum_{i=1}^{n} \lambda_i G_i(\lambda, \mu_i) + K(\lambda), \quad F = \sum_{i=1}^{n} G_i(\lambda, \mu_i), \tag{6.275}$$

where the functions $G_i = G_i(\lambda, \mu_i)$ and $K = K(\lambda)$ are arbitrary. Indeed, the solution of the first equation (6.273) for $i = 1$ is $H = (\lambda_1/\rho)F(\lambda, \mu) + \varphi_1(\lambda; \mu_2, ..., \mu_n)$, with φ_1 arbitrary. In regard to this result, the equation (6.273) for $i = 2$ has the solution

$$H = \frac{\lambda_1}{\rho} G_1(\lambda, \mu_1) + \frac{\lambda_2}{\rho} \varphi_1(\lambda; \mu_2, ..., \mu_n) + \varphi_2(\lambda; \mu_3, ..., \mu_n),$$

$$F = G_1(\lambda, \mu_1) + \psi_1(\lambda; \mu_2, ..., \mu_n)$$

with ψ_1 and φ_2 arbitrary. Iterating this procedure for $i = 3, ..., n$ one obtains the solution (6.275). Now inserting this solution into (6.274) we conclude that $K(\lambda)$ has to be a constant function, which can be taken equal to zero with no loss of generality. For ρ in the Pfaffian form (6.269) and H, F given by the formula (6.275), equations (6.274) can be written as

$$\frac{\partial}{\partial \lambda_i} \left(\sum_{j=1}^{n} \lambda_{ij} G_j \right) = \frac{1}{\lambda_i} \left(\sum_{j=1}^{n} \lambda_{ij} G_j \right) \quad i = 1, ..., n. \tag{6.276}$$

By integrating this equation and taking into account the dependence on μ, we obtain that

$$G_i(\lambda_i, \mu_i) = \frac{\rho}{\lambda_i} f_i(\lambda_i, \mu_i), \quad i = 1, ..., n, \tag{6.277}$$

where each f_i is an arbitrary function depending only on the pair of variables (λ_i, μ_i). Of course in addition to $H = \sum_{i=1}^{n} \frac{1}{\Delta_i} f_i$ an arbitrary function $g_H(\lambda)$

generates in $F = \sum_{i=1}^{n} \frac{\lambda_i}{\rho} \frac{1}{\Delta_i} f_i$ the addition of a function $g_F(\lambda)$ related to $g_H(\lambda)$ by the relation (6.272). $\square$

Theorem 6.3 *The Hamiltonian functions H and F, written in terms of the Nijenhuis coordinates (λ_i, μ_i) in the form (6.271), are separable for each n-tuple of functions $f_i(\lambda_i, \mu_i)$ if and only if*

$$g_H(\lambda) = 0 \quad \text{and} \quad g_F(\lambda) = 0,$$

$$g_H(\lambda) = d(\lambda_1 + \dots + \lambda_n) \quad \text{and} \quad g_F(\lambda) = d\rho,$$

$$g_H(\lambda) = \frac{d}{\rho} \quad \text{and} \quad g_F(\lambda) = d\left(\frac{1}{\lambda_1} + \dots + \frac{1}{\lambda_n}\right), \tag{6.278}$$

where d is an arbitrary constant.

Proof. The Hamilton–Jacobi equation for H is separable iff H verifies the Levi–Civita condition. It is in the form of a set of nonlinear partial differential equations

$$\frac{\partial H}{\partial \lambda_i} \frac{\partial H}{\partial \lambda_j} \frac{\partial^2 H}{\partial \mu_i \partial \mu_j} + \frac{\partial H}{\partial \mu_i} \frac{\partial H}{\partial \mu_j} \frac{\partial^2 H}{\partial \lambda_i \partial \lambda_j} - \frac{\partial H}{\partial \lambda_i} \frac{\partial H}{\partial \mu_j} \frac{\partial^2 H}{\partial \mu_i \partial \lambda_j} - \frac{\partial H}{\partial \mu_i} \frac{\partial H}{\partial \lambda_j} \frac{\partial^2 H}{\partial \lambda_i \partial \mu_j} = 0 \tag{6.279}$$

for all pairs (i, j), $i \neq j$. Applying the relations

$$\frac{\partial}{\partial \lambda_i} \frac{1}{\Delta_i} = -\frac{1}{\Delta_i} \sum_{\alpha \neq i} \frac{1}{\lambda_{i\alpha}},$$

$$\frac{\partial}{\partial \lambda_\beta} \frac{1}{\Delta_i} = \frac{1}{\Delta_i} \frac{1}{\lambda_{i\beta}}, \quad \beta \neq i, \tag{6.280}$$

one directly verifies the relation (6.279) for $H(\lambda, \mu)$ and $F(\lambda, \mu)$ in the case $g_H(\lambda) = g_F(\lambda) = 0$ (this was proved for the first time in [140]). The Levi–Civita condition (6.279) gives the following equation for a function $g_H(\lambda)$ different from zero

$$\frac{\partial^2 g_H}{\partial \lambda_i \partial \lambda_j}(\lambda_i - \lambda_j) - \frac{\partial g_H}{\partial \lambda_i} + \frac{\partial g_H}{\partial \lambda_j} = 0, \tag{6.281}$$

which, up to a constant factor, has only two solutions given by (6.278). In the same way one finds the admissible form of the function $g_F(\lambda)$.$\square$

Theorem 6.4 *An arbitrary bi-Hamiltonian finite dimensional system with two degenerate Poisson tensors of the form (6.244), where θ_{nc} takes the form (6.246), whose chain starts with a Casimir of the canonical Poisson structure and terminates with a Casimir of a noncanonical Poisson structure, is equivalent to a quasi-bi-Hamiltonian system, which in the Nijenhuis coordinates takes a Pfaffian form.*

Proof. Let

$$\bar{\theta}_0 \circ \overline{\nabla} \bar{h}_0 = 0$$

$$\bar{\theta}_0 \circ \overline{\nabla} \bar{h}_1 = \overline{K}_1 = \bar{\theta}_1 \circ \overline{\nabla} \bar{h}_0$$

$$\bar{\theta}_0 \circ \overline{\nabla} \bar{h}_2 = \overline{K}_2 = \bar{\theta}_1 \circ \overline{\nabla} \bar{h}_1 \qquad (6.282)$$

$$\vdots$$

$$\bar{\theta}_0 \circ \overline{\nabla} \bar{h}_n = \overline{K}_n = \bar{\theta}_1 \circ \overline{\nabla} \bar{h}_{n-1}$$

$$0 = \bar{\theta}_1 \circ \overline{\nabla} \bar{h}_n$$

be a bi-Hamiltonian chain on $U \subset \overline{M} \ni (q, p, c)$ with two degenerate Poisson structures, each with one Casimir, so that $\dim \overline{M} = 2n + 1$ and $\operatorname{rank} \bar{\theta}_0(\bar{\theta}_1) = 2n$, where

$$\bar{\theta}_0 = \begin{pmatrix} 0 & I & 0 \\ -I & 0 & 0 \\ 0 & 0 & 0 \end{pmatrix}, \qquad \bar{\theta}_1 = \begin{pmatrix} 0 & A(q) & \frac{\partial \bar{h}_1}{\partial p} \\ -A^{\mathrm{T}}(q) & B(q, p) & -\frac{\partial \bar{h}_1}{\partial q} \\ -\left(\frac{\partial \bar{h}_1}{\partial p}\right)^{\mathrm{T}} & -\left(\frac{\partial \bar{h}_1}{\partial q}\right)^{\mathrm{T}} & 0 \end{pmatrix}$$

$$(6.283)$$

which are compatible, where $\bar{h}_r(q, p, c) = \bar{h}_r(q, p) + c\rho_r(q)$ and $\overline{\nabla} = (\partial/\partial q, \partial/\partial p, \partial/\partial c)^{\mathrm{T}}$.

In order to have a bi-Hamiltonian hierarchy also in the original phase space $M \ni (q, p)$ one can try to project the chain (6.282) onto M. If $\pi : \overline{M} \to M : (q, p, c) \to (q, p)$ is the projection map, the Hamiltonians $\bar{h}_r$ and the vector fields $\overline{K}_r$ cannot be projected onto M, because they depend on the fibre coordinate c. Instead, the Poisson tensors $\bar{\theta}_0$ and $\bar{\theta}_1$ are projected onto

$$\theta_0 \equiv \theta_c = \pi' \circ \bar{\theta}_0 \circ \pi'^{\mathrm{T}} = \begin{pmatrix} 0 & I \\ -I & 0 \end{pmatrix},$$

$$\theta_1 \equiv \theta_{nc} = \pi' \circ \bar{\theta}_1 \circ \pi'^{\mathrm{T}} = \begin{pmatrix} 0 & A \\ -A^{\mathrm{T}} & B \end{pmatrix} \qquad (6.284)$$

with $A = A(q)$ and $B = B(q, p)$. Because these Poisson tensors are compatible and invertible, one obtains the following Nijenhuis tensor

$$\Phi = \theta_1 \circ \theta_0^{-1} = \begin{pmatrix} A & 0 \\ B & A^{\mathrm{T}} \end{pmatrix}. \qquad (6.285)$$

Nevertheless, the Poisson tensor θ_1 is not invariant along the flows of vector fields K_r, where $\overline{K}_r = (K_r, 0)^{\mathrm{T}}$, so Φ is not a recursion operator for any K_r. However, the vector fields K_r are still Hamiltonian with respect to θ_0, i.e.

$$K_r = \theta_0 \circ \nabla h_r, \tag{6.286}$$

where $h_r = \overline{h}_r$, c is treated as a parameter and $\nabla = (\partial/\partial q, \partial/\partial p)^{\mathrm{T}}$.

Consider the last equation from the hierarchy (6.282): $0 = \overline{\theta}_1 \overline{\nabla} \overline{h}_n$. On M it reads

$$\theta_1 \circ \nabla h_n = -\rho_n K_1, \qquad \rho_n = \frac{\partial h_n}{\partial c}. \tag{6.287}$$

As $h_0 = c$, hence on M $\nabla h_0 = 0$ and the hierarchy (6.282) turns into the form

$$\theta_0 \circ \nabla h_1 = K_1 = -\frac{1}{\rho_n}\theta_1 \circ \nabla h_n$$

$$\theta_0 \circ \nabla h_2 = K_2 = -\frac{\rho_1}{\rho_n}\theta_1 \circ \nabla h_n + \theta_1 \circ \nabla h_1$$

$$\vdots$$

$$\theta_0 \circ \nabla h_n = K_n = -\frac{\rho_{n-1}}{\rho_n}\theta_1 \circ \nabla h_n + \theta_1 \circ \nabla h_{n-1} \tag{6.288}$$

where $\rho_r = \frac{\partial h_r}{\partial c}$. The last equation terminates the sequence of vector fields K_r in the hierarchy as for the next equation from the chain we have

$$\theta_0 \circ \nabla h_{n+1} = K_{n+1} = -\theta_1 \circ \nabla h_n + \theta_1 \circ \nabla h_n = 0. \tag{6.289}$$

Because of the form of the first equation from the hierarchy (6.288), the vector field $K \equiv K_1$ is called quasi-bi-Hamiltonian. Now, from the relation

$$\Phi \circ K_r = \theta_1 \circ \nabla h_r = K_{r+1} + \frac{\rho_r}{\rho_n}\theta_1 \circ \nabla h_n = K_{r+1} - \rho_r K_1 \tag{6.290}$$

one finds that

$$K_{r+1} = \Phi_r K_1, \qquad \Phi_r = \sum_{i=0}^{r} \rho_{r-i}\Phi^i, \qquad \rho_0 = 1, \ r = 0, ..., n-1, \tag{6.291}$$

and for $r = n$

$$\Phi_n \circ K_1 = K_{n+1} = 0, \tag{6.292}$$

hence

$$\Phi^n + \rho_1\Phi^{n-1} + ... + \rho_{n-1}\Phi + \rho_n I = 0. \tag{6.293}$$

In Nijenhuis coordinates the operator Φ is diagonal, so (6.293) is equivalent to the following system

$$\begin{pmatrix} \lambda_1^{n-1} & \cdots & \lambda_1 & 1 \\ \lambda_2^{n-1} & \cdots & \lambda_2 & 1 \\ \vdots & & \vdots & \vdots \\ \lambda_n^{n-1} & \cdots & \lambda_n & 1 \end{pmatrix} \begin{pmatrix} \rho_1 \\ \rho_2 \\ \vdots \\ \rho_n \end{pmatrix} = \begin{pmatrix} -\lambda_1^n \\ -\lambda_2^n \\ \vdots \\ -\lambda_n^n \end{pmatrix}. \tag{6.294}$$

For all λ_i different there is a unique solution to (6.294) in the form

$$\rho_r(q) = (-1)^r \sum_{\substack{j_1, \ldots, j_r \\ j_1 < \ldots < j_r}} \lambda_{j_1} \cdot \ldots \cdot \lambda_{j_r} := \rho_r(\lambda) \qquad r = 1, \ldots, n. \qquad (6.295)$$

So, the quasi-bi-Hamiltonian representation (6.288) for $K_1 \equiv K$ takes the Pfaffian form (6.268) where now $H = h_1$, $F = h_n$ and $\rho = -\rho_n$.□

The explicit form of the canonical transformation $(q, p) \to (\lambda, \mu)$ to the Nijenhuis coordinates is given in what follows. From relation (6.295) we get $q_k = \varphi_k(\lambda)$. Then, we introduce the generating function $S = \sum_{i=1}^{n} p_i \varphi_i(\lambda)$ such that

$$q_i = \frac{\partial S}{\partial p_i}, \quad \mu_i = \frac{\partial S}{\partial \lambda_i}. \qquad (6.296)$$

The first equation reconstructs $q_k = \varphi_k(\lambda)$ and solving the second one with respect to p_k we get the missing part of the canonical transformation in the form $p_k = \psi_k(\lambda, \mu)$.

Theorem 6.5

(i)   *An arbitrary quasi-bi-Hamiltonian system, which in Nijenhuis coordinates has the Pfaffian form (6.268), has n independent constants of the motion of the form*

$$h_r(\lambda, \mu, c) = \sum_{i=1}^{n} \frac{\rho_{r-1}^i}{\Delta_i} f_i(\lambda_i, \mu_i) + c\rho_r(\lambda), \qquad r = 1, \ldots, n, \qquad (6.297)$$

where

$$\Delta_i = \prod_{j \neq i} \lambda_{ij}, \quad \lambda_{ij} = \lambda_i - \lambda_j,$$

$$\rho_r^i := (-1)^r \sum_{\substack{j_1, \ldots, j_r \\ j_1 < \ldots < j_r \\ j_1 \neq i, \ldots, j_r \neq i}} \lambda_{j_1} \cdot \ldots \cdot \lambda_{j_r}, \qquad (6.298)$$

ρ_r are defined by (6.295) and the n functions f_i are arbitrary smooth functions.

(ii)   *An arbitrary function h_r is separable for each n-tuple of functions $f_i(\lambda_i, \mu_i)$, that is an appropriate Hamilton–Jacobi equation can be solved.*

(iii) *On the phase space $M \ni (\lambda, \mu)$ the functions h_k, $k = 1, \ldots, n$, are in involution with respect to the Poisson tensors*

$$\theta_0 = \begin{pmatrix} 0 & I \\ -I & 0 \end{pmatrix}, \quad \theta_1 = \begin{pmatrix} 0 & \Lambda \\ -\Lambda & 0 \end{pmatrix}.$$

(iv)   *On the extended phase space $\overline{M} \ni (\lambda, \mu, c)$, the functions h_r with $c \neq 0$ form a bi-Hamiltonian Nijenhuis chain with respect to the degenerate Poisson tensors*

$$\bar{\theta}_0 = \begin{pmatrix} 0 & I & 0 \\ -I & 0 & 0 \\ 0 & 0 & 0 \end{pmatrix}, \quad \bar{\theta}_1 = \begin{pmatrix} 0 & \Lambda & \frac{\partial h_1}{\partial \mu} \\ -\Lambda & 0 & -\frac{\partial h_1}{\partial \lambda} \\ -\left(\frac{\partial h_1}{\partial \mu}\right)^{\mathrm{T}} & \left(\frac{\partial h_1}{\partial \lambda}\right)^{\mathrm{T}} & 0 \end{pmatrix}.$$

The chain starts with the Casimir $h_0 = c$ of the first Poisson tensor $\bar{\theta}_0$ and terminates with the Casimir h_n of the second Poisson tensor $\bar{\theta}_1$.

Proof.

(iii) Let $\gamma_r = \nabla h_r$ Then by direct calculations one can verify that

$$\gamma_{r+1} = \Phi^\dagger \gamma_r + \rho_r \gamma_1, \tag{6.299}$$

where

$$\Phi^\dagger = \theta_0^{-1} \circ \theta_1 = \begin{pmatrix} \Lambda & 0 \\ 0 & \Lambda \end{pmatrix}$$

and ρ_r is given by (6.295). In the calculations the following relations are helpful

$$\frac{\partial \rho_r}{\partial \lambda_i} = -\rho_{r-1}^i, \quad \lambda_i \frac{\partial \rho_r}{\partial \lambda_i} = \rho_r - \rho_r^i, \quad \lambda_i \rho_{r-1}^i = -\rho_r + \rho_r^i,$$

$$\frac{\partial \rho_r^k}{\partial \lambda_i} = -\rho_{r-1}^{ki}, \quad \lambda_i \frac{\partial \rho_r^k}{\partial \lambda_i} = \rho_r^k - \rho_r^{ki}, \quad \lambda_i \rho_{r-1}^{ki} = -\rho_r^k + \rho_r^{ki},$$

$$\rho_r^{ki} := (-1)^r \sum_{\substack{j_1, \ldots, j_r \\ j_1 < \ldots < j_r \\ j_1 \neq i, k \ldots, j_r \neq i, k}} \lambda_{j_1} \cdot \ldots \cdot \lambda_{j_r}, \tag{6.300}$$

as well as relations (6.277). From (6.299) it follows that

$$\gamma_{r+1} = \Phi_r^\dagger \gamma_1 = \sum_{i=0}^r \rho_{r-i} \left(\Phi^\dagger\right)^i \gamma_1$$

and thus, because $\theta_0 \circ \Phi^\dagger = \Phi \circ \theta_0$, we have

$$\{h_l, h_k\}_{\theta_0} = < \nabla h_k, \theta_0 \circ \nabla h_l >$$

$$= \sum_{i=0}^l \sum_{j=0}^k \rho_{r-i} \rho_{k-j} < \Phi^{\dagger j} \circ \gamma_1, \theta_0 \circ \Phi^{\dagger i} \circ \gamma_1 >$$

$$= \sum_{i=0}^l \sum_{j=0}^k \rho_{r-i} \rho_{k-j} < \gamma_1, \Phi^{i+j} \circ \theta_0 \circ \gamma_1 >= 0,$$

where the last equality follows from the skew-symmetry of the tensor $\Phi^m \circ \theta_0$ for any m. As $\theta_1 = \Phi \circ \theta_0$ we have

$$\{h_l, h_k\}_{\theta_1} = < \nabla h_k, \theta_1 \circ \nabla h_l >$$

$$= \sum_{i=0}^{l} \sum_{j=0}^{k} \rho_{r-i} \rho_{k-j} < \Phi^{\dagger j} \circ \gamma_1, \theta_1 \circ \Phi^{\dagger i} \circ \gamma_1 >$$

$$= \sum_{i=0}^{l} \sum_{j=0}^{k} \rho_{r-i} \rho_{k-j} < \gamma_1, \Phi^{i+j} \circ \theta_1 \circ \gamma_1 >$$

$$= \sum_{i=0}^{l} \sum_{j=0}^{k} \rho_{r-i} \rho_{k-j} < \gamma_1, \Phi^{i+j+1} \circ \theta_0 \circ \gamma_1 > = 0. \square$$

(i)   follows immediately from (iii) as

$$L_K h_{r+1} = < \nabla \gamma_{r+1}, \theta_0 \circ \gamma_1 > = 0. \ \square$$

(ii) We prove the statement by direct verification of the Levi–Civita condition (6.276) for each h_r. Notice that $h_1 = H$ and $h_n = F$ where H and F are given in Lemma 6.6, as $\rho_0^i = 1$ and $\rho_{n-1}^i = -\frac{\rho_n}{\lambda_i} = \frac{\rho}{\lambda_i}$. $\square$

(iv) The relation $\bar{\theta}_0 \circ \nabla h_{r+1} = \bar{\theta}_1 \circ \nabla h_r$ is equivalent to the following equalities

$$\frac{\partial h_{r+1}}{\partial \mu_i} = \lambda_i \frac{\partial h_r}{\partial \mu_i} + \rho_r \frac{\partial h_1}{\partial \mu_i},$$

$$\frac{\partial h_{r+1}}{\partial \lambda_i} = \lambda_i \frac{\partial h_r}{\partial \lambda_i} + \rho_r \frac{\partial h_1}{\partial \lambda_i}, \qquad i = 1, ..., n,$$

$$\left(\frac{\partial h_1}{\partial \lambda} \right)^{\mathrm{T}} \frac{\partial h_r}{\partial \mu} - \left(\frac{\partial h_1}{\partial \mu} \right)^{\mathrm{T}} \frac{\partial h_r}{\partial \lambda} = 0.$$

The first two are equivalent to the relation (6.299) while the last one is equal to $\{h_r, h_1\}_{\theta_0} = 0$. $\square$

Theorem 6.6 *An arbitrary bi-Hamiltonian finite dimensional system with two degenerate Poisson tensors of the form (6.244), where θ_{nc} takes the form (6.246), whose chain starts with a Casimir of noncanonical Poisson structure and terminates with a Casimir of the canonical structure, is equivalent to a quasi-bi-Hamiltonian system, which in the Nijenhuis coordinates takes an inverse Pfaffian form.*

Proof. Let

$$\bar{\theta}_{-1} \circ \overline{\nabla} \bar{h}_1 = 0$$

$$\bar{\theta}_{-1} \circ \overline{\nabla} \bar{h}_2 = \overline{K}_1 = \bar{\theta}_0 \circ \overline{\nabla} \bar{h}_1$$

$$\bar{\theta}_{-1} \circ \overline{\nabla} \bar{h}_3 = \overline{K}_2 = \bar{\theta}_0 \circ \overline{\nabla} \bar{h}_2 \qquad (6.301)$$

$$\vdots$$

$$\bar{\theta}_{-1} \circ \overline{\nabla} \bar{h}_{n+1} = \overline{K}_n = \bar{\theta}_0 \circ \overline{\nabla} \bar{h}_n$$

$$0 = \bar{\theta}_0 \circ \overline{\nabla} \bar{h}_{n+1}$$

be a bi-Hamiltonian chain on $U \subset \overline{M} \ni (q, p, c)$ with two degenerate Poisson tensors each with one Casimir, so that $\dim \overline{M} = 2n+1$ and $\operatorname{rank} \bar{\theta}_0(\bar{\theta}_{-1}) = 2n$, where

$$\bar{\theta}_{-1} = \begin{pmatrix} 0 & A(q) & \frac{\partial \bar{h}_n}{\partial p} \\ -A^{\mathrm{T}}(q) & B(q, p) & -\frac{\partial \bar{h}_n}{\partial q} \\ -\left(\frac{\partial \bar{h}_n}{\partial p}\right)^{\mathrm{T}} & -\left(\frac{\partial \bar{h}_n}{\partial q}\right)^{\mathrm{T}} & 0 \end{pmatrix}, \quad \bar{\theta}_0 = \begin{pmatrix} 0 & I & 0 \\ -I & 0 & 0 \\ 0 & 0 & 0 \end{pmatrix},$$

$$(6.302)$$

which are compatible, and where $\bar{h}_r(q, p, c) = \bar{h}_r(q, p) + c\bar{p}_r(q)$. As in the previous case (Theorem 6.4), in the original phase space $M \ni (q, p), \theta_c = \theta_0$ and $\theta_{nc} = \theta_{-1}$ constitutes a compatible pair of Poisson tensors but the tensor θ_{-1} is not invariant along any flow K_r. Because θ_0 is invertible, then $\Phi^{-1} = \theta_{-1} \circ \theta_0^{-1}$ is the inverse of the Nijenhuis operator Φ which is also not a recursion operator for K_r.

Consider the first equation from the hierarchy (6.301): $\bar{\theta}_{-1} \circ \bar{h}_1 = 0$. On M it reads

$$\theta_{-1} \circ \nabla \bar{h}_1 = -\bar{p}_1 K_n, \qquad \bar{p}_1 = \frac{\partial \bar{h}_1}{\partial c}, \qquad (6.303)$$

and the whole hierarchy turns into

$$-\frac{1}{\bar{p}_1} \theta_{-1} \circ \nabla \bar{h}_1 = K_n = \theta_0 \circ \nabla \bar{h}_n$$

$$\theta_{-1} \circ \nabla \bar{h}_n - \frac{\bar{p}_n}{\bar{p}_1} \theta_{-1} \circ \nabla \bar{h}_1 = K_{n-1} = \theta_0 \circ \nabla \bar{h}_{n-1}$$

$$\vdots$$

$$\theta_{-1} \circ \nabla \bar{h}_2 - \frac{\bar{p}_2}{\bar{p}_1} \theta_{-1} \circ \nabla \bar{h}_1 = K_1 = \theta_0 \circ \nabla \bar{h}_1, \qquad (6.304)$$

where $\bar{p}_r = \frac{\partial \bar{h}_r}{\partial c}$, $\bar{p}_{n+1} = 1$, $\overline{K}_r = (K_r, 0)^{\mathrm{T}}$. The last equation terminates the sequence of vector fields K_r in the hierarchy as for the next equation from the chain we have

$$\theta_{-1} \circ \nabla \overline{h}_1 - \theta_{-1} \circ \nabla \overline{h}_1 = K_0 = 0. \tag{6.305}$$

Because of the form of the first equation from the hierarchy (6.304), the vector field $K \equiv K_n$ is called quasi-bi-Hamiltonian. Now, from the relation

$$\Phi^{-1} \circ K_r = \theta_{-1} \circ \nabla \overline{h}_r = K_{r-1} + \frac{\overline{p}_r}{\overline{p}_1} \theta_{-1} \circ \nabla \overline{h}_1 = K_{r-1} - \overline{p}_r K_n \tag{6.306}$$

one finds that

$$K_{n-r} = \Phi_r^{-1} \circ K_n = \sum_{i=0}^{r} \overline{p}_{r+1-i} \Phi^{-r+i} \circ K_n, \tag{6.307}$$

and for $r = n$

$$K_0 = \Phi_n^{-1} K_n = \sum_{i=0}^{n} \overline{p}_{n+1-i} \Phi^{-n+i} \circ K_n = 0, \tag{6.308}$$

hence

$$\Phi^{-n} + \overline{p}_n \Phi^{-n+1} + \dots + \overline{p}_2 \Phi^{-1} + \overline{p}_1 I = 0. \tag{6.309}$$

In Nijenhuis coordinates we have

$$\theta_0 = \begin{pmatrix} 0 & I \\ -I & 0 \end{pmatrix}, \quad \theta_{-1} = \begin{pmatrix} 0 & \Lambda^{-1} \\ -\Lambda^{-1} & 0 \end{pmatrix}, \quad \Phi^{-1} = \begin{pmatrix} \Lambda^{-1} & 0 \\ 0 & \Lambda^{-1} \end{pmatrix}, \tag{6.310}$$

where $\Lambda^{-1} = \mathrm{diag}(\frac{1}{\lambda_1}, \dots, \frac{1}{\lambda_n})$, and for all λ_i different there is a unique solution to (6.309) in the form

$$\overline{p}_k(q) = (-1)^k \sum_{\substack{j_1, \dots, j_k \\ j_1 < \dots < j_k}} \frac{1}{\lambda_{j_1} \cdot \dots \cdot \lambda_{j_k}} := \overline{p}_{n+1-k}(\lambda), \quad k = 1, \dots, n. \tag{6.311}$$

So, the first equation in (6.304) takes the inverse Pfaffian form (6.270) where now $\overline{H} = \overline{h}_n$, $\overline{F} = \overline{h}_1$ and $\rho = -\overline{p}_1$. $\square$

Theorem 6.7 [38]

(i)   *An arbitrary quasi-bi-Hamiltonian system, which in Nijenhuis coordinates has the inverse Pfaffian form (6.270), has n independent constants of the motion of the form*

$$\overline{h}_r(\lambda, \mu, c) = \sum_{i=1}^{n} \frac{\rho_{r-1}^i}{\Delta_i} \overline{f}_i(\lambda_i, \mu_i) + c\overline{p}_r(\lambda), \quad r = 1, \dots, n, \tag{6.312}$$

   where Δ_i, ρ_r^i and $\overline{f}_i$ are the same as in Theorem 6.5.

(ii) *An arbitrary function $\overline{h}_r$ is separable for each n-tuple of functions $\overline{f}_i(\lambda_i, \mu_i)$, that is an appropriate Hamilton–Jacobi equation can be solved.*

(iii) *On the phase space* $M \ni (\lambda, \mu)$ *the functions* $\bar{h}_k$, $k = 1, ..., n$, *are in involution with respect to the Poisson tensors*

$$\theta_0 = \begin{pmatrix} 0 & I \\ -I & 0 \end{pmatrix}, \quad \theta_{-1} = \begin{pmatrix} 0 & \Lambda^{-1} \\ -\Lambda^{-1} & 0 \end{pmatrix}. \tag{6.313}$$

(iv) *On the extended phase space* $\overline{M} \ni (\lambda, \mu, c)$, *functions* $\bar{h}_r$ *with* $c \neq 0$ *form an inverse bi-Hamiltonian Nijenhuis chain with respect to the degenerate Poisson tensors*

$$\bar{\theta}_0 = \begin{pmatrix} 0 & I & 0 \\ -I & 0 & 0 \\ 0 & 0 & 0 \end{pmatrix}, \quad \bar{\theta}_{-1} = \begin{pmatrix} 0 & \Lambda^{-1} & \frac{\partial \bar{h}_n}{\partial \mu} \\ -\Lambda^{-1} & 0 & -\frac{\partial \bar{h}_n}{\partial \lambda} \\ -\left(\frac{\partial \bar{h}_n}{\partial \mu}\right)^{\mathrm{T}} & \left(\frac{\partial \bar{h}_n}{\partial \lambda}\right)^{\mathrm{T}} & 0 \end{pmatrix}. \tag{6.314}$$

The chain starts with the Casimir $\bar{h}_1$ *of the noncanonical Poisson tensor* $\bar{\theta}_{-1}$ *and terminates with the Casimir* $\bar{h}_{n+1} = c$ *of the canonical Poisson tensor* $\bar{\theta}_0$.

Proof. The proof is analogous to the one of Theorem 6.4, where additionally we use the relation $\bar{\rho}_r = -\rho_{r-1}/\rho_n$.

Finally, let us relate both the Pfaffian and inverse Pfaffian representations.

Lemma 6.7 *The canonical map* $\phi : \lambda_i = \frac{1}{\bar{\lambda}_i}$, $\mu_i = -\bar{\lambda}_i^2 \bar{\mu}_i$, $i = 1, ..., n$, *relates the Nijenhuis and inverse Nijenhuis bi-Hamiltonian chains from Theorems 6.6 and 6.7.*

Proof. From the definitions of ρ_r and $\bar{\rho}_r$ as well as from relations (6.300) one gets

$$\rho_r(\lambda) = \bar{\rho}_{n+1-r}(\bar{\lambda}) = -\frac{\rho_{n-r}(\bar{\lambda})}{\rho_n(\bar{\lambda})},$$

$$\rho_{r-1}^i(\lambda) = -\bar{\lambda}_i \frac{\rho_{n-r}^i(\bar{\lambda})}{\rho_n(\bar{\lambda})}, \quad \Delta_i(\lambda) = -\frac{\Delta_i(\bar{\lambda})}{\rho_n(\bar{\lambda})\bar{\lambda}_i^{n-2}}. \tag{6.315}$$

Hence we have

$$\bar{h}_{n-r+1}(\bar{\lambda}, \bar{\mu}, c) = h_r(\lambda, \mu, c) \cdot \phi, \quad r = 0, ..., n, \tag{6.316}$$

where

$$h_r(\lambda, \mu, c) = \sum_{i=1}^n \frac{\rho_{r-1}^i(\lambda)}{\Delta_i(\lambda)} f_i(\lambda_i, \mu_i) + c \rho_r(\lambda), \tag{6.317}$$

$$\bar{h}_{n-r+1}(\bar{\lambda}, \bar{\mu}, c) = \sum_{i=1}^n \frac{\rho_{n-r}^i(\bar{\lambda})}{\Delta_i(\bar{\lambda})} \bar{f}_i(\bar{\lambda}_i, \bar{\mu}_i) + c \bar{\rho}_{n-r+1}(\bar{\lambda}). \tag{6.318}$$

and $\overline{f}_i(\overline{\lambda}_i, \overline{\mu}_i) = \overline{\lambda}_i^{n-1} f_i(\frac{1}{\overline{\lambda}_i}, -\overline{\lambda}_i^2 \overline{\mu}_i)$. The relation $\overline{\theta}_1(\lambda) = \phi' \circ \overline{\theta}_{-1}(\overline{\lambda}) \circ \phi'^T$ can be verified by direct calculations. $\square$

As a consequence of Lemma 6.7 each of the bi-Hamiltonian systems considered can be transformed into both Pfaffian and inverse Pfaffian quasi-bi-Hamiltonian forms and hence the Pfaffian form is sufficient for the Nijenhuis representation. Nevertheless, the explicit inverse Pfaffian form seems to be necessary to complete the whole picture. For example, this form is crucial to understanding the last case of Theorem 6.3 and the related hierarchy. Moreover, as we shall see below, in many examples it is easier to calculate the inverse Nijenhuis chain then the Nijenhuis chain.

Example 6.20 Separability of the Henon–Heiles system and its generalizations.
For the first generalization (6.255)–(6.259) we find

$$\frac{\partial h_1}{\partial c} = \frac{\partial \overline{H}}{\partial c} = -q_1 = -\lambda_1 - \lambda_2,$$

$$\frac{\partial h_2}{\partial c} = \frac{\partial(-\frac{1}{4}\overline{F})}{\partial c} = -\frac{1}{4}q_2^2 = \lambda_1 \lambda_2, \tag{6.319}$$

and hence

$$q_1 = \lambda_1 + \lambda_2, \quad q_2 = 2\sqrt{-\lambda_1 \lambda_2}. \tag{6.320}$$

From the generating function $S = p_1(\lambda_1 + \lambda_2) + 2p_2\sqrt{-\lambda_1 \lambda_2}$ we get the conjugate momenta

$$p_1 = \sqrt{-\lambda_1 \lambda_2}\left(\frac{\mu_1}{\Delta_1} + \frac{\mu_2}{\Delta_2}\right), \quad p_2 = \frac{\lambda_1 \mu_1}{\Delta_1} + \frac{\lambda_2 \mu_2}{\Delta_2}, \tag{6.321}$$

where $\Delta_1 = -\Delta_2 = \lambda_1 - \lambda_2$. In the Nijenhuis coordinates, the conserved functions read

$$h_1 = \frac{\frac{1}{2}\lambda_1(\mu_1^2 + 2\lambda_1^3)}{\Delta_1} + \frac{\frac{1}{2}\lambda_2(\mu_2^2 + 2\lambda_2^3)}{\Delta_2} - c(\lambda_1 + \lambda_2),$$

$$h_2 = -\frac{\frac{1}{2}\lambda_2\lambda_1(\mu_1^2 + 2\lambda_1^3)}{\Delta_1} - \frac{\frac{1}{2}\lambda_1\lambda_2(\mu_2^2 + 2\lambda_2^3)}{\Delta_2} + c\lambda_1 \lambda_2. \tag{6.322}$$

The Hamilton–Jacobi equations $h_i(\lambda, \frac{\partial W_i}{\partial \lambda}) = E_i$, $i = 1, 2$, are separable and have the complete integrals $W_i = \sum_{k=1}^{2} W_i^{(k)}(\lambda_k; c_1, c_2)$, where $W_i^{(k)}$ are solutions of the following equations

$$\frac{dW_i^{(k)}}{d\lambda_k} = \left[\frac{2}{\lambda_k}\left(-\lambda_k^4 + c\lambda_k^2 + c_2\lambda_k + c_1\right)\right]^{1/2}, \quad E_i = c_{3-i}, \quad i, k = 1, 2. \tag{6.323}$$

For the second generalized Henon–Heiles representation (6.260)–(6.264) from the relations

$$\frac{\partial h_1}{\partial d} = \frac{\partial(-2\overline{H})}{\partial d} = -\frac{1}{q_2^2} = \frac{1}{\lambda_1\lambda_2},$$

$$\frac{\partial h_2}{\partial d} = \frac{\partial \overline{F}}{\partial d} = \frac{2q_1}{q_2^2} = -\frac{1}{\lambda_1} - \frac{1}{\lambda_2}, \tag{6.324}$$

we find the following Nijenhuis coordinates

$$q_1 = \frac{1}{2}(\lambda_1 + \lambda_2), \quad q_2 = \sqrt{-\lambda_1\lambda_2},$$

$$p_1 = 2\sqrt{-\lambda_1\lambda_2}\left(\frac{\mu_1}{\Delta_1} + \frac{\mu_2}{\Delta_2}\right), \quad p_2 = 2\left(\frac{\lambda_1\mu_1}{\Delta_1} + \frac{\lambda_2\mu_2}{\Delta_2}\right), \tag{6.325}$$

and the following conserved functions

$$h_1 = \frac{\frac{1}{4}\lambda_1(16\mu_1^2 + \lambda_1^3)}{\Delta_1} + \frac{\frac{1}{4}\lambda_2(16\mu_2^2 + \lambda_2^3)}{\Delta_2} - \frac{d}{\lambda_1\lambda_2},$$

$$h_2 = -\frac{\frac{1}{4}\lambda_2\lambda_1(16\mu_1^2 + \lambda_1^3)}{\Delta_1} - \frac{\frac{1}{4}\lambda_1\lambda_2(16\mu_2^2 + \lambda_2^3)}{\Delta_2} + d\left(\frac{1}{\lambda_1} + \frac{1}{\lambda_2}\right). \tag{6.326}$$

Again the Hamilton–Jacobi equations $h_i(\lambda, \frac{\partial W_i}{\partial \lambda}) = E_i$, $i = 1, 2$, are separable and have the complete integrals $W_i = \sum_{k=1}^2 W_i^{(k)}(\lambda_k; c_1, c_2)$, where $W_i^{(k)}$ are solutions of the following equations

$$\frac{dW_i^{(k)}}{d\lambda_k} = \left[\frac{4}{\lambda_k}\left(-14\lambda_k^4 + c_2\lambda_k + c_1 + \frac{d}{\lambda_k}\right)\right]^{1/2}, \quad E_i = c_{3-i}, \quad i, k = 1, 2. \tag{6.327}$$

Example 6.21 Separability of the second Newton representation for the stationary flow of the Harry–Dym equation.
Newton system (6.138) with $\gamma = -2$ reads

$$q_{1xx} = 8q_1^{-5}q_2, \tag{}$$

$$q_{2xx} = -2q_1^{-4} - d, \tag{6.328}$$

where $q_1 = \overline{r}_1$, $q_2 = \overline{p}_2$ and $d = \overline{d}$. With $p_1 = q_{1x}$ and $p_2 = q_{2x}$ the related bi-Hamiltonian chain (6.140) is given by the Hamiltonians

$$h_0 = d,$$

$$h_1 = \frac{1}{2}p_1^2 + \frac{1}{2}p_2^2 + 2q_1^{-4}q_2 + dq_2,$$

$$h_2 = \frac{1}{2}q_2p_1^2 - \frac{1}{2}q_1p_1p_2 + \frac{1}{2}q_1^{-2} + 2q_1^{-4}q_2^2 - \frac{1}{4}dq_1^2, \tag{6.329}$$

the canonical Poisson tensor θ_0 and the noncanonical tensor given by (6.142)

$$\theta_1 = \begin{pmatrix} 0 & 0 & 0 & -\frac{1}{2}q_1 & p_1 \\ 0 & 0 & -\frac{1}{2}q_1 & -q_2 & p_2 \\ 0 & \frac{1}{2}q_1 & 0 & \frac{1}{2}q_1 & -8q_1^{-5}q_2 \\ \frac{1}{2}q_1 & q_2 & -\frac{1}{2}q_1 & 0 & 2q_1^{-4}+d \\ -p_1 & -p_2 & 8q_1^{-5}q_2 & -2q_1^{-4}-d & 0 \end{pmatrix}. \tag{6.330}$$

Hence we have

$$\rho_1 = \frac{\partial h_1}{\partial d} = q_2 = -\lambda_1 - \lambda_2,$$

$$\rho_2 = \frac{\partial h_2}{\partial d} = -\frac{1}{4}q_1^2 = \lambda_1\lambda_2, \tag{6.331}$$

and then

$$q_1 = 2\sqrt{-\lambda_1\lambda_2}, \qquad q_2 = -\lambda_1 - \lambda_2. \tag{6.332}$$

Introducing the generating function $S = 2p_2\sqrt{-\lambda_1\lambda_2} - p_2(\lambda_1 + \lambda_2)$ and inverting the system $\mu = \frac{\partial S}{\partial \lambda}$ we get the second part of the transformation

$$p_1 = \sqrt{-\lambda_1\lambda_2}\left(\frac{\mu_1}{\Delta_1} + \frac{\mu_2}{\Delta_2}\right), \quad p_2 = -\lambda_1\frac{\mu_1}{\Delta_1} - \lambda_2\frac{\mu_2}{\Delta_2}, \tag{6.333}$$

and hence

$$h_1 = \frac{\frac{1}{8}\lambda_1(4\mu_1^2 + \lambda_1^{-3})}{\Delta_1} + \frac{\frac{1}{8}\lambda_2(4\mu_2^2 + \lambda_2^{-3})}{\Delta_2} - d(\lambda_1 + \lambda_2),$$

$$h_2 = -\frac{\frac{1}{8}\lambda_2\lambda_1(4\mu_1^2 + \lambda_1^{-3})}{\Delta_1} - \frac{\frac{1}{8}\lambda_1\lambda_2(4\mu_2^2 + \lambda_2^{-3})}{\Delta_2} + d\lambda_1\lambda_2. \tag{6.334}$$

The Hamilton–Jacobi equations $h_i(\lambda, \frac{\partial W_i}{\partial \lambda}) = E_i$, $i = 1, 2$, are separable and have the complete integrals $W_i = \sum_{k=1}^{2} W_i^{(k)}(\lambda_k; c_1, c_2)$, where $W_i^{(k)}$ are solutions of the following equations

$$\frac{dW_i^{(k)}}{d\lambda_k} = \left[\frac{2}{\lambda_k}\left(-\frac{1}{8}\lambda_k^{-2} + d\lambda_k^2 + c_2\lambda_k + c_1\right)\right]^{1/2}, \qquad E_i = c_{3-i}, \ i, k = 1, 2. \tag{6.335}$$

Example 6.22 Separability of the first Newton representation of the seventh order stationary flow of the KdV.

We consider the Newton system (6.111) from Example 6.13 with related bi-Hamiltonian chain given by formulas (6.120)–(6.125). Hence we have

$$\rho_1 = \frac{\partial h_1}{\partial c} = -r_1 = -\lambda_1 - \lambda_2 - \lambda_3,$$

$$\rho_2 = \frac{\partial h_2}{\partial c} = \frac{1}{4}r_1^2 + \frac{1}{2}r_2 = \lambda_1\lambda_2 + \lambda_1\lambda_3 + \lambda_2\lambda_3,$$

$$\rho_3 = \frac{\partial h_3}{\partial c} = -\frac{1}{4}r_1r_2 - \frac{1}{4}r_3 = -\lambda_1\lambda_2\lambda_3 \tag{6.336}$$

and then

$$r_1 = \lambda_1 + \lambda_2 + \lambda_3,$$

$$r_2 = (\lambda_1\lambda_2 + \lambda_1\lambda_3 + \lambda_2\lambda_3) - \frac{1}{2}(\lambda_1^2 + \lambda_2^2 + \lambda_3^2),$$

$$r_3 = 4\lambda_1\lambda_2\lambda_3 + \frac{1}{3}(\lambda_1 + \lambda_2 + \lambda_3)[4(\lambda_1\lambda_2 + \lambda_1\lambda_3 + \lambda_2\lambda_3)$$

$$- \frac{1}{3}(\lambda_1^2 + \lambda_2^2 + \lambda_3^2)]. \tag{6.337}$$

From the appropriate generating function we calculate

$$s_1 = -\frac{1}{8}\left\{[(-3\lambda_1^2 + \lambda_2^2 + \lambda_3^2) + 2(\lambda_1\lambda_2 + \lambda_1\lambda_3 - \lambda_2\lambda_3)]\frac{\mu_1}{\Delta_1} + \text{c.p.}\right\},$$

$$s_2 = \frac{1}{4}\left[(-\lambda_1 + \lambda_2 + \lambda_3)\frac{\mu_1}{\Delta_1} + (\lambda_1 - \lambda_2 + \lambda_3)\frac{\mu_2}{\Delta_2} + (\lambda_1 + \lambda_2 - \lambda_3)\frac{\mu_3}{\Delta_3}\right],$$

$$s_3 = \frac{1}{4}\left(\frac{\mu_1}{\Delta_1} + \frac{\mu_2}{\Delta_2} + \frac{\mu_3}{\Delta_3}\right), \tag{6.338}$$

where $\Delta_1 = (\lambda_1 - \lambda_2)(\lambda_1 - \lambda_3), ...,$ and c.p. stands for cyclic permutations of the indices $1, 2, 3$, in λ and μ. The three Hamiltonians in the Nijenhuis coordinates take the form

$$h_1 = \frac{f_1}{\Delta_1} + \frac{f_2}{\Delta_2} + \frac{f_3}{\Delta_3} - c(\lambda_1 + \lambda_2 + \lambda_3),$$

$$h_2 = -\frac{(\lambda_2 + \lambda_3)f_1}{\Delta_1} - \frac{(\lambda_1 + \lambda_3)f_2}{\Delta_2} - \frac{(\lambda_1 + \lambda_2)f_3}{\Delta_3} + c(\lambda_1\lambda_2 + \lambda_1\lambda_3 + \lambda_2\lambda_3),$$

$$h_3 = \frac{\lambda_2\lambda_3 f_1}{\Delta_1} + \frac{\lambda_1\lambda_3 f_2}{\Delta_2} + \frac{\lambda_1\lambda_2 f_3}{\Delta_3} - c\lambda_1\lambda_2\lambda_3,$$

$$\tag{6.339}$$

where $f_i(\lambda_i, \mu_i) = \frac{1}{8}(\mu_i^2 + 128\lambda_i^7)$. The appropriate Hamilton–Jacobi equations $h_i(\lambda, \frac{\partial W_i}{\partial \lambda}) = E_i$ are separable and have the complete integrals

$$W_i = \sum_{k=1}^{3} W_i^{(k)}(\lambda_k; c_1, c_2, c_3), \quad i = 1, 2, 3 \tag{6.340}$$

with $W_i^{(k)}$ being solutions of the following equations

$$\frac{dW_i^{(k)}}{d\lambda_k} = [8(-16\lambda_k^7 + c\lambda_k^3 + c_3\lambda_k^2 + c_2\lambda_k + c_1)]^{1/2}, \quad E_i = c_{4-i}. \tag{6.341}$$

Example 6.23 Separability of the second Newton representation of the seventh order stationary flow of the KdV.

Let us consider the Newton system (6.112) from Example 6.13 ($\bar{r}_i = q_i$)

$$0 = q_{1_{xx}} + 10q_1^2 - 4q_2,$$

$$0 = q_{2_{xx}} + 20q_1q_2 - 10q_1^3 + 2q_3,$$

$$0 = q_{3_{xx}} + 4q_1q_3 - dq_3^{-3}. \tag{6.342}$$

The related bi-Hamiltonian chain takes the form (6.301); h_r and $\bar{\theta}_{-1}$ are [164]

$$h_1 = -2p_1p_2 - 2p_3^2 - 20q_1^2q_2 + 4q_2^2 + 5q_1^4 - 4q_1q_3^2 - dq_3^{-2},$$

$$h_2 = -p_1^2 - 2q_1p_1p_2 + 2q_2p_2^2 - 2q_1p_3^2 + 2q_3p_2p_3 + 2q_1^2q_3^2$$
$$\qquad -4q_2q_3^2 - 16q_1q_2^2 + 4q_1^5 - 2dq_2q_3^{-2},$$

$$h_3 = -q_3^2p_2^2 + 2q_3p_1p_3 + 2q_1q_3p_2p_3 - q_1^2p_3^2 + 8q_2p_3^2 + q_3^4$$
$$\qquad +8q_1q_2q_3^2 - 4q_1^3q_3^2 - dq_1^2q_3^{-2} - 2dq_2q_3^{-2},$$

$$h_4 = d, \tag{6.343}$$

and

$$\bar{\theta}_{-1} = \begin{pmatrix} 0 & 0 & 0 & 0 & 0 & -1/q_3 & A_1 \\ 0 & 0 & 0 & 1 & 0 & -q_1/q_3 & A_2 \\ 0 & 0 & 0 & -q_1/q_3 & -1/q_3 & B_1 & A_3 \\ 0 & -1 & q_1/q_3 & 0 & 0 & B_2 & A_4 \\ 0 & 0 & 1/q_3 & 0 & 0 & -p_3/q_3^2 & A_5 \\ 1/q_3 & q_1/q_3 & -B_1 & -B_2 & p_3/q_3^2 & 0 & A_6 \\ -A_1 & -A_2 & -A_3 & -A_4 & -A_5 & -A_6 & 0 \end{pmatrix}, \tag{6.344}$$

where

$$p_1 = q_{2_x}, \ p_2 = q_{1_x}, \ p_3 = q_{3_x}, \ (A_1, ..., A_6, 0)^{\mathrm{T}} = \bar{\theta}_0 \circ \nabla h_3,$$

$$B_1 = (q_1^2 + 2q_2)/q_3^2, \ B_2 = (q_3p_2 - q_1p_3)/q_3^2.$$

Hence we find

$$\bar{\rho}_1 = -q_3^{-2} = -\frac{1}{\lambda_1\lambda_2\lambda_3},$$

$$\bar{\rho}_2 = -2q_1q_3^{-2} = \frac{1}{\lambda_1\lambda_2} + \frac{1}{\lambda_1\lambda_3} + \frac{1}{\lambda_2\lambda_3},$$

$$\bar{\rho}_3 = -q_1^2q_3^{-2} - 2q_2q_3^{-2} = -\frac{1}{\lambda_1} - \frac{1}{\lambda_2} - \frac{1}{\lambda_3}, \tag{6.345}$$

and then the Nijenhuis map

$$q_1 = -\frac{1}{2}(\lambda_1 + \lambda_2 + \lambda_3),$$

$$q_2 = \frac{1}{2}(\lambda_1\lambda_2 + \lambda_1\lambda_3 + \lambda_2\lambda_3) - \frac{1}{8}(\lambda_1 + \lambda_2 + \lambda_3)^2,$$

$$q_3 = (\lambda_1\lambda_2\lambda_3)^{1/2},$$

$$p_1 = \lambda_1(-\lambda_1 + \lambda_2 + \lambda_3)\frac{\mu_1}{\Delta_1} + \lambda_2(\lambda_1 - \lambda_2 + \lambda_3)\frac{\mu_2}{\Delta_2}$$

$$+\lambda_3(\lambda_1 + \lambda_2 - \lambda_3)\frac{\mu_3}{\Delta_3},$$

$$p_2 = -2\left(\lambda_1\frac{\mu_1}{\Delta_1} + \lambda_2\frac{\mu_2}{\Delta_2} + \lambda_3\frac{\mu_3}{\Delta_3}\right),$$

$$p_3 = 2(\lambda_1\lambda_2\lambda_3)^{1/2}\left(\frac{\mu_1}{\Delta_1} + \frac{\mu_2}{\Delta_2} + \frac{\mu_3}{\Delta_3}\right). \tag{6.346}$$

The three Hamiltonians (6.343) in the Nijenhuis coordinates take the form

$$h_1 = \frac{f_1}{\Delta_1} + \frac{f_2}{\Delta_2} + \frac{f_3}{\Delta_3} - d\frac{1}{\lambda_1\lambda_2\lambda_3},$$

$$h_2 = -\frac{(\lambda_2 + \lambda_3)f_1}{\Delta_1} - \frac{(\lambda_1 + \lambda_3)f_2}{\Delta_2} - \frac{(\lambda_1 + \lambda_2)f_3}{\Delta_3}$$

$$+d\left(\frac{1}{\lambda_1\lambda_2} + \frac{1}{\lambda_1\lambda_3} + \frac{1}{\lambda_2\lambda_3}\right),$$

$$h_3 = \frac{\lambda_2\lambda_3 f_1}{\Delta_1} + \frac{\lambda_1\lambda_3 f_2}{\Delta_2} + \frac{\lambda_1\lambda_2 f_3}{\Delta_3} - d\left(\frac{1}{\lambda_1} + \frac{1}{\lambda_2} + \frac{1}{\lambda_3}\right), \tag{6.347}$$

where $f_i(\lambda_i, \mu_i) = \lambda_i(\lambda_i^5 - 4\mu_i^2)$. The appropriate Hamilton–Jacobi equations $h_i(\lambda, \frac{\partial W_i}{\partial \lambda}) = E_i$ are separable and have the complete integrals

$$\frac{dW_i^{(k)}}{d\lambda_k} = \left[-\frac{1}{4\lambda_k}\left(-\lambda_k^6 + c_3\lambda_k^2 + c_2\lambda_k + c_1 + \frac{d}{\lambda_k}\right)\right]^{1/2}, \quad E_i = c_{4-i}. \tag{6.348}$$

Notice that for all the above examples the functions $f_i(\lambda_i, \mu_i)$ are quadratic with respect to the momenta μ_i

$$f_i(\lambda_i, \mu_i) = \frac{1}{2}\varphi(\lambda_i)\mu_i^2 + \psi(\lambda_i), \tag{6.349}$$

which allows us to find a general solution for such a Nijenhuis chains. First, let us consider a Nijenhuis chain with f_i functions given by (6.349). The involutive hierarchy of Hamiltonians (6.297) reads

$$h_i(\lambda, \mu, c) = \frac{1}{2}\sum_{k=1}^{n}\gamma_i^k(\lambda)g^{kk}(\lambda)\mu_k^2 + V_i(\lambda) + c\rho_i(\lambda), \tag{6.350}$$

where

$$g^{kk}(\lambda) = \frac{\varphi(\lambda_k)}{\Delta_k(\lambda)}, \quad \gamma_i^k(\lambda) = \rho_{i-1}^k(\lambda), \quad V_i(\lambda) = \sum_{k=1}^{n}\rho_{i-1}^k(\lambda)\frac{\psi(\lambda_k)}{\Delta_k(\lambda)}. \tag{6.351}$$

According to the Hamilton–Jacobi theory, we look for a canonical transformation $(\lambda_i, \mu_i) \to (b_i, c_i)$ in the form $b_i = \partial W/\partial c_i$, $\mu_i = \partial W/\partial \lambda_i$, where

$W(\lambda_1, ..., \lambda_n, c_1, ..., c_n)$ is the generating function, satisfying the Hamilton–Jacobi equation

$$h_1\left(\lambda, \frac{\partial W}{\partial \lambda}, c\right) = \frac{1}{2}\sum_{k=1}^{n} \frac{\varphi(\lambda_k)(\partial W/\partial \lambda_k)^2}{\Delta_k(\lambda)} + V_1(\lambda) - c\,(\lambda_1 + ... + \lambda_n) = \text{const.}$$

(6.352)

The integrability of (6.352) with $c = 0$ was derived in [65]. Here, we include a Casimir variable (c parameter) different from zero. Since (6.352) is separable we can write

$$W(\lambda_1, ..., \lambda_n, c_1, ..., c_n) = W_1(\lambda_1, c_1, ..., c_n) + ... + W_n(\lambda_n, c_1, ..., c_n) \quad (6.353)$$

and hence (6.352) turns to the form

$$\sum_{k=1}^{n} \frac{\frac{1}{2}\varphi(\lambda_k)(\partial W_k/\partial \lambda_k)^2 + \psi(\lambda_k)}{\Delta_k(\lambda)} - c\,(\lambda_1 + ... + \lambda_n) = \text{const.} \quad (6.354)$$

with the solution

$$\frac{1}{2}\varphi(\lambda_k)(\partial W_k/\partial \lambda_k)^2 + \psi(\lambda_k) = \nu(q^k), \quad (6.355)$$

where

$$\nu(\xi) = c_1 + c_2\xi + ... + c_n\xi^{n-1} + c\xi^n. \quad (6.356)$$

Thus, we obtain

$$W(\lambda_1, ..., \lambda_n, c_1, ..., c_n) = \sum_{k=1}^{n} \int^{\lambda_k} \sqrt{\frac{\nu(\xi) - \psi(\xi)}{\frac{1}{2}\varphi(\xi)}}\,d\xi, \quad (6.357)$$

which result stays in excellent agreement with the considered examples.

In the new canonical variables c_i, $b_i = \partial W/\partial c_i$, the Hamiltonians h_i become

$$h_i = c_{n+1-i} \quad (6.358)$$

with

$$b^i = \frac{\partial W}{\partial c_i} = \frac{1}{2}\sum_{k=1}^{n} \int^{\lambda_k} \frac{\xi^{i-1}}{\sqrt{\frac{1}{2}\varphi(\xi)[\nu(\xi) - \psi(\xi)]}}\,d\xi. \quad (6.359)$$

As in the new coordinates each h_i generates the trivial flow

$$(c_j)_{t_i} = -\frac{\partial h_i}{\partial b^j} = 0, \quad (b^j)_{t_i} = \frac{\partial h_i}{\partial c_j} = \frac{1}{2}\delta^j_{n+1-i}, \quad c_{t_i} = 0 \quad (6.360)$$

hence

$$b^i = t_{n+1-i} + \text{const.} \quad (6.361)$$

Combining (6.359) with (6.361) we arrive at implicit solutions for the trajectories $q^i(t_k)$, with respect to the evolution parameter t_k, in the form

$$\frac{1}{2}\int^{\lambda_1}\frac{\xi^i\mathrm{d}\xi}{f(\xi)}+...+\frac{1}{2}\int^{\lambda_n}\frac{\xi^i\mathrm{d}\xi}{f(\xi)}=\delta_i^{n-k}t_k+\text{const},\quad i=1,...,n,\quad(6.362)$$

where $f(\xi)=\sqrt{\frac{1}{2}\varphi(\xi)[\nu(\xi)-\psi(\xi)]}$.

In an analogous way one can integrate the inverse bi-Hamiltonian Nijenhuis chain where now the involutive set of Hamiltonians (6.312) reads

$$\overline{h}_i(\lambda,\mu,c)=\frac{1}{2}\sum_{k=1}^{n}\gamma_i^k(\lambda)\overline{g}^{kk}(\lambda)\mu_k^2+\overline{V}_i(\lambda)+c\overline{p}_i(\lambda),\qquad(6.363)$$

and

$$\overline{g}^{kk}(\lambda)=\frac{\overline{\varphi}(\lambda_k)}{\Delta_k(\lambda)},\qquad \overline{V}_1(\lambda)=\sum_{k=1}^{n}\rho_{i-1}^k(\lambda)\frac{\overline{\psi}(\lambda_k)}{\Delta_k(\lambda)}.\qquad(6.364)$$

Again the Hamilton–Jacobi equation is separable and takes the form

$$\overline{h}_1\left(\lambda,\frac{\partial\overline{W}}{\partial\lambda},c\right)=\sum_{k=1}^{n}\frac{\frac{1}{2}\overline{\varphi}(\lambda_k)(\partial\overline{W}_k/\partial\lambda_k)^2+\overline{\psi}(\lambda_k)}{\Delta_k(\lambda)}+\frac{(-1)^n}{\lambda_1\cdot...\cdot\lambda_n}=\text{const},$$
$$(6.365)$$

with the solution

$$\frac{1}{2}\overline{\varphi}(\lambda_k)(\partial\overline{W}_k/\partial\lambda_k)^2+\overline{\psi}(\lambda_k)=\overline{\nu}(\lambda_k),\qquad(6.366)$$

where

$$\overline{\nu}(\xi)=\frac{c}{\xi}+c_1+c_2\xi+...+c_n\xi^{n-1}.\qquad(6.367)$$

The remaining calculations and general formulas are the same as in the case of the bi-Hamiltonian Nijenhuis chain.

Present results allow us to re-examine the theory of classical Hamiltonian systems with separable potentials in the framework of the bi-Hamiltonian theory. In this respect we follow a remarkable paper [165] with some modifications according to the results of this section. In [188],[131] there has been formulated and proved a theorem which says that the Hamilton–Jacobi equation for a natural Hamiltonian system $H=T+V(q)$ is separable in generalized elliptic coordinates if it has N functionally independent and commuting integrals of motion of the form

$$K_i(q,p)=\frac{1}{2}\sum_{j=1,j\neq i}^{N}(\alpha_i-\alpha_j)^{-1}l_{ij}^2+p_i^2+V_i(q)=P_i(q,p)+V_i(q),\quad(6.368)$$

where q_i,p_i are canonically conjugated variables, $l_{ij}=q_ip_j-q_jp_i$ $(i,j=1,...,N)$ are angular momenta, α_i are different, positive constants and $V_i(q)$ are functions of $q=(q_1,...,q_N)$ such that the Poisson brackets $\{H,K_i\}=0, i=1,...,N$. The Hamiltonian is given as

$$H = \frac{1}{2} \sum_{i=1}^{N} K_i(q,p) = \frac{1}{2}(p,p) + \sum_{i=1}^{N} \frac{1}{2} V_i(q) = T + V(q). \qquad (6.369)$$

Theorem 6.8 *A bi-Hamiltonian formulation for elliptic separable potentials. Every natural Hamiltonian system $H = T + V(q)$ separable in generalized elliptic coordinates admits, in the extended phase space of variables (q, p, c), the bi-Hamiltonian formulation*

$$\begin{pmatrix} q \\ p \\ c \end{pmatrix}_t = \theta_0 \circ \nabla h_1 = \theta_1 \circ \nabla h_0, \qquad \theta_0 = \begin{pmatrix} 0 & I & 0 \\ -I & 0 & 0 \\ 0 & 0 & 0 \end{pmatrix},$$

$$\theta_1 = \begin{pmatrix} 0 & A - \frac{1}{2} q \otimes q & q_t \\ -A + \frac{1}{2} q \otimes q & \frac{1}{2} p \otimes q - \frac{1}{2} q \otimes p & p_t \\ -q_t & -p_t & 0 \end{pmatrix}, \qquad (6.370)$$

where $A = \text{diag}(\alpha_1, ..., \alpha_N)$, $h_0 = c$, $h_1 = H + \frac{1}{2} c(q,q)$ and in the remaining column and row of θ_1, $q_t = \partial h_1 / \partial p = p$, $p_t = -\partial h_1 / \partial q = -\partial(V + \frac{1}{2} cq^2) / \partial q$ denote the components of the Hamiltonian vector field. Conversely, every natural Hamiltonian system $H = T + V(q)$ admitting the bi-Hamiltonian formulation (6.370) is separable in generalized elliptic coordinates. The potential V satisfies the equations

$$0 = (\alpha_i - \alpha_r) \frac{\partial^2 V}{\partial q_i \partial q_r} + \frac{3}{2} \left(q_r \frac{\partial V}{\partial q_i} - q_i \frac{\partial V}{\partial q_r} \right)$$

$$+ \frac{1}{2} \sum_{j=1}^{N} \left(q_j q_r \frac{\partial^2 V}{\partial q_i \partial q_j} - q_i q_j \frac{\partial^2 V}{\partial q_j \partial q_r} \right), \qquad i, r = 1, ..., N, \qquad (6.371)$$

which are the iff condition for θ_1 to fulfil the Jacobi identity. This bi-Hamiltonian formulation generates an infinite chain of commuting bi-Hamiltonian vector fields

$$\theta_0 \circ \nabla h_0 = 0, ..., \theta_0 \circ \nabla h_{r+1} = \theta_1 \circ \nabla h_r, ... \qquad r = 0, 1, 2, ... \qquad (6.372)$$

with $h_0 = c$ and

$$h_r = \frac{1}{2} \sum_{i=1}^{N} \alpha_i^{r-1} \widetilde{K}_i = \frac{1}{2} \sum_{i=1}^{N} \alpha_i^{r-1} K_i + \frac{1}{2} c(q, A^{r-1} q), \qquad r = 1, 2, ... \qquad (6.373)$$

where $\widetilde{K}_i = K_i + c q_i^2$ and K_i are given by (6.368). Moreover we can define new Hamiltonians

$$\widetilde{h}_r = h_r + \sum_{k=1}^{r} \beta_{N-k} h_{r-k}, \qquad r = 0, ..., N, \qquad (6.374)$$

where β_r are solutions of the set of linear equations

$$\alpha_i^N + \sum_{r=0}^{N-1} \beta_r \alpha_i^r = 0, \tag{6.375}$$

so that the new chain starts with $\widetilde{h}_0 = h_0 = c$, a Casimir of θ_0, and terminates with

$$\widetilde{h}_N = \beta_0 \left(\frac{1}{2} \sum_{i=1}^{N} \alpha_i^{-1} \widetilde{K}_i - c \right), \tag{6.376}$$

which is a Casimir of θ_1.

Proof. One can check that (6.370) holds and that θ_1 satisfies the Jacobi identity due to condition (6.371). The bi-Hamiltonian chain (6.372),(6.373) has to be calculated directly. The kinetic (p-dependent) part of $\theta_0 \circ \nabla h_{r+1} = \theta_1 \circ \nabla h_r$ holds due to the identity

$$\sum_{\text{pairs}(i,j)} \left(\alpha_i^r + \alpha_i^{r-1} \alpha_j + \ldots + \alpha_i \alpha_j^{r-1} + \alpha_j^r \right)$$

$$= (q, A^r q)(p, p) + (q A^{r-1}, q)(p, Ap) + \ldots + (q, q)((p, A^r p)$$
$$- (q, A^r p)(q, p) - (q, A^{r-1} p)(q, Ap) - \ldots - (q, p)(q, A^r p). \tag{6.377}$$

The remaining 'potential part' of (6.372),(6.373) follows from the equality

$$\left(A - \frac{1}{2} q \otimes q \right) \frac{\partial}{\partial q} \left(\frac{1}{2} \sum_{i=1}^{N} \alpha_i^r V_i \right) + \frac{1}{2}(q, A^r q) \frac{\partial V}{\partial q} = \frac{\partial}{\partial q} \left(\frac{1}{2} \sum_{i=1}^{N} \alpha_i^{r+1} V_i \right), \tag{6.378}$$

where $\frac{\partial}{\partial q}$ denotes the vector derivative. This equality can be verified by expressing the derivatives $\partial V_i / \partial q_j$ through the first derivatives of the potential V as follows from $0 = \{H, K_i\} = \{V, P_i\} + \{\frac{1}{2}(p, p), V_i\}$. The possibility of redefining the chain (6.372) is due to the nonuniqueness of h_{r+1} as defined by $\theta_0 \circ \nabla h_{r+1} = \theta_1 \circ \nabla h_r$: every next h_{r+1} is determined up to the additive term const $\times h_0$ which is a Casimir of θ_0. Of course, only a finite number of Hamiltonians h_r can be functionally independent but the special property of this chain is that h_{r+1} can be expressed through the previous h_r as their linear combination with constant coefficients.

Finally, let us construct the Nijenhuis coordinates. The solution to the set of equations (6.375) takes the form

$$\beta_{N-r} = (-1)^r \sum_{\substack{j_1, \ldots, j_r \\ j_1 < \ldots < j_r}} \alpha_{j_1} \cdot \ldots \cdot \alpha_{j_r} := \rho_r(\alpha), \quad \beta_N = 1. \tag{6.379}$$

Hence, the c-dependent part of

$$\widetilde{h}_r(q, p, c; \alpha) = \widetilde{h}_r(q, p; \alpha) + c\rho_r(q; \alpha) \tag{6.380}$$

reads

$$\rho_r(q;\alpha) = \beta_{N-r} + \frac{1}{2}\sum_{k=0}^{r-1}\beta_{N-k}(q, A^{r-k-1}q) \tag{6.381}$$

and defines the Nijenhuis coordinates (6.295) through the relation

$$\rho_r(\alpha) + \frac{1}{2}\sum_{k=0}^{r-1}\rho_k(\alpha)(q, A^{r-k-1}q) = \rho_r(\lambda). \tag{6.382}$$

But the so defined Nijenhuis coordinates are just the generalized elliptic coordinates. Indeed, the generalized elliptic coordinates $\lambda_1, ..., \lambda_N$ defined by the relation [188]

$$1 + \frac{1}{2}\sum_{k=1}^{N}\frac{q_k^2}{z - \alpha_k} = \frac{\prod_{j=1}^{N}(z - \lambda_j)}{\prod_{k=1}^{N}(z - \alpha_k)}, \tag{6.383}$$

are equivalent to the Nijenhuis coordinates defined by relations (6.391). To see this let

$$A(z) := \prod_{k=1}^{N}(z - \alpha_k) = \sum_{r=0}^{N}\rho_r(\alpha)z^{N-r},$$

$$\Lambda(z) := \prod_{k=1}^{N}(z - \lambda_k) = \sum_{r=0}^{N}\rho_r(\lambda)z^{N-r},$$

$$A_k(z) := \frac{A(z)}{z - \alpha_k} = \sum_{r=1}^{N}\rho_{r-1}^{k}(\alpha)z^{N-r}, \tag{6.384}$$

where

$$\rho_r^i(\alpha) = (-1)^r \sum_{\substack{j_1, \ldots, j_r \\ j_1 < \cdots < j_r \\ j_1 \neq i, \ldots, j_r \neq i}} \alpha_{j_1} \cdot \ldots \cdot \alpha_{j_r}.$$

From (6.392) it follows that

$$\Lambda(z) = A(z) + \sum_{k=1}^{N}A_k(z)q_k^2$$

$$= \sum_{r=0}^{N}\rho_r(\alpha)z^{N-r} + \frac{1}{2}\sum_{k=1}^{N}\sum_{r=1}^{N}\rho_{r-1}^{k}(\alpha)z^{N-r}q_k^2$$

$$= z^N + \sum_{r=1}^{N}\left[\rho_r(\alpha) + \frac{1}{2}\sum_{k=1}^{N}\rho_{r-1}^{k}(\alpha)q_k^2\right]z^{N-r}. \tag{6.385}$$

Comparing the coefficients of the same powers of z and applying the relation

$$\sum_{l=0}^{r}\rho_l(\alpha)\alpha_i^{r-l} = \rho_r^i(\alpha), \tag{6.386}$$

we get

$$\rho_r(\lambda) = \rho_r(\alpha) + \frac{1}{2}\sum_{k=1}^{N}\rho_{r-1}^k(\alpha)q_k^2$$

$$= \rho_r(\alpha) + \frac{1}{2}\sum_{k=1}^{N}\sum_{l=0}^{r-1}\rho_l(\alpha)\alpha_k^{r-l-1}q_k^2$$

$$= \rho_r(\alpha) + \frac{1}{2}\sum_{l=0}^{r-1}\rho_l(\alpha)(q, A^{r-l-1}q), \qquad (6.387)$$

which is relation (6.382). □

Example 6.24 The basic example here is the Garnier system, considered in detail in the previous section, with the potential $V^{(2)}(q) = \frac{1}{4}(q,q)^2 + \frac{1}{2}(q, Aq)$. Its bi-Hamiltonian formulation (6.178),(6.179) with the bi-Hamiltonian infinite chain (6.180),(6.181) is a special case of the system (6.370) and the chain (6.372),(6.373). This potential is a member of an infinite family of permutationally symmetric potentials separable in generalized elliptic coordinates which were found and characterized recursively in [188]. This family contains for instance $V^{(3)}(q) = \frac{1}{8}(q,q)^3 - \frac{1}{2}(q,q)(q, Aq) + \frac{1}{2}(q, A^2q)$ and $V^{(-1)}(q) = \left[1 - \frac{1}{2}(q, A^{-1}q)\right]^{-1}$. All these potentials come up as constraint spectral problems of the coupled KdV systems and its bi-Hamiltonian formulation has been initially derived in [13]. Nevertheless, all potentials, not necessarily related to the coupled KdV hierarchies, which have the bi-Hamiltonian formulation given in (6.370) are separable in generalized elliptic coordinates (6.383).

The generalized elliptic and the generalized parabolic coordinates are fundamental for the generation of all separable systems of coordinates on $\mathbb{R}^N$. Thus, the bi-Hamiltonian formulation of the parabolic separable potentials is of primary interest for understanding the bi-Hamiltonian formulation of all separable potentials. In order to derive this formulation one may use the fact that the generalized parabolic coordinates can be derived from the generalized elliptic coordinates by taking the improper limit $a_N \to \infty$ if one sets $\alpha_N = \frac{1}{2}a_N^2$, $\alpha_k = m_k a_N$, $q_N = \bar{q}_N - a_N$. Then the elliptic equations (6.370),(6.371) go to the limiting equations (6.389),(6.390) below. The elliptic integral K_N goes to the new, parabolic Hamiltonian $H = T + V(q)$ with the potential separable in the generalized parabolic coordinates and the other integrals K_i ($i = 1, ..., N - 1$) go to the integrals of motion of the parabolic Hamiltonian

$$K_i(q,p) = -\frac{1}{4}\sum_{j=1,j\neq i}^{N-1}(m_i - m_j)^{-1}l_{ij}^2 + m_i p_i^2 - l_{Ni}p_i + V_i(q) = \overline{P}_i(q,p) + V_i(q),$$

$$(6.388)$$

with $V_i(q)$ satisfying $\{\frac{1}{2}p^2, V_i\} + \{V(q), \overline{P}_i\} = 0(i = 1, ..., N)$. The elliptic Poisson operator θ_1 provides in the limit $a_N \to \infty$ the Poisson operator θ_1 given below.

Theorem 6.9 A bi-Hamiltonian formulation for parabolic separable potentials.

Every natural Hamiltonian system $H = T + V(q)$ separable in generalized parabolic coordinates admits, in the extended phase space of variables (q, p, c), the bi-Hamiltonian formulation

$$\begin{pmatrix} q \\ p \\ c \end{pmatrix}_t = \theta_0 \circ \nabla h_1 = \theta_1 \circ \nabla h_0,$$

$$\theta_1 = \begin{pmatrix} 0 & M + \frac{1}{2}\mathbf{1} \otimes q + \frac{1}{2}q \otimes \mathbf{1} & q_t \\ -M - \frac{1}{2}\mathbf{1} \otimes q - \frac{1}{2}q \otimes \mathbf{1} & \frac{1}{2}\mathbf{1} \otimes p - \frac{1}{2}p \otimes \mathbf{1} & p_t \\ -q_t & -p_t & 0 \end{pmatrix}, \quad (6.389)$$

where $h_0 = c$, $h_1 = H - c\overline{q}_N$, θ_0 as before, $M = \mathrm{diag}(m_1, ..., m_{N-1}, 0)$, $\mathbf{1} = (0, ..., 0, 1)$, $q = (q_1, ..., q_{N-1}, \overline{q}_N)$, $p = (p_1, ..., p_N)$ are N-vectors and in the remaining column and row of θ_1 the $q_t = \partial h_1/\partial p = p$, $p_t = -\partial h_1/\partial q = -\partial V/\partial q$ denote the components of the Hamiltonian vector field. Conversely, every natural Hamiltonian system $H = T + V(q)$ admitting the bi-Hamiltonian formulation (6.389) is separable in generalized parabolic coordinates. The potential V satisfies the equations

$$0 = \frac{\partial^2 V}{\partial q_i \partial q_j}(m_j - m_i) + \frac{1}{2}\left(\frac{\partial^2 V}{\partial q_i \partial \overline{q}_N}q_j - \frac{\partial^2 V}{\partial q_j \partial \overline{q}_N}q_i\right), \quad i, j = 1, ..., N-1,$$

$$(6.390)$$

and

$$0 = \frac{3}{2}\frac{\partial V}{\partial q_i} + \frac{\partial^2 V}{\partial q_i \partial \overline{q}_N}(\overline{q}_N - m_i) + \frac{1}{2}\left(\frac{\partial^2 V}{\partial q_i \partial q_i} - \frac{\partial^2 V}{\partial \overline{q}_N \partial \overline{q}_N}\right)q_i$$

$$+ \frac{1}{2}\sum_{r \neq i}^{N-1}\frac{\partial^2 V}{\partial q_i \partial q_r}q_r, \quad i = 1, ..., N-1, \quad (6.391)$$

which are the iff condition for θ_1 to fulfil the Jacobi identity. This bi-Hamiltonian formulation generates an infinite chain of commuting bi-Hamiltonian vector fields

$$\theta_0 \circ \nabla h_0 = 0, ..., \theta_0 \circ \nabla h_{r+1} = \theta_1 \circ \nabla h_r, ... \quad r = 0, 1, 2, ... \quad (6.392)$$

with $h_0 = c, h_1 = H - c\rho_N$, and the higher Hamiltonians

$$h_r = \frac{1}{2}\sum_{i=1}^{N}m_i^{r-2}\overline{K}_i = \frac{1}{2}\sum_{i=1}^{N}m_i^{r-2}K_i - \frac{1}{4}c(q, M^{r-2}q), \quad r = 2, 3, ..., \quad (6.393)$$

where $\overline{K}_i = K_i - \frac{1}{2}cq_i^2$ are integrals of $h_1 = H - c\bar{q}_N$. Moreover, we can define new Hamiltonians

$$\tilde{h}_r = h_r + \sum_{k=1}^{r} \beta_{N-k-1}h_{r-k}, \quad r = 0, ..., N-1, \tag{6.394}$$

where β_r are solutions of the set of linear equations

$$m_i^{N-1} + \sum_{r=0}^{N-2} \beta_r m_i^r = 0, \quad i = 1, ..., N-1, \tag{6.395}$$

so that the new chain starts with $\tilde{h}_0 = h_0 = c$ and terminates with

$$\tilde{h}_N = \beta_0 \left(\frac{1}{2} \sum_{i=1}^{N} m_i^{-1}\overline{K}_i - h_1 \right), \tag{6.396}$$

which is a Casimir of θ_1. For all m_i different, the Hamiltonians $\tilde{h}_r$ ($r = 0, ..., N$) are functionally independent.

Proof. In order to prove the structure (6.389), (6.392), (6.393) of bi-Hamiltonian chains we proceed similarly as for the elliptic case. The kinetic part of $\theta_0 \circ \nabla h_{r+1} = \theta_1 \circ \nabla h_r$, for $r = 2, 3, ...$ holds due to the identity (6.377). The remaining potential part relies on the equality

$$\left(M + \frac{1}{2}\mathbf{1} \otimes q + \frac{1}{2}q \otimes \mathbf{1} \right) \frac{\partial}{\partial q} \left(\frac{1}{2} \sum_{i=1}^{N} m_i^r V_i \right) - \frac{1}{4}(q, M^r q) \frac{\partial V}{\partial q}$$

$$= \frac{\partial}{\partial q} \left(\frac{1}{2} \sum_{i=1}^{N} m_i^{r+1} V_i \right), \tag{6.397}$$

which is valid due to $\{H, K_i\} = \{\frac{1}{2}p^2, V_i\} + \{V(q), \overline{P}_i\} = 0$.

Now let us construct the Nijenhuis coordinates. The solution to the set of equations (6.395) takes the form

$$\beta_{N-r-1} = \sum_{\substack{i_1, ..., i_r \\ i_1 < ... < i_r}} m_{i_1} \cdot ... \cdot m_{i_r} := \rho_r(m), \quad \beta_{N-1} = 1. \tag{6.398}$$

Hence, the c-dependent part of

$$\tilde{h}_r(q, p, c; m) = \tilde{h}_r(q, p; m) + c\rho_r(q; m) \tag{6.399}$$

reads

$$\rho_r(q; m) = \beta_{N-r-1} - \bar{q}_N \beta_{N-r} - \frac{1}{4} \sum_{k=0}^{r-2} \beta_{N-k-1}(q, M^{r-k-2}q) \tag{6.400}$$

and defines the Nijenhuis coordinates (6.295) through the relation

$$\rho_r(m) - \bar{q}_N \rho_{r-1}(m) - \frac{1}{4} \sum_{k=0}^{r-2} \rho_k(m)(q, M^{r-k-2}q) = \rho_r(\lambda). \qquad (6.401)$$

The Nijenhuis coordinates given by formula (6.401) are just the generalized parabolic coordinates $\lambda_1, ..., \lambda_N$ defined by the relation [131]

$$\frac{1}{4} \sum_{k=1}^{N-1} \frac{q_k^2}{z - m_k} + (\bar{q}_N - z) = - \frac{\prod_{j=1}^{N}(z - \lambda_j)}{\prod_{k=1}^{N-1}(z - m_k)}, \qquad (6.402)$$

where $0 < m_1 < ... < m_{N-1}$. To see this let

$$M(z) := \prod_{k=1}^{N-1}(z - m_k) = \sum_{r=0}^{N-1} \rho_r(m) z^{N-r-1},$$

$$\Lambda(z) := \prod_{k=1}^{N}(z - \lambda_k) = \sum_{r=0}^{N} \rho_r(\lambda) z^{N-r},$$

$$M_k(z) := \frac{M(z)}{z - m_k} = \sum_{r=1}^{N-1} \rho_{r-1}^k(m) z^{N-r-1}. \qquad (6.403)$$

From (6.402) it follows that

$$\Lambda(z) = M(z)(z - \bar{q}_N) - \frac{1}{4} \sum_{r=1}^{N-1} M_k(z) q_k^2$$

$$= \sum_{r=0}^{N-1} \rho_r(m) z^{N-r} - \sum_{r=0}^{N-1} \bar{q}_N \rho_r(m) z^{N-r-1} - \frac{1}{4} \sum_{r=1}^{N-1} \sum_{r=1}^{N-1} \rho_{r-1}^k(m) q_k^2 z^{N-r-1}$$

$$= z^N + \sum_{r=1}^{N-1} \rho_r(m) z^{N-r} - \sum_{r=1}^{N} \bar{q}_N \rho_{r-1}(m) z^{N-r} - \frac{1}{4} \sum_{r=1}^{N-1} \sum_{r=2}^{N} \rho_{r-2}^k(m) q_k^2 z^{N-r}.$$

$$\qquad (6.404)$$

Comparing coefficients of the same powers of z and applying the formula

$$\sum_{l=0}^{r} \rho_l(m) m_i^{r-l} = \rho_r^i(m), \qquad (6.405)$$

we get relation (6.401). □

Example 6.25 The most important example here is the Kepler problem $V(q) = \gamma(q,q)^{-1/2}$ which is separable in parabolic coordinates. Substitution of V into (6.390),(6.391) implies that all $m_i = 0$. Thus, it admits the bi-Hamiltonian formulation (6.389) and the chain consists of four functional independent Hamiltonians

$$\tilde{h}_0 = c, \quad \tilde{h}_1 = \frac{1}{2}(p,p) + \gamma(q,q)^{-1/2} - c\bar{q}_N,$$

$$\tilde{h}_2 = \frac{1}{2}\sum_{i=1}^{N-1} p_i l_{iN} - \frac{1}{2}\gamma(q,q)^{-1/2}\bar{q}_N - \frac{1}{4}c\left(\sum_{i=1}^{N-1} q_i^2\right),$$

$$\tilde{h}_3 = -\frac{1}{8}\sum_{i=1}^{N-1}\sum_{j=1}^{N-1} l_{ij}^2, \tag{6.406}$$

where $\tilde{h}_3$ is a Casimir of θ_1. The other integrals of motion come up here as additional Casimirs

$$C_m = \sum_{i=1}^{m}\sum_{j=1}^{m} l_{ij}^2, \quad m = 2, ..., N-1, \tag{6.407}$$

of θ_1 with $M = 0$. Note that it is the Kepler potential in the external homogeneous force which admits the bi-Hamiltonian formulation (6.389).

Example 6.26 Another interesting class of potentials admitting (6.392), (6.393) is the so-called parabolic family of permutationally symmetric potentials [188]. These potentials are characterized by the recursion relation

$$V_i^{(m+1)} = -q_i^2 V^{(m)} - 2m_i V_i^{(m)},$$

$$V^{(m+1)} = -2\bar{q}_N V^{(m)} - \sum_{j=1}^{N-1} V_j^{(m)}, \tag{6.408}$$

$$V_i^{(0)} = 0, \quad V^{(0)} = 1, \quad i = 1, ..., N-1.$$

To this class belong such potentials as

$$V^{(1)} = -2\bar{q}_N,$$

$$V^{(3)} = -8\bar{q}_N^3 - 4\bar{q}_N\sum_{j=1}^{N-1} q_j^2 - 2\sum_{j=1}^{N-1} m_j q_j^2,$$

$$V^{(-1)} = \left(\sum_{j=1}^{N-1} q_j^2 - 2\bar{q}_N\right). \tag{6.409}$$

At the other extreme of the whole family of separation coordinates is the spherical system of coordinates in which spherically symmetric potentials separate. A bi-Hamiltonian formulation of these follows directly from the elliptic one (6.370) if $\alpha_1 = ... = \alpha_N = \alpha$.

Theorem 6.10 *A bi-Hamiltonian formulation for spherically symmetric potentials.*

Every natural Hamiltonian system $H = T + V(q)$ separable in spherical coordinates $V(q) = U((q,q))$ admits, in the extended phase space of variables (q, p, c), the bi-Hamiltonian formulation (6.370) where $A = \alpha I, h_0 = c, h_1 = H + u((q,q)) + \frac{1}{2}c(q,q)$. The potential V satisfies the equations

$$q_r \frac{\partial V}{\partial q_i} - q_i \frac{\partial V}{\partial q_r} = 0, \qquad i, r = 1, ..., N, \tag{6.410}$$

which are the iff condition for θ_1 to satisfy the Jacobi identity. Both Hamiltonian structures θ_0 and θ_2 generate one chain of length 3

$$\theta_0 \circ \nabla \tilde{h}_0 = 0, \; \theta_0 \circ \nabla \tilde{h}_1 = \theta_1 \circ \nabla \tilde{h}_0, \; \theta_0 \circ \nabla \tilde{h}_2 = \theta_1 \circ \nabla \tilde{h}_1, \; 0 = \theta_1 \circ \nabla \tilde{h}_2, \tag{6.411}$$

where $\tilde{h}_0 = c, \tilde{h}_1 = h_1 - 2\alpha c$, and $\tilde{h}_2 = \frac{1}{4}[(q,q)(p,p) - (q,p)^2] - \alpha h_1 + \alpha^2 c$ is a Casimir of θ_1. Moreover for $\mu = -\alpha$ the operator θ_μ has $N - 2$ additional Casimirs

$$C_m = -\frac{1}{4} \sum_{i=1}^{m} \sum_{j=1}^{m} l_{ij}^2, \qquad m = 2, ..., N - 1, \tag{6.412}$$

which Poisson commute with each other and with $\tilde{h}_k$ ($k = 0, 1, 2$). The commutation holds with respect to all Poisson brackets in the pencil since these Casimirs generate $N - 2$ periodic chains of length 1: $\theta_1 \circ \nabla C_m = \lambda \theta_0 \circ \nabla C_m$.

For two degrees of freedom there are only four separation systems: elliptic, parabolic spherical and Cartesian.

Lemma 6.8 *Every separable natural Hamiltonian system $H = T + V(q)$ of two degrees of freedom admits, in the extended phase space of variables (q, p, c), a bi-Hamiltonian formulation with a chain of the form (6.411).*

Proof. For Hamiltonians separable in elliptic, parabolic or spherical coordinates we have to specialize the previous theorems. For a Hamiltonian of the form $H = \frac{1}{2}(p_1^2 + p_2^2) + V(q, q_2)$ with $V(q_1, q_2) = f_1(q_1) + f_2(q_2)$ we take θ_0 as before and

$$\theta_1 = \begin{pmatrix} 0 & \Lambda & q_t \\ -\Lambda & 0 & p_t \\ -q_t & -p_t & 0 \end{pmatrix}, \tag{6.413}$$

where $\Lambda = \text{diag}(1, -1)$, $q_t = \partial H / \partial p = p, p_t = -\partial H / \partial q = -\partial V / \partial q$. The potential $V(q_1, q_2) = f_1(q_1) + f_2(q_2)$ satisfies the equation $\partial^2 V / \partial q_1 \partial q_2 = 0$ which is the iff condition for θ_1 to fulfil the Jacobi identity. The bi-Hamiltonian chain has the form (6.411) with $\tilde{h}_0 = c, \tilde{h}_1 = H, \tilde{h}_2 = \frac{1}{2}[p_1^2 + f_1(q_1)] - \frac{1}{2}[p_2^2 + f_2(q_2)] - c$ which is a Casimir of θ_1.

Example 6.27 The Euler problem of two centres of gravitation is defined by $H = \frac{1}{2}(p, p) + \mu_1 r_1^{-1} + \mu_2 r_2^{-1}$ where $r_1 = [q_1^2 + (q_2 + \gamma)^2]^{1/2}, r_2 = [q_1^2 +$

$(q_2 - \gamma)^2]^{1/2}$. Take $h_1 = H + \frac{1}{2}c(q,q)$; then the Hamiltonian equations have the bi-Hamiltonian formulation (6.370) with $A = \mathrm{diag}(\alpha, \alpha + \frac{1}{2}\gamma^2)$. The bi-Hamiltonian chain (6.372),(6.373) consists of

$$h_0 = c, \quad h_1 = \frac{1}{2}(p,p) + \mu_1 r_1^{-1} + \mu_2 r_2^{-1} + \frac{1}{2}c(q,q),$$

$$h_2 = \frac{1}{2}\alpha(p_1^2 + cq_1^2) + \frac{1}{2}(\alpha + \frac{1}{2}\gamma^2)(p_2^2 + cq_2^2) + \frac{1}{4}(q_1 p_2 - q_2 p_1)$$

$$+ \mu_1 \left(\alpha - \frac{1}{2}q_2\gamma^2\right) r_1^{-1} + \mu_2 \left(\alpha + \frac{1}{2}q_2\gamma^2\right) r_2^{-1}, \quad \dots . \tag{6.414}$$

The Casimir of θ_1

$$\tilde{h}_2 = h_2 - \left(2\alpha + \frac{1}{2}\gamma^2\right) h_1 + \alpha \left(2\alpha + \frac{1}{2}\gamma^2\right) h_0, \tag{6.415}$$

generates a finite chain of length 3. In the limit $\gamma = 0$ one recovers the spherical bi-Hamiltonian formulation for the Kepler problem while the parabolic limit (one centre goes to zero and the other one to infinity) leads to the parabolic bi-Hamiltonian formulation of the Kepler problem.

Finally in this section let us reconsider bi-Hamiltonian Nijenhuis chains which are quadratic in momenta, from the point of view of Riemannian (pseudo-Riemannian) geometry. Let us consider a Riemannian or pseudo-Riemannian manifold, with coordinates $q^1, ..., q^n$ and metric components g_{ij}, with inverse g^{ij}, satisfying: $\sum_{j=1}^{n} g_{ij}g^{jk} = \delta_i^k$. The Levi–Civita connection components are defined by

$$\Gamma_{jk}^i = \frac{1}{2}\sum_{l=1}^{n} g^{il}(\partial_k g_{lj} + \partial_j g_{kl} - \partial_l g_{jk}), \quad \partial_i \equiv \frac{\partial}{\partial q^i}. \tag{6.416}$$

The equations

$$q_{tt}^i + \Gamma_{jk}^i q_t^j q_t^k = g^{ik}\partial_k V(q), \quad i = 1, ..., n, \quad q_t \equiv \frac{dq}{dt} \tag{6.417}$$

describe the motion of a particle in curved space with the metric g_{ij}. Moreover (6.417) can be obtained by varying the Lagrangian

$$L = \frac{1}{2}\sum_{i,j} g_{ij}q_t^i q_t^j - V(q). \tag{6.418}$$

The corresponding Hamiltonian reads

$$H(q,p) = \sum_{i=1}^{n} q_t^i \frac{\partial L}{\partial q_t^i} - L = \frac{1}{2}\sum_{i,j=1}^{n} g^{ij}p_i p_j + V(q), \quad p_i = \frac{\partial L}{\partial q_t^i}. \tag{6.419}$$

Now, let us consider the Hamiltonian $H = h_1$ from the Nijenhuis chain (6.297) with the f_i functions given by (6.349), i.e.

$$H = h_1(q, p, c) = \frac{1}{2} \sum_{k=1}^{n} g^{kk}(q)p_k^2 + V_1(q) - c(q^1 + \ldots + q^n), \qquad (6.420)$$

where

$$g^{kk}(q) = \frac{\varphi(q^k)}{\Delta^k(q)} \qquad (6.421)$$

is the kk-component of the inverse of the diagonal metric $\mathrm{d}s = \sum g_{kk}(\mathrm{d}q^k)^2$ on some Riemannian manifold,

$$V_1(q) = \sum_{k=1}^{n} \frac{\psi(q^k)}{\Delta^k(q)} \qquad (6.422)$$

and φ, ψ are arbitrary smooth functions. Notice that here we identify the Nijenhuis coordinates (μ_i, λ_i) with the Riemannian coordinates (q^i, p_i).

As H is separable for each pair of (φ, ψ) we deal with the motion of a particle in the so called Stäckel space [63] under the potential $V_1(q) + c\rho_1(q)$. Actually, the Stäckel space is a Riemannian space with diagonal metric of such a form that the corresponding geodesic equations are separable. Hence, from the point of view of Riemannian geometry, bi-Hamiltonian Nijenhuis systems, which are quadratic in momenta, are nothing else but separable bi-Hamiltonian Stäckel systems. Moreover, geodesics of an arbitrary Stäckel space, with the following components of the metric tensor:

$$g_{ii}(q) = \frac{\Delta^i(q)}{\varphi(q^i)}, \qquad (6.423)$$

are trajectories of free particle, i.e. integral curves of a quasi-bi-Hamiltonian system with Hamiltonian function

$$H = \frac{1}{2} \sum_{k=1}^{n} g^{kk}(q)p_k^2. \qquad (6.424)$$

The simplest potential which makes the dynamics of (6.424) a bi-Hamiltonian one is

$$V(q) = -c \sum_{k=1}^{n} q^k. \qquad (6.425)$$

Additing to the potential (6.425) terms of the form (6.422) preserves the bi-Hamiltonian structure.

From the point of view of Riemannian geometry all examples of dynamical systems presented in this chapter are equivalent to the bi-Hamiltonian dynamics of a single particle in a flat Stäckel space. Here we illustrate our considerations by integrable dynamics in a curved Stäckel space.

Example 6.28 Integrable dynamics on the unit sphere S^n.
For the particular choice

$$\varphi(q^k) = 4 \prod_{j=1}^{n+1} (a^j - q^k), \qquad a^j = \text{const.}, \quad j = 1, ..., n+1, \qquad (6.426)$$

the metric (6.423) is that of the n-dimensional unit sphere S^n:

$$(x^1)^2 + ... + (x^{n+1})^2 = 1, \qquad (6.427)$$

written down in the spherical-conical coordinates q^i

$$x^1 = \sqrt{\frac{\prod_{k=1}^{n}(a^1 - q^k)}{\prod_{j \neq 1}(a^1 - a^j)}}, ..., x^{n+1} = \sqrt{\frac{\prod_{k=1}^{n}(a^{n+1} - q^k)}{\prod_{j \neq n+1}(a^{n+1} - a^j)}}, \qquad (6.428)$$

which in our terminology are the Nijenhuis coordinates. The Hamiltonian for the geodesics reads

$$H \equiv h_1(q, p) = 2 \sum_{k=1}^{n} \frac{\prod_{j=1}^{n+1}(a^j - q^k)}{\prod_{j \neq k}(q^k - q^j)} p_k^2. \qquad (6.429)$$

In the phase space $M \ni (q, p)$ it is a quasi-bi-Hamiltonian system in Pfaffian form

$$\begin{pmatrix} q \\ p \end{pmatrix}_t = \begin{pmatrix} 0 & I \\ -I & 0 \end{pmatrix} dH = \frac{(-1)^{n+1}}{\prod_{j=1}^{n} q^j} \begin{pmatrix} 0 & \Lambda \\ -\Lambda & 0 \end{pmatrix} dF, \qquad (6.430)$$

where $q = (q^1, ..., q^n)^T, p = (p_1, ..., p_n)^T, I = \text{diag}(1, ..., 1), \Lambda = \text{diag}(q^1, ..., q^n)$, H is given by (6.429) and F takes the form

$$F \equiv h_n(q, p) = 2(-1)^{n+1} \sum_{k=1}^{n} \left(\prod_{j \neq k} q^j \right) \frac{\prod_{j=1}^{n+1}(a^j - q^k)}{\prod_{j \neq k}(q^k - q^j)} p_k^2. \qquad (6.431)$$

Now, let us consider the simplest potential (6.425), which, up to an irrelevant constant term, is just the restriction of the quadratic potential $V(x) = c[a^1(x^1)^2 + ... + a^{n+1}(x^{n+1})^2]$ from the ambient Euclidean space onto S^n and turns the free motion (6.429) into the bi-Hamiltonian form. Here c will be treated as the additional Casimir variable. In extended phase space $\overline{M} \ni (q, p, c)$ the perturbed geodesic motion turns out to be bi-Hamiltonian

$$\begin{pmatrix} q \\ p \\ c \end{pmatrix}_t = \begin{pmatrix} 0 & I & 0 \\ -I & 0 & 0 \\ 0 & 0 & 0 \end{pmatrix} dH = \begin{pmatrix} 0 & \Lambda & \frac{\partial H}{\partial p} \\ -\Lambda & 0 & -\frac{\partial H}{\partial q} \\ -(\frac{\partial H}{\partial p})^T & (\frac{\partial H}{\partial q})^T & 0 \end{pmatrix} dF, \qquad (6.432)$$

where

$$H \equiv h_1(q, p, c) = 2 \sum_{k=1}^{n} \frac{\prod_{j=1}^{n+1}(a^j - q^k)}{\prod_{j \neq k}(q^k - q^j)} p_k^2 - c(q^1 + ... + q^n), \quad F \equiv h_0 = c. \qquad (6.433)$$

The bi-Hamiltonian representation (6.432) is sufficient for generating the whole bi-Hamiltonian chain. Hence, we have proved the bi-Hamiltonian property of the famous integrable problem of the motion of a particle on the unit sphere under the action of a quadratic potential, discussed for the first time by C. Neumann. Of course system (6.432) is solvable by quadratures, where the implicit form of the integral curves $q^i(t)$ $(t \equiv t_1)$ is given by a set of equations (6.362) with

$$f(\xi) = 2 \sqrt{(c_1 + c_2\xi + ... + c_n\xi^{n-1} + c\xi^n) \prod_{j=1}^{n+1} (a^j - \xi)}. \qquad (6.434)$$

Obviously the addition of an extra potential in the form (6.422) preserves the bi-Hamiltonian structure of the dynamics as well as the solvability by quadratures.

6.6 Nonstandard Multi-Hamiltonian Structures of Field Systems and Their Finite Dimensional Reductions

In the following section of this chapter we analyse the relation between all the types of finite dimensional bi-Hamiltonian systems considered in previous sections and the so-called nonstandard representations of field systems [11],[28],[89]. Developing the idea from [28], we investigate algebraic properties of field systems constructed directly from the restricted (stationary) flows. Such systems are related with what we refer to as the t-type Hamiltonian formulation of dynamics. Let

$$u_t = K[u] \qquad (6.435)$$

be a given Hamiltonian evolution equation, where t is the evolution parameter and K is a differential function of u. Then, in many cases, there exists an equivalent Hamiltonian formulation of (6.435) taking the form

$$\begin{pmatrix} q \\ p \\ c \end{pmatrix}_x = G[q, p, c], \qquad (6.436)$$

where now x plays the role of the evolution parameter and G is a t-differential function (i.e. it depends on q, p, c and their t-derivatives). The algebraic structure of (6.436) is as rich as that of (6.435) and the suppression of the t-dependence in (6.436) results in the classical Hamiltonian dynamics of stationary flows of (6.435) in the canonical variables.

We will consider here the evolution of a (vector) function $u = (u_1, \ldots, u_N)^T$ of one space variable x and the evolution parameter t:

$$u_t = \theta \circ \delta H, \tag{6.437}$$

where θ is a regular (Definition 6.1) x-type Hamiltonian structure. This restriction imposed on a Hamiltonian operator means that

$$\ker \theta = c_1 \delta F_1 + \ldots + c_N \delta F_N, \tag{6.438}$$

where c_i are arbitrary constants and we confine ourselves to the case where

$$F_i = F_i(u_1, ..., u_N) \quad i = 1, ..., N. \tag{6.439}$$

Stationary (i.e. t-independent) solutions of (6.437) satisfy

$$\theta \circ \delta H = 0 \tag{6.440}$$

which is equivalent to

$$\delta H = \ker \theta = \delta R \tag{6.441}$$

where $R = c_1 F_1 + \ldots + c_N F_N$. Note that because c_i are constants we have

$$\delta R = \ker \theta = \begin{pmatrix} \frac{\delta F_1}{\delta u_1} & \cdots & \frac{\delta F_N}{\delta u_1} \\ \vdots & \vdots & \vdots \\ \frac{\delta F_1}{\delta u_N} & \cdots & \frac{\delta F_N}{\delta u_N} \end{pmatrix} \begin{pmatrix} c_1 \\ \vdots \\ c_N \end{pmatrix} = F'^\dagger \begin{pmatrix} c_1 \\ \vdots \\ c_N \end{pmatrix}, \tag{6.442}$$

where $F = (F_1, \ldots, F_N)^{\mathrm{T}}$. Equation (6.441) is the Lagrange equation (with respect to u)

$$\delta L = 0 \tag{6.443}$$

with the Lagrangian $L = H - R$ depending on N parameters $c_1, \ldots, c_N$. If L is nondegenerate we can perform the (generalized) Legendre transformation given by formulas (6.8) and (6.9). This brings the stationary equation (6.441), in the Ostrogradsky representation, into the canonical Hamiltonian form

$$\begin{pmatrix} q \\ p \end{pmatrix}_x = \begin{pmatrix} 0 & I \\ -I & 0 \end{pmatrix} \begin{pmatrix} \delta_q h \\ \delta_p h \end{pmatrix}, \tag{6.444}$$

where c_i are treated as some constant parameters. Of course the Newton representation can be used as well.

Let us now consider $c = (c_1, \ldots, c_N)^{\mathrm{T}}$ as differential functions of the u variable defined by the relation

$$\delta H = (F')^\dagger c. \tag{6.445}$$

Then the original flow (6.437) reads

$$u_t = \theta \circ \delta H = \theta \circ (F')^\dagger c. \tag{6.446}$$

On the other hand

$$F' \circ u_t = (F' \circ \theta \circ F'^\dagger) c = b[D_x] c, \tag{6.447}$$

where $b[D_x]$ is a normal form (i.e. ker $b[D_x] \in R^N$) of a regular Hamiltonian operator θ. Here we confine ourselves to the case where

$$b[D_x] = bD_x, \quad b - \text{constant matrix}, \tag{6.448}$$

i.e. a first order, constant coefficient, differential operator. So the original flow (6.437) is equivalent to the system

$$\delta H = (F')^\dagger c,$$

$$c_x = b^{-1} F' u_t, \tag{6.449}$$

where we have assumed that the constant matrix b is invertible.

According to the assumption (6.439), the $p_{i,k}$ variables do not depend on the constants c_i, so the Hamiltonian (6.9) depends on c only through the term R in L and hence

$$\delta_c h = \delta_c(-L) = \delta_c R = (F_1, \ldots, F_N)^{\mathrm{T}} = F. \tag{6.450}$$

Moreover the following relation holds

$$b^{-1} \partial_t (\delta_c h) = b^{-1} \partial_t F = b^{-1} F' \circ u_t. \tag{6.451}$$

Thus considering h as a function of q, p and c and bearing in mind that the first equation of (6.449) is equivalent to (6.444) and the second equation of (6.449) is equivalent to (6.451), we can rewrite (6.449) as

$$\begin{pmatrix} q \\ p \\ c \end{pmatrix}_x = \begin{pmatrix} 0 & I & 0 \\ -I & 0 & 0 \\ 0 & 0 & b^{-1} D_t \end{pmatrix} \delta h, \tag{6.452}$$

where $\delta = (\delta_q, \delta_p, \delta_c)^{\mathrm{T}}$. Equation (6.452) represents a Hamiltonian system with the Poisson operator

$$\pi = \begin{pmatrix} 0 & I & 0 \\ -I & 0 & 0 \\ 0 & 0 & b^{-1} D_t \end{pmatrix} \tag{6.453}$$

acting in the algebra of t-differential functions of q, p, c. The variable x now plays the role of an evolution parameter while t is a new 'space' variable. Thus, we have proven that any nondegenerate Hamiltonian flow in the space of x-differential functions with a regular Poisson operator θ can be equivalently described as a Hamiltonian flow in the space of t-differential functions with a constant Poisson operator (6.453). The latter formulation, which will be referred to as the t-type one, is especially convenient when considering stationary solutions of (6.437).

The complicated passage from an infinite dimensional Hamiltonian system (6.437) to its finite dimensional Hamiltonian stationary flow is superseded by a simple and natural elimination of a t-dependence in the system (6.452). The Hamiltonian operator π reduces to a degenerate canonical operator

$$\pi_s \equiv \theta_0 = \begin{pmatrix} 0 & I & 0 \\ -I & 0 & 0 \\ 0 & 0 & 0 \end{pmatrix}, \tag{6.454}$$

since D_t acts trivially on the space of t-independent functions. Superiority of the t-type Hamiltonian description is even more visible when considering second (or higher) Hamiltonian formulations of the starting equation (and the corresponding stationary flow). We do not know any direct connection between the higher Hamiltonian structures of (2.1) and their stationary counterparts. In fact the only method of constructing the latter involves stationary versions of Miura maps (if they exist) considered in previous sections. Higher Hamiltonian operators of the t-type can be constructed using, for example, t-representations of standard Miura maps $u = M[\bar{u}]$.

Let us assume that our original equation $u_t = \theta \circ \delta H$ is Miura related to a (modified) equation $\bar{u}_t = \bar{\theta} \circ \delta \bar{H}$. We can translate the x-differential relation $u = M[\bar{u}]$ (M is an x-differential function of $\bar{u}$) into the language of canonical variables arriving at the t-differential relation $(q, p, c) = M[\bar{q}, \bar{p}, \bar{c}]$ (i.e. M is now a t-differential function of modified canonical variables). Now, using this t-type Miura map, one constructs the second (t-type) Hamiltonian structure of the flow (6.452)

$$\pi_1 = M' \circ \bar{\pi} \circ M'^\dagger \big|_{q=M[\bar{q}]} \tag{6.455}$$

where $\bar{\pi}$ is a t-type Hamiltonian structure of the modified equation

$$\bar{\pi} = \begin{pmatrix} 0 & I & 0 \\ -I & 0 & 0 \\ 0 & 0 & \bar{b}^{-1}D_t \end{pmatrix} \tag{6.456}$$

and M' (M'^*) denotes the directional derivative of M (its adjoint). Further development, provided $\pi_0 \equiv \pi$ and π_1 are compatible, is standard for the theory of bi-Hamiltonian systems.

The next problem, concerning the t-type representation of integrable systems, is Lax representation. Here, the question arises: how do we construct the appropriate Lax hierarchy? The recipe would be the following. Let us consider a standard Lax hierarchy in the form

$$\varphi_{t_n} = L_n \varphi \qquad n = 0, 1, \ldots, \tag{6.457}$$

where $L_n = L_n(\alpha; u, u_x, \ldots)$ are differential functions of field variables $u_1, \ldots, u_N$ and polynomials with respect to the parameter α. The case $n = 0$, $t_0 = x$, represents the principal Lax equation, whereas the other cases, i.e. $n > 0$, represent the auxiliary Lax equations. The hierarchy of evolution equations in standard form is reconstructed from the zero curvature condition

$$(L_0)_{t_n} - (L_n)_{t_0=x} + [L_0, L_n] = 0, \quad n = 0, 1, \ldots. \tag{6.458}$$

Now, let us pass to the representation where $t_k = \xi$ plays the role of the space variable. In the first step, we have to transform the hierarchy (6.457) in such a way that the x-differential functions $L_n = L_n(\alpha; u, u_x, \ldots)$ transform into the ξ-differential ones $L_n = L_n(\alpha; q, p, c, q_\xi, p_\xi, c_\xi, \ldots)$. This transformation is straightforward as (6.436) provides the desired relationships. In the second step, we consider the kth Lax equation as the principal one and the others as the auxiliary ones. The hierarchy of evolution equations with $\xi = t_k$ as the space variable is generated from the zero curvature condition in the following form

$$(L_k)_{t_n} - (L_n)_{t_k=\xi} + [L_k, L_n] = 0, \quad n = 0, 1, \ldots. \tag{6.459}$$

It is interesting to note that the stationary reduction of (6.457) and (6.459) with respect to the variable $t_k = \xi$ gives just the Lax representation of the stationary flows of the kth equation from the standard hierarchy, considered in previous sections.

First we shall illustrate the considerations by examples of field counterparts of stationary flows in the Ostrogradsky representation.

Example 6.29 The t-type multi-Hamiltonian structure of the fifth order KdV. Here, we consider the fifth order KdV equation

$$
\begin{aligned}
u_{t_2} &= K_2 = (\frac{1}{16}u_{4x} + \frac{5}{8}uu_{xx} + \frac{5}{16}u_x^2 + \frac{5}{8}u^3)_x \\
&= D_x \delta \int_R (\frac{1}{32}u_{xx}^2 - \frac{5}{8}uu_x^2 + \frac{5}{32}u^4)dx
\end{aligned}
\tag{6.460}
$$

where now $\theta = D_x$ is a constant x-type Hamiltonian structure. According to (6.8) and (6.449), its t_2-type Hamiltonian formulation reads

$$
\begin{pmatrix} q_1 \\ q_2 \\ p_1 \\ p_2 \\ c \end{pmatrix}_{t_0} = G_0 =
\begin{pmatrix} q_2 \\ 16p_2 \\ \frac{5}{8}q_1^3 - \frac{5}{16}q_2^2 - c \\ -p_1 - \frac{5}{8}q_1q_2 \\ (q_1)_\xi \end{pmatrix} = \pi_0 \circ \delta H_1
$$

$$
= \begin{pmatrix}
0 & 0 & 1 & 0 & 0 \\
0 & 0 & 0 & 1 & 0 \\
-1 & 0 & 0 & 0 & 0 \\
0 & -1 & 0 & 0 & 0 \\
0 & 0 & 0 & 0 & D_\xi
\end{pmatrix}
\delta \int_R (8p_2^2 + q_2p_1 + \frac{5}{16}q_1q_2^2 - \frac{5}{32}q_1^4 + q_1c)d\xi
$$

$$\tag{6.461}$$

where

$$t_0 = x, \quad t_2 = \xi, \quad q_1 = u, \quad q_2 = u_x, \quad p_1 = -\frac{1}{16}u_{3x} - \frac{5}{8}uu_x,$$

$$p_2 = \frac{1}{16}u_{xx}, \quad c = \frac{1}{16}u_{4x} + \frac{5}{8}uu_{xx} + \frac{5}{16}u_x^2 + \frac{5}{8}u^3.$$

The x-type and the t_2-type Hamiltonian formulations of the modified KdV system are

$$v_{t_2} = (\frac{1}{16}v_{4x} - \frac{5}{8}vv_x^2 - \frac{5}{8}v^2v_{xx} + \frac{3}{8}v^5)_x$$

$$= (-D_x)\delta \int_R (-\frac{1}{32}v_{xx}^2 - \frac{5}{16}v^2v_x^2 - \frac{1}{16}v^6)d\xi \qquad (6.462)$$

and

$$\begin{pmatrix} \bar{q}_1 \\ \bar{q}_2 \\ \bar{p}_1 \\ \bar{p}_2 \\ \bar{c} \end{pmatrix}_{t_0}$$

$$= \bar{G}_0 = \begin{pmatrix} \bar{q}_2 \\ -4\bar{p}_2 \\ -\frac{3}{2}\bar{q}_1^5 - \frac{5}{2}\bar{q}_1\bar{q}_2^2 - \bar{c} \\ -\frac{5}{2}\bar{q}_1^2\bar{q}_2 - \bar{p} \\ -4(\bar{q}_1)_\xi \end{pmatrix} = \bar{\pi}_0 \circ \delta\bar{H}_1$$

$$= \begin{pmatrix} 0 & 0 & 1 & 0 & 0 \\ 0 & 0 & 0 & 1 & 0 \\ -1 & 0 & 0 & 0 & 0 \\ 0 & -1 & 0 & 0 & 0 \\ 0 & 0 & 0 & 0 & -4D_\xi \end{pmatrix} \delta \int_R (\frac{1}{4}\bar{q}_1^6 + \frac{5}{4}\bar{q}_1^2\bar{q}_2^2 + \bar{q}_2\bar{p}_1 - 2\bar{p}_2^2 + \bar{q}_1\bar{c})d\xi$$

$$(6.463)$$

where

$$t_0 = x, \quad t_2 = \xi, \quad \bar{q}_1 = v, \quad \bar{q}_2 = v_x, \quad \bar{p}_1 = -\frac{5}{8}v^2v_x + \frac{1}{16}v_{3x},$$

$$\bar{p}_2 = -\frac{1}{16}vv_{xx}, \quad \bar{c} = -\frac{1}{16}v_{4x} + \frac{5}{8}vv_x^2 + \frac{5}{8}v^2v_{xx} - \frac{3}{8}v^5.$$

The t_2-type Miura map M relating (6.461) and (6.463) is

$$\begin{aligned}
q_1 &= -\bar{q}_1^2 - \bar{q}_2,\\
q_2 &= 4\bar{p}_2 - 2\bar{q}_1\bar{q}_2,\\
p_1 &= \bar{q}_2\bar{p}_2 + \tfrac{1}{2}\bar{q}_1\bar{p}_1 - \tfrac{5}{8}\bar{q}_1\bar{q}_2^2 - \tfrac{3}{8}\bar{q}_1^5 + \tfrac{1}{4}\bar{c},\\
p_2 &= -\tfrac{1}{4}\bar{p}_1 + \tfrac{1}{2}\bar{q}_1\bar{p}_2 - \tfrac{5}{8}\bar{q}_2^2 - \tfrac{5}{8}\bar{q}_1^2\bar{q}_2,\\
c &= -\bar{p}_2^2 + \tfrac{1}{2}\bar{q}_2\bar{p}_1 + \tfrac{5}{8}\bar{q}_1^2\bar{q}_2^2 + \tfrac{1}{8}\bar{q}_1^6 - \tfrac{1}{2}\bar{q}_1\bar{c} - (\bar{q}_1)_\xi.
\end{aligned} \tag{6.464}$$

Now, according to the standard procedure, the Miura map (6.464) generates the second Hamiltonian structure of G_0

$$\pi_1 = M' \circ \bar{\pi}_0 \circ M'^\dagger$$

$$= \begin{pmatrix}
0 & -4 & q_1 & 0 & \tfrac{1}{2}q_2 \\
* & 0 & -q_2 & -\tfrac{3}{2}q_1 & 8p_2 \\
* & * & -\tfrac{1}{4}D_\xi & \tfrac{1}{2}p_2 + \tfrac{15}{32}q_1^2 & -\tfrac{5}{32}q_2^2 + \tfrac{5}{16}q_1^3 - \tfrac{1}{2}c \\
* & * & * & 0 & -\tfrac{1}{2}p_1 - \tfrac{5}{16}q_1q_2 + \tfrac{1}{4}\partial_\xi \\
* & * & * & * & \tfrac{1}{2}D_\xi q_1 + \tfrac{1}{2}q_1 D_\xi
\end{pmatrix}, \tag{6.465}$$

where ($*$) represents the lower triangular part of π_1 chosen so as to make it skew adjoint. Thus, the second Hamiltonian description of G_0, generated from the Casimir h_0 of π_0, reads

$$G_0 = \pi_1 \circ \delta h_0, \qquad h_0 = 2c. \tag{6.466}$$

One can check that both Hamiltonian operators π_0 and π_1 are compatible. This implies the heredity of the recursion operator

$$\Phi = \pi_1 \circ \pi_0^{-1}$$

$$= \begin{pmatrix}
q_1 & 0 & 0 & 4 & \tfrac{1}{2}q_2 D_\xi^{-1} \\
-q_2 & -\tfrac{3}{2}q_1 & -4 & 0 & 8p_2 D_\xi^{-1} \\
-\tfrac{1}{4}\partial_\xi & \tfrac{1}{2}p_2 + \tfrac{15}{32}q_1^2 & q_1 & -q_2 & AD_\xi^{-1} \\
-\tfrac{1}{2}p_2 - \tfrac{15}{32}q_1^2 & 0 & 0 & -\tfrac{3}{2}q_1 & \tfrac{1}{4} - BD_\xi^{-1} \\
-A & B + \tfrac{1}{4}D_\xi & \tfrac{1}{2}q_2 & 8p_2 & \tfrac{1}{2}q_1 + \tfrac{1}{2}D_\xi q_1 D_\xi^{-1}
\end{pmatrix}, \tag{6.467}$$

where $A = -\tfrac{5}{32}q_2^2 + \tfrac{5}{16}q_1^3 - \tfrac{1}{2}c$, $B = \tfrac{1}{2}p_1 + \tfrac{5}{16}q_1q_2$ and we deal with the infinite chain of local, commuting bi-Hamiltonian flows $G_n = \Phi^n \circ G_0$. We list the first few:

$$G_0 = \begin{pmatrix}
q_2 \\
16p_2 \\
\tfrac{5}{8}q_1^3 - \tfrac{5}{16}q_2^2 - c \\
-\tfrac{5}{8}q_1q_2 - p_1 \\
(q_1)_\xi
\end{pmatrix},$$

$$G_1 = \begin{pmatrix} -q_1 q_2 - 4p_1 \\ \frac{1}{4}q_2^2 - \frac{5}{2}q_1^3 - 16q_1 p_2 + 4c \\ 8p_2^2 + \frac{5}{32}q_1 q_2^2 + \frac{15}{2}q_1^2 p_2 + q_2 p_1 + \frac{15}{16}q_1^4 - \frac{3}{2}q_1 c - \frac{1}{4}q_2 \xi \\ \frac{5}{32}q_1^2 q_2 + q_1 p_1 - \frac{1}{2}q_2 p_2 + \frac{1}{4}q_1 \xi \\ \frac{3}{2}q_1 (q_1)_\xi + 4(p_2)_\xi \end{pmatrix},$$

$$G_2 = \begin{pmatrix} q_1 \\ q_2 \\ p_1 \\ p_2 \\ c \end{pmatrix}_\xi,$$

$$G_3 =$$
$$\begin{pmatrix} \frac{1}{2}q_2 c + q_1 q_{1\xi} + 4p_{2\xi} \\ 8p_2 c - q_2 q_{1\xi} - \frac{3}{2}q_1 q_{2\xi} - 4p_{1\xi} \\ c(-\frac{5}{32}q_2^2 + \frac{5}{16}q_1^3 - \frac{1}{2}c) - \frac{1}{4}q_{1\xi} + (\frac{1}{2}p_2 + \frac{15}{32}q_1^2)q_{2\xi} + q_1 p_{1\xi} - q_2 p_{2\xi} \\ -(\frac{1}{2}p_2 + \frac{15}{32}q_1^2)q_{1\xi} - \frac{3}{2}q_1 p_{2\xi} + \frac{1}{4}c_\xi - c(\frac{1}{2}p_1 + \frac{5}{16}q_1 q_2) \\ (\frac{5}{32}q_2^2 - \frac{5}{16}q_1^3 + \frac{1}{2}c)q_{1\xi} + (\frac{1}{2}p_1 + \frac{5}{16}q_1 q_2)q_{2\xi} + \frac{1}{4}q_{2\xi\xi} + \frac{1}{2}q_2 p_{1\xi} \\ \qquad\qquad +8p_2 p_{2\xi} + \frac{1}{2}q_1 c_\xi + \frac{1}{2}(q_1 c)_\xi \end{pmatrix},$$

$$(6.468)$$

and the corresponding Hamiltonians

$$H_0 = 2c,$$

$$H_1 = 8p_2^2 + q_2 p_1 + \frac{5}{16}q_1 q_2^2 - \frac{5}{32}q_1^4 + q_1 c,$$

$$H_2 = -2p_1^2 - 8q_1 p_2^2 + \frac{1}{4}q_2^2 p_2 - \frac{5}{2}q_1^3 p_2 - q_1 q_2 p_1 - \frac{5}{64}q_1^2 q_2^2$$
$$\qquad - \frac{3}{16}q_1^5 + 4p_2 c + \frac{3}{4}q_1^2 c + \frac{1}{4}q_1 (q_2)_\xi,$$

$$H_3 = \frac{1}{2}c^2 - q_1 (p_1)_\xi - q_2 (p_2)_\xi, \; \cdots$$

$$(6.469)$$

On the other hand, the first few Lax matrices L_n are the following

$$L_0 = \begin{pmatrix} 0 & 1 \\ \lambda - q_1 & 0 \end{pmatrix},$$

$$L_1 = \begin{pmatrix} -\frac{1}{4}q_2 & \lambda + \frac{1}{2}q_1 \\ \lambda^2 - \frac{1}{2}\lambda q_1 - 4p_2 - \frac{1}{2}q_1^2 & -\frac{1}{4}q_2 \end{pmatrix},$$

$$L_2 = \begin{pmatrix} -\frac{1}{4}\lambda q_2 + p_1 + \frac{1}{4}q_1 q_2 & \lambda^2 + \frac{1}{2}\lambda q_1 + 2p_2 + \frac{3}{8}q_1^2 \\ \lambda^3 - \frac{1}{2}\lambda^2 q_1 - \lambda(\frac{1}{8}q_1^2 + 2p_2) & \frac{1}{4}\lambda q_2 - p_1 - \frac{1}{4}q_1 q_2 \\ +2q_1 p_2 - \frac{1}{16}q_2^2 + \frac{1}{4}q_1^3 - c \end{pmatrix},$$

$$
L_3 = \begin{pmatrix}
-\frac{1}{4}\lambda^2 q_2 + \frac{1}{2}\lambda(2p_1 & \lambda^3 + \frac{1}{2}\lambda^2 q_1 + \lambda(2p_2 \\
+\frac{1}{2}q_1 q_2) + \frac{1}{4}(q_1)_\xi & +\frac{3}{8}q_1^2) + \frac{1}{2}c \\
& \\
\lambda^4 - \frac{1}{2}\lambda^3 q_1 - \lambda^2(2p_2 - \frac{1}{8}q_1^2) & \frac{1}{4}\lambda^2 q_2 - \frac{1}{2}\lambda(2p_1 \\
+\lambda(\frac{1}{4}q_1^3 - \frac{1}{16}q_2^2 + 2q_1 p_2 - \frac{1}{2}c) & +\frac{1}{2}q_1 q_2) + \frac{1}{4}(q_1)_\xi \\
-\frac{1}{2}q_1 c - \frac{1}{4}(q_2)_\xi &
\end{pmatrix}, \qquad (6.470)
$$

where L_2 plays the role of the principal one and the others are auxiliary ones. Moreover, we have

$$
(L_2)_{t_n} - (L_n)_\xi + [L_2, L_n] = 0 \quad \Longleftrightarrow \quad \begin{pmatrix} q \\ p \\ c \end{pmatrix}_{t_n} = G_n. \qquad (6.471)
$$

Now, let us reduce our systems to the stationary subspace, i.e. let us assume that q, p, c do not depend on the variable $t_2 = \xi$. The natural projections exist for differential functions, functionals and operators through $D_\xi \to 0$. There are no such reductions for nonlocal objects including D_ξ^{-1}. Thus, in the reduced phase space, which is just the phase space of the fifth KdV stationary flow in the canonical representation, there are two degenerate Hamiltonian operators

$$
\pi_0^s \equiv \theta_0 = \begin{pmatrix}
0 & 0 & 1 & 0 & 0 \\
0 & 0 & 0 & 1 & 0 \\
-1 & 0 & 0 & 0 & 0 \\
0 & -1 & 0 & 0 & 0 \\
0 & 0 & 0 & 0 & 0
\end{pmatrix},
$$

$$
\pi_1^s \equiv \theta_1 = \begin{pmatrix}
0 & -4 & q_1 & 0 & \frac{1}{2}q_2 \\
* & 0 & -q_2 & -\frac{3}{2}q_1 & 8p_2 \\
* & * & 0 & \frac{1}{2}p_2 + \frac{15}{32}q_1^2 & -\frac{15}{32}q_2^2 + \frac{5}{16}q_1^3 - \frac{1}{2}c \\
* & * & * & 0 & -\frac{1}{2}p_1 - \frac{5}{16}q_1 q_2 \\
* & * & * & * & 0
\end{pmatrix} \qquad (6.472)
$$

and no recursion operator at all. The Hamiltonian flows (6.468) and respective Hamiltonians (6.469) take the form

$$
G_0^s = \begin{pmatrix}
q_2 \\
16p_2 \\
\frac{5}{8}q_1^3 - \frac{5}{16}q_2^2 - c \\
-\frac{5}{8}q_1 q_2 - p_1 \\
0
\end{pmatrix},
$$

$$G_1^s = \begin{pmatrix} -q_1 q_2 - 4p_1 \\ \frac{1}{4}q_2^2 - \frac{5}{2}q_1^3 - 16q_1 p_2 + 4c \\ 8p_2^2 + \frac{5}{32}q_1 q_2^2 + \frac{15}{2}q_1^2 p_2 + q_2 p_1 + \frac{15}{16}q_1^4 - \frac{3}{2}q_1 c \\ \frac{5}{32}q_1^2 q_2 + q_1 p_1 - \frac{1}{2}q_2 p_2 \\ 0 \end{pmatrix},$$

$$G_2^s = 0, \quad G_3^s = \frac{1}{2}cG_0^s, \dots, \tag{6.473}$$

$$H_0^s \equiv h_0 = 2c,$$

$$H_1^s \equiv h_1 = 8p_2^2 + q_2 p_1 + \frac{5}{16}q_1 q_2^2 - \frac{5}{32}q_1^4 + q_1 c,$$

$$H_2^s \equiv h_2 = -2p_1^2 - 8q_1 p_2^2 + \frac{1}{4}q_2^2 p_2 - \frac{5}{2}q_1^3 p_2 - q_1 q_2 p_1 - \frac{5}{64}q_1^2 q_2^2$$
$$- \frac{3}{16}q_1^5 + 4p_2 c + \frac{3}{4}q_1^2 c,$$

$$H_3^s \equiv h_3 = \frac{1}{2}c^2 = \frac{1}{4}ch_0, \quad \dots \tag{6.474}$$

familiar from Sect. 6.1. Now, the chain $\pi_1 \delta H_k = G_k = \pi_0 \delta H_{k+1}$ terminates with the H_2^s, i.e. the Casimir of π_1^s

$$\pi_0^s \circ \delta H_0^s = 0$$
$$\pi_0^s \circ \delta H_1^s = G_0^s \quad = \pi_1^s \circ \delta H_0^s$$
$$\pi_0^s \circ \delta H_2^s = G_1^s \quad = \pi_1^s \circ \delta H_1^s$$
$$0 \quad = \pi_1^s \circ \delta H_2^s. \tag{6.475}$$

Simultaneously, the stationary reduction of Lax matrices reads

$$L_0^s = L_0, \quad L_1^s = L_1, \quad L_2^s = L_2, \quad L_3^s = \lambda L_2^s + \frac{1}{2}cL_0^s, \quad \dots, \tag{6.476}$$

hence, both nontrivial stationary flows can be obtained from the stationary reduction of the zero curvature condition (6.471)

$$(L_2^s)_{t_n} + [L_2^s, L_n^s] = 0 \quad \Longleftrightarrow \quad \begin{pmatrix} q \\ p \\ c \end{pmatrix}_{t_n} = G_n^s, \quad n = 0, 1. \tag{6.477}$$

Example 6.30 The t-type multi-Hamiltonian structure of the DWW hierarchy. Let us consider the t_2-flow of the dispersive water waves, which is represented by the following two-component dynamical system

$$\begin{pmatrix} u \\ v \end{pmatrix}_{t_2} = \begin{pmatrix} \frac{1}{4}u_{xxx} + \frac{3}{2}uu_x + \frac{3}{8}v^2u_x + \frac{3}{2}uvv_x + \frac{9}{8}v_xv_{xx} + \frac{3}{8}vv_{xxx} \\ \frac{1}{4}v_{xxx} + \frac{3}{2}uv_x + \frac{3}{2}vu_x + \frac{15}{8}v^2u_x \end{pmatrix}$$

$$= \begin{pmatrix} -\frac{1}{2}vD_x - \frac{1}{2}D_xv & D_x \\ D_x & 0 \end{pmatrix} \delta \int_R \left(-\frac{1}{4}u_xv_x - \frac{5}{16}vv_x^2 + \frac{5}{8}uv^3 \right.$$

$$\left. + \frac{3}{4}u^2v + \frac{7}{64}v^5 \right) dx. \qquad (6.478)$$

The appropriate modified version has the form

$$\begin{pmatrix} \bar{u} \\ \bar{v} \end{pmatrix}_{t_2} = \begin{pmatrix} \frac{1}{4}\bar{u}_{xxx} - \frac{3}{2}\bar{u}^2\bar{u}_x - \frac{3}{8}\bar{v}_x^2 - \frac{3}{8}\bar{v}\bar{v}_{xx} + \frac{3}{8}\bar{u}_x\bar{v}^2 - \frac{3}{4}\bar{u}\bar{v}\bar{v}_x \\ -\frac{3}{2}\bar{u}_{xx}\bar{v} - \frac{3}{2}\bar{u}_x\bar{v}_x - 3\bar{u}\bar{u}_x\bar{v} - \frac{3}{2}\bar{u}^2\bar{v}_x + \frac{15}{8}\bar{v}^2\bar{v}_x + \frac{1}{4}\bar{v}_{xxx} \end{pmatrix}$$

$$= \begin{pmatrix} -\frac{1}{4}D_x & 0 \\ 0 & D_x \end{pmatrix} \delta \int_R \left(\frac{1}{2}\bar{u}_x^2 + \frac{1}{2}\bar{u}^4 - \frac{3}{4}\bar{u}_x\bar{v}^2 - \frac{3}{4}\bar{u}^2\bar{v}^2 \right.$$

$$\left. + \frac{5}{32}\bar{v}^4 - \frac{1}{8}\bar{v}_x^2 \right) dx \qquad (6.479)$$

related to (6.478) by the x-type Miura map

$$u = -\bar{u}_x - \bar{u}^2, \quad v = \bar{v}. \qquad (6.480)$$

As θ_0 is a regular Hamiltonian operator and $\ker\theta_0 = c_1\delta(\frac{1}{2}v) + c_2\delta(\frac{1}{2}u + \frac{1}{8}v^2)$, according to the general scheme, the $t_2 = \xi$-type Hamiltonian formulation of (6.478) reads

$$\begin{pmatrix} q_1 \\ q_2 \\ p_1 \\ p_2 \\ c_1 \\ c_2 \end{pmatrix}_{t_0=x} = G_0$$

$$= \begin{pmatrix} 10q_2p_1 - 4p_2 \\ -4p_1 \\ \frac{5}{8}q_2^3 + \frac{3}{2}q_1q_2 - \frac{1}{2}c_2 \\ \frac{15}{8}q_1q_2^2 + \frac{3}{4}q_1^2 + \frac{35}{64}q_2^4 - 5p_1^2 - \frac{1}{2}c_1 - \frac{1}{4}q_2c_2 \\ 2(q_1)_\xi + q_2(q_2)_\xi \\ 2(q_2)_\xi \end{pmatrix} = \pi_0 \circ \delta H_1 \qquad (6.481)$$

where

$$\pi_0 = \begin{pmatrix} 0 & 0 & 1 & 0 & 0 & 0 \\ 0 & 0 & 0 & 1 & 0 & 0 \\ -1 & 0 & 0 & 0 & 0 & 0 \\ 0 & -1 & 0 & 0 & 0 & 0 \\ 0 & 0 & 0 & 0 & 0 & 4D_\xi \\ 0 & 0 & 0 & 0 & 4D_\xi & 0 \end{pmatrix} \tag{6.482}$$

and

$$q_1 = u, \; p_1 = -\frac{1}{4}v_x, \; q_2 = v, \; p_2 = -\frac{1}{4}u_x - \frac{5}{8}vv_x, \; b = \begin{pmatrix} 0 & 1 \\ 1 & 0 \end{pmatrix},$$

$$c_1 = \frac{1}{2}u_{xx} + \frac{5}{8}v_x^2 + vv_{xx} + \frac{9}{4}uv_2 + \frac{3}{2}u^2 + \frac{15}{32}v^4,$$

$$c_2 = \frac{1}{2}v_{xx} + \frac{5}{4}v^3 + 3uv,$$

$$h_1 = -4p_1 p_2 + 5q_2 p_1^2 - \frac{5}{8}q_1 q_2^3 - \frac{3}{4}q_1^2 q_2 - \frac{7}{64}q_2^5$$

$$+ \frac{1}{2}q_2 c_1 + \frac{1}{2}q_1 c_2 + \frac{1}{8}q_2^2 c_2. \tag{6.483}$$

The second t_2-type Hamiltonian structure of the flow (6.481), being the $(t_2$-type) Miura image of the first Hamiltonian structure of the modification of system (6.481),

$$\overline{\pi}_0 = \begin{pmatrix} 0 & 0 & 1 & 0 & 0 & 0 \\ 0 & 0 & 0 & 1 & 0 & 0 \\ -1 & 0 & 0 & 0 & 0 & 0 \\ 0 & -1 & 0 & 0 & 0 & 0 \\ 0 & 0 & 0 & 0 & -4D_\xi & 0 \\ 0 & 0 & 0 & 0 & 0 & D_\xi \end{pmatrix}, \tag{6.484}$$

is given as

$$\pi_1 = \begin{pmatrix} 0 & 0 & -\frac{3}{2}q_2 & -\frac{1}{2}q_1 - \frac{15}{8}q_2^2 & G_0^1 + 2D_\xi & 0 \\ * & 0 & 1 & q_2 & G_0^2 & 0 \\ * & * & 0 & -p_1 & G_0^3 & 0 \\ * & * & * & -\frac{1}{4}D_\xi & G_0^4 & 0 \\ * & * & * & * & 2q_1 D_\xi + 2D_\xi q_1 + q_2 D_\xi q_2 & 2q_2 D_\xi \\ * & * & * & * & * & 4D_\xi \end{pmatrix},$$

$$\tag{6.485}$$

where $(G_0^1, \ldots, G_0^6)^{\mathrm{T}} = G_0$ are given in (6.481).

The further development is straightforward. One constructs the recursion operator

$$\Phi = \pi_1 \circ (\pi_0)^{-1}$$

$$
= \begin{pmatrix}
-\frac{3}{2}q_2 & -\frac{1}{2}q_1 - \frac{15}{8}q_2^2 & 0 & 0 & 0 & \frac{1}{2} + \frac{1}{4}G_0^1 D_\xi^{-1} \\
1 & q_2 & 0 & 0 & 0 & \frac{1}{4}G_0^2 D_\xi^{-1} \\
0 & -p_1 & -\frac{3}{2}q_2 & 1 & 0 & \frac{1}{4}G_0^3 D_\xi^{-1} \\
p_1 & -\frac{1}{4}D_\xi & -\frac{1}{2}q_1 - \frac{15}{8}q_2^2 & q_2 & 0 & \frac{1}{4}G_0^4 D_\xi^{-1} \\
-G_0^3 & -G_0^4 & G_0^1 - 2D_\xi & G_0^2 & \frac{1}{2}q_2 & \frac{1}{2}q_1 + B D_\xi^{-1} \\
0 & 0 & 0 & 0 & 1 & \frac{1}{2}\partial q_2 \partial^{-1}
\end{pmatrix},
$$

$$\tag{6.486}$$

where $B = \frac{1}{2}D_\xi q_1 + \frac{1}{4}q_2 D_\xi q_2$ and the infinite hierarchy of flows $G_n = \Phi^n \circ G_0$

$$
G_0 = \begin{pmatrix}
10q_2 p_1 - 4p_2 \\
-4p_1 \\
\frac{5}{8}q_2^3 + \frac{3}{2}q_1 q_2 - \frac{1}{2}c_2 \\
\frac{15}{8}q_1 q_2^2 + \frac{3}{4}q_1^2 + \frac{35}{64}q_2^2 - 5p_1 - \frac{1}{2}c_1 - \frac{1}{4}q_2 c_2 \\
2q_{1\xi} + q_2 q_{2\xi} \\
2q_{2\xi}
\end{pmatrix},
$$

$$
G_1 = \Phi \circ G_0 = \begin{pmatrix}
-\frac{5}{2}q_2^2 p_1 + 4q_2 p_2 + 2q_1 p_1 + q_{2\xi} \\
4q_2 p_1 - 4p_2 \\
-p_1^2 + \frac{3}{8}q_1 q_2^2 + \frac{3}{4}q_1^2 - \frac{5}{64}q_2^4 - \frac{1}{2}c_1 + \frac{1}{4}q_2 c_2 \\
\frac{5}{2}q_2 p_1^2 - 4p_1 p_2 - \frac{5}{16}q_1 q_2^3 + \frac{3}{8}q_1^2 q_2 - \frac{45}{128}q_2^5 + \frac{1}{4}q_1 c_2 \\
\qquad\qquad - \frac{3}{4}q_2 c_1 + \frac{9}{16}q_2^2 c_2 + p_{1\xi} \\
-(q_1 q_2)_\xi - \frac{9}{4}q_2^2 q_{2\xi} + c_{2\xi} \\
-2q_{1\xi} + 3q_2 q_{2\xi}
\end{pmatrix},
$$

$$
G_2 = \Phi \circ G_1 = \begin{pmatrix}
q_1 \\
q_2 \\
p_1 \\
p_2 \\
c_1 \\
c_2
\end{pmatrix}_\xi, \quad \dots . \tag{6.487}
$$

The hierarchy G_n is not only bi-Hamiltonian but even tri-Hamiltonian, just the same as its x-type counterpart [9]. Its third (local) Hamiltonian structure is given as

π_2

$$
=
\begin{pmatrix}
0 & 0 & \frac{3}{8}q_2^2 - \frac{1}{2}q_1 & -\frac{1}{4}q_1q_2 - \frac{15}{16}q_2^3 & G_1^1 - 2q_2 D_\xi & G_0^1 + 2D_\xi \\
* & 0 & -\frac{1}{2}q_2 & -\frac{1}{2}q_1 - \frac{7}{8}q_2^2 & G_1^2 + 2D_\xi & G_0^2 \\
* & * & 0 & \frac{1}{2}q_2 p_1 - \frac{1}{4}D_\xi & G_1^3 & G_0^3 \\
* & * & * & -\frac{1}{4}(D_\xi q_2 + q_2 D_\xi) & G_1^4 + 3p_1 D_\xi & G_0^4 \\
* & * & * & * & A & B \\
* & * & * & * & * & 2q_2 D_\xi + 2 D_\xi q_2
\end{pmatrix},
$$
(6.488)

where

$$
\begin{aligned}
A & = (-2q_1 q_2 - \frac{5}{4}q_2^3 + c_2)D_\xi + D_\xi(-2q_1 q_2 - \frac{5}{4}q_2^3 + c_2) + q_2 D_\xi q_1 \\
& \quad + q_1 D_\xi q_2 + \frac{1}{2}q_2^2 D_\xi q_2 + \frac{1}{2}q_2 D_\xi q_2^2, \\
B & = 2q_1 D_\xi + 2D_\xi q_1 + q_2^2 D_\xi + q_2 D_\xi q_2.
\end{aligned}
$$

The first few Lax matrices in the t_2-representation are the following

$$
L_0 =
\begin{pmatrix}
0 & 1 \\
\lambda^2 - \lambda q_2 - q_1 & 0
\end{pmatrix},
$$

$$
L_1 =
\begin{pmatrix}
p_1 & \lambda \frac{1}{2} q_2 \\
\lambda^3 - \frac{1}{2}\lambda^2 q_2 - \lambda(q_1 + \frac{1}{8}q_2^2) & -p_1 \\
+ (q_1 q_2 + \frac{5}{8}q_2^3 - \frac{1}{2}c_2) &
\end{pmatrix},
$$

$$
L_2 =
\begin{pmatrix}
\lambda p_1 + p_2 - q_2 p_1 & \lambda^2 + \frac{1}{2}\lambda q_2 + \frac{3}{8}q_2^2 + \frac{1}{2}q_1 \\
\lambda^4 - \frac{1}{2}\lambda^3 q_2 - \lambda^2(\frac{1}{2}q_1 + \frac{1}{8}q_2^2) & -\lambda p_1 - p_2 + q_2 p_1 \\
-\lambda(\frac{1}{2}c_2 - \frac{1}{4}q_2^3 - \frac{1}{2}q_1 q_2) + (-p_1^2 & \\
- \frac{5}{64}q_2^4 + \frac{7}{4}q_1^2 + \frac{1}{4}q_2 c_2 - \frac{1}{2}c_1) &
\end{pmatrix},
$$
(6.489)

where L_2 is the principal one.

The reduction of this hierarchy to the $t_2 = \xi$-independent functions gives us the DWW stationary flows in the Ostrogradsky representation. It is of interest to note that the reduced π_2 operator gives us the third Hamiltonian structure $\pi_2^s \equiv \theta_2$ of the stationary DWW equations, not found in Sect. 6.1.

We shall illustrate the field counterparts of restricted flows on the example of the so called KdV hierarchy with sources, in order to compare it with the results of the previous sections. The m-th KdV equation with sources is defined as the set of nonlinear partial differential equations of the form

$$u_{t_m} + \sum_{k=1}^{N} \left(\varphi_k^2 \right)_x = K_m \left[u \right],$$

$$\varphi_{k_{xx}} = (\alpha_k - u)\varphi_k, \qquad k = 1, ..., N. \tag{6.490}$$

Here, we examine some algebraic properties of the system (6.490). Actually, we derive its t-type bi-Hamiltonian structure, construct the appropriate recursion operator, and finally consider its stationary reduction.

Let us rewrite the system under consideration in the more convenient form

$$\begin{pmatrix} \varphi_{1k} \\ \varphi_{2k} \end{pmatrix}_x = \begin{pmatrix} 0 & 1 \\ \lambda_k - u & 0 \end{pmatrix} \begin{pmatrix} \varphi_{1k} \\ \varphi_{2k} \end{pmatrix}, \qquad k = 1, ..., N$$

$$\gamma_m \left[u \right] = \sum_{k=1}^{N} \left(\varphi_{1k}^2 \right) + c, \tag{6.491}$$

$$c_x = u_t,$$

where $\varphi_{1k} = \varphi_k$, $\varphi_{2k} = \varphi_{k_x}$, $\gamma_m \left[u \right]$ are cosymmetries (conserved one-forms) of the KdV, such that $(\gamma_m \left[u \right])_x = K_m \left[u \right]$, c is a new field variable and for simplicity we dropped the index m on the evolution parameter t. Notice that the stationary flow of system (6.491) is equivalent to the m-th restricted flow of the KdV (6.218). Hence, applying the results (6.218)–(6.224) from Sect. 6.4, the first Hamiltonian formulation of the m-th KdV equation with sources (6.491) can be presented in the following form

$$\begin{pmatrix} q \\ r \\ p \\ s \\ c \end{pmatrix}_x = \begin{pmatrix} p \\ K_{rm} \\ (A - s_1 I_N)Q \\ K_{sm} + K_{\text{int}} \\ s_{1_t} \end{pmatrix} \equiv K_{m,1} = \pi_{m,0} \circ \delta H_{m,1}$$

$$\equiv \begin{pmatrix} 0 & 0 & I_N & 0 & 0 \\ 0 & 0 & 0 & I_m & 0 \\ -I_N & 0 & 0 & 0 & 0 \\ 0 & -I_m & 0 & 0 & 0 \\ 0 & 0 & 0 & 0 & 2D_t \end{pmatrix} \delta H_{m,1}, \tag{6.492}$$

where

$$K_{\text{int}} = \left(-\frac{1}{2} q^{\mathrm{T}} q, 0, ..., 0 \right)^{\mathrm{T}}, \qquad H_{m,1} = H_L + H_m + H_{\text{int}}.$$

Let us consider now the following system of nonlinear partial differential equations

$$v_t + \frac{1}{2} \sum_{k=1}^{N} (\psi_{1k}, \psi_{2k}) = \overline{K}_m[u],$$

$$\begin{pmatrix} \psi_{1k} \\ \psi_{2k} \end{pmatrix}_x = \begin{pmatrix} v & \alpha_k \\ 1 & -v \end{pmatrix} \begin{pmatrix} \psi_{1k} \\ \psi_{2k} \end{pmatrix}, \qquad k = 1, \dots, N, \qquad (6.493)$$

where $\overline{K}_m [v]$ is the m-th vector field of the modified KdV hierarchy and the lower of (6.493) are N copies of the MKdV Lax equation. We call this system the m-th MKdV with sources.

First, let us rewrite the system (6.493) in a more convenient form. Integrating the first equation once, we get

$$\begin{pmatrix} \psi_{1k} \\ \psi_{2k} \end{pmatrix}_x = \begin{pmatrix} v & \alpha_k \\ 1 & -v \end{pmatrix} \begin{pmatrix} \psi_{1k} \\ \psi_{2k} \end{pmatrix}, \qquad k = 1, \dots, N$$

$$2\overline{\gamma}_m [v] = 2\overline{c} + \sum_{k=1}^{N} \psi_{1k}\psi_{2k}, \qquad (6.494)$$

$$\overline{c}_x = -v_t,$$

where $\overline{\gamma}_m [v]$ are cosymmetries of the MKdV, such that $\overline{K}_m [v] = -(\overline{\gamma}_m [v])_x$, and $\overline{c}$ is a new field variable. Repeating the procedure presented in the case of the KdV with sources and applying relations (6.225)–(6.231) we find the following Hamiltonian representation of (6.494):

$$\begin{pmatrix} \overline{q} \\ \overline{r} \\ \overline{p} \\ \overline{s} \\ \overline{c} \end{pmatrix}_x = \begin{pmatrix} A\overline{p} + \overline{s}_1\overline{q} \\ K_{\overline{r}m} \\ \overline{q} - \overline{s}_1\overline{p} \\ K_{\overline{s}m} + K_{\text{int}} \\ -\overline{s}_{1t} \end{pmatrix} \equiv \overline{K}_{m,1} = \overline{\pi}_{m,0} \circ \delta\overline{H}_{m,1}$$

$$= \begin{pmatrix} 0 & 0 & I_N & 0 & 0 \\ 0 & 0 & 0 & I_m & 0 \\ -I_N & 0 & 0 & 0 & 0 \\ 0 & -I_m & 0 & 0 & 0 \\ 0 & 0 & 0 & 0 & -\frac{1}{2}D_t \end{pmatrix} \delta\overline{H}_{m,1}, \qquad (6.495)$$

where

$$\overline{H}_{\text{int}} = \overline{q}_1\overline{Q}^{\mathrm{T}}\overline{P}, \qquad \overline{H}_{m,1} = \overline{H}_L + \overline{H}_m + \overline{H}_{\text{int}}.$$

To derive the second Hamiltonian formulation of the KdV hierarchy with sources again it is convenient to find an explicit form of the Miura map relating the modified system to the given one. Its general form is given by Lemma 6.5. Now, applying the map M (6.233) to the first Hamiltonian structure of the MKdV hierarchy we generate the second Hamiltonian structure of the KdV hierarchy with sources:

$$\pi_{m,1} = M' \circ \overline{\pi}_{m,0} \circ M'^{\dagger}. \qquad (6.496)$$

On the other hand, because the Poisson operator $\pi_{m,0}$ is invertible, we can immediately construct a recursion operator

$$\Phi_m = \pi_{m,1} \circ (\pi_{m,0})^{-1}, \tag{6.497}$$

which has the hereditary property.

Thus, we have at our disposal the necessary tool for the construction for the system (6.490) an infinite number of symmetries, cosymmetries and conserved functionals. Actually, applying the operator Φ_m to the vector field $K_{m,1}$ (6.492) we can generate the hierarchy of Hamiltonian commuting vector fields (symmetries) $K_{m,n+1} = (\Phi_m)^n \circ K_{m,1}$. Then, applying the operator $\Phi_m^\dagger$ to the closed one-form $\gamma_{m,1} = (\pi_{m,0})^{-1} \circ K_{m,1} = \delta H_{m,1}$, we generate the hierarchy of invariant closed one-forms (cosymmetries) $\gamma_{m,n+1} = (\Phi_m^\dagger)^n \circ \gamma_{m,1} = \delta H_{m,n+1}$ and hence a hierarchy of conserved functionals $H_{m,n}$ which are all in involution, and such that $\pi_{m,0} \circ \delta H_{m,n} = K_{m,n} = \pi_{m,1} \circ \delta H_{m,n-1}$. Moreover $\pi_{m,n} = (\phi_m)^n \circ \pi_{m,0}$ are all Poisson operators of (6.490) although only the first two are purely differential objects.

Let us illustrate the considerations by an example of the field Garnier system [32] and the Melnikov system [36].

Example 6.31 Bi-Hamiltonian field Garnier system.
The simplest case $K_0[u] = u_x$ in (6.490) leads directly to the so called field Garnier system

$$\varphi_{kxx} = (\alpha_k - u)\varphi_k, \quad k = 1, \ldots, N$$

$$u_x = u_t + \sum_{i=1}^{N} (\varphi_i^2)_x. \tag{6.498}$$

Let

$$c_x = u_t \Rightarrow u = c + \sum_{i=1}^{N} \varphi_i^2 \Rightarrow u_t = c_t + \sum_{i=1}^{N} (\varphi_i^2)_t, \tag{6.499}$$

where now the variable c is considered as a field variable $c = c(x,t)$. So, the system (6.499) takes the form

$$\varphi_{kxx} = \left(\alpha_k - c - \sum_{i=1}^{N} \varphi_i^2 \right) \varphi_k, \quad k = 1, \ldots, N$$

$$c_x = c_t + \sum_{i=1}^{N} (\varphi_i^2)_t, \tag{6.500}$$

and its stationary reduction gives just the Garnier system (6.151). According to our previous considerations the Hamiltonian representation of system (6.500) reads

$$\begin{pmatrix} q \\ p \\ c \end{pmatrix}_x = \begin{pmatrix} p \\ [A - c - (q,q)]q \\ c_t + 2(q,q_t) \end{pmatrix} = K(q,p,c)$$

$$= \begin{pmatrix} 0 & I_N & 0 \\ -I_N & 0 & 0 \\ 0 & 0 & 2D_t \end{pmatrix} \delta \int_R \left[\frac{1}{2}(p,p) - \frac{1}{2}(q,Aq) + \frac{1}{4}(q,q)^2 \right.$$

$$\left. + \frac{1}{2}c(q,q) + \frac{1}{4}c^2 \right] dt = \pi_0 \circ \delta H_1.$$

(6.501)

We construct the second Hamiltonian structure of (6.500) by applying the Miura map to the first Hamiltonian structure of the modified system.

Let us consider the following modified system

$$\begin{pmatrix} \psi_{1k} \\ \psi_{2k} \end{pmatrix}_x = \begin{pmatrix} v & \alpha_k \\ 1 & -v \end{pmatrix} \begin{pmatrix} \psi_{1k} \\ \psi_{2k} \end{pmatrix}, \quad k = 1, \dots, N,$$

$$v_x = v_t - \frac{1}{2} \sum_{i=1}^N (\psi_{1i}\psi_{2i})_x,$$

(6.502)

where (6.502) represents the N replicas of the Lax spectral problem for the modified KdV (MKdV) equation with respective Bergman type restriction of the MKdV field variable v. The stationary version of (6.502) was considered in Sect. 6.4. Then, substituting

$$v_t = \bar{c}_x \Rightarrow v = \bar{c} - \frac{1}{2} \sum_{i=1}^N \psi_{1i}\psi_{2i} \Rightarrow v_t = \bar{c}_t - \frac{1}{2} \sum_{i=1}^N (\psi_{1i}\psi_{2i})_t$$

(6.503)

and introducing new variables

$$\varphi_{1k} = \bar{q}_k, \quad \varphi_{2k} = \bar{p}_k, \quad \bar{q} = (\bar{q}_1, \dots, \bar{q}_N)^T, \quad \bar{p} = (\bar{p}_1, \dots, \bar{p}_N)^T,$$

we can put the system (6.502) into the following Hamiltonian form

$$\begin{pmatrix} \bar{q} \\ \bar{p} \\ \bar{c} \end{pmatrix}_x = \begin{pmatrix} A\bar{p} + [\bar{c} - \frac{1}{2}(\bar{q},\bar{p})]\bar{q} \\ \bar{q} - [\bar{c} - \frac{1}{2}(\bar{q},\bar{p})]\bar{p} \\ \bar{c}_t - \frac{1}{2}(\bar{q},\bar{p})_t \end{pmatrix}$$

$$= \begin{pmatrix} 0 & I_N & 0 \\ -I_N & 0 & 0 \\ 0 & 0 & -\frac{1}{2}D_t \end{pmatrix} \delta \int_R \left[\frac{1}{2}(\bar{p},A\bar{p}) - \frac{1}{2}(\bar{q},\bar{q}) + \bar{c}(\bar{q},\bar{p}) \right.$$

$$\left. - \frac{1}{4}(\bar{q},\bar{p})^2 - \bar{c}^2 \right] dt = \bar{\pi}_0 \circ \delta \bar{H}_1$$

(6.504)

which is the Hamiltonian formulation of the modified field Garnier system. The relating Miura map $M : (\bar{q}, \bar{p}, \bar{c}) \to (q, p, c)$ is of the form

$$q = \bar{q},$$
$$p = (\bar{c} - \tfrac{1}{2}\bar{q}^T\bar{p})\bar{q} + A\bar{p}, \tag{6.505}$$
$$c = \tfrac{1}{2}A\bar{p}^T\bar{p} - \tfrac{1}{2}\bar{q}^T\bar{q} + \bar{c}\bar{q}^T\bar{p} - \tfrac{1}{4}(\bar{q}^T\bar{p})^2 - \bar{c}^2 + \tfrac{1}{2}(\bar{q}^T\bar{p})_t - \bar{c}_t.$$

Now, following the standard procedure, the second Hamiltonian structure of the field Garnier system has the form

$$\pi_1 = M' \circ \bar{\pi}_0 \circ M'^\dagger$$

$$= \begin{pmatrix} 0 & A - \tfrac{1}{2}q \otimes q & p - \tfrac{1}{2}qD_t \\[1em] -A + \tfrac{1}{2}q \otimes q & \begin{matrix}\tfrac{1}{2}p \otimes q - \tfrac{1}{2}q \otimes p \\ -\tfrac{1}{2}q \otimes D_t q\end{matrix} & \begin{matrix}[A - c - (q,q)]q \\ -\tfrac{1}{2}pD_t - \tfrac{1}{2}qD_t^2\end{matrix} \\[2em] -p^T - \tfrac{1}{2}D_t q^T & \begin{matrix}[(-A + c + q^T q)q]^T \\ -\tfrac{1}{2}D_t p^T + \tfrac{1}{2}D_t^2 q^T\end{matrix} & \begin{matrix}(c + q^T q)D_t \\ +D_t(c + q^T q) + \tfrac{1}{2}D_t^3\end{matrix} \end{pmatrix}$$
$$\tag{6.506}$$

and is compatible with the first Hamiltonian structure π_0. This leads to the bi-Hamiltonian formulation of the field Garnier dynamics $\pi_0 \circ \delta H_1 = \pi_1 \circ \delta H_0$, where $H_0 = \int_R c \, dt$. As π_0 is invertible, we can construct the recursion operator

$$\Phi = \pi_1 \circ \pi_0^{-1}$$

$$= \begin{pmatrix} A - \tfrac{1}{2}q \otimes q & 0 & -\tfrac{1}{4}q + \tfrac{1}{2}pD_t^{-1} \\[1em] \begin{matrix}\tfrac{1}{2}p \otimes q - \tfrac{1}{2}q \otimes p \\ -\tfrac{1}{2}q \otimes D_t q\end{matrix} & A - \tfrac{1}{2}q \otimes q & \begin{matrix}\tfrac{1}{2}[A - c - (q \otimes q)]qD_t^{-1} \\ -\tfrac{1}{4}p - \tfrac{1}{4}qD_t\end{matrix} \\[2em] \begin{matrix}[(-A + c + (q,q))q]^T \\ -\tfrac{1}{2}D_t p^T + \tfrac{1}{2}D_t^2 q^T\end{matrix} & p^T + \tfrac{1}{2}D_t q^T & \begin{matrix}+\tfrac{1}{2}[c_t + 2(q, q_t)]D_t^{-1} \\ c + (q,q) + \tfrac{1}{4}D_t^2\end{matrix} \end{pmatrix}$$
$$\tag{6.507}$$

which has the heredity property. Applying this operator to the Garnier vector field K (6.501) we can generate the hierarchy of Hamiltonian commuting symmetries $K_n = \Phi^n \circ K$. Then, applying the $\Phi^\dagger$ operator to the gradient of the conserved functional H_1 we generate the hierarchy of closed one-forms $\delta H_{n+1} = (\Phi^\dagger)^n \circ \delta H_1$ and hence a hierarchy of conserved functionals H_n which are all in involution and such that $\pi_0 \circ \delta H_n = K_n = \pi_1 \circ \delta H_{n+1}$. Moreover, $\pi_n = \Phi^n \circ \pi_0$ are all Hamiltonian structures of (6.501) although only the first two are purely differential objects.

Let us reduce the system (6.501) to the stationary subspace, i.e. we assume that q_k, p_k and c do not depend on the variable t. The natural projections exist for differential functions, functionals and operators through $D_t \to 0$. There are no such reductions for nonlocal objects including D_t^{-1}. Thus, in the reduced phase space, which is the phase space of the finite-dimensional Garnier system, there are only two degenerate Hamiltonian operators

$$
\pi_0^s = \begin{pmatrix} 0 & I_N & 0 \\ -I_N & 0 & 0 \\ 0 & 0 & 0 \end{pmatrix},
$$

$$
\pi_1^s = \begin{pmatrix} 0 & A - \frac{1}{2}q \otimes q & p \\ -A + \frac{1}{2}q \otimes q & \frac{1}{2}p \otimes q - \frac{1}{2}q \otimes p & [A - c - (q,q)]q \\ -p^{\mathrm{T}} & [-A + c + (q,q)]q^{\mathrm{T}} & 0 \end{pmatrix} \tag{6.508}
$$

familiar from previous sections, and no recursion operator.

Example 6.32 Bi-Hamiltonian Melnikov system.
The case $K_1[u] = \frac{1}{4}u_{xxx} + \frac{3}{4}uu_x$ in (6.490) leads directly to the so called
Melnikov system

$$
\varphi_{kxx} = (\alpha_k - u)\varphi_k, \quad k = 1, \dots, N
$$

$$
\frac{1}{4}u_{xxx} + \frac{3}{4}uu_x = u_t + \sum_{i=1}^{N}(\varphi_i^2)_x. \tag{6.509}
$$

Passing to the Ostrogradsky variables

$$
r = u, \quad s = -\frac{1}{4}u_x, \quad c = \frac{1}{4}u_{xx} + \frac{3}{4}u^2 - q^{\mathrm{T}}, \tag{6.510}
$$

the first Hamiltonian representation of (6.509) is the following

$$
\begin{pmatrix} q \\ r \\ p \\ s \\ c \end{pmatrix}_x = \begin{pmatrix} p \\ -8s \\ (A - sI_N)q \\ \frac{3}{8}r^2 - \frac{1}{2}c - \frac{1}{2}(q,q) \\ r_t \end{pmatrix}
$$

$$
= \begin{pmatrix} 0 & 0 & I_N & 0 & 0 \\ 0 & 0 & 0 & 1 & 0 \\ -I_N & 0 & 0 & 0 & 0 \\ 0 & -1 & 0 & 0 & 0 \\ 0 & 0 & 0 & 0 & 2D_t \end{pmatrix}
$$

$$
\times \delta \int_R [\frac{1}{2}(p,p) - \frac{1}{2}(q, Aq) + \frac{1}{2}r(q,q) - 4s^2 - \frac{1}{8}r^3 + \frac{1}{2}cr] dt.
$$

$$
\tag{6.511}
$$

As $\overline{K}_1[v] = \frac{1}{4}v_{xxx} - \frac{3}{4}v^2v_x$, the appropriate modified system (6.493) has the
Hamiltonian form

$$
\begin{pmatrix} \bar{q} \\ \bar{r} \\ \bar{p} \\ \bar{s} \\ \bar{c} \end{pmatrix}_x = \begin{pmatrix} A\bar{p} + \bar{r}\bar{q} \\ 2\bar{s} \\ \bar{q} - \bar{r}\bar{p} \\ \bar{r}^3 - 2\bar{c} - (\bar{q},\bar{p}) \\ -\bar{r}_t \end{pmatrix}
$$

$$
= \begin{pmatrix} 0 & 0 & I_N & 0 & 0 \\ 0 & 0 & 0 & 1 & 0 \\ -I_N & 0 & 0 & 0 & 0 \\ 0 & -1 & 0 & 0 & 0 \\ 0 & 0 & 0 & 0 & -\frac{1}{2}D_t \end{pmatrix}
$$

$$
\times \delta \int_R \left[\frac{1}{2}(\bar{p}, A\bar{p}) - \frac{1}{2}(\bar{q},\bar{q}) + \bar{r}(\bar{q},\bar{p}) + \bar{s}^2 - \frac{1}{4}\bar{r}^4 + 2\bar{c}\bar{r} \right] dt,
$$

$$
(6.512)
$$

where $\bar{r} = v$, $\bar{s} = \frac{1}{2}v_x$, $\bar{c} = -\frac{1}{4}v_{xx} + \frac{1}{2}v^3 - \frac{1}{2}(\bar{q},\bar{p})$. The Miura map relating systems (6.512) to (6.511) reads

$$
M: \quad
\begin{aligned}
q &= \bar{q}, \\
r &= -2\bar{s} - \bar{r}^2, \\
p &= A\bar{p} + \bar{r}\bar{q}, \\
s &= \frac{1}{4}\bar{r}^3 + \frac{1}{2}\bar{r}\bar{s} - \frac{1}{2}\bar{c} - \frac{1}{4}(\bar{q},\bar{p}), \\
c &= \bar{s}^2 - \frac{1}{4}\bar{r}^4 + 2\bar{r}\bar{c} - \bar{r}_t + \frac{1}{2}(\bar{p}, A\bar{p}) - \frac{1}{2}(\bar{q},\bar{q}) + \bar{r}(\bar{q},\bar{p}).
\end{aligned}
\qquad (6.513)
$$

This leads to the second Hamiltonian structure of (6.511):

$$
\pi_{1,1} = M' \circ \bar{\pi}_{1,0} \circ M'^\dagger
$$

$$
= \begin{pmatrix}
0 & 0 & A & -\frac{1}{4}q & p \\
0 & 0 & -2q^{\mathrm{T}} & -\frac{1}{2}r & -8s + 2D_t \\
-A & 2q & 0 & \frac{1}{4}p & (A - rI_N)q \\
\frac{1}{4}q & \frac{1}{2}r & -\frac{1}{4}p & -\frac{1}{8}D_t & \frac{3}{8}r^2 - \frac{1}{2}c + \frac{1}{2}(q,q) \\
* & * & * & * & rD_t + D_t r
\end{pmatrix}
\qquad (6.514)
$$

and to the recursion operator of the form

$$
\phi_1 = \pi_{1,1} \circ (\pi_{1,0})^{-1}
$$

$$
= \begin{pmatrix}
A & -\frac{1}{4}q & 0 & 0 & \frac{1}{2}pD_t^{-1} \\
-2q^{\mathrm{T}} & -\frac{1}{2}r & 0 & 0 & -4sD_t^{-1} + 1 \\
0 & \frac{1}{4}p & A & -2q & \frac{1}{2}(A - rI_N)qD_t^{-1} \\
-\frac{1}{4}p^{\mathrm{T}} & -\frac{1}{8}D_t & -\frac{1}{4}q^{\mathrm{T}} & -\frac{1}{2}r & \frac{1}{2}BD_t^{-1} \\
(rI_N - A)q^{\mathrm{T}} & -B & p^{\mathrm{T}} & -8s - 2D_t & \frac{1}{2}r + \frac{1}{2}D_t r D_t^{-1}
\end{pmatrix},
$$

$$
(6.515)
$$

where $B = \frac{3}{8}r^2 - \frac{1}{2}c + \frac{1}{2}(q,q)$. Hence, the bi-Hamiltonian formulation of dynamics (6.509) reads

$$K_{1,1} = \pi_{1,0} \circ \delta H_{1,1} = \pi_{1,1} \circ \delta H_{1,0}, \qquad H_{1,0} = c. \qquad (6.516)$$

The higher order KdV flows with sources can be illustrated in the same way. The stationary reduction gives exactly the KdV restricted flows studied in Sect. 6.4.

The results of this section confirm the suggestion that a study of field representatives of finite-dimensional integrable systems provides a simple method of generating symmetries, cosymmetries and related constants of motion of a given finite-dimensional system involving recursive generation of the respective counterparts on the field level and then projection of the results onto the stationary submanifold.

We will complete this section with some comments on the change of the evolution variable for integrable systems. The idea of interchanging the standard evolution parameter t and the space variable x is not new. It was considered, for example, in the paper [89]. But already in [69] Flaschka et al. made a point of the fact that the integrability of the AKNS hierarchy is a strictly algebraic feature and that we are free to pick out any t_n (in an infinite series $t_1, t_2, t_3, \ldots$) as the space variable x. In the KdV reduction of the general AKNS scheme the choice $t_1 = x$ corresponds to the standard KdV hierarchy. Choosing $t_3 = x$ we get our t_2-type KdV hierarchy. We will not go into details of the Flaschka-Newell-Ratiu construction and remark only that it can be used to construct directly t-type hierarchies. This changes the perspective completely since the Hamiltonian structure of a stationary flow is obtained as a natural and straightforward reduction of its (t-type) Hamiltonian structure rather than being used to construct the hierarchy in question. Although in this section we have confined the discussion to Hamiltonian systems with regular Poisson structures, nevertheless the possibility of interchange of the evolution variable seems nowadays a more general property and is applicable to other (perhaps all) multi-Hamiltonian systems.

6.7 Bi-Hamiltonian Chains
on Poisson–Nijenhuis Manifolds

The last section of this chapter is devoted to case different from that considered in previous sections. Actually, we present a few examples of bi-Hamiltonian dynamical systems on a Poisson–Nijenhuis manifold. This means that at least one of the Hamiltonian structures is not degenerate and the related system has a hereditary recursion operator.

A comparison of the results of Sects 6.5 and 5.2 indicates that the Nijenhuis coordinates (μ, λ) can be identified with the linear coordinates

(q, c) considered in the previous chapter. That is, the Poisson structures θ_i, $i = -1, 0, 1$, and the Nijenhuis operator Φ (6.267) are identical to $\bar{\theta}_i$, $\bar{\Phi}$ (5.10),(5.11) with a unit scaling factor $\rho = 1$ when $q_i \equiv \mu_i$ and $c_i \equiv \lambda_i$.

Let (M, ω_0) be a $2N$ dimensional symplectic manifold spanned by general coordinates $\{x^i\}_{i=1}^{2N}$. In this basis the closed two-form ω_0 reads

$$\omega_0 = \sum_{i,j} J_{0ij}(x) \mathrm{d}x^i \wedge \mathrm{d}x^j, \tag{6.517}$$

where J_0 is a nonsingular symplectic matrix. Its inverse defines a bi-vector

$$V_0 = \sum_{i,j} \theta_0^{ij}(x) \frac{\partial}{\partial x^i} \wedge \frac{\partial}{\partial x^j}, \tag{6.518}$$

where the implectic matrix $\theta_0 = J_0^{-1}$ is the inverse of the symplectic one, J_0. If there exists another closed two-form

$$\omega_1 = \sum_{i,j} J_{1ij}(x) \mathrm{d}x^i \wedge \mathrm{d}x^j \tag{6.519}$$

such that the symplectic matrix J_1 is compatible with the implectic one θ_0, then the symplectic manifold (M, ω_0) turns into a Poisson–Nijenhuis manifold (M, θ_0, J_1).

Let the vector field $K(x)$ be bi-Hamiltonian with respect to the pair (θ_0, J_1), i.e.

$$K(x) = \theta_0(x) \nabla H(x), \qquad J_1(x) K(x) = \nabla G(x). \tag{6.520}$$

Then the Nijenhuis (hereditary) operator

$$\Phi = \sum_{i,j} \Phi_j^i(x) \mathrm{d}x^j \otimes \frac{\partial}{\partial x^i}, \qquad \Phi_j^i = \sum_k \theta_0^{ik} J_{1kj}, \tag{6.521}$$

is the recursion operator for symmetries and its adjoint

$$\Psi = \Phi^\dagger = \sum_{i,j} \Psi_j^i(x) \mathrm{d}x^j \otimes \frac{\partial}{\partial x^i}, \qquad \Psi_j^i = \sum_k J_{1jk} \theta_0^{ki} \tag{6.522}$$

is the recursion operator for cosymmetries.

A natural Lax representation for the dynamics $x_t = K(x)$ reads

$$\Psi_t = [\Psi, B], \tag{6.523}$$

where

$$B_i^j(x) = (K'^T)_i^j = \partial_i K^j = \partial_i \left(\sum_k \theta_0^{jk} \partial_k H \right), \qquad \partial_j := \frac{\partial}{\partial x^j}. \tag{6.524}$$

There also exists a natural set of constants of motion in the form

$$H_n = \frac{1}{2n}\mathrm{Tr}\Psi^n, \quad n = 1, 2, \ldots \tag{6.525}$$

Indeed

$$\frac{\mathrm{d}H_k}{\mathrm{d}t} = \frac{1}{2}\mathrm{Tr}\Psi^{n-1}\Psi_t = \frac{1}{2}\mathrm{Tr}(\Psi B - B\Psi) = 0. \tag{6.526}$$

As presented by Okubo [157], if $H_1 = G$ and $H_2 = H$ then the hierarchy of commuting bi-Hamiltonian vector fields reads

$$K_n(x) = \theta_0(x)\nabla H_n(x) = \theta_1(x)\nabla H_{n-1}(x), \quad \theta_1 = \theta_0 \circ J_1 \circ \theta_0. \tag{6.527}$$

We recall that according to the Darboux theorem in a neighbourhood of any point of a symplectic manifold there exist local coordinates (q^i, p_i) which give the symplectic two-form ω_0 in the canonical form

$$\omega_0 = -\sum_i \mathrm{d}q^i \wedge \mathrm{d}p_i. \tag{6.528}$$

These coordinates are defined up to a canonical transformation. On a Poisson–Nijenhuis manifold we can analyse another problem of finding, inside the class of canonical coordinates, a subclass of Nijenhuis coordinates (μ_i, λ_i) transforming Φ as well into the canonical form (6.267). The reader can find more information on the bi-Hamiltonian dynamics on Poisson–Nijenhuis manifolds in the literature [39],[61],[67],[116],[134],[178],[179].

Nonperiodic Toda Lattice

The finite Toda lattice describes the motion of N point masses on the line under the influence of exponential forces. The Hamiltonian in terms of the canonical coordinates (q^i, p_i) is given by

$$H = \frac{1}{2}\sum_{i=1}^{N}p_i^2 + \sum_{i=1}^{N-1}f_i e^{q^i - q^{i+1}}, \tag{6.529}$$

and the related equations of motion with respect to the canonical Poisson matrix are

$$(q^i)_t = p_i, \quad (p_i)_t = f_{i-1}e^{q^{i-1}-q^i} - f_i e^{q^i - q^{i+1}} \tag{6.530}$$

with the understanding that $f_0 = f_N = 0$, where $f_1, \ldots, f_{N-1}$ are real coupling constants. Let us note that the canonical coordinates can be simply thought of as a particular choice of the coordinates x^i such that

$$x^i = q^i, \quad x^{N+i} = p_i, \quad i = 1, \ldots, N. \tag{6.531}$$

The canonical two-form

$$\omega_0 = -\sum_{i=1}^{N}\mathrm{d}q^i \wedge \mathrm{d}p_i \tag{6.532}$$

gives rise to the symplectic structure

$$J_0 = \begin{pmatrix} 0 & -I \\ I & 0 \end{pmatrix}, \tag{6.533}$$

and hence to the canonical Poisson structure

$$\theta_0 = J_1^{-1} = \begin{pmatrix} 0 & I \\ -I & 0 \end{pmatrix}. \tag{6.534}$$

What is important is that there exists a another invariant symplectic form [55]

$$\omega_1 = \sum_{i=1}^{N} \left[f_i e^{q^i - q^{i+1}} dq^i \wedge dq^{i+1} - p_i\, dq^i \wedge dp_i \right] + \sum_{i,j}^{N} \epsilon_{ij}\, dp_i \wedge dp_j. \tag{6.535}$$

where

$$\epsilon_{ij} = \left\{ \begin{array}{cc} 1 & i > j \\ 0 & i = j \\ -1 & i < j \end{array} \right\}. \tag{6.536}$$

Thus, the second symplectic structure J_1 can be represented in $N \times N$ block form as

$$J_1 = \begin{pmatrix} A & -B \\ B & E \end{pmatrix}, \tag{6.537}$$

where the block matrices have the form

$$A_{ij} = f_i \delta_{i+1,j} e^{q^i - q^{i+1}} - f_j \delta_{i,j+1} e^{q^j - q^{j+1}}, \quad B_{ij} = p_i \delta_{ij}, \quad E_{ij} = \epsilon_{ij}. \tag{6.538}$$

Moreover, one can verify the compatibility of the pair (θ_0, J_1). Hence, we find the following forms of the recursion hereditary operator, its adjoint and the second implectic structure

$$\Phi = \theta_0 \circ J_1 = \begin{pmatrix} B & E \\ -A & B \end{pmatrix}, \quad \Psi = J_1 \circ \theta_0 = \begin{pmatrix} B & A \\ -E & B \end{pmatrix}$$

$$\theta_1 = \theta_0 \circ J_1 \circ \theta_0 = \begin{pmatrix} -E & B \\ -B & -A \end{pmatrix}. \tag{6.539}$$

The bi-Hamiltonian formulation for Toda dynamics (6.530) reads

$$u_t = \theta_0 \circ \nabla H_2 = \theta_1 \circ \nabla H_1, \quad u = (q,p)^{\mathrm{T}},$$

$$H_1 = \frac{1}{2}\mathrm{Tr}\Psi = \sum_{i=1}^{N} p_i, \quad H_2 = \frac{1}{4}\mathrm{Tr}\Psi^2 = \frac{1}{2}\sum_{i=1}^{N} p_i^2 + \sum_{i=1}^{N-1} f_i e^{q^i - q^{i+1}}. \tag{6.540}$$

The reader can find more information on algebraic structure of finite dimensional Toda chains in the literature [53],[54],[66].

The Calogero–Moser (CM) System

This system consists of n identical particles moving on a line and repulsing each other by forces proportional to the cube of the mutual distances. Denoting by $(q^1, ..., q^n)$ the positions of the particles, by $(p_1, ..., p_n)$ the corresponding momenta, and by

$$q_{ij} := (q^i - q^j)^{-1} \tag{6.541}$$

the inverse of the distance between any pair of particles, the Hamiltonian of the system is written in the form

$$H = \frac{1}{2} \sum_{i=1}^{n} p_i^2 + \sum_{j \neq k}^{n} q_{ij}^2, \tag{6.542}$$

and therefore the equations of motion are

$$(q^i)_t = p_i, \quad (p_i)_t = -\frac{\partial H}{\partial q^i}. \tag{6.543}$$

This system was first solved by Calogero [42] in the quantum case. In the classical case Moser [141] have found the Lax pair L, A:

$$L_{ik} = \delta_{jk} p_j + \mathrm{i}(1 - \delta_{jk}) q_{jk},$$

$$A_{jk} = \delta_{jk} \sum_{k \neq j} \mathrm{i} q_{jk}^2 + \mathrm{i}(1 - \delta_{jk}) q_{jk}^2, \quad \mathrm{i}^2 = -1. \tag{6.544}$$

The hierarchy of constants of motion in involution is given by

$$H_k = \frac{1}{k+1} \mathrm{Tr} L^{k+1}, \quad k = 0, 1, ..., \tag{6.545}$$

where Tr = matrix trace. The CM system is the second member of the hierarchy, since the Hamiltonian H coincides with H_1. A second remarkable property of the CM system is that all members of its hierarchy admit a Lax representation. This means that there exist matrices A_k with $k = 0, 1, ...,$ such that the Hamiltonian equations associated with H_k may be written in the Lax form

$$\frac{\mathrm{d}L}{\mathrm{d}t_k} = [L, A_k] \tag{6.546}$$

with the same Lax matrix L.

Let us introduce the matrix

$$X = \mathrm{diag}(q_1, ..., q_n) \tag{6.547}$$

representing the positions of the particles of the CM system at a certain time. Then X evolves to [103]

$$\frac{\mathrm{d}X}{\mathrm{d}t_k} = [X, A_k] + L^{k-1}. \tag{6.548}$$

Introducing the functions H_l and

$$T_l = \mathrm{Tr}\, X L^{l+1} \tag{6.549}$$

we verify that they obey the equations

$$\frac{\mathrm{d}H_l}{\mathrm{d}t_k} = 0, \qquad \frac{\mathrm{d}T_l}{\mathrm{d}t_k} = (l+k+1)H_{l+k+1}. \tag{6.550}$$

Hence, the change of variables from the canonical coordinates $(q^1, ..., q^n, p_1, ..., p_n)$ to the new noncanonical coordinates $(H_1, ..., H_n, T_1, ..., T_n)$ allows a transformation of the equations of motion into the readily solvable form (6.550). Moreover, noncanonical coordinates fulfil the following commutator relations with respect to the canonical Poisson operator θ_0

$$\{H_n, H_m\}_{\theta_0} = 0, \quad \{T_n, H_m\}_{\theta_0} = (n+m+1)H_{n+m},$$

$$\{T_n, T_m\}_{\theta_0} = (m-n)T_{n+m}. \tag{6.551}$$

Comparing this result with the noncanonical action/angle algebra (5.34) with $\rho = 1$ and $p = 0$, one concludes that we deal with the same scalar fields as in (5.33), now expressed by canonical coordinates (q, p) instead of linearizing Nijenhuis coordinates $(\mu, \lambda) \equiv (q, c)$. So, according to the general relation (5.34), the commutators between the coordinates (H, T) with respect to non-canonical Poisson operator θ_1 take the form

$$\{H_n, H_m\}_{\theta_1} = 0, \quad \{T_n, H_m\}_{\theta_1} = (n+m+2)H_{n+m+1},$$

$$\{T_n, T_m\}_{\theta_1} = (m-n)T_{n+m+1}. \tag{6.552}$$

Now, from (6.552), the explicit form of the second Poisson structure in canonical coordinates (q, p) can be reconstructed.

Let us consider the particular case of the two-particle CM system. The noncanonical coordinates constructed from matrices L and X read

$$H_0 = p_1 + p_2,$$

$$H_1 = \tfrac{1}{2}p_1^2 + \tfrac{1}{2}p_2^2 + q_{12}^2,$$

$$T_{-1} = q^1 + q^2, \tag{6.553}$$

$$T_0 = q^1 p_1 + q^2 p_2.$$

If we use the relations (6.551) and we pass from the $(H_0, H_1 T_{-1} T_0)$ coordinates to the (q^1, q^2, p_1, p_2) coordinates by means of the transformation (6.553), we recover the canonical Poisson bracket $\{.,.\}_{\theta_0}$ on the cotangent bundle of the CM system. If we perform the same computation for the second Poisson bracket (6.552), we arrive at a new Poisson bracket [128] defined by

$$\{q^1, q^2\}_{\theta_1} = 2q_{12}/(4q_{12}^2 + p_{12}^2),$$

$$\{q^1, p_1\}_{\theta_1} = p_1 + p_{12}q_{12}^2/(4q_{12}^2 + p_{12}^2),$$

$$\{q^1, p_2\}_{\theta_1} = -p_{12}q_{12}^2/(4q_{12}^2 + p_{12}^2),$$

$$\{q^2, p_1\}_{\theta_1} = p_{12}q_{12}^2/(4q_{12}^2 + p_{12}^2),$$

$$\{q^2, p_2\}_{\theta_1} = p_2 - p_{12}q_{12}^2/(4q_{12}^2 + p_{12}^2),$$

$$\{p_1, p_2\}_{\theta_1} = 2q_{12}^3.$$

(6.554)

Thus, the CM system may be written in the bi-Hamiltonian form

$$u_t = \theta_0 \circ \nabla H = \theta_1 \circ \nabla P, \tag{6.555}$$

where $u = (q, p)^{\mathrm{T}}$, $H = H_1$ is the total energy of the system and $P = H_0$ is the total momentum. The same procedure is true for the system with three and more particles, even if the explicit form of the second Poisson bracket becomes prohibitively complicated.

7. Multi-Hamiltonian Lax Dynamics in (1+1)-Dimensions

In the previous chapters we presented a variety of multi-Hamiltonian integrable dynamical systems of finite and infinite dimensions and their common geometric and algebraic features. What we still lack, is a systematic method for the construction of multi-Hamiltonian infinite dimensional systems. This chapter will present an approach to a unified description of the integrable equations, based on the use of a simple and powerful algebraic tool, the so called $\mathcal{R}$-structure (or $\mathcal{R}$-matrix)[62],[175],[176]. This approach is formulated in a rather abstract algebraic way, but as an advantage one gets a simple and effective method for analysis of the multi-Hamiltonian structure of integrable systems. The crucial point of this approach is the observation that the Lax equation (1.2) can be treated as an abstract dynamical system from which the 'physical' dynamical systems are obtained by introducing suitable charts. Hence, the phase space for most of these equations can be regarded as given by the set of Lax operators taking values from some Lie algebra. This abstract representation of integrable dynamics is referred to as the *Lax dynamics*. The multi-Hamiltonian construction of the integrable equations becomes quite transparent when using the terminology of $\mathcal{R}$-structures. Based on results of Gelfand and Dikii [91], Adler [4] used a Lie algebraic setting to describe integrable partial differential equations via their Lax representations. As an important consequence it turned out that integrable systems of a different nature (discrete lattice systems or differential equations) may be constructed in a similar manner using Lie algebra techniques. The celebrated Adler–Gelfand–Dikii (AGD) scheme starts with a dual Lie algebra as a natural phase space for integrable equations. The Lie–Poisson bracket associated with the Lie algebra structure provides a natural Hamiltonian structure, and the invariant functions provide a natural set of functions in involution on the algebra. In order to obtain a nontrivial integrable dynamics for these functions, only a few additional structures have to be provided, which are again of a purely Lie algebraic nature. As the simplest example, a decomposition of the original algebra into proper subalgebras gives rise to a hierarchy of integrable Hamiltonian equations. It turn out that this construction may be regarded as a special case of a yet larger picture. Following Drinfeld's ideas [62] Semenov showed that the notion of classical $\mathcal{R}$-structures leads to an algebraic construction of integrable systems generalizing the AGD scheme [175]. In [117],[150] it was shown that there are in fact three natural Poisson brackets associated with such classical $\mathcal{R}$–structures. They lead to an abstract tri-Hamiltonian formulation of the Lax equations describing the nonlinear integrable systems. Applications of this general construction to the particular algebra of pseudo-differential operators or algebra of shift operators leads to a compact formulation of the multi-Hamiltonian structures for certain classes of integrable hierarchies.

In the first section of this chapter we collect the definitions and results regarding Lie groups and Lie algebras, necessary to clarify the concept of Lax dynamics. For proofs of statements of this section given without proof we refer the reader to the literature [3],[118]. In the second section we present basic concepts of the $\mathcal{R}$–structure and related multi-Hamiltonian formalism. The application of the theory to the infinite algebra of pseudo-differential operators and the infinite algebra of shift operators is presented in Sects 3 and 4, respectively. In the first case we obtain of unified theory of multi-Hamiltonian field systems; in the second case we get a unified multi-Hamiltonian theory of lattice systems.

7.1 Hamiltonian Dynamics on Lie Algebras

Definition 7.1 *A Lie group isa finite dimensional smooth manifold G which is a group and for which the group operation of multiplication*

$$G \times G \to G : \quad (g, h) \to gh \tag{7.1}$$

and inversion

$$G \to G : \quad g \to g^{-1} \tag{7.2}$$

are smooth.

Definition 7.2 *For every $g \in G$ the maps $L_g : G \to G : h \to gh$ and $R_g : G \to G : h \to hg$ are, respectively, left and right translations by g.*

Since $L_g \circ L_h = L_{gh}$ and $R_h \circ R_g = R_{gh}$ so $(L_g)^{-1} = L_{g^{-1}}$ and $(R_g)^{-1} = R_{g^{-1}}$, thus, both L_g and R_g are diffeomorphisms and moreover commute: $L_g \circ R_h = R_h \circ L_g$.

Recall that by $T_g G$ we denote a tangent space to a manifold G at a point g. All tangent spaces $T_g G$, $g \in G$ constitute a $2n$-dimensional $(n = \dim G)$ tangent bundle TG. The smooth mappings

$$X : G \to TG, \quad g \to X(g), \quad g \in G \tag{7.3}$$

are called vector fields on G. These vector fields form in a natural way a linear (infinite dimensional) space which we denote by $\mathcal{G}$. The linear space $\mathcal{G}$ is also a Lie algebra with respect to the commutator $[.,.]$.

Let $\phi : G \to G$ be a smooth mapping. Its differential (derivative)

$$d\phi = \phi' := T\phi : T_g G \to T_{\phi(g)} G \qquad (7.4)$$

constitutes a smooth mapping as well. If ϕ is a diffeomorphism, then for any vector fields $X, Y \in \mathcal{G}$ the direct image (push-forward) under ϕ of a vector field X is the vector field $Y = \phi_* X$ which satisfies

$$(\phi_* X)(\phi(g)) = T_g \phi \cdot X(g) = (d\phi)_g \cdot X(g)$$

$$\Leftrightarrow \phi_* X \cdot \phi = T\phi \circ X = d\phi \circ X \Leftrightarrow \phi_* X = (d\phi) \circ X \cdot \phi^{-1}. \qquad (7.5)$$

The inverse image (pull-back) under ϕ of Y is the vector field $\phi^* Y$, the direct image of Y by ϕ^{-1} that is $\phi^* Y = (\phi^{-1})_* Y$, which satisfies

$$
\begin{aligned}
(\phi^* Y)(g) &= ((\phi^{-1})_* Y)(g) = T_{\phi(g)} \phi^{-1} \cdot Y(\phi(g)) = (d\phi^{-1})_{\phi(g)} \cdot Y(\phi(g)) \\
&= ((d\phi)^{-1})_g \circ Y \cdot \phi(g) \Leftrightarrow \phi^* Y = T\phi^{-1} \circ Y \cdot \phi = (d\phi)^{-1} \circ Y \cdot \phi.
\end{aligned}
$$
$$(7.6)$$

Definition 7.3 *A vector field $X \in \mathcal{G}$ is said tobe left-invariant (resp., right-invariant) if, for every $g \in G$, the direct image $L_{g*} X$ (resp., $R_{g*} X$) of X by the left translation L_g (resp., by the right translation R_g), is equal to X*

$$L_{g*} X = X, \qquad R_{g*} X = X. \qquad (7.7)$$

Lemma 7.1 *Let G be a Lie group. The set $\mathcal{G}^L$ (resp., $\mathcal{G}^R$) of left-invariant (resp., right-invariant) vector fields on G is a Lie subalgebra of the Lie algebra of vector fields $\mathcal{G}$. The map $X^L \to X^L(e)$ (resp., $X^R \to X^R(e)$), which associates with each $X^L \in \mathcal{G}^L$ (resp., $X^R \in \mathcal{G}^R$) the value of this vector field at the neutral element e, is an isomorphism of $\mathcal{G}^L$ (resp., $\mathcal{G}^R$) onto the vector space $T_e G$. The dimension of $\mathcal{G}^L$ (resp., $\mathcal{G}^R$) is equal to the dimension of G.*

Proof. The set $\mathcal{G}^L$ is obviously a vector space, and an element X^L of $\mathcal{G}^L$ is completely determined by its value at the neutral element. Indeed, for every $g \in G$

$$X^L(g) = (dL_g)_e \cdot X^L(e). \qquad (7.8)$$

Conversely, once an element ξ in $T_e G$ is given, we define the vector field X^L by setting, for every $g \in G$,

$$X^L(g) = (dL_g)_e \cdot \xi. \qquad (7.9)$$

We can see that this vector field is left-invariant. The map $X^L \to X^L(e) = \xi$ is thus an isomorphism from $\mathcal{G}^L$ onto $T_e G$, and the dimension of $\mathcal{G}^L$ is indeed equal to that of G. Lastly, if $X, Y \in \mathcal{G}$, then, for every $g \in G$,

$$L_{g*}[X, Y] = [L_{g*} X, L_{g*} Y]. \qquad (7.10)$$

Hence $\mathcal{G}^L$ is indeed a Lie subalgebra of the Lie algebra $\mathcal{G}$. The same arguments may of course be applied to $\mathcal{G}^R$. $\square$

Let us introduce the following notation

$$X^L(g) \equiv X_\xi^L(g) = (L_{g*}\xi)(g), \quad X_\xi^L(e) = \xi, \quad \xi \in T_eG$$

$$X^R(g) \equiv X_\xi^R(g) = (R_{g*}\xi)(g), \quad X_\xi^R(e) = \xi, \tag{7.11}$$

Lemma 7.2 *Let $\chi : G \to G$ be the map*

$$g \to \chi(g) = g^{-1}. \tag{7.12}$$

It satisfies

$$T_e\chi = (d\chi)_e = -\mathrm{id}_{T_eG} \Leftrightarrow T_e\chi(\xi) = -\xi. \tag{7.13}$$

The mapping $X_\xi^L \to \chi_*X_\xi^L = -X_\xi^R$, which maps each left-invariant vector field X_ξ^L onto its direct image by χ, is an isomorphism from the Lie algebra $\mathcal{G}^L$ onto the Lie algebra $\mathcal{G}^R$, whose inverse is $X_\xi^R \to \chi_*X_\xi^R = -X_\xi^L$.

Defining a Lie bracket in T_eG by

$$[\xi,\eta] := [X_\xi^L, X_\eta^L](e) = [X_\eta^R, X_\xi^R](e), \qquad \xi,\eta \in T_eG, \tag{7.14}$$

makes T_eG a Lie algebra. Note that

$$[X_\xi^L, X_\eta^L] = X_{[\xi,\eta]}^L, \quad [X_\xi^R, X_\eta^R] = X_{[\eta,\xi]}^R. \tag{7.15}$$

Definition 7.4 *The vector space T_eG with the Lie algebra structure (7.14) is called the Lie algebra of G and is denoted by g or g_G.*

Lemma 7.3 *Let H and G be Lie groups and $f : H \to G$ a smooth homomorphism. Then $T_ef = (df)_e : g_H \to g_G$ is a Lie algebra homomorphism.*

Definition 7.5 *Let M be a smooth manifold.*

(i) *A left-action of a Lie group G on M is a differentiable map $\phi : G \times M \to M$ which for every $x \in M$ and $g_1, g_2 \in G$ satisfies*

$$\phi(e, x) = x, \qquad \phi(g_1, \phi(g_2, x)) = \phi(g_1g_2, x). \tag{7.16}$$

(ii) *A right-action of a Lie group G on M is a differentiable map $\psi : G \times M \to M$ which for every $x \in M$ and $g_1, g_2 \in G$ satisfies*

$$\psi(x, e) = x, \qquad \psi(\psi(x, g_1), g_2) = \psi(x, g_1g_2). \tag{7.17}$$

Observe that with each right-action ψ of G on M, we may associate the left-action ϕ defined by

$$\phi(g, x) = \psi(x, g^{-1}). \tag{7.18}$$

Definition 7.6 *Let ϕ be a left-action of G on M. For $x \in M$ the orbit of x isgiven by*

$$G \cdot x = \{\phi(g, x),\ g \in G\}. \tag{7.19}$$

Definition 7.7 *Suppose $\phi : G \times M \to M$ is a smooth left-action and $\xi \in g$. Then*

$$\phi^\xi : \mathbb{R} \times M \to M : \quad (t, x) \to \phi^\xi(t, x) \tag{7.20}$$

is an $\mathbb{R}-action$ on M or a flow on M.

The above definition relates the results of Sect. 2.6 to the considerations of this section. Applying the exponential representation of the flow

$$\bar{x} = \phi^\xi(t, x) = \phi_t^\xi(x) = \phi_t^\xi \cdot x = \exp t\xi \cdot x \tag{7.21}$$

we immediately find that the corresponding vector field on M is given by

$$X_\xi(x) = \frac{d}{dt}\phi^\xi(t, x)_{|t=0} = (dR_x)_e\xi = X_\xi^R(x). \tag{7.22}$$

It is called the infinitesimal generator of the $\mathbb{R}-$action corresponding to ξ and is a right-invariant vector field.

For each $g \in G$, let $\alpha_g : G \to G$ be the diffeomorphism of G such that for every $x \in G$

$$\alpha_g(x) = gxg^{-1} = (R_{g^{-1}} \circ L_g)(x). \tag{7.23}$$

Definition 7.8 *The map*

$$Ad_g = d(\alpha_g)_e = d(R_{g^{-1}} \circ L_g)_e : \ g \to g \tag{7.24}$$

is called the adjoint mapping associated with g.

For each g, g_1 and $g_2 \in G$, ξ and $\eta \in g$ the following relations hold :

$$\alpha_g(g_1 g_2) = \alpha_g(g_1)\alpha_g(g_2),\ \alpha_{g_1} \circ \alpha_{g_2} = \alpha_{g_1 g_2},\ \alpha_e = id_G,$$
$$Ad_g([\xi, \eta]) = [Ad_g\xi, Ad_g\eta],\ Ad_{g_1} \circ Ad_{g_2} = Ad_{g_1 g_2},\ Ad_e = id g. \tag{7.25}$$

For every $g \in G$ and $\xi \in g$, $Ad_g\xi$ is the image of ξ by α_g. The flows of vector fields X_ξ and $X_{Ad_g\xi}$ are thus brought into correspondence by α_g, so for every $t \in \mathbb{R}$,

$$\exp(tAd_g\xi) = \alpha_g(\exp(t\xi)) = g\exp(t\xi)g^{-1}. \tag{7.26}$$

Lemma 7.4 *The adjoint mapping Ad_g relates the right and left-invariant vector fields in the following way*

$$X_\xi^L(g) = X_{Ad_g\xi}^R(g), \tag{7.27}$$

$$X_\xi^R(g) = X_{Ad_{g^{-1}}\xi}^L(g). \tag{7.28}$$

Proof. We only prove the first relation as the proof of the second one is analogous:

$$X_\xi^L(g) = (dL_g)_e\xi = (dR_g)_e(dR_{g^{-1}})_g(dL_g)_e\xi$$

$$= (dR_g)_e d(R_{g^{-1}} \circ L_g)_e\xi = (dR)_e Ad_g\xi = X_{Ad_g\xi}^R(g). \square$$

The map

$$Ad : G \times g \to g : (g, \xi) \to Ad_g\xi \tag{7.29}$$

is a smooth action called the *adjoint action of G on g*.

Definition 7.9 *Let*

$$ad : g \times g \to g : (\xi, \eta) \to ad_\xi\eta \tag{7.30}$$

where

$$ad_\xi\eta := \frac{d}{dt}(Ad_{\exp t\xi}\eta)_{|t=0}. \tag{7.31}$$

The map ad is called the adjoint action of g on g.

Lemma 7.5 *Let $\xi, \eta \in g$, then*

$$ad_\xi\eta = [\xi, \eta]. \tag{7.32}$$

Proof. We apply the definition of the Lie derivative from Sect. 2.7. Let ϕ_t be a flow whose generator is the vector field X, so $X(g)$ is a tangent vector in $\phi_t(g)$. Then we have

$$[X, Y](g) = (L_X Y)(g) = \frac{d}{dt}(\phi_t^* Y)_{|t=0} = \frac{d}{dt}(\phi_{-t*}Y)(g)_{|t=0}$$

$$= \frac{d}{dt}(d\phi_{-t}(\phi_t(g)) \cdot Y(\phi_t(g)))_{|t=0}.$$

So, from (7.14), Definition 7.7 and the fact that $\phi_t(g) = \exp t\xi \cdot g = L_{\exp t\xi}g$, we find

$$[\xi, \eta] = -[X_\xi^R, Y_\eta^R](e) = -\frac{d}{dt}(d\phi_{-t}(\phi_t(e)) \cdot Y_\eta^R(\phi_t(e)))_{|t=0}$$

$$= -\frac{d}{dt}((dL_{\exp(-t\xi)})_{\exp t\xi} \cdot (dR_{\exp t\xi})_e\eta)_{|t=0}$$

$$= -\frac{d}{dt}((d(L_{\exp(-t\xi)} \circ R_{\exp t\xi}))_e\eta)_{|t=0}$$

$$= \frac{d}{dt}((d(R_{\exp(-t\xi)} \circ L_{\exp t\xi}))_e\eta)_{|t=0} = \frac{d}{dt}(Ad_{\exp t\xi}\eta)_{|t=0}$$

$$= ad_\xi\eta. \square$$

Let g^* be a dual algebra related to g by a duality map

$$g^* \times g \to \mathbb{R} : \quad (\beta, \xi) \to < \beta, \xi >, \quad \xi \in g, \ \beta \in g^*. \tag{7.33}$$

Definition 7.10 *Let* $\rho(g)$ *be a linear operator in* g, *i.e.* $\rho(g) : G \times g \to g$. *Its dual counterpart* $\rho^*(g)$ *in* g^* *is defined as follows*

$$< \rho^*(g)\beta, \xi > = < \beta, \rho(g^{-1})\xi >. \tag{7.34}$$

From Definition 7.10 we find that the dual to the Ad_g is $Ad^*_{g^{-1}}$ and the action

$$Ad^* : G \times g^* \to g^* \quad : \quad (g, \beta) \to Ad^*_{g^{-1}}\beta \tag{7.35}$$

is called the *co-adjoint action of G on* g^*. Now we may introduce the *co-adjoint action of g on* g^* as follows

$$ad^* : g \times g^* \to g^* \quad : \quad (\xi, \beta) \to ad^*_\xi \beta, \tag{7.36}$$

where

$$ad^*_\xi \beta = \frac{d}{dt} \left(Ad^*_{\exp t\xi} \beta \right)_{|t=0}. \tag{7.37}$$

Lemma 7.6 *For arbitrary* $\xi, \eta \in g$ *and* $\beta \in g^*$

$$< ad^*_\xi \beta, \eta > = - < \beta, ad_\xi \eta > = - < \beta, [\xi, \eta] >. \tag{7.38}$$

Proof.

$$< ad^*_\xi \beta, \eta > \ = \ ad^*_\xi \beta(\eta) = \frac{d}{dt} \left(Ad^*_{\exp t\xi} \beta(\eta) \right)_{|t=0}$$

$$= \ \frac{d}{dt} \left(\beta \left(Ad_{\exp(-t\xi)} \eta \right) \right)_{|t=0}$$

$$= \ \beta \left(\frac{d}{dt} \left(Ad_{\exp(-t\xi)} \eta \right)_{|t=0} \right)$$

$$= \ \beta([-\xi, \eta]) = -\beta([\xi, \eta]) = - < \beta, [\xi, \eta] >. \ \square$$

Let g be a Lie algebra, g^* the dual algebra related to g by a duality map

$$g^* \times g \to \mathbb{R} : \quad (\beta, \xi) \to < \beta, \xi >, \quad \xi \in g, \ \beta \in g^*. \tag{7.39}$$

Then, let

$$ad : g \times g \to g : (\xi, \eta) \to ad_\xi \eta, \quad \xi, \eta \in g \tag{7.40}$$

be the adjoint action of g on g and

$$ad^* : g \times g^* \to g^* \quad : \quad (\xi, \beta) \to ad^*_\xi \beta, \quad \xi \in g, \ \beta \in g^* \tag{7.41}$$

be the co-adjoint action of g on g^*. For arbitrary $\xi, \eta \in g$ and $\beta \in g^*$ the following relations hold

$$ad_\xi \eta = [\xi, \eta], \quad < ad_\xi^* \beta, \eta > = - < \beta, ad_\xi \eta > = - < \beta, [\xi, \eta] > . \quad (7.42)$$

Equipped with the knowledge of this section we can now pass to the concept of Hamiltonian dynamics on a Lie algebra. Let G be a Lie group, g its Lie algebra, g^* the dual algebra and $F(g^*) := C^\infty(g^*)$ the space of C^∞–functions on g^*. There exists a natural Lie–Poisson bracket in $F(g^*)$. Let $L \in g^*$, $f, h \in F(g^*)$, $\nabla f, \nabla h \in g$. Then

$$\{h, f\}(L) := < L, [\nabla h, \nabla f] >, \quad (7.43)$$

where $< .,. >: g^* \times g \rightarrow \mathbb{R}$ is a duality map between g^* and g, respectively. So, according to Lemma 7.6, the Hamiltonian dynamical system on g^* can be defined by the equation

$$\frac{d}{dt} L = ad_{\nabla f}^* L = -P(L)\nabla f, \quad L \in g^*, \ \nabla f \in g, \quad (7.44)$$

where ad^* is the co-adjoint action of g on g^*, P is a Poisson tensor $P : g \rightarrow g^*$ and the minus sign is for convention.

Assumption. We confine our further considerations to such algebras for which g^* can be identified with g. So, we assume the existence of an invariant metrix, i.e. a non-degenerate symmetric product on g

$$(a, b) = (b, a), \quad a, b \in g, \quad (7.45)$$

invariant under the adjoint action

$$(ad_a b, c) + (b, ad_a c) = 0 \Leftrightarrow ([a, b], c) = (a, [b, c]), \ a, b, c \in g. \quad (7.46)$$

Let us assume that this is the *trace form* tr $: g \rightarrow \mathbb{R}$

$$(a, b) = \text{tr}(ab) = \text{tr}(ba) = (b, a). \quad (7.47)$$

Then, we can identify g^* with g by setting

$$(\eta, \xi) = < \eta^*, \xi >, \quad (7.48)$$

where $\eta, \xi \in g$, $\eta^* \in g^*$ and η^* is identified with $\eta, \in g$.

Thus, now we can write the Lie–Poisson bracket as

$$\{f, h\}(L) := < L, [\nabla f, \nabla h] > = (L, [\nabla f, \nabla h]) = ([L, \nabla f], \nabla h)$$

$$= (\nabla h, [L, \nabla f]) \equiv (\nabla h, P(L)\nabla f). \quad (7.49)$$

Hence, the equation of motion (7.44) takes the form

$$\frac{\mathrm{d}}{\mathrm{d}t} L = ad^*_{\nabla f} L = -P(L)\nabla f = [\nabla f, L] = ad_{\nabla f} L. \tag{7.50}$$

We notice that under the above identification of g^* and g, the co-adjoint action goes over into the adjoint one.

So far we have proved the equivalence between a hierarchy of dynamical systems and the respective hierarchy of Lax equations

$$u_{t_n} = K_n(u) \Leftrightarrow L_{t_n} = [A_n, L] \tag{7.51}$$

as well as the equivalence between algebras (3.107), (3.92) of symmetries and algebras (4.97), (4.107) of respective Lax operators. Now, we can identify the dynamic equation (7.50) and the Lax equation (1.2) with a natural Hamiltonian structure

$$[L, A] = [L, \nabla f] = P(L)\nabla f. \tag{7.52}$$

So, the alternative description of the Hamiltonian dynamics on a manifold M

$$u_t = K(u) = \theta(u)\nabla_u H(u), \quad u \in M \tag{7.53}$$

is the Hamiltonian dynamics on a Lie algebra of Lax operators

$$L_t = [A, L] = -P(L)A = \Theta(L)\nabla_L H(L), \quad L \in g. \tag{7.54}$$

As will be demonstrated in the following sections, constants of the motion $H_n(u)$ of dynamics (7.53) come from the Casimirs of the Lie–Poisson structure $\Theta(L)$, while the Poisson structure $\theta(u)$ is related with another Hamiltonian representation of operator dynamics (7.54) coming from a so called R-structures. This alternative abstract approach to integrable systems profits from the deeper understanding of the nature of integrability as well as equips us with a very general and efficient tool for the construction of multi-Hamiltonian systems from scratch.

7.2 Basic Facts About R-Structures

Definition 7.11
An R-structure is a Lie algebra g equipped with a linear map $R : g \rightarrow g$ (called the R-matrix) such that the bracket

$$[a, b]_R := [R\, a, b] + [a, Rb] \qquad a, b \in g \tag{7.55}$$

is a second Lie product on g.

Lemma 7.7 *A sufficient condition for R to be an R-matrix is*

$$[R(a), R(b)] - R([a, b]_R) = -\alpha[a, b], \tag{7.56}$$

where α is some real number.

Proof. One easily checks that the Jacobi identity for $[,]_{\mathcal{R}}$ can be rewritten as a Jacobi identity of the expression $[[\mathcal{R}a, \mathcal{R}b] - \mathcal{R}[a, b]_{\mathcal{R}}, c]$. Hence claiming the first entry to be just a scalar multiple of the original Lie bracket $[,]$ is a sufficient condition for $\mathcal{R}$ to be an $\mathcal{R}$-matrix. $\square$

Equation (7.56) is called the Yang–Baxter equation, and we will refer to it as YB(α). It is clear that there are only two relevant cases, namely, $\alpha = 0$ and $\alpha = 1$, as for $\alpha \neq 0$ the dilatation $\mathcal{R} \to (1/\sqrt{\alpha})\mathcal{R}$ maps the solution of YB(α) to solutions of YB(1). The case $\alpha = 1$ is also called the modified Yang–Baxter equation.

Any solution $\mathcal{R}$ of YB(α) for the Lie bracket $[.,.]$ also solves YB(α) for the Lie bracket $[.,.]_{\mathcal{R}}$. As a result of this observation one can iterate the construction of a 'modified' Lie bracket by starting with $[.,.]_{\mathcal{R}}$ instead of $[.,.]$, thus obtaining a third Lie bracket $[a, b]_{\mathcal{R}\mathcal{R}} = [\mathcal{R}a, b]_{\mathcal{R}} + [a, \mathcal{R}b]_{\mathcal{R}}$, etc. Hence an $\mathcal{R}$-structure equips a Lie algebra with a hierarchy of Lie brackets, where the n-th iterated bracket is given by

$$[a, b]_{\mathcal{R}\mathcal{R}...\mathcal{R}} = \sum_{k=0}^{n} \binom{n}{k} [\mathcal{R}^k a, \mathcal{R}^{n-k} b]. \tag{7.57}$$

Each bracket induces a Lie–Poisson structure on the dual g^*, given by

$$\{f_1, f_2\}(L) = < L, [\nabla f_1, \nabla f_2]_{\mathcal{R}\mathcal{R}...\mathcal{R}} > . \tag{7.58}$$

Here the f_i are scalar-valued functions on g^* and their gradients are interpreted as elements of g. L is chosen for a point in g^* since in the applications L will be the Lax operator for the considered integrable system.

For both cases ($\alpha = 0, 1$) a systematic scheme of solving the YB equations is given in [94]. There is a special class of solutions to YB(1) that arises in a very simple manner, relevant for our further applications. Assume that the Lie algebra g can be split into a direct sum of Lie subalgebras g_+ and g_-, i.e.

$$g = g_+ \oplus g_- , \qquad [g_\pm, g_\pm] \subset g_\pm. \tag{7.59}$$

Denoting the projections onto these subalgebras by $P_\pm$, it is easy to verify that

$$\mathcal{R} = \frac{1}{2}(P_+ - P_-) \tag{7.60}$$

solves YB(1/4) and hence defines an $\mathcal{R}$-structure on g. In this case the hierarchy (7.57) of Lie brackets generated by $\mathcal{R}$ reduces to just three different brackets

$$[a, b],$$
$$[a, b]_{\mathcal{R}} = [P_+(a), P_+(b)] - [P_-(a), P_-(b)], \tag{7.61}$$
$$[a, b]_{\mathcal{R}\mathcal{R}} = [P_+(a), P_+(b)] + [P_-(a), P_-(b)],$$

as $[a, b]_{\mathcal{R}\mathcal{R}\mathcal{R}} = [a, b]_{\mathcal{R}}$. For YB(0) there does not seem to be such a natural class of solutions.

Now, having in mind the Lax equations we start with an abstract asso-ciative algebra g (of Lax operators in our applications) as the natural phase space of integrable dynamical systems. We assume that g bears a symmetric, non-degenerate trace form tr:$g \to \mathbb{R}$ (7.47), so that g can be identified with its dual g^* via the pairing (7.49). Considering the natural Lie algebra structure $[a, b] = ab - ba$ on g, we thus have an invariant metric satisfying (7.46). Let R:$g \to g$ be some linear map. We will be interested in the investigation of the following three brackets on the space $C^\infty(g^*)$ of smooth functions on $g^* = g$

$$\{f_1, f_2\}_1(L) := (ad_L \nabla f_1, R \nabla f_2) - (ad_L \nabla f_2, R \nabla f_1) \qquad (7.62)$$

$$\{f_1, f_2\}_2(L) := (ad_L \nabla f_1, R \, ad_L^+ \nabla f_2) - (ad_L \nabla f_2, R \, ad_L^+ \nabla f_1), \qquad (7.63)$$

$$\{f_1, f_2\}_3(L) := (ad_L \nabla f_1, R(L \nabla f_2 L)) - (ad_L \nabla f_2, R(L \nabla f_1 L)), \qquad (7.64)$$

evaluated at a point $L \in g^* = g$, where ad_ξ^+ is the anticommutator, i.e. $ad_\xi^+ \eta = \xi\eta + \eta\xi$. Referring to the dependence on the point L they are called the 'linear', the 'quadratic' and the 'cubic' bracket.

Theorem 7.1

(i)   For any R-matrix on g, the linear bracket is a Poisson bracket.

(ii) If R and its skew symmetric part $\widetilde{R} = \frac{1}{2}(R - R^*)$ (where R^* is the adjoint of R with respect to the trace duality : $(Ra, b) = (a, R^* b)$) both satisfy $YB(\alpha)$ (with the same α), then the quadratic bracket is a Poisson bracket.

(iii) If R solves $YB(\alpha)$, then the cubic bracket is a Poisson bracket.

Proof.

(i)   From the invariance (7.46) of the metric (7.47) it follows that the linear bracket coincides with the Lie–Poisson bracket (7.58) arising from the modified Lie product $[,]_R$

$$\{f_1, f_2\}_1(L) = ([L, \nabla f_1], R \nabla f_2) - ([L, \nabla f_2], R \nabla f_1)$$

$$= ([\nabla f_1, R \nabla f_2], L) - ([\nabla f_2, R \nabla f_1], L)$$

$$= (L, [\nabla f_1, R \nabla f_2] + [R \nabla f_1, \nabla f_2]) = (L, [\nabla f_1, \nabla f_2]_R),$$

i.e. the linear bracket will become a Poisson bracket if R is an R-matrix on g. $\square$

(ii-iii) For the other two brackets the proof consists of some tedious but straightforward calculations, whose essential steps the reader can find in [150]. We remark that for the special case of a unitary R-matrix $(R^* = -R)$, $\widetilde{R}$ coincides with R and the condition (ii) for a quadratic bracket reduces to the condition (iii) for a cubic bracket.

For further applications we will rather work with the Poisson tensors $\Theta_i(L) : g \to g$ associated to the brackets $\{,\}_i$ via

$$\{f_1, f_2\}_i(L) = (\nabla f_1, \Theta_i(L)\nabla f_2), \qquad i = 1, 2, 3, \tag{7.65}$$

than with the brackets themselves.

Lemma 7.8 *The following Poisson operators are related to the Poisson brackets (7.62)-(7.64)*

$$\Theta_1(L) : \nabla f \to -ad_L \mathcal{R}(\nabla f) - \mathcal{R}^* ad_L \nabla f, \tag{7.66}$$

$$\Theta_2(L) : \nabla f \to -ad_L \mathcal{R} \, ad_L^+ \nabla f - L\mathcal{R}^* ad_L \nabla f - \mathcal{R}^*(ad_L \nabla f)L, \tag{7.67}$$

$$\Theta_3(L) : \nabla f \to -ad_L \mathcal{R}(L\nabla f L) - L\mathcal{R}^*(ad_L \nabla f)L. \tag{7.68}$$

Proof. We only prove the relations for the linear bracket as the proofs for quadratic and cubic brackets are analogous. Using the invariance (7.46) of the metric (7.47) we find

$$\begin{aligned}
\{g, f\}_1(L) &= ([L, \nabla g], \mathcal{R}(\nabla f)) - ([L, \nabla f], \mathcal{R}(\nabla g)) \\
&= ([\mathcal{R}(\nabla f), L], \nabla g) + (\mathcal{R}^*[\nabla f, L], \nabla g) \\
&= (\nabla g, [\mathcal{R}(\nabla f), L]) + (\nabla g, \mathcal{R}^*[\nabla f, L]) \\
&= (\nabla g, [\mathcal{R}(\nabla f), L] + \mathcal{R}^*[\nabla f, L]) \\
&= (\nabla g, -ad_L \mathcal{R}(\nabla f) - \mathcal{R}^* ad_L \nabla f) = (\nabla g, \Theta_1(L)\nabla f). \square
\end{aligned}$$

We remark that all three brackets are interrelated by the simple deformations

$$\Theta_2(L + \epsilon \mathbf{1}) = \Theta_2(L) + 2\epsilon\Theta_1(L),$$
$$\Theta_3(L + \epsilon \mathbf{1}) = \Theta_3(L) + \epsilon\Theta_2(L) + \epsilon^2\Theta_1(L), \tag{7.69}$$

where $\mathbf{1}$ is the identity element of the algebra. As was presented in Chap. 3, an important property of the multi-Hamiltonian formulation for a given dynamics is the compatibility of the Poisson operators. We recall that two Poisson operators are called compatible if their sum is again a Poisson operator. From (7.69) it is readily seen that all three brackets will be compatible, automatically, if $\mathcal{R}$ is such that all three operators are indeed Poisson operators. The problem arises of how to deal with a quadratic operator in the case when $\mathcal{R}$ does not satisfy the YB(α) equation. In such a case the more general formulation of a quadratic bracket was suggested by Suris [180].

Theorem 7.2 *On any algebra g equipped with a non-degenerate invariant metric $(a, bc) = (ab, c) = (c, ab)$ the quadratic tensor*

$$\overline{\Theta}(L) : \nabla f \to A_1(L\nabla f)L - LA_2(\nabla f L) + S(\nabla f L)L - LS^*(L\nabla f) \tag{7.70}$$

defines a Poisson structure on $g = g^*$, if the linear maps $A_{1,2} : g \to g$ are skew-symmetric solutions of the YB(1) equation (7.56) and the linear map $S : g \to g$ with the adjoint S^* satisfies

$$S([A_2(\xi), \eta] + [\xi, A_2(\eta)]) = [S(\xi), S(\eta)],$$

$$S^*([A_1(\xi), \eta] + [\xi, A_1(\eta)]) = [S^*(\xi), S^*(\eta)], \qquad \xi, \eta \in g. \qquad (7.71)$$

We notice that in the special case when $\widetilde{R}$ satisfies the YB(1) equation, under the substitution $A_1 = A_2 = 2\widetilde{R}$, $S = S^* = R + R^*$, the quadratic Poisson operator (7.70) reduces to (7.67) and the conditions (7.71) are equivalent to YB(1) for the operator R.

For many important examples the maps A_1, A_2, S and S^* originate from decompositions of the R−matrix

$$R = A_1 + S = A_2 + S^*. \qquad (7.72)$$

Hence, the following theorem is useful for further considerations.

Theorem 7.3 [155] *Let $R = A_1 + S = A_2 + S^*$ satisfy the modified YB(1) equation. If $A_{1,2}$ are skew symmetric w.r.t. the underlying invariant metric, then conditions (7.71) imply that both A_1 and A_2 satisfy (7.56), i.e.*

$$[A(\xi), A(\eta)] + [\xi, \eta] = A([A(\xi), \eta] + [\xi, A(\eta)]). \qquad (7.73)$$

Proof. From the modified YB(1) equation for $R = A_1 + S$, using (7.71), one derives

$$Y(\xi, \eta) := [A_1(\xi), A_1(\eta)] + [\xi, \eta] - A_1([A_1(\xi), \eta] + [\xi, A_1(\eta)])$$

$$= S([S^*(\xi), \eta] + [\xi, S^*(\eta)]) + A_1([S(\xi), \eta] + [\xi, S(\eta)])$$

$$- [S(\xi), A_1(\eta)] - [A_1(\xi), S(\eta)].$$

The invariance of the metric and skew symmetry of A_1 leads to the identity

$$(Y(\xi, \eta), \zeta) = ([S^*(\eta), S^*(\zeta)] - S^*([A_1(\eta), \zeta] + [\eta, A_1(\zeta)]), \xi)$$

$$+ ([S^*(\zeta), S^*(\xi)] - S^*([A_1(\zeta), \xi] + [\zeta, A_1(\xi)]), \eta)$$

with arbitrary $\xi, \eta, \zeta \in g$. The corresponding identity holds for A_2 and S^*. □

This observation reduces the verification of the condition (7.73) for $A_{1,2}$ to (7.71). The latter relations are often more easily checked.

Thus the introduction of an R-matrix on g turns the algebra into a Hamiltonian phase space. The compatibility results are a strong hint that the Hamiltonian structures (7.66)–(7.68) should turn out to be interesting in the context of the integrable multi-Hamiltonian equations considered in Chap. 3. We will look for a natural set of functions in involution w.r.t. the

Poisson brackets (7.62)–(7.64). Of course, on any dual Lie algebra g^* the invariant functions (Casimir functions), i.e. those functions $C(L) \in C^\infty(g^*)$ satisfying

$$[\nabla C(L), L] = 0 \qquad (7.74)$$

are in involution relative to the natural Lie–Poisson bracket (7.43) related with the Lie–Poisson operator $\Theta(L)$. Here an invariant metric on g is assumed. But this certainly does not lead to any interesting dynamics on $g \cong g^*$, as the Poisson bracket of a Casimir function with any other function vanishes. The essential observation is given by the following theorem.

Theorem 7.4

(i) *The Casimir functions of the natural Lie–Poisson bracket on $g \cong g^*$ are in involution with respect to all brackets (7.62)-(7.64). The Hamiltonian equations associated with a Casimir function $C(L)$ are Lax equations given by*

$$L_t = \Theta_1(L)\nabla C = [\mathcal{R}(\nabla C), L],$$

$$L_t = \Theta_2(L)\nabla C = [\mathcal{R}(2L\nabla C), L],$$

$$L_t = \Theta_3(L)\nabla C = [\mathcal{R}(L^2\nabla C), L], \qquad (7.75)$$

(ii) *The orbits of these equations are restricted to the symplectic leaves of the natural Lie–Poisson structure on $g \cong g^*$.*

Proof. The proof of (i) is obvious. For (ii) one just has to note that the tangent space of the symplectic leaf at a point L is given by the image of the Lie–Poisson tensor $\Theta(L)$, i.e. consists of elements of the form $[a, L]$ with arbitrary $a \in g$. But obviously all the Hamiltonian vector fields $\Theta_i(L)\nabla C$ in (7.75) are of this form and hence are tangent to the symplectic leaves. □

As we have assumed a nondegenerate trace form tr on g, we will consider the Casimir functions given by the trace of powers of L, that is,

$$C_q(L) = \frac{1}{q}\mathrm{tr}(L^q), \qquad \nabla C_q(L) = L^{q-1}. \qquad (7.76)$$

Taking these $C_q(L)$ as Hamiltonian functions one finds a hierarchy of equations

$$\frac{d}{dt_q}L = [\mathcal{R}(L^q), L] = \Theta_1\nabla C_{q+1} = \frac{1}{2}\Theta_2\nabla C_q = \Theta_3\nabla C_{q-1}, \qquad (7.77)$$

which are evidently tri-Hamiltonian with respect to the three brackets, provided that $\mathcal{R}$ is such that all three brackets are indeed Poisson brackets. In the Lax equations (7.77) we have introduced an evolution parameter t_q for each power q. For any $\mathcal{R}$-matrix each two evolution equations in the hierarchy (7.77) commute due to the involutivity of the Casimir functions C_q. Each equation admits all the Casimir functions as a set of conserved quantities in involution. In this sense we will regard (7.77) as a hierarchy of 'integrable' evolution equations. In this construction we have defined the equation based on a Hamiltonian framework involving the $\mathcal{R}$-matrix. The Lax form (7.77) owes its simplicity to the assumption of the trace-duality on the algebra g.

For the special $\mathcal{R}$-matrices (7.60) originating from a Lie algebra decomposition of the algebra g, we can use the identity

$$0 = [L^q, l] = [P_+(L^q) + P_-(L^q), L] \Rightarrow [P_+(L^q), L] = -[P_-(L^q), L] \quad (7.78)$$

to rewrite the integrable hierarchy as

$$\frac{\mathrm{d}}{\mathrm{dt_q}} L = [\mathcal{R}(L^q), L] = \frac{1}{2}[P_+(L^q) - P_-(L^q), L]$$

$$= [P_+(L^q), L] = -[P_-(L^q), L]. \quad (7.79)$$

The commutativity of these equations for different values of q is also reflected in the following compatibility equations for the projected powers $P_+(L^q)$. As $g_\pm$ form subalgebras in g, it is readily verified that they satisfy the zero-curvature equations

$$\frac{\mathrm{d}}{\mathrm{dt_q}}(P_+(L^q)) - \frac{\mathrm{d}}{\mathrm{dt_r}}(P_+(L^r)) + [P_+(L^q), P_+(L^r)] = 0. \quad (7.80)$$

The abstract algebraic setting presented above, although interesting in itself, can be applied to a variety of particular algebras g. Such applications create a lot of new questions and technical problems. For example, in our applications, if we define g to be the algebra of pseudo-differential operators or the algebra of discrete shift operators, then (7.77) involve infinitely many fields u_i. Our natural task is then to define a proper restriction of the original algebra to a smaller subspace, to obtain the coupled system of equations involving a finite number of fields. The best situation is when we choose a subspace in such a way that images of Poisson operators $\Theta_i(L), i = 1, 2, 3$, lie in the space tangent to the subspace for each element of the subspace, because then, for any given Hamiltonian, the dynamics is restricted to the subspace. If this is not the case the *Dirac reduction* can be invoked for restriction of Poisson tensors to suitable subspaces.

Lemma 7.9 *For two linear subspaces U and V spanned by $u \in U$ and $v \in V$ let*

$$\Theta(u, v) = \begin{pmatrix} \Theta_{uu}(u, v) & \Theta_{uv}(u, v) \\ \Theta_{vu}(u, v) & \Theta_{vv}(u, v) \end{pmatrix} \quad (7.81)$$

be a Poisson tensor on $U \oplus V$. Also let Θ_{vv} be invertible. Then, for arbitrary $c \in V$ the tensor

$$\Theta(u; c) = \Theta_{uu}(u, c) - \Theta_{uv}(u, c)[\Theta_{vv}(u, c)]^{-1}\Theta_{vu}(u, c) \quad (7.82)$$

is a Poisson tensor on the affine space $c + U$.

The *proof* consists of a tedious, yet straightforward calculations, investigating the condition (3.5) for tensor (7.81) on $U \oplus V$ and eliminating derivatives w.r.t. the v-variables from (3.5). Of course the demand for invertibility of Θ_{vv} is a strong one, so Dirac reduction is not always possible and we may lose the Hamiltonian structure in the reduction process from generic $L \in g$ to specific Lax operators. However, we can weaken this condition, namely, it is enough to demand invertibility of Θ_{vv} on the image of Θ_{vu} . Observe that the inversion of Θ_{vv} creates a strong nonlocality in $\Theta(u; c)$.

7.3 Multi-Hamiltonian Dynamics
of Pseudo-Differential Lax Operators

Now we shall apply the formalism presented in the previous section to the associative algebra of pseudo-differential operators. We shall follow in general the results of [108]. Let us consider the operator algebra

$$g = \left\{ L = \sum_{-\infty < i << +\infty} u_i(x)\partial^i \right\},\qquad(7.83)$$

where $\partial = \partial/\partial x$ and the coefficients u_i are to be functions of the "space variable" x. The negative powers of ∂ are to be understood in the following way. One introduces a formal integration symbol ∂^{-1} , for which algebraic multiplication with a multiplication operator a (given by a function $a = a(x)$) is defined as the formal series

$$\partial^{-1}a = a\partial^{-1} - a_x\partial^{-2} + a_{xx}\partial^{-3} - ... \quad .\qquad(7.84)$$

In fact, this algebraic rule is just a special case of the general Leibniz rule

$$\partial^k a = \sum_{i \geq 0} C_i^k \frac{\partial^i a}{\partial x^i}\partial^{k-i},\qquad(7.85)$$

where

$$C_i^k = \begin{cases} \begin{pmatrix} k \\ i \end{pmatrix} & \text{for} \quad k \geq i \geq 0 \\ 0 & \text{for} \quad i \geq k \geq 0 \ \text{or} \ i < 0 \\ \begin{pmatrix} |k| + i - 1 \\ i \end{pmatrix}(-1)^i & \text{for} \ k < 0, \ i \geq 0 \end{cases}$$

and

$$\begin{pmatrix} k \\ i \end{pmatrix} = \frac{k(k-1) \cdot ... \cdot (k-i+1)}{1 \cdot 2 \cdot ... \cdot i},$$

for differential operators, now extended to negative powers of k. With this assumption we have fixed the algebraic structure of the set (7.83). The crucial observation (made for the first time by Adler [4]) is given by the following lemma.

Lemma 7.10 *The trace form*

$$\text{tr}(L) = \text{tr}\left(\sum_i u_i(x)\partial^i\right) = \int_R u_{-1}(x)\mathrm{d}x \qquad (7.86)$$

yields a symmetric and non-degenerate pairing on g

$$(L_1, L_2) = \text{tr}(L_1 L_2). \qquad (7.87)$$

In (7.86) the integration denotes the equivalence class of differential expressions modulo total derivatives.

Proof. It is sufficient to prove (7.87) for $L_1 = A\partial^k$, $L_2 = B\partial^l$. By the definition of the multiplicative law (7.85), we have

$$A\partial^k \cdot B\partial^l = \sum_i C_i^k A(D^i B)\partial^{k+l-i}.$$

Comparing this expression with

$$B\partial^l \cdot A\partial^k = \sum_i C_i^l B(D^i A)\partial^{k+l-i}$$

we find that

$$\text{tr}(A\partial^k \cdot B\partial^l) = C_{k+l+1}^k \int_R A(D^{k+l+1}B)\mathrm{d}x,$$

$$\text{tr}(B\partial^l \cdot A\partial^k) = C_{k+l+1}^l \int_R B(D^{k+l+1}A)\mathrm{d}x.$$

The coincidence of the two expressions follows from the skew symmetry of D and the definition of C_i^k. $\square$

As a consequence, for operators $L = \sum u_i \partial^i$, the vector fields $\frac{\mathrm{d}}{\mathrm{dt}}L$ and gradients $\nabla H(L)$ are conveniently parametrized by

$$\frac{\mathrm{d}}{\mathrm{dt}}L = \sum_i (u_i)_t \partial^i, \qquad \nabla H(L) = \sum_i \partial^{-1-i}\frac{\delta H}{\delta u_i}, \qquad (7.88)$$

where $\delta H / \delta u_i$ is the usual variational derivative of a functional $H(L) = H(u_i, u_{i_x}, u_{i_{xx}}, ...)$ in terms of the Euler operator. In these frames the trace duality assumes the usual Euclidean form

$$\left\langle \nabla H, \frac{\mathrm{d}}{\mathrm{dt}}L \right\rangle = \left\langle \frac{\mathrm{d}}{\mathrm{dt}}L, \nabla H \right\rangle = \sum_i \int_R \frac{\delta H}{\delta u_i}(u_i)_t \, \mathrm{d}x. \qquad (7.89)$$

As mentioned before, natural $\mathcal{R}$-matrices on the algebra (7.83) come from the decomposition of this algebra into a direct sum of its subalgebras. We consider simple decompositions of g of the form

$$
g_+ = g_{\geq k} = \left\{ \sum_{i \geq k} u_i \partial^i \right\}, \qquad g_- = g_{<k} = \left\{ \sum_{i<k} u_i \partial^i \right\}, \quad k = \ldots - 1, 0, 1 \ldots.
$$
(7.90)

With $P_+ = P_{\geq k}$ and $P_- = P_{<k}$ we denote the projections to these subalgebras. In order to ensure that the maps $\mathcal{R}_k = \frac{1}{2}(P_{\geq k} - P_{<k})$ define classical $\mathcal{R}$-matrices, we have to look for those cases of k, for which both $g_{\geq k}$ and $g_{<k}$ are Lie subalgebras of g. It is readily verified that $g_{\geq k}$ constitutes a closed subalgebra of differential operators for any integer values $k \geq 0$. On the other hand, due to (7.84) and (7.85), it is also easily checked that $g_{<k}$ constitutes a closed subalgebra of integro-differential symbols for any integer values $k < 2$. The common cases $k = 0, 1, 2$ give the only admissible decompositions of the algebra (7.83) and hence the three admissible Lax hierarchies (7.79)

$$
\frac{\mathrm{d}}{\mathrm{d}t_q} L = [P_{\geq k}(L^q), L] \equiv [(L^q)_{\geq k}, L]
$$

$$
= -[P_{<k}(L^q), L] \equiv -[(L^q)_{<k}, L], \qquad k = 0, 1, 2. \tag{7.91}
$$

Let us point out that Lax dynamics of the form (7.91) with $k = 0$ was first considered by Gelfand and Dikii [91] while the cases $k = 1, 2$ were introduced by Kupershmidt [113].

Before entering a systematic discussion of the integrable nonlinear systems hidden in (7.91), we look at the three simplest examples of equations related to the three choices of k in (7.91). We consider the operators

$$
k = 0: \qquad L_{KdV} = \partial^2 + u,
$$

$$
k = 1: \qquad L_{MKdV} = \partial^2 + 2v\partial,
$$

$$
k = 2: \qquad L_{HD} = w^2 \partial^2, \tag{7.92}
$$

and try to evaluate the hierarchies of Lax equations (7.91) with these operators and the corresponding values for k as indicated above. Here u, v and w are the dynamical fields satisfying the nonlinear equations given by the Lax representation. The subscripts are implied by the fact that these operators constitute the well-known spectral operators associated to the KdV equation, the modified KdV equation and the Harry–Dym equation, respectively. The crucial point about the following analysis is that in all cases the isospectral hierarchy of equations associated with a given operator L can be calculated from L in a straightforward way. The recipe to obtain the second operator $P_{\geq k}(L^q) = (L^q)_{\geq k}$ needed for the dynamical equations (7.91) is simply given by choosing a power q of L and applying one of the projections $P_{\geq k}$ to this operator. It should be noted, however, that integer powers q of L will not lead to any interesting dynamics, when we start with purely differential operators. In the case $k = 0$, for instance, we would have $(L^q)_{\geq 0} = L^q$ for integer q, leading to the trivial dynamics $L_{t_q} = [L^q, L] = 0$. As shown by Gelfand and Dikii [91] fractional powers of the differential operators will lead to interesting results. In the context of algebra (7.83) the relevant procedure can be summarized as follows. We first consider the Schrödinger operator L_{KdV} of (7.92) and review the calculations of fractional powers in a purely algebraic way. Using the formal integration symbol ∂^{-1} we consider operators of the form

$$L_{KdV}^{1/2} = \partial + a_0 + a_1\partial^{-1} + a_2\partial^{-2} + a_3\partial^{-3} + \dots \ . \tag{7.93}$$

Trying to turn the ansatz (7.93) into a formal square root of the Schrödinger operator, we identify the coefficients $a_0, a_1, a_2, \dots$ by calculating

$$L_{KdV}^{1/2} \cdot L_{KdV}^{1/2} = \partial^2 + 2a_0\partial + (a_{0_x} + a_0^2 + 2a_1) + (a_{1_x} + 2a_0a_1 + 2a_2)\partial^{-1} + (\dots)\partial^{-2} + \dots \tag{7.94}$$

using (7.85) and $\partial\partial^{-1} = \partial^{-1}\partial = 1$. Hence, requiring $(L_{KdV}^{1/2})^2 = \partial^2 + u$, we can calculate all the coefficients $a_0, a_1, \dots$, recursively in terms of the field u. In particular, one finds the formal expansion

$$L_{KdV}^{1/2} = \partial + \frac{1}{2}u\partial^{-1} - \frac{1}{4}u_x\partial^{-2} + \frac{1}{8}(u_{xx} - u)\partial^{-3} + \dots \ . \tag{7.95}$$

Calculating the third power of (7.95) (or multiplying with $\partial^2 + u$) one finds the operator

$$L_{KdV}^{3/2} = \partial^3 + \frac{3}{2}u\partial + \frac{3}{4}u_x + (\dots)\partial^{-1} + (\dots)^{-2} + \dots, \tag{7.96}$$

which, by construction, commutes with $\partial^2 + u$. Applying the projection $P_{\geq 0}$ we have extracted the second operator of the Lax formulation of the KdV equation as the purely differential part of (7.96)

$$L_{KdV} = \partial^2 + u, \qquad A_1 = (L_{KdV}^{3/2})_{\geq 0} = \partial^3 + \frac{3}{2}u\partial + \frac{3}{4}u_x,$$

$$\frac{d}{dt}L_{KdV} = [(L_{KdV}^{3/2})_{\geq 0}, L_{KdV}] \Leftrightarrow u_t = \frac{1}{4}u_{3x} + \frac{3}{2}uu_x. \tag{7.97}$$

In fact, the entire hierarchy of higher KdV equations can be obtained in the same fashion. For instance, including further terms in the expressions (7.93) one can easily calculate $L_{KdV}^{5/2}$ as

$$L_{KdV}^{5/2} = L_{KdV}^{3/2} \cdot L_{KdV} = \partial^5 + \frac{5}{2}u\partial^3 + \frac{15}{4}u_x^2 + \frac{5}{8}(3u^2 + 5u_{xx})\partial$$

$$+ \frac{15}{16}u_{3x} + \frac{15}{8}uu_x + (\dots)\partial^{-1} + \dots \tag{7.98}$$

and consider its differential part. This operator will supplement the Schrödinger operator to form a Lax pair for the fifth order KdV flow

$$L_{KdV} = \partial^2 + u,$$

$$A_2 = (L_{KdV}^{5/2})_{\geq 0} = \partial^5 + \frac{5}{2}u\partial^3 + \frac{15}{4}u_x^2 + \frac{5}{8}(3u^2 + 5u_{xx})\partial + \frac{15}{16}u_{3x} + \frac{15}{8}uu_x,$$

$$\frac{d}{dt}L_{KdV} = [(L_{KdV}^{5/2})_{\geq 0}, L_{KdV}] \Leftrightarrow u_t = \frac{1}{16}(u_{5x} + 10uu_{3x} + 20u_xu_{xx} + 30u^2u_x).$$

$$(7.99)$$

In general the n-th Lax operator $A_n = \left(L_{KdV}^{(2n+1)/2}\right)_{\geq 0}$ covers the result of Chap. 4 (4.45) in an alternative way. Nevertheless, the above approach is more fundamental in the sense that it allows the construction of the appropriate hierarchy of commuting flows while the method of Chap. 4 allows only a reconstruction of a given hierarchy.

In a similar way one can extract the Lax operators for the Lax pairs of the modified KdV and the HD equations out of the given spectral operators L_{MKdV} and L_{HD} in (7.92). Using the same ansatz (7.93) for $L_{MKdV}^{1/2}$ and a modified ansatz $L_{HD}^{1/2} = w\partial + a_0 + a_1\partial^{-1} + ..$ for the HD case, one calculates

$$L_{MKdV}^{1/2} = \partial + v + \frac{1}{2}(-v_x - v^2)\partial^{-1} + (...)\partial^{-2} + ...,$$

$$L_{HD}^{1/2} = w\partial - \frac{1}{2}w_x + \frac{1}{8}\left(2w_{xx} - \frac{w_x^2}{w}\right)\partial^{-1} + (...)^{-2} + ..., \qquad (7.100)$$

and

$$L_{MKdV}^{3/2} = \partial^3 + 3v\partial^2 + \frac{3}{2}(v_xv^2)\partial + (...) + (...)\partial^{-1} + ...,$$

$$L_{HD}^{3/2} = w^3\partial^3 + \frac{3}{2}w^2w_x\partial^2 + (...)\partial + (...) + (...)\partial^{-1} + ... \qquad (7.101)$$

Using the projections $P_{\geq k}$ with $k = 1$ for the MKdV and $k = 2$ for the HD equation, one obtains the Lax pairs for these equations

$$L_{MKdV} = \partial^2 + 2v\partial, \quad (L_{MKdV}^{3/2})_{\geq 1} = \partial^3 + 3v\partial^2 + \frac{3}{2}(v_xv^2)\partial,$$

$$\frac{d}{dt}L_{MKdV} = 2v_t\partial = [(L_{MKdV}^{3/2})_{\geq 1}, L_{MKdV}] = \frac{1}{2}(v_{3x} - 6v^2v_x)\partial$$

$$\updownarrow$$

$$v_t = \frac{1}{4}v_{3x} - \frac{3}{2}v^2v_x \qquad (7.102)$$

and

$$L_{HD} = w^2\partial^2, \quad (L_{HD}^{3/2})_{\geq 2} = w^3\partial^3 + \frac{3}{2}w^2w_x\partial^2,$$

$$\frac{d}{dt}L_{HD} = 2ww_t\partial^2 = [(L_{HD}^{3/2})_{\geq 2}, L_{HD}] = \frac{1}{2}w^4w_{3x}\partial^2$$

$$\updownarrow$$

$$w_t = \frac{1}{4} w^3 w_{3x}. \tag{7.103}$$

Generally, the hierarchies of Lax operators $A_n = \left(L_{MKdV}^{(2n+1)/2} \right)_{\geq 1}$ and $A_n = \left(L_{HD}^{(2n+1)/2} \right)_{\geq 2}$ generate the well-known hierarchies of MKdV and HD vector fields.

As the first step in a more general discussion of the integrable equations (7.91) we have to explain what type of Lax operator may be used in (7.91) to obtain a consistent operator evolution equivalent to some nonlinear integrable equation. We start with looking for spectral Lax operators L in the general form

$$L = u_N \partial^N + u_{N-1} \partial^{N-1} + \dots + u_1 \partial + u_0 + u_{-1} \partial^{-1} + \dots \tag{7.104}$$

of N-th order, parametrized by infinitely many fields $u_N, u_{N-1}, \dots$. To obtain a consistent Lax equation we have to ensure that the commutator in (7.91) yields an integro-differential operator of an order not exceeding the order N of (7.92). Observing $[(L^q)_{\geq k}, L] = -[(L^q)_{<k}, L]$ with some $(L^q)_{<k} = a_{k-1}\partial^{k-1} + a_{k-2}\partial^{k-2} + \dots$ one immediately obtains the highest order of the commutator as

$$\frac{d}{dt_q} L = -[(L^q)_{<k}, L] = -[a_{k-1}\partial^{k-1} + \text{lower}, u_N \partial^N + \text{lower}]$$

$$= (N u_N (a_{k-1})_x - (k-1) a_{k-1} (u_N)_x) \partial^{N+k-2} + \text{lower}, \tag{7.105}$$

where lower represents lower differential orders. Hence, for the cases $k = 0, 1, 2$ under consideration here, the form of the commutator (7.105) matches the form (7.104) of the Lax operator and the corresponding time evolution for the fields $u_N, u_{N-1}, \dots$ is obtained from the coefficients of each power of ∂ in (7.105). However, we note that for $k = 0$ the two highest orders N and $N-1$ are not present in the commutator. As a result the fields u_N and u_{N-1} of (7.104) will not inherit any dynamics from the Lax equation (7.105). In other words, the two highest fields will be time independent functions (of the space variable x) which can be chosen arbitrarily. For $k = 1$ only the highest field u_N will inherit a trivial dynamics; for $k = 2$ all fields in (7.104) will be dynamical fields. Hence, as the first step, we have obtained some information on the highest orders of the operators admissible for the Lax equations (7.91). They are given in the form

$$k = 0: \quad L = c_N \partial^N + c_{N-1} \partial^{N-1} + u_{N-2} \partial^{N-2} + \dots + u_0 + u_{-1} \partial^{-1} + \dots,$$

$$k = 1: \quad L = c_N \partial^N + u_{N-1} \partial^{N-1} + u_{N-2} \partial^{N-2} + \dots + u_0 + u_{-1} \partial^{-1} + \dots,$$

$$k = 2: \quad L = u_N \partial^N + u_{N-1} \partial^{N-1} + u_{N-2} \partial^{N-2} + \dots + u_0 + u_{-1} \partial^{-1} + \dots,$$

$$\tag{7.106}$$

where the u_i are dynamical fields and c_N, c_{N-1} are arbitrary time independent functions of x. The three hierarchies of equations (7.91) in the general form (7.106) are to be interpreted as nonlinear hierarchies of coupled systems for the parametrizing fields u_i. In this sense (7.91) represents three hierarchies of (1+1)-dimensional equations involving the time variable t_q and the space variable x for an infinite number of fields u_i. Obviously, we are interested in extracting closed systems of nonlinear integrable equations for a finite number of fields. Not every possible restriction will work, of course. The restriction is valid if the commutator on the right hand side of (7.91) does not produce terms noncontained in L_{t_q}. Simple computation leads to the conclusion that there are three basic types of restriction :

$$k = 0: \quad L = c_N \partial^N + c_{N-1}\partial^{N-1} + u_{N-2}\partial^{N-2} + ... + u_0, \qquad (7.107)$$

$$k = 1: \quad L = c_N \partial^N + u_{N-1}\partial^{N-1} + ... + u_0 + \partial^{-1}u_{-1}, \qquad (7.108)$$

$$k = 2: \quad L = u_N \partial^N + u_{N-1}\partial^{N-1} + ... + u_0 + \partial^{-1}u_{-1} + \partial^{-2}u_{-2}. \quad (7.109)$$

We point out that here the integration symbols ∂^{-1} and ∂^{-2} turn up on the left of the fields u_{-1} and u_{-2}, hence, rewriting these symbols using (7.84), one encounters an 'infinite tail' of integration symbols in these operators. It is quite easy to see from a direct computation that operators of the above form do indeed lead to consistent Lax equations of the form (7.91). For instance, for the case $k = 0$ the commutator $[(L^q)_{\geq 0}, L]$ will be a purely differential operator, if L is a purely differential operator. As we will see later the reduction to operators of the form (7.95) can be best understood from the underlying Hamiltonian concept. Operators (7.106) lie in the dual of the subspace $g_{<k}$. It will be argued that the linear Poisson bracket associated with a Lie algebra decomposition can always be properly restricted to the dual subspaces. Hence, all Hamiltonian equations (in particular the equations (7.91) associated with the Casimir functions) can be restricted to the dual subalgebras. As a result, all the nonlinear equations (7.91) with Lax operators of the form (7.107)–(7.109) will inherit a Hamiltonian formulation from the abstract linear bracket.

We remark that for $k = 1$ further admissible reductions of the Lax form (7.108) are given by

$$L = c_N \partial^N + u_{N-1}\partial^{N-1} + u_{N-2}\partial^{N-2} + ... + u_1\partial + u_0, \qquad (7.110)$$

$$L = c_N \partial^N + u_{N-1}\partial^{N-1} + u_{N-2}\partial^{N-2} + u_1\partial + \lambda, \qquad (7.111)$$

where λ is an arbitrary constant parameter. Again, it is easy to see this from a direct computation. In the reduction to $u_{-1} = 0$ the operator L becomes purely differential and the commutator with $(L^q)_{\geq 1}$ will be again purely differential. In the last reduction the parameter λ has no effect on the commutator. $[(L^q)_{\geq 1}, L] = [a_1\partial + \text{higher}, \lambda + u_1\partial + \text{higher}] = (...)\partial + \text{higher}$, and the resulting operator matches the form of the Lax operator. As a result, one can immediately predict the following property of the nonlinear equations for $u_{-1}, ..., u_{N-1}$ given by the choice $L = c_N \partial^N + ... + u_0 + \partial^{-1}u_{-1}$. These equations are such that they will admit the reductions $u_{-1} = 0$ and $(u_{-1} =$

$0, u_0 = \lambda = \text{const}$). We will see, however, that only the choice in (7.108) and (7.111) stem from the Hamiltonian reduction mentioned above. The reduction to $u_{-1} = 0$ (7.110) will leave the particular equations (7.91) invariant, but do not necessarily admit a proper restriction of the Hamiltonian structure.

In a similar way, the following reductions: $(u_{-2} = 0), (u_{-2} = u_{-1} = 0), (u_{-2} = u_{-1} = 0, u_0 = \lambda_1 + \lambda_2 x), (u_{-2} = u_{-1} = 0, u_0 = \lambda_3, u_1 = \lambda_1 + \lambda_2 x)$, are admissible for the case $k = 2$ of (7.109), where all λ's are arbitrary constant parameters. In all these cases it is readily verified that the lowest order of the commutator $[(L^q)_{\geq 2}, L]$ matches the lowest order of L_{t_q} for arbitrary operators $(L^q)_{\geq 2} = a_M \partial^M + ... + a_2 \partial^2$. Again, only the choice in (7.109) is a Hamiltonian for the general linear Poisson bracket. All other reductions leave the equations (7.91), $k = 2$, invariant, but will not necessarily admit a reduction of the Poisson structure.

Now we will display a list of the simplest nonlinear integrable equations encoded in (7.91) by specializing the spectral Lax operator as in (7.106) and its admissible reductions. All equations are obtained in exactly the same way as the three 'prototype' equations: KdV, MKdV and HD. We consider an operator of the N-th order $L = u_N \partial^N +$ lower orders. Here, the highest coefficient will be chosen as $u_N = c_N = 1$ for the cases $k = 0, 1$ and $u_N = u^N$ (with some dynamical field u) for the case $k = 2$, respectively. Then, following the procedure of the algebraic calculation of the 'N-th root', one constructs $L^{\frac{1}{N}} = A\partial + a_0 + a_1\partial^{-1} + ...$, where $A = 1$ in the cases $k = 0, 1$ and $A = u$ in the case $k = 2$, respectively. Calculating the powers $L^{\frac{n}{N}}$ and applying the projections $P_{\geq k}$ one considers the Lax equations

$$\frac{d}{dt_n} L = [(L^{\frac{n}{N}})_{\geq k}, L], \qquad n = 1, 2, 3, ... \, . \tag{7.112}$$

For $k = 0, 1$ the choice $n = 1$ will always lead to the dynamics $(u_i)_{t_1} = (u_i)_x$ for the fields u_i in L, so that we may identify $t_1 = x$ in these cases. For $k = 0$ and purely differential L the equations become trivial for integer values of $\frac{n}{N}$, because then $(L^{\frac{n}{N}})_{\geq 0} = L^{\frac{n}{N}}$. In the calculations leading to the nonlinear equations the only difficulty arising is of a technical nature. In order to obtain $(L^{\frac{n}{N}})_{\geq k}$ one has to calculate sufficiently many coefficients of the expression $L^{\frac{1}{N}} = A\partial + \sum_{i \leq 0} a_i \partial^i$ in terms of the fields u_i parametrizing the Lax operators. In particular, the coefficients up to a_{n-1-k} will contribute to the n-th equation in the hierarchy (7.112). Hence, for large n and N, this calculation will become very cumbersome.

Example 7.1 The case $k = 0$.
The second order spectral Lax operator $L = \partial^2 + u$ has been already discussed and leads to the well known KdV hierarchy. For the third order Lax operator one finds

$$L = \partial^3 + u\partial + v, \qquad (L^{\frac{2}{3}})_{\geq 0} = \partial^2 + \frac{2}{3}u, \tag{7.113}$$

where the fractional power was calculated according to the procedure given previously. The resulting equation is

$$
\begin{pmatrix} u \\ v \end{pmatrix}_{t_2} = \begin{pmatrix} -u_{xx} + 2v_x \\ -\frac{2}{3}u_{3x} + v_{xx} - \frac{2}{3}uu_x \end{pmatrix}. \tag{7.114}
$$

As a consequence, the field u satisfies the Boussinesq equation $u_{t_2 t_2} + \frac{1}{3}(u_{3x} + 4uu_x)_x = 0$. For the fourth order spectral operator one finds

$$
L = \partial^4 + u\partial^2 + v\partial + w, \qquad (L^{\frac{2}{4}})_{\geq 0} = \partial^2 + \frac{1}{2}u. \tag{7.115}
$$

The resulting equation is

$$
\begin{pmatrix} u \\ v \\ w \end{pmatrix}_{t_2} = \begin{pmatrix} -2u_{xx} + 2v_x \\ -2u_{3x} + v_{xx} + 2w_x - uu_x \\ -\frac{1}{2}u_{4x} + w_{xx} - \frac{1}{2}uu_{xx} - \frac{1}{2}u_x v \end{pmatrix}. \tag{7.116}
$$

Example 7.2 The case $k = 1$.
For the first order Lax operator one obtains

$$
L = \partial + u + \partial^{-1}v, \qquad (L^2)_{\geq 1} = \partial^2 + 2u\partial, \tag{7.117}
$$

so that (7.112) is equivalent to the system of equations

$$
\begin{pmatrix} u \\ v \end{pmatrix}_{t_2} = \begin{pmatrix} u_{xx} + 2v_x + 2uu_x \\ -v_{xx} + 2(uv)_x \end{pmatrix} \tag{7.118}
$$

of Kaup and Broer discussed in [113]. According to the additional reductions we can restrict the operators and equations to $v = 0$, leading to Burgers equation

$$
L = \partial + u, \qquad (L^2)_{\geq 1} = \partial^2 + 2u\partial, \qquad u_{t_2} = u_{xx} + 2uu_x. \tag{7.119}
$$

For the second order Lax operator one finds

$$
L = \partial^2 + u\partial + v + \partial^{-1}w, \qquad (L^{\frac{2}{2}})_{\geq 1} = \partial^2 + u\partial, \tag{7.120}
$$

leading to the equation

$$
\begin{pmatrix} u \\ v \\ w \end{pmatrix}_{t_2} = \begin{pmatrix} 2v_x \\ v_{xx} + 2w_x + uv_x \\ -w_{xx} + (uw)_x \end{pmatrix}. \tag{7.121}
$$

The reduction $w = 0$ leads to

$$
\begin{pmatrix} u \\ v \end{pmatrix}_{t_2} = \begin{pmatrix} 2v_x \\ v_{xx} + uv_x \end{pmatrix}. \tag{7.122}
$$

The next reduction to $v = 0$ leaves (7.122) trivial. However, for the next equation in the hierarchy one calculates

$$L = \partial^2 + u\partial + v + \partial^{-1}w, \qquad (L^{\frac{3}{2}})_{\geq 1} = \partial^3 + \frac{3}{2}u\partial^2 + \frac{3}{8}(2u_x + u^2 + 4v)\partial. \quad (7.123)$$

The resulting equation is

$$\begin{pmatrix} u \\ v \\ w \end{pmatrix}_{t_3} = \begin{pmatrix} \frac{1}{4}u_{3x} + \frac{3}{2}v_{xx} + 3w_x - \frac{3}{8}u^2u_x + \frac{3}{2}(uv)_x \\ v_{3x} + \frac{3}{2}uv_{xx} + 6u_xv_x + \frac{3}{2}u_xw + 3uw_x + \frac{3}{2}vv_x + \frac{3}{8}u^2v_x \\ w_{3x} - \frac{3}{2}uw_{xx} - \frac{9}{4}u_xw_x - \frac{3}{4}u_{xx}w + \frac{3}{2}(vw)_x + \frac{3}{8}(u^2w)_x \end{pmatrix},$$

$$(7.124)$$

for which the reduction $w = v = 0$ leads to the modified KdV equation

$$L = \partial^2 + u\partial, \quad (L^{\frac{3}{2}})_{\geq 1} = \partial^3 + \frac{3}{2}u\partial^2 + \frac{3}{8}(2u_x + u^2)\partial,$$

$$u_{t_3} = \frac{1}{4}u_{3x} - \frac{3}{8}u^2u_x \qquad (7.125)$$

in accordance with the previous results.

For the third order Lax operators one finds

$$L = \partial^3 + u\partial^2 + v\partial + w + \partial^{-1}z, \quad (L^{\frac{2}{3}})_{\geq 1} = \partial^2 + \frac{2}{3}u\partial. \qquad (7.126)$$

The resulting equation is

$$\begin{pmatrix} u \\ v \\ w \\ z \end{pmatrix}_{t_2} = \begin{pmatrix} -u_{xx} + 2v_x - \frac{2}{3}uu_x \\ -\frac{2}{3}u_{3x} + v_{xx} + 2w_x + \frac{2}{3}uv_x - \frac{2}{3}u_xv - \frac{2}{3}uu_{xx} \\ w_{xx} + 2z_x + \frac{2}{3}uw_x \\ -z_{xx} + \frac{2}{3}(uz)_x \end{pmatrix}. \qquad (7.127)$$

For the reduction $z = 0, w = $ const we obtain a modified Boussinesq equation

$$\begin{pmatrix} u \\ v \end{pmatrix}_{t_2} = \begin{pmatrix} -u_{xx} + 2v_x - \frac{2}{3}uu_x \\ -\frac{2}{3}u_{3x} + v_{xx} + \frac{2}{3}uv_x - \frac{2}{3}u_xv - \frac{2}{3}uu_{xx} \end{pmatrix}. \qquad (7.128)$$

The way of constructing a Miura transformation to the Boussinesq equation (7.114) will be explained later.

Example 7.3 The case $k = 2$.
For first order Lax operator one obtains

$$L = u\partial + v + \partial^{-1}w + \partial^{-2}z, \quad (L^2)_{\geq 2} = u^2\partial^2, \qquad (7.129)$$

so that (7.112) is equivalent to the system of equations

$$\begin{pmatrix} u \\ v \\ w \\ z \end{pmatrix}_{t_2} = \begin{pmatrix} u^2u_{xx} + 2u^2v_x \\ u^2v_{xx} + 2u(uw)_x \\ -(u^2w)_{xx} + 2(u^2z)_x \\ -(u^2z)_{xx} \end{pmatrix}. \qquad (7.130)$$

According to the additional reductions we can restrict the operators and equations to

$$\begin{pmatrix} u \\ v \\ w \end{pmatrix}_{t_2} = \begin{pmatrix} u^2 u_{xx} + 2u^2 v_x \\ u^2 v_{xx} + 2u(uw)_x \\ -(u^2 w)_{xx} \end{pmatrix}, \quad \begin{pmatrix} u \\ v \end{pmatrix}_{t_2} = \begin{pmatrix} u^2 u_{xx} + 2u^2 v_x \\ u^2 v_{xx} \end{pmatrix},$$

$$u_{t_2} = u^2 u_{xx} + 2\lambda u^2. \tag{7.131}$$

For the last reduction we have used $z = w = 0, v = \lambda x$.
For the second order Lax operator one finds

$$L = u^2 \partial^2 + v\partial + w + \partial^{-1} z + \partial^{-2} r, \quad (L)_{\geq 2} = u^2 \partial^2, \tag{7.132}$$

leading to the equation

$$\begin{pmatrix} u \\ v \\ w \\ z \\ r \end{pmatrix}_{t_2} = \begin{pmatrix} uv_x - vu_x \\ u^2 v_{xx} + 2u^2 w_x \\ u^2 w_{xx} + 2u(uz)_x \\ -(u^2 z)_{xx} + 2(u^2 r)_x \\ -(u^2 r)_{xx} \end{pmatrix}. \tag{7.133}$$

The admissible reductions lead to

$$\begin{pmatrix} u \\ v \\ w \\ z \end{pmatrix}_{t_2} = \begin{pmatrix} uv_x - vu_x \\ u^2 v_{xx} + 2u^2 w_x \\ u^2 w_{xx} + 2u(uz)_x \\ -(u^2 z)_{xx} \end{pmatrix}, \quad \begin{pmatrix} u \\ v \\ w \end{pmatrix}_{t_2} = \begin{pmatrix} uv_x - vu_x \\ u^2 v_{xx} + 2u^2 w_x \\ u^2 w_{xx} \end{pmatrix},$$

$$\tag{7.134}$$

for $r = 0$ and $r = z = 0$. Choosing $w = \lambda x$ and $(v = \lambda x, w = \text{const})$, respectively, one can reduce (7.134) to the following equations

$$\begin{pmatrix} u \\ v \end{pmatrix}_{t_2} = \begin{pmatrix} uv_x - vu_x \\ u^2 v_{xx} + 2\lambda u^2 \end{pmatrix}, \quad u_{t_2} = \lambda(u - xu_x). \tag{7.135}$$

For the very last reduction the evolution equation for u becomes linear. It is the first member of the Harry–Dym hierarchy related to the scaling symmetry. Indeed, for the next equation in the hierarchy one calculates

$$L = u^2 \partial^2 + v\partial + w + \partial^{-1} z + \partial^{-2} r, \quad (L^{\frac{3}{2}})_{\geq 2} = u^3 \partial^3 + \frac{3}{2}(u^2 u_x + uv)\partial^2. \tag{7.136}$$

The reduction $r = z = w = v = 0$ leads to the HD equation $u_{t_3} = \frac{1}{4} u^3 u_{3x}$ in accordance with the previous results.

For any given Lax formulation $L_t = [A, L]$ encoding a nonlinear integrable equation one can use an equivalent Lax formulation in terms of the formally transposed pair $L^\dagger$ and $-A^\dagger$, which again satisfy $L_t^\dagger = [-A^\dagger, L^\dagger]$. In terms of Lax operators the transposition is given by

$$L^\dagger = \left(\sum_i u_i \partial^i\right)^\dagger = \sum_i (-1)^i \partial^i u_i. \tag{7.137}$$

We now discuss the invariance of the Lax equations (7.91),(7.112) with respect to transposition of L. It is readily seen that

$$((B)_{\geq 0})^\dagger = (B^\dagger)_{\geq 0}, \qquad ((B)_{<0})^\dagger = (B^\dagger)_{<0} \tag{7.138}$$

for any operator $B \in g$. Further, observing that $(B)_{\geq k} = (B\partial^{-k})_{\geq 0}\partial^k$, one derives the identity

$$\partial^{-k}((B)_{\geq k})^\dagger \partial^k = (\partial^{-k} B^\dagger \partial^k)_{\geq k}. \tag{7.139}$$

For a given N-th order operator $L = B^N \partial^N +$ lower orders all integrable equations discussed in this section were constructed via the N-th root

$$L^{\frac{1}{N}} = B\partial + a_0 + a_1 \partial^{-1} + ... \tag{7.140}$$

from the condition $\left(L^{\frac{1}{N}}\right)^N = L$. All coefficients $a_0, a_1, ...$ in this ansatz are determined uniquely from this condition, once the leading order term B is fixed. The isospectral equations were then obtained using (integer) powers of $L^{\frac{1}{N}}$. Let us consider the transposed operators

$$\tilde{L}^{\frac{1}{N}} = -\partial^{-k}\left(L^{\frac{1}{N}}\right)^\dagger \partial^k = -\partial^{-k}\left(B\partial + \text{lower}\right)^\dagger \partial^k = -\partial^{-k}\left(\partial B + \text{lower}\right)\partial^k$$

$$= B\partial + \text{lower}, \tag{7.141}$$

having the same leading order term B as the original $L^{\frac{1}{N}}$ and representing the N-th root of the operator

$$\tilde{L} = (-1)^N \partial^{-k} L^\dagger \partial^k = (-1)^N \partial^{-k}\left(B^N \partial^N + \text{lower}\right)^\dagger \partial^k = B^N \partial^N + \text{lower}. \tag{7.142}$$

Applying identity (7.139), the time evolution of $\tilde{L}$ is obtained from (7.112) as

$$\frac{d}{dt_n}\tilde{L} = (-1)^N \partial^{-k}\left([((L^{\frac{1}{N}})^n)_{\geq k}, L]\right)^\dagger \partial^k = (-1)^{n+1}[((\tilde{L}^{\frac{1}{N}})^n)_{\geq k}, \tilde{L}]. \tag{7.143}$$

Absorbing the sign into the time variable we find that all equations (7.112) are invariant under the discrete transformation

$$L \to \tilde{L} = (-1)^N \partial^{-k} L^\dagger \partial^k,$$

$$L^{\frac{1}{N}} \to \tilde{L}^{\frac{1}{N}} = -\partial^{-k}\left(L^{\frac{1}{N}}\right)^\dagger \partial^k,$$

$$t_n \to \tilde{t}_n = (-1)^{n+1} t_n \tag{7.144}$$

Applying transformation (7.144), a certain invariance for the nonlinear equations can be extracted directly from the Lax operator. For example, the Boussinesq equation (7.114) is associated with the third order operator $L = \partial^3 + u\partial + v$, constructed via $(L^{\frac{2}{3}})_{\geq 0}$. Considering

$$\widetilde{L} = \partial^3 + \widetilde{u}\partial + \widetilde{v} = -L^\dagger = \partial^3 + \partial u - v = \partial^3 + u\partial + u_x - v \qquad (7.145)$$

we immediately obtain the invariance

$$u \to \widetilde{u} = u, \quad v \to \widetilde{v} = -v + u_x, \quad t_2 \to \widetilde{t}_2 = -t_2 \qquad (7.146)$$

for the dynamical equation (7.114). The same analysis for the modified Boussinesq equation (7.128), associated with the operators $L = \partial^3 + u\partial^2 + v\partial$, $(L^{\frac{2}{3}})_{\geq 1}$, leads to

$$\widetilde{L} = \partial^3 + \widetilde{u}\partial^2 + \widetilde{v}\partial = -\partial^{-1}L^\dagger\partial = \partial^3 - u\partial^2 + (v - u_x)\partial. \qquad (7.147)$$

Hence, we find the invariance

$$u \to \widetilde{u} = -u, \quad v \to \widetilde{v} = v - u_x, \quad t_2 \to \widetilde{t}_2 = -t_2 \qquad (7.148)$$

for the dynamical equation (7.128).

An important consequence of these symmetry transformations is the conclusion that for odd n, that is $\widetilde{t}_n = t_n$, the fixed points $L = \widetilde{L}$ of the transformation (7.144) are invariant with respect to the dynamics (7.112). Hence, we may restrict each second equation in the three hierarchies (7.112) (with odd n) to Lax operators satisfying the constraint

$$L = (-1)^N \partial^{-k} L^\dagger \partial^k. \qquad (7.149)$$

This constraint was discussed for the first time by Kupershmidt [113] and hence is referred to as *Kupershmidt reduction*.

Example 7.4 The Kupershmidt constraint for the case $k = 0$.
Imposing the constraint (7.149) on the third order operator $L = \partial^3 + u_1\partial + u_0$ one finds the condition $u_0 = \frac{1}{2}u_{1_x}$. This reduction cannot be imposed on the Boussinesq equation (7.114), because it represents the flow associated with the even time $t = t_2$. The next admissible index $n = 3$ leads to a trivial flow, hence, we have chosen $n = 5$ to find the first nontrivial flow associated with a third order operator constrained by (7.149). Putting $u_1 = 2u$ one finds

$$L = \partial^3 + \partial u + u\partial, \quad (L^{\frac{5}{3}})_{\geq 0} = \partial^5 + \frac{10}{9}(\partial^3 u + u\partial^3) + \frac{5}{9}(\partial^2 u\partial + \partial u\partial^2)$$

$$+ \frac{10}{9}(\partial u^2 + u^2\partial), \qquad (7.150)$$

leading to the Kaup–Kupershmidt (KK) equation [78]

$$9u_{t_5} + u_{5x} + 10uu_{3x} + 25u_x u_{xx} + 20u^2 u_x = 0. \qquad (7.151)$$

This represents a reduction of the t_5-flow in the Boussinesq hierarchy.

Example 7.5 The Kupershmidt constraint for the case $k = 1$.
Imposing restriction (7.149) on the first order operator $L = \partial + u_0 + \partial^{-1}u_{-1}$ one finds the constraint $u_0 = 0$. With $u = u_{-1}$ one calculates

$$L = \partial + \partial^{-1}u, \quad (L^3)_{\geq 1} = \partial^3 + 3u\partial, \tag{7.152}$$

leading to the KdV equation

$$u_{t_3} = u_{3x} + 6uu_x. \tag{7.153}$$

Hence, the pair (7.152) is one more alternative Lax representation of the KdV.

For the second order operator $L = \partial^2 + u_1\partial + u_0 + \partial^{-1}u_{-1}$ the Kupershmidt constraint leads to $u_1 = 0, u_{-1} = -\frac{1}{2}u_{0_x}$. Putting $u_0 = 2u$ one finds

$$L = \partial^2 + u + \partial^{-1}u\partial, \quad (L^{\frac{3}{2}})_{\geq 1} = \partial^3 + 3u\partial, \tag{7.154}$$

again leading to the KdV equation

$$u_{t_3} = u_{3x} + 3uu_x. \tag{7.155}$$

In this case the spectral Lax operator is just the recursion operator for cosymmetries of the KdV hierarchy, when identifying the pseudo-differential symbol ∂^{-1} with the integration operator.

For the third order operator $L = \partial^3 + u_2\partial^2 + u_1\partial + u_0 + \partial^{-1}u_{-1}$ the Kupershmidt reduction leads to the condition $u_2 = u_0 = 0$. With $u = u_1, v = u_{-1}$, one finds

$$L = \partial^3 + u\partial + \partial^{-1}v, \quad (L)_{\geq 1} = \partial^3 + u\partial, \tag{7.156}$$

leading to

$$\begin{pmatrix} u \\ v \end{pmatrix}_{t_3} = \begin{pmatrix} 3v_x \\ v_{3x} + (uv)_x \end{pmatrix}. \tag{7.157}$$

We can intersect the Kupershmidt constraint with the additional reductions. Assuming $v = 0$ equation (7.157) becomes trivial. However, passing to the next nontrivial flow admitting the Kupershmidt constraint, one finds

$$L = \partial^3 + u\partial, \quad (L^{\frac{5}{3}})_{\geq 1} = \partial^5 - \frac{5}{9}\partial u\partial^2 + \frac{10}{9}(\partial^2 u\partial + u\partial^3) + \frac{5}{9}u^2\partial, \tag{7.158}$$

leading to the Sawada–Kotera equation

$$9u_{t_5} + u_{5x} + 5uu_{3x} + 5u_x u_{xx} + 5u^2 u_x = 0, \tag{7.159}$$

considered in Example 3.12. Let us point out that stationary flows of the KK equation (7.151) and the SK equation (7.159) are equivalent to the remaining two integrable cases of the Henon–Heiles system [79],[33]. The integrable case of the Henon–Heiles system related to the fifth order KdV stationary flow was considered in the previous chapter.

Example 7.6 The Kupershmidt constraint of the $k = 2$.

Imposing restriction (7.149) on the first order operator $L = u_1\partial + u_0 + \partial^{-1}u_{-1} + \partial^{-2}u_{-2}$ one finds the constraints $u_0 = -\frac{1}{2}u_{-1}$, $u_{-2} = \frac{1}{4}u_{1xxx} - \frac{1}{2}u_{-1x}$. With $u_1 = 2u$, $u_{-1} = u_{xx} + 2v$, one calculates

$$L = u\partial + \partial^{-1}u\partial^2 + \partial^{-1}v + \partial^{-2}v\partial, \quad (L^3)_{\geq 2} = 4u^3\partial^3 + 4\partial u^3\partial^2, \quad (7.160)$$

leading to the equation

$$\begin{pmatrix} u \\ v \end{pmatrix}_{t_3} = 4\begin{pmatrix} 2u^3u_{3x} + 6u^2(uv)_x + 3u^2u_xu_{xx} \\ 2u^3v_{3x} + 6u^2u_x^2v_x + 3u^2u_{xx}v_x + 9u^2u_xv_{xx} \end{pmatrix}. \quad (7.161)$$

Intersecting the Kupershmidt constraint with the additional reduction of the form $v = \frac{1}{2}\lambda = $ const one finds

$$L = u\partial + \partial^{-1}u\partial^2 + \lambda\partial^{-1}, \quad (L^3)_{\geq 2} = 4u^3\partial^3 + 4\partial u^3\partial^2,$$

$$u_{t_3} = 8u^3u_{3x} + 12u^2u_xu_{xx} + 12\lambda u^2u_x. \quad (7.162)$$

The reader can find other examples of the Kupershmidt reduction in [108].

Before we pass to relations between various Lax hierarchies we shall define the notion of an eigenfunction of the Lax equation (7.91).

Definition 7.12

(i)   *For a given $k = 0, 1, 2$ a function $\psi = \psi(x, t_q)$ satisfying the linear equations*

$$\psi_{t_q} = (L^q)_{\geq k}\psi \quad (7.163)$$

is called an eigenfunction for the hierarchy of Lax equations $\frac{d}{dt_q}L = [(L^q)_{\geq k}, L]$.

(ii) *For a given $k = 0, 1, 2$ a function $\varphi = \varphi(x, t_q)$ satisfying the linear equations*

$$\varphi_{t_q} = -((L^q)_{\geq k})^\dagger\varphi \quad (7.164)$$

is called an adjoint eigenfunction for the hierarchy of Lax equations $\frac{d}{dt_q}L^\dagger = [-((L^q)_{\geq k})^\dagger, L^\dagger]$.

The Lax equations $\frac{d}{dt_q}L = [(L^q)_{\geq k}, L]$ and $\frac{d}{dt_q}L^\dagger = [-((L^q)_{\geq k})^\dagger, L^\dagger]$ guarantee the compatibility between the linear problems (7.163), (7.164) and the isospectral problems

$$L\psi = \lambda\psi, \qquad L^\dagger\varphi = \mu\varphi \quad (7.165)$$

with eigenvalues λ and μ. We again remember, in order to avoid confusion, that an application of a pseudo-differential operator A to a function ϕ can be described as the zero order term $P_0(A\phi) = (A\phi)_0$ of the operator $A\phi$. Thus, relations (7.163), (7.164) should be understood as

$$\psi_{t_q} = ((L^q)_{\geq k}\psi)_0, \qquad (L\psi)_0 = \lambda\psi, \tag{7.166}$$

when now $(L^q)_{\geq k}\psi$ and $L\psi$ are operators instead of differential functions.

Remark 7.1 For the case $k = 0$, besides the Gelfand–Dikii restriction (7.107) (with $c_N = 1, c_{N-1} = 0$) of a spectral operator, there exists another important restriction of L to the form

$$L = \partial^N + u_{N-2}\partial^{N-2} + \ldots + u_0 + \psi\partial^{-1}\varphi, \tag{7.167}$$

where ψ and φ satisfy

$$\psi_{t_q} = ((L^q)_{\geq 0}\psi)_0, \qquad \varphi_{t_q} = -\left(((L^q)_{\geq 0})^\dagger\varphi\right)_0. \tag{7.168}$$

Note that the simplest case of the Lax operator (7.167) with $N = 1$ reads

$$L = \partial + \psi\partial^{-1}\varphi, \qquad (L^2)_{\geq 0} = \partial^2 + 2\psi\varphi. \tag{7.169}$$

As a result, equation (7.112) gives

$$\begin{pmatrix} \psi \\ \varphi \end{pmatrix}_{t_2} = \begin{pmatrix} \psi_{xx} + 2\varphi\psi^2 \\ -\varphi_{xx} - 2\psi\varphi^2 \end{pmatrix}, \tag{7.170}$$

that is the first nontrivial equation of the AKNS hierarchy. Hence, $L = \partial + q\partial^{-1}r$ is the alternative (to the one considered in Chap. 4) spectral operator for the AKNS hierarchy. The reader can find more examples of nonlinear integrable equations encoded in (7.112) ($k = 0$) by specializing the Lax operator as in (7.167) in [151].

Remark 7.2 For the case $k = 1$ and the restriction to $L = \partial^N + u_{N-1}\partial^{N-1} + \ldots + u_0 + \partial^{-1}u_{-1}$, the fields u_0 and u_{-1} are expressible by the respective eigenfunction ψ and adjoint eigenfunction φ of hierarchy (7.112) in the following way

$$u_0 = \psi - D_x^{-1}\varphi, \qquad u_{-1} = \varphi. \tag{7.171}$$

The reader can find proof in [151]. Now, further reductions of L to the forms (7.110) and (7.111) appear in a very natural way. Actually, the restriction of u_{-1} to the trivial eigenfunction $\varphi = 0$ generates the operator (7.110) with $u_0 = \psi$ and further restriction of ψ to trivial solutions $\psi = \lambda = \text{const}$ (as $(L^q)_{\geq 1}$ has no zero order term) generates the operator (7.111).

The three cases $k = 0, 1, 2$ of the Lax hierarchies (7.91), we have considered in this Section, are interrelated in the following way.

Theorem 7.5

(i) *Gauge $k = 0 \rightarrow k = 1$.*
    Let $L \in \mathfrak{g}$ *satisfy the hierarchy of Lax equations* $\frac{d}{dt_q}L = [(L^q)_{\geq 0}, L]$ *and the function* $\psi \neq 0$ *be an eigenfunction of this hierarchy:* $\psi_{t_q} = ((L^q)_{\geq 0}\psi)_0$
    Then $\overline{L} = \psi^{-1}L\psi$ *satisfies the hierarchy* $\frac{d}{dt_q}\overline{L} = [(\overline{L}^q)_{\geq 1}, \overline{L}]$.

(ii) Link $k = 1 \to k = 2$.

Let $\overline{L} = \overline{L}(x, t_q)$ *satisfy* $\frac{d}{dt_q}\overline{L} = [(\overline{L}^q)_{\geq 1}, \overline{L}]$ *and* $\phi(x, t_q)$, $\phi_x \neq 0$, *be an eigenfunction of this hierarchy satisfying* $\phi_{t_q} = ((L^q)_{\geq 1}\phi)_0$. *Then $L'(x', t'_q) = \overline{L}(x, t_q)$, linked by the transformation $x' = \phi(x, t_q)$, $t'_q = t_q$, satisfies the hierarchy* $\frac{d}{dt'_q}L' = [(L'^q)_{\geq 2}, L']$.

The *proof* does not require any assumptions on the explicit form of L other than those implied by the Lax equation $L_t = [(L^q)_{\geq k}, L]$. For example, in order to prove (i) one establishes the operator identity

$$(\psi^{-1}L\psi)_t = [((\psi^{-1}L\psi)^q)_{\geq 1}, \psi^{-1}L\psi]$$

$$= \psi^{-1}(L - [(L^q)_{\geq 0}, L])\psi - [\psi^{-1}((L^q)_{\geq 0}\psi)_0, \psi^{-1}L\psi], \qquad (7.172)$$

which holds for an arbitrary operator L and any arbitrary function ψ. Obviously, from (7.172) the result (i) of Theorem 7.4 is obtained trivially.

In particular, starting with L of the form (7.167) the gauge transformed operator $\overline{L} = \psi^{-1}L\psi$ has the form

$$\overline{L} = \partial^N + N\psi^{-1}\psi_x\partial^{N-1} + \left(u_{N-2} + \frac{1}{2}N(N-1)\psi^{-1}\psi_{xx}\right)\partial^{N-2}$$

$$+ \ldots + v_1\partial + v_0 + \partial^{-1}\varphi\psi, \qquad (7.173)$$

where the highest field $v_{N-1} = N\psi^{-1}\psi_x$ is given in terms of ψ only. The other fields $v_{N-2}, , \ldots, v_0$ are given as differential expressions of ψ and the fields $u_{N-2}, \ldots, u_0$ parametrizing L. In particular, the zero order term is given by $v_0 = \psi^{-1}(L\psi)_0$ and the v_{-1} term is given as the 'squared eigenfunction' $v_{-1} = \varphi\psi$. Gauge transformation with the choice $\varphi = 0$ and the additional assumption that ψ is an eigenfunction of L, i.e. $(L\psi)_0 = \lambda\psi$, relates the $k = 0$ spectral problem (7.107) with the $k = 1$ spectral problem (7.111) as $v_{-1} = 0$ and $v_0 = \lambda$. Eliminating ψ from the transformation one can obtain a Miura map between the fields $(u_{N-2}, \ldots, u_0)$ and $(v_{N-1}, \ldots, v_1)$.

We shall demonstrate this procedure for the simplest cases of $N = 2, 3$. For $N = 2$ one starts with the Schrödinger operator $L = \partial^2 + u$ giving rise to the KdV hierarchy. Taking a special ψ related to the $\lambda = 0$ eigenvalue of L, i.e. $(L\psi)_0 = 0$, the gauge transformed operator is calculated as $\overline{L} = \psi^{-1}L\psi = \partial^2 + 2\psi^{-1}\psi_x\partial$, giving rise to the hierarchy of the modified KdV equation for the field $v = \psi^{-1}\psi_x$. Hence, the celebrated Miura map is encoded in the transformation

$$v = \psi^{-1}\psi_x, \quad \psi_{xx} + u\psi = 0. \qquad (7.174)$$

Indeed, eliminating ψ from this transformation one finds $u = -v_x - v^2$, i.e. the Miura relation (3.124) mapping solutions v of the modified KdV equation to solutions u of the KdV equation.

For $N = 3$ let $L = \partial^3 + u\partial + v$ be a Boussinesq spectral operator, $\overline{L} = \partial^3 + \overline{u}\partial^2 + \overline{v}\partial$ be a modified Boussinesq spectral operator and the eigenfunction ψ be such that $(L\psi)_0 = 0$. Then we find

$$\overline{L} = \psi^{-1}L\psi \Rightarrow \overline{u} = 3\psi^{-1}\psi_x, \quad \overline{v} = 3\psi^{-1}\psi_{xx} + u, \tag{7.175}$$

and elimination of ψ from (7.175) gives

$$u = \overline{v} - \overline{u}_x - \frac{1}{3}\overline{u}^2, \quad v = -\frac{1}{3}\overline{u}_{xx} - \frac{1}{3}\overline{u}\,\overline{v} + \frac{2}{27}\overline{u}^3, \tag{7.176}$$

i.e. the Miura map between the modified Boussinesq equation (7.128) and Boussinesq equation (7.114).

Similarly, considering the transformations $k = 1 \rightarrow k = 2$, one starts with an operator $\overline{L}$ of the form (7.111,$k = 1$). Using an eigenfunction ϕ satisfying $\overline{L}\phi = \mu\phi$, one performs the transformation $x' = \phi$. Using $\partial = \phi_x\partial'$, the linked operator $L' = \overline{L}$ is calculated to be of the form

$$L' = \phi_x^N\partial'^N + \left(v_{N-1}\phi_x^{N-1} + \frac{1}{2}N(N-1)\phi_x^{N-2}\phi_{xx}\right)\partial'^{N-1} + ... + (\mu-\lambda)x'\partial' + \lambda. \tag{7.177}$$

It is readily seen that the zero order term w_0 of $L' = \sum_i w_i\partial'^i$ coincides with the zero order term of $\overline{L}$, that is, $w_0 = \lambda$. The first order term w_1 of $\overline{L}$ is calculated from $w_1 = (L' - \lambda)x' = (\overline{L} - \lambda)\phi = (\mu - \lambda)x'$. Identifying $w_N = \phi_x^N$, $w_{N-1} = v_{N-1}\phi_x^{N-1} + \frac{1}{2}N(N-1)\phi_x^{N-2}\phi_{xx}, ..., w_1 = (\mu - \lambda)x', w_0 = \lambda$, the operator L' is of the form

$$k = 2: \quad L' = w_N\partial'^N + w_{N-1}\partial'^{N-1} + ... + w_2\partial'^2 + \lambda_1 x'\partial' + \lambda_2. \tag{7.178}$$

If $\overline{L}$ satisfies the $k = 1$ hierarchy, then according to Theorem 7.3 (ii) the operator L' satisfies the $k = 2$ hierarchy in the coordinates x'. Thus we have the so called *reciprocal transformation* from the fields $(v_{N-1}, ..., v_1)$ to the fields $(w_N, ..., w_2)$ satisfying the isospectral hierarchies associated with the operators $\overline{L}$ and L'. This transformation involves the eigenfunction ϕ of $\overline{L}$, which can be eliminated using the additional constraint $(\overline{L}\phi)_0 = \mu\phi$.

We demonstrate this for the simplest case $N = 2$. Starting with the operator $\overline{L} = \partial^2 + 2v\partial$ giving rise to the modified KdV hierarchy one introduces $x' = \phi$, where ϕ is of the special form $(\overline{L}\phi)_0 = 0$, and calculates the linked operator $L' = \phi_x^2\partial'^2$. Hence, the field $w = \phi_x$ satisfies the hierarchy of the Harry–Dym equation (7.103) in the coordinate x' if v satisfies the modified KdV hierarchy and ϕ satisfies the corresponding linear equations. The transformation is encoded in the linking relation

$$x' = \phi, \quad w = \phi_x, \quad \phi_{xx} + 2v\phi_x = 0, \tag{7.179}$$

from which the eigenfunction can be eliminated. Thus, the well known reciprocal links [185] between the hierarchies of the modified KdV and the Harry–Dym equation are found to be characterized by

$$\frac{\partial x}{\partial x'} = \frac{1}{w}, \quad t'_q = t_q, \quad 2v = -w_{x'}, \tag{7.180}$$

mapping solutions of the HD hierarchy to solutions of the MKdV hierarchy.

Finally, we want to remark that for the Lax operators considered in this section, other types of gauge transformations and reciprocal links may be established as well. As these results go beyond the scope of this book we refer the reader to the literature [108],[152].

As discussed in the previous section the equation (7.91) originates from a Hamiltonian concept making use of the algebraic properties of the $\mathcal{R}$-matrix. Hence, the Lax equations (7.91) as evolutionary systems on the algebra g are Hamiltonian. In fact, they are multi-Hamiltonian according to representations (7.66)–(7.68) involving the three natural brackets (7.62)–(7.64). Both the linear and cubic brackets will be Poisson structures for any $\mathcal{R}$-matrix satisfying the modified Yang–Baxter equation (7.56). As the $\mathcal{R}$-matrices of this section

$$\mathcal{R}_k = \frac{1}{2}(P_{\geq k} - P_{<k}), \qquad k = 0, 1, 2 \tag{7.181}$$

stem from Lie algebra decompositions of the algebra of pseudo-differential symbols, they do indeed satisfy the Yang–Baxter condition and hence lead to two Hamiltonian formulations involving Θ_1 and Θ_3. The quadratic bracket is more delicate. As we know a sufficient condition for this bracket to be a Poisson structure involves the skew-symmetric part $\widetilde{\mathcal{R}} = \frac{1}{2}(\mathcal{R} - \mathcal{R}^*)$ of the $\mathcal{R}$-matrix. In particular, if this skew-symmetric part again satisfies the modified Yang–Baxter equation, then this will guarantee the Poisson properties of the quadratic bracket. For the case $k = 0$ the $\mathcal{R}$-matrix $\mathcal{R}_0$ turns out to be skew symmetric. Hence $\widetilde{\mathcal{R}}_0 = \mathcal{R}_0$ thus $\widetilde{\mathcal{R}}_0$ satisfies the Yang–Baxter relation and leads to the quadratic Poisson structure. For the cases $k = 1$ and $k = 2$ the $\mathcal{R}$-matrices are no longer skew-symmetric. Even worse, as $\widetilde{\mathcal{R}}_1$ and $\widetilde{\mathcal{R}}_2$ do not satisfy the Yang–Baxter relation, in these cases the quadratic bracket does not define a Poisson bracket for generic points in the algebra. As a consequence, on the general abstract level, for the cases $k = 1$ and $k = 2$ the Poisson brackets associated with Θ_1 and Θ_3 are not compatible. For the case $k = 1$ we will overcome this difficulty by constructing new quadratic Poisson tensors.

For a detailed discussion of the reduction properties of the tensors (7.66)–(7.68) we first rewrite them in terms of the particular $\mathcal{R}$-matrices (7.181) under consideration here. The transpose $\mathcal{R}_k^*$ entering the definitions are given by $\mathcal{R}_k^* = P_{\geq k}^* - P_{<k}^*$, where $P_{\geq k}^*$ and $P_{<k}^*$ are the projections to the dual subalgebras given by the dual decomposition $g^* = g = g_{\geq k}^* \oplus g_{<k}^*$ with

$$g_{<k}^* = \left\{ \sum_{i \geq -k} \partial^i u_i \right\}, \qquad g_{\geq k}^* = \left\{ \sum_{i < -k} \partial^i u_i \right\}. \tag{7.182}$$

This dual decomposition can be verified observing that $(g_{<k}^*, g_{\geq k}) = (g_{\geq k}^*, g_{<k})$ $= 0$, when using the trace duality (7.86),(7.87). Thus, the tensors are defined in terms of the four projections $P_{\geq k}, P_{<k}, P_{\geq k}^*, P_{<k}^*$. Parametrizing each element $A \in /,g$ by $A = \sum_{i \geq 0}^{N} a_i \partial^i + \sum_{i < 0} \partial^i a_i$ these projections are given by

$P_{\geq k}(A) = (A)_{\geq k}$, $P_{<k}(A) = (A)_{<k}$, $P^*_{\geq k}(A) = (A)_{<-k}$, $P^*_{<k}(A) = (A)_{\geq -k}$.
For example $P^*_{<1}(A) = a_N \partial^N + ... + a_1 \partial + a_0 + \partial^{-1} a_{-1}$. Using $P_{\geq k} + P_{<k} = P^*_{\geq k} + P^*_{<k} = 1$ we can either eliminate $P_{<k} = 1 - P_{\geq k}, P^*_{\geq k} = 1 - P^*_{<k}$ or $P^*_{<k} = 1 - P^*_{\geq k}, P_{\geq k} = 1 - P_{<k}$ and obtain two equivalent representations for each of the tensors (7.66)–(7.68)

$$\Theta_1(L)\nabla H = -ad_L(\nabla H)_{\geq k} + (ad_L \nabla H)_{\geq -k}$$

$$= ad_L(\nabla H)_{<k} - (ad_L \nabla H)_{<-k},$$

$$\Theta_2(L)\nabla H = -ad_L(ad_L^+ \nabla H)_{\geq k} + ad_L^+(ad_L \nabla H)_{\geq -k}$$

$$= ad_L(ad_L^+ \nabla H)_{<k} - ad_L^+(ad_L \nabla H)_{<-k},$$

$$\Theta_3(L)\nabla H = -ad_L(L\nabla H L)_{\geq k} + L(ad_L \nabla H)_{\geq -k}L$$

$$= ad_L(L\nabla H L)_{<k} - L(ad_L \nabla H)_{<-k}L. \tag{7.183}$$

It turns out that for each tensor the first representation yields direct access to the lowest differential order of the operators $\Theta_i(L)\nabla H$, $i = 1, 2, 3$, whereas the second representation yields information about the highest orders present.

The Case of $k = 0$.

Let us consider operators $L \in g^*_{<0} = g_{\geq 0}$ of the form

$$L = \partial^N + u_{N-2}\partial^{N-2} + u_{N-3}\partial^{N-3} + ... + u_1 \partial + u_0 \tag{7.184}$$

with dynamical fields $u_{N-2}, ..., u_0$. These are the Gelfand-Dikii spectral problems, for which some of the Lax dynamics and related evolution equations were exhibited in Example 7.1. Inserting such an operator into (7.183) it becomes clear from the first representation of the linear tensor that $\Theta_1(L)\nabla H$ will be a purely differential operator. From the second representation it is evident that the highest differential order will be at most $N - 2$. Hence, for L of the form (7.184) the Hamiltonian vector field $\Theta_1(L)\nabla H$ is tangent to the space of operators (7.184) spanned by the coordinates $u_{N-2}, ..., u_0$. As a result, this space is a proper Poisson subspace of the linear bracket on g. Because of the form (7.184) of the L operator, the linear Poisson tensor (7.183) takes the form

$$\Theta_1(L) : \nabla H \to [L, \nabla H]_{\geq 0} \tag{7.185}$$

as $\nabla H = \partial^{-N+1}\gamma_{N-2} + ... + \partial^{-1}\gamma_0$. The quadratic tensor defines a Poisson bracket on g, as the underlying $\mathcal{R}$-matrix is skew-symmetric for the present case $k = 0$. From the first representation in (7.183) it is clear that Θ_2 yields a purely differential operator, if L is a purely differential operator. Hence, the quadratic bracket can be properly restricted to the subalgebra $g_{\geq 0}$ of differential operators. Inserting an operator L of the form (7.184) into the second representation in (7.183) one sees that the highest differential order of the Hamiltonian vector field $\Theta_2(L)\nabla H$ will be $N - 1$. Hence, operators of the form (7.184) spanned by the variables $u_{N-2}, ..., u_0$ do not form a proper

Poisson subspace. Instead, the quadratic bracket can be properly restricted to a subspace of the form

$$L = \partial^N + u_{N-1}\partial^{N-1} + u_{N-2}\partial^{N-2} + u_{N-3}\partial^{N-3} + ... + u_1\partial + u_0 \quad (7.186)$$

spanned by the coordinates $u_{N-1}, ..., u_0$. Again, for the particular form (7.186) of L, the quadratic Poisson tensor (known as the Adler map), up to an irrelevant factor, takes the form

$$\Theta_2(L) : \nabla H \to (L\nabla H)_{\geq 0}L - L(\nabla HL)_{\geq 0}. \quad (7.187)$$

As the space (7.184) sits inside (7.186), it is simple to invoke Dirac reduction to reduce the quadratic bracket on (7.186) by the constraint $u_{N-1} = 0$. The reduced structure is given by the Poisson tensor

$$\Theta_2^{(red)}(L) : \nabla H \to (L\nabla H)_{\geq 0}L - L(\nabla HL)_{\geq 0} + \frac{1}{N}[D_x^{-1}\mathrm{res}([\nabla H, L]), L], \quad (7.188)$$

where res represents the usual residue, i.e. the coefficient of the differential order ∂^{-1}. The reduced bracket is again local, as the residue of a commutator is in the image of the differential operator [4]. It provides the second Hamiltonian structure of the Gelfand–Dikii hierarchies.

The Suris result presented in Theorem 7.2 leads to a direct construction of (7.188) with

$$A_1 : \xi \to \xi_{\geq 0} - \xi_{<0} - \frac{1}{N}D_x^{-1}\mathrm{res}(\xi), \quad S = \frac{1}{N}D_x^{-1}\mathrm{res}(\xi),$$

$$A_2 : \xi \to \xi_{\geq 0} - \xi_{<0} + \frac{1}{N}D_x^{-1}\mathrm{res}(\xi), \quad S = -\frac{1}{N}D_x^{-1}\mathrm{res}(\xi). \quad (7.189)$$

The tensor $\overline{\Theta}_2(L)$ produces the reduced Poisson structure (7.188). Moreover one has

$$\mathcal{R} = A_1 + S = A_2 + S^* : \xi \to \xi_{\geq 0} - \xi_{<0}, \quad (7.190)$$

where the $\mathcal{R}$−matrix $\mathcal{R}$ obviously satisfies the modified YB(1) equation, since it comes from a Lie algebra decomposition. The assumptions (7.71) are easily checked.

For the cubic tensor the first representation in (7.183) shows that Θ_3 yields a purely differential operator if L is a purely differential operator. Hence, also the cubic bracket can be properly restricted to the subalgebra $g_{\geq 0}$ of differential operators. Inserting an operator L of the form (7.184) into the second representation in (7.183) one finds that the highest differential order of the Hamiltonian vector field $\Theta_3(L)\nabla H$ will be $2N - 1$. Hence, operators of the form (7.184) spanned by the variables $u_{N-2}, ..., u_0$, do not form a proper Poisson subspace. In fact, there are no obvious proper Poisson subspaces for the cubic bracket, apart from the trivial case of the first order operators. Nevertheless, Dirac reduction can be invoked to restrict the cubic bracket on the differential operators to the subspace of the form (7.184). As was demonstrated in [150] the Dirac reduction of the cubic bracket leads to $\Theta_3 = \frac{1}{4}\Theta_2\Theta_1^{-1}\Theta_2$, involving the recursion operator $\Theta_2\Theta_1^{-1}$ of the bi-Hamiltonian formulation given by Θ_1 and Θ_2.

Example 7.7 The KdV Hamiltonian Lax dynamics.
The first Lax Hamiltonian structure reads

$$L_t = \Theta_1(L)\nabla H,$$

where $L = \partial^2 + u$, $\nabla H = \partial^{-1}\frac{\delta H}{\delta u}$ (compare with (7.88)) and $\Theta_1(L)$ is given by formula (7.185). Thus, applying the general Leibniz rule (7.85), we obtain

$$
\begin{aligned}
L_t = 1_t\partial^2 + u_t &= \left[\partial^2 + u, \left(\frac{\delta H}{\delta u}\right)\partial^{-1} - \left(\frac{\delta H}{\delta u}\right)_x\partial^{-2} + \dots\right]_{\geq 0} \\
&= 2\left(\frac{\delta H}{\delta u}\right)_x + \left(\frac{\delta H}{\delta u}\right)\partial - \left(\frac{\delta H}{\delta u}\right)_x - \left(\frac{\delta H}{\delta u}\right)\partial + \left(\frac{\delta H}{\delta u}\right)_x \\
&= 0\cdot\partial^2 + 2\left(\frac{\delta H}{\delta u}\right)_x \Leftrightarrow u_t = 2D\frac{\delta H}{\delta u},
\end{aligned}
$$

which, up to a constant factor, covers the first Hamiltonian structure of the field KdV representation.
The second Lax Hamiltonian structure is given by

$$L_t = \Theta_2^{(\mathrm{red})}(L)\nabla H,$$

where now $\Theta_2^{(\mathrm{red})}(L)$ is given by (7.188). The reader can verify that

$$(L\nabla H)_{\geq 0} = \left(\frac{\delta H}{\delta u}\right)_x + \left(\frac{\delta H}{\delta u}\right)\partial,$$

$$(\nabla H L)_{\geq 0} = -\left(\frac{\delta H}{\delta u}\right)_x + \left(\frac{\delta H}{\delta u}\right)\partial,$$

$$(L\nabla H)_{\geq 0}L - L(\nabla H L)_{\geq 0} = \left(\frac{\delta H}{\delta u}\right)_{xx}\partial + 2u\left(\frac{\delta H}{\delta u}\right)_x + u_x\left(\frac{\delta H}{\delta u}\right) + \left(\frac{\delta H}{\delta u}\right)_{3x},$$

$$\mathrm{res}[\nabla H, L] = \left(\frac{\delta H}{\delta u}\right)_{xx},$$

$$\frac{1}{2}\left[\left(\frac{\delta H}{\delta u}\right)_x, L\right] = -\frac{1}{2}\left(\frac{\delta H}{\delta u}\right)_{3x} - \left(\frac{\delta H}{\delta u}\right)_{xx}\partial.$$

Hence,

$$L_t = \Theta_2^{(\mathrm{red})}(L)\nabla H \Leftrightarrow u_t = \left(\frac{1}{2}D^3 + 2uD + u_x\right)\frac{\delta H}{\delta u},$$

which, up to a constant factor, covers the second Hamiltonian structure of the field KdV representation.

Example 7.8 The Boussinesq Hamiltonian Lax dynamics.

The Lax operator $L = \partial^3 + u\partial + v$ gives rise to the Boussinesq equation (7.114) via $L_t = [(L^{\frac{2}{3}})_{\geq 0}, L]$. With $\nabla H = \partial^{-1}\frac{\delta H}{\delta v} + \partial^{-2}\frac{\delta H}{\delta u}$ the linear and quadratic structures (7.187) and (7.188) produce, up to a constant factor, the well known Boussinesq Hamiltonian operators

$$\theta_0 = \begin{pmatrix} 0 & D \\ D & 0 \end{pmatrix},$$

$$\theta_1 =$$
$$\begin{pmatrix} 2D^3 + Du + uD & -D^4 - D^2u + 2Dv + vD \\ D^4 + uD^2 + Dv + 2vD & -\frac{2}{3}(D^5 + D^3u + uD^3 + uDu) + D^2v - vD^2 \end{pmatrix}.$$

The rider can find two Poisson brackets for the restriction (7.167), as well as many particular examples (including AKNS), in [151].

The Case of k = 1

Let us consider operators $L \in g^*_{<1}$ of the form

$$L = c_N\partial^N + u_{N-1}\partial^{N-1} + u_{N-2}\partial^{N-2} + u_{N-3}\partial^{N-3} + ... + u_1\partial + u_0 + \partial^{-1}u_{-1} \tag{7.191}$$

with fixed c_N. These are the spectral problems for which some of the related dynamical systems were exhibited previously. Inserting such operators into (7.183), one finds from the first representation for the linear tensor ($k = 1$) that they do again lie in $g^*_{<1}$. From the second representation one finds that the highest differential order will be at most $N - 1$. Hence, for L of the form (7.189), according to (7.183) and the fact that $\nabla H = \partial^{-N}\gamma_{N-1} + ... + \gamma_{-1}$, the Hamiltonian vector field

$$\Theta_1(L) : \nabla H \to [L, \nabla H]_{\geq -1} \tag{7.192}$$

is tangent to the affine spaces of operators (7.189) spanned by the coordinates $u_{N-1}, ..., u_{-1}$. As a result, these spaces are proper Poisson subspaces of the linear bracket (7.192) on g. Upon the reduction to $u_{-1} = 0$ one loses the Hamiltonian structure. However, Dirac reduction will lead to Hamiltonian operators for the equations associated with the second admissible reduction to $u_{-1} = u_0 = 0$, i.e. when

$$L = c_N\partial^N + u_{N-1}\partial^{N-1} + u_{N-2}\partial^{N-2} + u_{N-3}\partial^{N-3} + ... + u_1\partial. \tag{7.193}$$

The quadratic tensor Θ_2 in (7.67) does not define a Poisson bracket on g for the present choice $k = 1$, as it is neither skew-symmetric nor does its skew-symmetric part satisfy the Yang–Baxter equation. Yet, it was pointed out in [108] that this tensor is still capable of producing suitable Poisson structures via Dirac reduction. We will check it for some of our examples. However, it is possible to transform the tensor Θ_2 into a proper quadratic Poisson structure for the case $k = 1$ simply by adding some extra terms.

Theorem 7.6 *The tensor*

$$\tilde{\Theta}_2(L) : \nabla H \rightarrow (L\nabla H)_{\geq 1}L - L(\nabla HL)_{\geq 0} + L(L\nabla H)_0$$

$$- \partial^{-1}\mathrm{res}([\nabla H, L])L + [D_x^{-1}\mathrm{res}([\nabla H, L]), L] \qquad (7.194)$$

defines a Poisson bracket on g *which is compatible with the linear tensor (7.183). It admits a proper restriction to Lax operators of the form (7.191).*

The proof can be found in [151]. Notice that it is a special case of the general quadratic tensor $\bar{\Theta}_2$ (7.70) with

$$A_1 : \xi \rightarrow \xi_{\geq 1} - \xi_0 + \xi_{-1} - \xi_{<-1} - 2D_x^{-1}(\mathrm{res}\,\xi),$$

$$A_2 : \xi \rightarrow \xi_{\geq 0} - \xi_{<0} + 2D_x^{-1}(\mathrm{res}\,\xi),$$

$$S : \xi \rightarrow -2\xi_{-1} + 2D_x^{-1}(\mathrm{res}\,\xi), \quad S^* : \xi \rightarrow -2\xi_0 - 2D_x^{-1}(\mathrm{res}\,\xi). \qquad (7.195)$$

We note that
$$\mathcal{R} = A_1 + S = A_2 + S^* : \xi \rightarrow \xi_{\geq 1} - \xi_{<1},$$

so, according to Theorem 7.3, only conditions (7.71) need to be checked.

No proper restriction to Lax operators of the form (7.193) is possible. Hence, Dirac reduction has to be invoked to obtain the quadratic bracket associated with the constraint $u_{-1} = u_0 = 0$. Fortunately, this reduction process can be carried out formally and yields a local Poisson structure which is naturally extended to the whole of g [153].

Theorem 7.7 *The tensor*

$$\hat{\Theta}_2(L) : \nabla H \rightarrow (L\nabla H)_{\geq 0}L - L(\nabla HL)_{\geq 1} - (\nabla HL)_0 L$$

$$- L\partial^{-1}\mathrm{res}([\nabla H, L]) - [D_x^{-1}\mathrm{res}([\nabla H, L]), L] \qquad (7.196)$$

defines a Poisson bracket on g *which is compatible with the linear tensor (7.183). It admits a proper restriction to Lax operators of the form (7.193), where it coincides with the Dirac reduction of (7.194).*

Both the Poisson property and the reduction result are checked by a straightforward, yet lengthy and cumbersome computation. The compatibility with (7.192) follows from the usual deformation argument that $L \rightarrow L + \epsilon \mathbf{1}$ produces the linear tensor as the term linear in ϵ.

The cubic tensor in (7.68) indeed defines a Poisson bracket, so that the following considerations are well motivated. An analysis similar to the case of the quadratic bracket shows that only for spectral operators of the restricted class (7.193) in $g_{\geq 1}$ will the operator $\Theta_3(L)$ $(k = 1)$ take values in $g_{\geq 1} \subset g_{<1}^*$. Thus, the cubic bracket can be properly restricted to the subalgebra $g_{\geq 1} \subset g_{<1}^*$. Considering the second representation for Θ_3 in (7.183), one finds that the highest differential order of $\Theta_3(L)\nabla H$ will be $2N - 1$. Hence, there is only one simple Poisson subspace for the cubic bracket in the class (7.193), given by the second order operators of the form $L = v\partial^2 + u\partial$. We insert $\nabla H = \partial^{-2}\frac{\delta H}{\delta v} + \partial^{-3}\frac{\delta H}{\delta u}$ and $L_t = \Theta_3(L)\nabla H$ translates to

$$
\begin{pmatrix} u \\ v \end{pmatrix}_t = 2 \begin{pmatrix} uvD^2v - vD^2vu & -vD^2v^2 + Duv^2 - 2uDv^2 \\ v^2D^2v + uv^2D - 2v^2Du & -v^3D - Dv^3 \end{pmatrix} \delta H.
$$
(7.197)

For $N > 2$ Dirac reduction can be performed to obtain further Hamiltonian formulations of hierarchies associated with isospectral operators of the form (7.193).

Example 7.9 The Kaup–Broer system.
According to Theorems 7.6 and 7.7 the bi-Hamiltonian structure of the field hierarchy associated with the operator $L = \partial + u + \partial^{-1}v$ can be obtained from tensors (7.192) and (7.194) . Indeed, with $\nabla H = \frac{\delta H}{\delta v} + \partial^{-1}\frac{\delta H}{\delta u}$ and the relation $\partial^{-1}a = a\partial^{-1} - \partial^{-1}a_x\partial^{-1}$, one derives the Hamiltonian operators

$$
\theta_0(u,v) : \begin{pmatrix} \delta H/\delta u \\ \delta H/\delta v \end{pmatrix} \rightarrow \begin{pmatrix} 0 & D \\ D & 0 \end{pmatrix} \begin{pmatrix} \delta H/\delta u \\ \delta H/\delta v \end{pmatrix},
$$
(7.198)

and

$$
\theta_1(u,v) : \begin{pmatrix} \delta H/\delta u \\ \delta H/\delta v \end{pmatrix} \rightarrow \begin{pmatrix} 2D & D^2 + Du \\ -D^2 + uD & Dv + vD \end{pmatrix} \begin{pmatrix} \delta H/\delta u \\ \delta H/\delta v \end{pmatrix}
$$
(7.199)

for the Kaup–Broer system (7.118). As this equation is obtained from $L_t = [(L^2)_{\geq 1}, L] = \Theta_1(L)\nabla tr(L^3)/6$, the Hamiltonian function for (7.118) is calculated as $H = tr(L^3)/6 = \int_R(u_xv + v^2 + u^2v)\,dx$ using (7.86). We had observed previously that the Kaup–Broer system can be restricted to Burgers equation (7.119) by the constraint $u_{-1} = v = 0$. However, the Dirac reduction cannot be invoked, hence, although Burgers equation is a restriction of the Hamiltonian equation (7.118), the Hamiltonian structure is lost in the reduction. Also, the Hamiltonian function $H = tr(L^3)/6$ becomes trivial for $v = 0$.

Example 7.10 For $N = 2$ we consider the spectral operator $L = \partial^2 + u\partial + v + \partial^{-1}w$. The linear Poisson structure (7.192) translates to

$$
\begin{pmatrix} u \\ v \\ w \end{pmatrix}_t = \begin{pmatrix} 0 & 0 & 2D \\ 0 & 2D & uD + D^2 \\ 2D & Du - D^2 & 0 \end{pmatrix} \begin{pmatrix} \delta H/\delta u \\ \delta H/\delta v \\ \delta H/\delta w \end{pmatrix}.
$$
(7.200)

This is the Hamiltonian formulation of equations (7.121) and (7.124). The Hamiltonian functionals can be calculated as $H = \frac{1}{2}tr(L^2)$ for (7.121) and $H = \frac{2}{5}tr(L^{\frac{5}{2}})$ for (7.124), that is for example $\frac{1}{2}tr(L^2) = \int_R vw\,dx$. Again, in the reduction to $w = 0$ we will lose the Hamiltonian structure of equation (7.122). However, in the reduction $(w = 0, v = 0)$ to the modified KdV equation (7.125) we can use the Dirac reduction on (7.200) to obtain

$$u_t = - \begin{pmatrix} 0 & 2D \end{pmatrix} \begin{pmatrix} 2D & uD + D^2 \\ Du - D^2 & 0 \end{pmatrix}^{-1} \begin{pmatrix} 0 \\ 2D \end{pmatrix} \frac{\delta H}{\delta u}$$

as a formal Hamiltonian formulation. With

$$\begin{pmatrix} 2D & uD + D^2 \\ Du - D^2 & 0 \end{pmatrix}^{-1}$$

$$= \begin{pmatrix} 0 & (u - D)^{-1}D^{-1} \\ D^{-1}(u + D)^{-1} & -2D^{-1}(u + D)^{-1}D(u - D)^{-1}D^{-1} \end{pmatrix}$$

one obtains a formal Hamiltonian formulation

$$u_t = 8(u + D)^{-1}D(u - D)^{-1}\frac{\delta H}{\delta u} \tag{7.201}$$

for the modified KdV equation (7.125). This, in fact, corresponds to the formal Poisson operator $\Phi^{-1} \circ \theta_0 = \theta_0 \circ \theta_1^{-1} \circ \theta_0$ obtained from the recursion operator $\Phi = \theta_1 \circ \theta_0^{-1}$. Here $\theta_0 = -D$ and $\theta_1 = D^3 - DuD^{-1}uD = D(u - D)D^{-1}(u + D)D$ are two Poisson operators constituting the well know bi-Hamiltonian formulation of the modified KdV. These operators are identified by the quadratic bracket and the reduction of the cubic bracket. Actually, the quadratic tensor of Theorem 7.7 produces the Poisson operator $-D$, while the Dirac reduction of tensor (7.197) to $v = 1$ yields the second Poisson operator $-D^3 + DuD^{-1}uD$.

Example 7.11 The modified Boussinesq equation.
The modified Boussinesq system (7.128) is associated with the spectral operator $L = \partial^3 + u\partial^2 + v\partial$ via $L_t = [(L^{\frac{2}{3}})_{\geq 1}, L]$. The quadratic tensor of Theorem 7.7 yields the Poisson operator

$$\theta_0 = \begin{pmatrix} -6D & 3D(D - u) \\ -3(D + u)D & 2D^3 + vD + Dv + 2uD^2 - 2D^2u - 2uDu \end{pmatrix}. \tag{7.202}$$

The Case of $k = 2$

We are interested in operators $L \in g_{<2}^*$ of the form

$$L = u_N\partial^N + u_{N-1}\partial^{N-1} + \ldots + u_0 + \partial^{-1}u_{-1} + \partial^{-2}u_{-2} \tag{7.203}$$

spanned by the fields $u_N, \ldots, u_{-2}$. Inserting such operators into (7.183) it is clear from the first representation of the linear tensor that $\Theta_1(L)\nabla H$ for $k = 2$ do again lie in $g_{<2}^*$. From the second representation one finds that the highest differential order of $\Theta_1(L)\nabla H$ will be at most N. Hence, for L of the form (7.203), the Hamiltonian vector field $\Theta_1(L)\nabla H$ is tangent to the space of operators (7.203). As a result, the spaces given by these operators are proper Poisson subspaces of the linear bracket on g. For further restrictions of (7.203) (presented in examples) one loses the Hamiltonian structure. Only for the final restriction $(u_{-2} = u_{-1} = 0, u_0 = \lambda_3, u_1 = \lambda_1 + \lambda_2 x)$ will Dirac reduction lead to the Hamiltonian formulation of the equations associated with these spectral operators.

The quadratic tensor Θ_2 in (7.67) does not define a Poisson bracket on g for the present choice $k = 2$. Although, contrary to the case $k = 1$, we still do not know the proper form of the quadratic tensor, nevertheless, it turns out that Hamiltonian formulations are hidden in tensor (7.67) for certain reductions. Considering the reduction properties we first look at the second representation for Θ_2 in (7.183). Inserting L of the form (7.203), the highest differential order of the operator $\Theta_2 \nabla H$ is at most N, matching the form of the highest differential order of the ansatz (7.203). To consider the lowest differential orders we insert L into the first representation for Θ_2 in (7.183). The second and third terms do not lie in the dual subalgebra $g_{<2}^*$, from which the Lax operators (7.203) were chosen. Hence, the quadratic bracket cannot be properly restricted to $g_{<2}^*$. However, choosing L to be of the form

$$L = u_N \partial^N + u_{N-1} \partial^{N-1} + ... + u_2 \partial^2, \qquad (7.204)$$

the operator $\Theta_2(L)\nabla H$ will take values in $g_{<2}^*$, but expressions of the form $(...)\partial + (...) + (...)\partial^{-1} + (...)\partial^{-2}$ will be present. Hence, the quadratic bracket cannot be properly restricted to operators (7.204) either. Nevertheless, such properties allow application of the Dirac reduction to operators (7.204) in a simple way.

The cubic tensor in (7.68) does define a Poisson bracket, so that the following considerations are well motivated. An analysis similar to that of the case of the quadratic bracket shows that only for Lax operators of the restricted class (7.204) will the operator $\Theta_3 \nabla H$ take values in $g_{\geq 2} \subset g_{<2}^*$. Hence the cubic bracket can be properly restricted to the subalgebra $g_{\geq 2}$. Considering the second representation for Θ_3 in (7.183), one finds that the highest differential order of $\Theta_3 \nabla H$ will be $\max(N, 2N - 3)$. Hence, there is only one simple Poisson subspace for the cubic bracket in the class (7.204), given by second order operators of the form $L = u^2 \partial^2$. In this case, we insert $\nabla H = \partial^{-3} \frac{1}{2u} \frac{\delta H}{\delta u}$ and $L_t = \Theta_3(L)\nabla H$ translates to the Hamiltonian formulation

$$u_t = u^2 D u^2 \frac{\delta H}{\delta u} \qquad (7.205)$$

of the Harry–Dym hierarchy. For $N > 2$ Dirac reduction can be performed to obtain further Hamiltonian formulations of the field hierarchies associated with the spectral Lax operators of the form (7.204). The reader can find examples of various Dirac reductions for the case $k = 2$ in [108].

Finally in this section let us briefly mention the possible extension of the operator algebra (7.83) to include the one built over the noncommutative matrix valued dynamical variables [37]. In fact, let

$$g = \left\{ L = \sum_{-\infty < i}^{N} U_i(x)\partial \right\}, \tag{7.206}$$

where now $\{(U_i)_{kl}\}$ are $n \times n$ matrices. The appropriate trace form reads

$$\mathrm{tr}(L) = \int_R \mathrm{Tr} U_{-1} \mathrm{d}x, \tag{7.207}$$

where $\mathrm{Tr} = $ matrix *trace*, and the gradients $\nabla H(L)$ are conveniently parametrized by

$$\nabla H(L) = \sum_i \partial^{-1-i} \frac{\delta H}{\delta U_i}, \quad \left\{ \frac{\delta H}{\delta U_i} \right\}_{kl} = \frac{\delta H}{\delta (U_i)_{lk}}. \tag{7.208}$$

The noncommutativity of the basic fields U_i induces some changes in the form of admissible Lax operators L. First of all one finds

$$\frac{\mathrm{d}}{\mathrm{d}t_q} L = -[(L^q)_{<k}, L] = -[A_{k-1}\partial^{k-1} + \mathrm{lower}, U_N\partial^N + \mathrm{lower}]$$

$$= [U_N, A_{k-1}]\partial^{N+k-1} + \mathrm{lower}, \tag{7.209}$$

hence, only for the cases $k = 0, 1$ does the form of the commutator match the form (7.206) of the Lax operator, i.e.

$$k = 0: \quad L = C_N\partial^N + U_{N-1}\partial^{N-1} + \dots + U_0 + U_{-1}\partial^{-1} + \dots,$$

$$k = 1: \quad L = U_N\partial^N + U_{N-1}\partial^{N-1} + \dots + U_0 + U_{-1}\partial^{-1} + \dots, \tag{7.210}$$

where C_N is an arbitrary constant coefficient matrix different from a unit matrix. The basic restrictions to a finite number of fields are the following

$$k = 0: \quad L = C_N\partial^N + U_{N-1}\partial^{N-1} + \dots + U_0, \tag{7.211}$$

$$k = 1: \quad L = U_N\partial^N + U_{N-1}\partial^{N-1} + \dots + U_0 + \partial^{-1}U_{-1}. \tag{7.212}$$

Passing to the Hamiltonian representation we consider the case $k = 0$. One can readily verify that for both, the linear Poisson structure (7.185) and the quadratic Poisson structure (7.187), the restricted operators (7.211) form the proper Poisson subspaces of g. Let us illustrate it on the simplest example of the first order differential operator (the generalized Zakharov–Shabat spectral problem)

$$L = C\partial + U, \tag{7.213}$$

where C is an $n \times n$ constant matrix and U is an $n \times n$ matrix of dynamical variables. The pair of compatible Poisson operators (7.185) and (7.187), where $\nabla H(L) = \partial^{-1}\delta H/\delta U$, takes the following form

$$\theta_1 = -\frac{1}{2}\mathrm{ad}_C, \quad \theta_2 = -\lambda_C \rho_C D_x - U\rho_C + C\rho_U, \tag{7.214}$$

when it acts on the gradient matrix $\delta H/\delta U$, where λ_A and ρ_A denote the operators of left multiplication and right multiplication by $A : \lambda_A F = AF$, $\rho_A F = FA$. Notice that for $C = I$, $\theta_1 = 0$ and $\theta_2 = -D_x + \mathrm{ad}_U$. We perform detailed calculations for the case $n = 2$. Let

$$C = \mathrm{i}\sigma_3, \quad U = \begin{pmatrix} a(x) & b(x) \\ c(x) & d(x) \end{pmatrix}, \tag{7.215}$$

where $\sigma_3 = \mathrm{diag}(1, -1)$ is the third Pauli matrix. Then we have

$$\frac{\delta H}{\delta U} = \begin{pmatrix} \delta H/\delta a & \delta H/\delta c \\ \delta H/\delta b & \delta H/\delta d \end{pmatrix} := \begin{pmatrix} \gamma_a & \gamma_c \\ \gamma_b & \gamma_d \end{pmatrix}, \tag{7.216}$$

and hence the first Hamiltonian representation reads

$$U_t = -\frac{1}{2}\mathrm{ad}_C \frac{\delta H}{\delta U} \Leftrightarrow \begin{pmatrix} a & b \\ c & d \end{pmatrix}_t = \mathrm{i}\begin{pmatrix} 0 & -\gamma_c \\ \gamma_b & 0 \end{pmatrix}. \tag{7.217}$$

So $a_t = d_t = 0$ and for $b = -\psi, c = \varphi$ one finds

$$\begin{pmatrix} \psi \\ \varphi \end{pmatrix}_t = \begin{pmatrix} 0 & \mathrm{i} \\ -\mathrm{i} & 0 \end{pmatrix}\begin{pmatrix} \delta H/\delta\psi \\ \delta H/\delta\varphi \end{pmatrix}, \tag{7.218}$$

which is exactly the first Hamiltonian representation of the AKNS equation (7.170).

Analogously, the second Hamiltonian representation reads

$$U_t = \theta_2 \frac{\delta H}{\delta U}$$

$$\Updownarrow$$

$$\begin{pmatrix} a & b \\ c & d \end{pmatrix}_t$$

$$= \mathrm{i}\begin{pmatrix} c\gamma_c - b\gamma_b - \mathrm{i}\gamma_{ax} & (a+d)\gamma_c + b(\gamma_a + \gamma_d) - \mathrm{i}\gamma_{cx} \\ -(a+d)\gamma_b - c(\gamma_a + \gamma_d) - \mathrm{i}\gamma_{bx} & c\gamma_c - b\gamma_b - \mathrm{i}\gamma_{ax} \end{pmatrix} \tag{7.219}$$

Under the constraint $a_t = d_t = 0$ one finds

$$\gamma_d = \gamma_a = \mathrm{i}D_x^{-1}b\gamma_b - \mathrm{i}D_x^{-1}c\gamma_c \tag{7.220}$$

and (7.219) turns into the form

$$\begin{pmatrix} b \\ c \end{pmatrix}_t = \begin{pmatrix} -2bD_x^{-1}b & D_x + 2bD_x^{-1}c + \mathrm{i}(a+d) \\ D_x + 2cD_x^{-1}b - \mathrm{i}(a+d) & -2cD_x^{-1}c \end{pmatrix}\begin{pmatrix} \gamma_b \\ \gamma_c \end{pmatrix}. \tag{7.221}$$

Now, substituting $a = d = 0$, $b = -\psi$ and $c = \varphi$, one finds

$$\begin{pmatrix} \psi \\ \varphi \end{pmatrix}_t = \begin{pmatrix} -2\psi D_x^{-1}\psi & -D_x + 2\psi D_x^{-1}\varphi \\ -D_x + 2\varphi D_x^{-1}\psi & -2\varphi D_x^{-1}\varphi \end{pmatrix}\begin{pmatrix} \delta H/\delta\psi \\ \delta H/\delta\varphi \end{pmatrix}, \tag{7.222}$$

i.e. the second Hamiltonian formulation of the AKNS system.

7.4 Multi-Hamiltonian Dynamics of Shift Lax Operators

In this section we develop a theory similar to that constructed in Sect. 3.3, but now we shall deal with lattice systems. We demonstrate that the $\mathcal{R}$-matrix formulation of the lattice system theory exhibits new features in spite of numerous similarities to the previous case.

Consider the space of suitable discrete fields $\{u : \mathbb{Z} \to \mathbb{R}\}$ and also the algebra g of shift operators of the form:

$$L = \sum_{i<\infty} u_i(n)\mathcal{E}^i \tag{7.223}$$

where $\mathcal{E}$ is the shift operator, satisfying the simple commutation rule

$$\mathcal{E}^\alpha u = (E^\alpha u)\mathcal{E}^\alpha \tag{7.224}$$

This is clearly an associative algebra, so it fits the framework of Sect. 3.2. Now we have to define the suitable trace form on g.

Lemma 7.11 *The trace form*

$$tr(L) \equiv tr\left(\sum_i u_i(n)\mathcal{E}^i\right) = \sum_{n=-\infty}^{+\infty} u_0(n) \tag{7.225}$$

yields a symmetric pairing on g $: (L_1, L_2) = tr(L_1 L_2)$.

Proof. Similarly to the continuous case, it is sufficient to prove (7.225) for $L_1 = A\mathcal{E}^k$, $L_2 = B\mathcal{E}^l$. From the commutation rule (7.224) it follows that

$$L_1 L_2 = A(E^k B)\mathcal{E}^{k+l},$$

$$L_2 L_1 = B(E^l A)\mathcal{E}^{k+l}.$$

But the zero order term of a commutator is in the image of the difference operation $\Delta = E - 1$

$$[L_1, L_2]_0 = A(E^k B) - B(E^{-k}A) = (E^k - 1)(E^{-k}A)B$$

$$= (E-1)(1 + E + ... + E^{k-1})(E^{-k}A)B = (E-1)F(A, B).$$

Consequently

$$tr(L_1 L_2 - L_2 L_1) = \sum_{n=-\infty}^{n=+\infty} [L_1, L_2] = \sum_{n=-\infty}^{n=+\infty} \Delta F(A, B) = 0.\square$$

Let us notice that L as given by (7.223) is uniquely determined through the set of coefficient fields u_i. Hence, for operators $L = \sum_{i<\infty} u_i(n)\mathcal{E}^i$, the vector fields $\frac{\mathrm{d}}{\mathrm{d}t}L$ and gradients $\nabla H(L)$ are conveniently parametrized by

$$\frac{\mathrm{d}}{\mathrm{d}t}L = \sum_i u_{i_t}\mathcal{E}^i, \qquad \nabla H(L) = \sum_i \mathcal{E}^{-i}\frac{\delta H}{\delta u_i}. \tag{7.226}$$

Now, exactly as before, we use the splitting of g into subalgebras to find natural $\mathcal{R}$-matrices in g. Put

$$g_{\geq k} = \left\{\sum_{i\geq k} u_i\mathcal{E}^i\right\} \quad , \quad g_{<k} = \left\{\sum_{i<k} u_i\mathcal{E}^i\right\} \quad k = 0,1. \tag{7.227}$$

Of course, $g = g_{\geq k}\oplus g_{<k}$ in the sense of linear subspaces for any integer k. Let us find the cases when both the above subspaces are also Lie subalgebras. Consider first $g_{\geq k}$. Each element of this subspace is of the form field $\cdot \mathcal{E}^k +$ higher so the commutator of two such elements has the form field$\cdot\mathcal{E}^{2k}+$higher, so it will remain in the subspace if $2k \geq k$, which yields $k \geq 0$. Similarly, every element of $g_{<k}$ has the form field $\cdot \mathcal{E}^{k-1} +$ lower, so the commutator of two such elements attains the form field$\cdot \mathcal{E}^{2k-2} +$ lower. It will remain in this subspace if $2k - 2 < k$, which yields $k < 2$. It follows that only for $k = 0,1$ is the splitting (7.227) into subalgebras. So in the lattice case with the algebra (7.223) we have only two $\mathcal{R}$-matrices at our disposal, namely

$$\mathcal{R}_k : g \to g$$
$$\mathcal{R}_k = \tfrac{1}{2}(P_{\geq k} - P_{<k}), \tag{7.228}$$

where $k = 0,1$. Now, using the general scheme (7.79) we obtain two hierarchies of flows on g

$$L_{t_q} = [\mathcal{R}(L^q), L] = [(L^q)_{\geq k}, L] \quad , \quad k = 0,1. \tag{7.229}$$

Formulas (7.229) generate dynamics on the whole algebra g, so we now have to find restrictions on the form of L (i.e. subspaces of g) such that the vector field given by the right hand side of (7.229) is tangent to these subspaces. Let us look for the n-component restriction of the form

$$L = u_{\alpha+n}\mathcal{E}^{\alpha+n} + u_{\alpha+n-1}\mathcal{E}^{\alpha+n-1} + \ldots + u_\alpha\mathcal{E}^\alpha. \tag{7.230}$$

Consider, for example, the case $k = 0$ (the case $k = 1$ is similar). For $\alpha \geq 0$ we have $(L^q)_{\geq 0} = L^q$, so the commutator will give zero, thus trivial dynamics. For $\alpha \leq -n$, $(L^q)_{\geq 0} = $ field $\cdot \mathcal{E}^0 \equiv$ field, so the commutator will again give zero. Consequently, the only non-trivial cases are when $-n < \alpha \leq -1$. Exactly the same condition can be obtained for the case $k = 1$. So the interesting reductions have the form

$$L = u_{\alpha+n}\mathcal{E}^{\alpha+n} + u_{\alpha+n-1}\mathcal{E}^{\alpha+n-1} + \ldots + u_\alpha\mathcal{E}^\alpha \ , \quad -n < \alpha \le -1. \quad (7.231)$$

By a simple analysis, similar to the above, one can immediately check that for the case $k = 0$ the above restriction will not produce the evolution of the highest field $u_{\alpha+n}$: $u_{\alpha+n} = $ const, so we can put $u_{\alpha+n} = 1$. In the case $k = 1$ the lowest field will remain constant: $u_\alpha = $ const, so we put $u_\alpha = 1$ in this case. In this way we finally arrive at the following sets of admissible reductions of the original algebra g

$$k = 0 \quad L = \mathcal{E}^{\alpha+n} + u_{\alpha+n-1}\mathcal{E}^{\alpha+n-1} + .. + u_\alpha\mathcal{E}^\alpha, \quad u_{\alpha+n} = 1$$

$$k = 1 \quad \bar{L} = \bar{u}_{\alpha+n}\mathcal{E}^{\alpha+n} + \bar{u}_{\alpha+n-1}\mathcal{E}^{\alpha+n-1} + .. + \bar{u}_{\alpha+1}\mathcal{E}^{\alpha+1} + E^\alpha, \quad \bar{u}_\alpha = 1$$
$$(7.232)$$

and the only non-trivial cases (when $(L^q)_{\ge 0,1} \ne 0, L^q$) are for $-n < \alpha \le -1$. (For $k = 1$ bars were introduced above the field variables to show a direct visual distinction between the cases $k = 0$ and $k = 1$. We will retain this notation.) This reduction leads to many lattice equations. Below we list a few examples up to four-field systems.

One-Field Systems

Do not exist in this scheme; there are only 1-field reductions of the 2-field case.

Two-Field Systems

1. $k = 0$, $\alpha = -1$.   Then Lax operator has the form

$$L = u_{-1}\mathcal{E}^{-1} + u_0 + \mathcal{E} := v\mathcal{E}^{-1} + p + \mathcal{E}. \quad (7.233)$$

This choice of L, through the formula (7.229), leads to the evolution of L. But now $L_t = v_t\mathcal{E}^{-1} + p_t$ so the first three equations of the hierarchy are

$$
\begin{aligned}
q = 1 \quad v(n)_{t_1} &= v(n)[p(n) - p(n-1)] \\
p(n)_{t_1} &= v(n+1) - v(n) \\
q = 2 \quad v(n)_{t_2} &= v(n)[p^2(n) - p^2(n-1) + v(n+1) - v(n-1)] \\
p(n)_{t_2} &= v(n+1)[p(n+1) + p(n)] - v(n)[p(n-1) + p(n)] \\
q = 3 \quad v(n)_{t_3} &= p(n)[v^2(n) + v(n)p^2(n) + 2v(n)v(n+1) \\
&\quad + p(n+1)v(n+1)] \\
&\quad - v(n)[v(n-1)p(n-2) + 2v(n-1)p(n-1) \\
&\quad + p^3(n-1) + v(n)p(n-1)] \\
p(n)_{t_3} &= v(n+1)[p^2(n) + p(n)p(n+1) + p^2(n+1) \\
&\quad + v(n+1) + v(n+2)] \\
&\quad - v(n)[v(n-1) + v(n) + p^2(n-1) + p(n)p(n-1) \\
&\quad + p^2(n)]
\end{aligned}
$$
$$(7.234)$$

producing the famous Toda lattice hierarchy [183].

2. $k = 1$, $\alpha = -1$. Then the Lax operator has the form

$$L = \mathcal{E}^{-1} + \bar{u}_0 + \bar{u}_1 \mathcal{E} := \mathcal{E}^{-1} + \bar{p} + \bar{v}\mathcal{E} \tag{7.235}$$

and the first equation ($q = 1$) of the hierarchy has the form

$$\begin{aligned}
\bar{p}(n)_{t_1} &= \bar{v}(n) - \bar{v}(n-1) \\
\bar{v}(n)_{t_1} &= \bar{v}(n)[\bar{p}(n+1) - \bar{p}(n)]
\end{aligned} \tag{7.236}$$

which is just the reparametrized Toda equation. So the choice $k = 1$, $\alpha = -1$ does not lead to any new two-component system.

Three-Field Systems

1. $k = 0$, $\alpha = -1$.   Then the Lax operator has the form

$$L = u_{-1}\mathcal{E}^{-1} + u_0 + u_1\mathcal{E} + \mathcal{E}^2 := u\mathcal{E}^{-1} + v + p\mathcal{E} + \mathcal{E}^2 \tag{7.237}$$

and ($q = 1$):

$$\begin{aligned}
u(n)_{t_1} &= u(n)[v(n) - v(n-1)] \\
v(n)_{t_1} &= p(n)u(n+1) - u(n)p(n-1) \\
p(n)_{t_1} &= u(n+2) - u(n).
\end{aligned} \tag{7.238}$$

2. $k = 0$, $\alpha = -2$. This leads to:

$$L = u_{-2}\mathcal{E}^{-2} + u_1\mathcal{E}^{-1} + u_0 + \mathcal{E} := w\mathcal{E}^{-2} + v\mathcal{E}^{-1} + u + \mathcal{E} \tag{7.239}$$

and ($q = 1$)

$$\begin{aligned}
w(n)_{t_1} &= w(n)[u(n) - u(n-2)] \\
v(n)_{t_1} &= v(n)[u(n) - u(n-1)] + w(n+1) - w(n) \\
u(n)_{t_1} &= v(n+1) - v(n).
\end{aligned} \tag{7.240}$$

3. $k = 1$, $\alpha = -1$. Then

$$L = \mathcal{E}^{-1} + \bar{u}_0 + \bar{u}_1\mathcal{E} + \bar{u}_2\mathcal{E}^2 := \mathcal{E}^{-1} + \bar{u} + \bar{v}\mathcal{E} + \bar{w}\mathcal{E}^2 \tag{7.241}$$

and ($q = 1$)

$$\begin{aligned}
\bar{u}(n)_{t_1} &= \bar{v}(n) - \bar{v}(n-1) \\
\bar{v}(n)_{t_1} &= \bar{v}(n)[\bar{u}(n+1) - \bar{u}(n)] + \bar{w}(n) - \bar{w}(n-1) \\
\bar{w}(n)_{t_1} &= \bar{w}(n)[\bar{u}(n+2) - \bar{u}(n)].
\end{aligned} \tag{7.242}$$

4. $k = 1$, $\alpha = -2$

$$L = \mathcal{E}^{-2} + \bar{u}_{-1}\mathcal{E}^{-1} + \bar{u}_0 + \bar{u}_1\mathcal{E} := \mathcal{E}^{-2} + \bar{p}\mathcal{E}^{-1} + \bar{v} + \bar{u}\mathcal{E} \tag{7.243}$$

and ($q = 1$)

$$\begin{aligned}
\bar{p}(n)_{t_1} &= \bar{u}(n) - \bar{u}(n-2) \\
\bar{v}(n)_{t_1} &= \bar{u}(n)\bar{p}(n+1) - \bar{u}(n-1)\bar{p}(n) \\
\bar{u}(n)_{t_1} &= \bar{u}(n)[\bar{v}(n+1) - \bar{v}(n)].
\end{aligned} \tag{7.244}$$

Four-Field Systems

The reader can find all admissible examples of four-field systems in [34]. Here we quote only two examples.

1. $k = 0, \alpha = -2$

$$L = p\mathcal{E}^{-2} + u\mathcal{E}^{-1} + v + w\mathcal{E} + \mathcal{E}^2 \qquad (7.245)$$

and $(q = 1)$

$$
\begin{aligned}
p(n)_{t_1} &= p(n)[v(n) - v(n-2)] \\
u(n)_{t_1} &= u(n)[v(n) - v(n-1)] + w(n)p(n+1) - p(n)w(n-2) \\
v(n)_{t_1} &= w(n)u(n+1) - u(n)w(n-1) + p(n+2) - p(n) \\
w(n)_{t_1} &= u(n+2) - u(n).
\end{aligned}
$$

$$\qquad (7.246)$$

2. $k = 1, \alpha = -2$

$$L = \mathcal{E}^{-2} + \bar{u}\mathcal{E}^{-1} + \bar{v} + \bar{w}\mathcal{E} + \bar{q}\mathcal{E}^2 \qquad (7.247)$$

and $(q = 1)$

$$
\begin{aligned}
\bar{u}(n)_{t_1} &= \bar{w}(n) - \bar{w}(n-2) \\
\bar{v}(n)_{t_1} &= \bar{w}(n)\bar{u}(n+1) - \bar{u}(n)\bar{w}(n-1) + \bar{q}(n) - \bar{q}(n-2) \\
\bar{w}(n)_{t_1} &= \bar{w}(n)[\bar{v}(n+1) - \bar{v}(n)] + \bar{q}(n)\bar{u}(n+2) - \bar{u}(n)\bar{q}(n-1) \\
\bar{q}(n)_{t_1} &= \bar{q}(n)[\bar{v}(n+2) - \bar{v}(n)].
\end{aligned}
$$

$$\qquad (7.248)$$

Now we investigate the Hamiltonian structures of the lattice hierarchies presented above. First let us calculate adjoints of projection operators, which enter into (7.66)–(7.68) :

$$P^*_{\geq 0} = P_{\leq 0}, \quad P^*_{\geq 1} = P_{\leq -1}, \quad P^*_{<0} = P_{>0}, , \quad P^*_{<1} = P_{>-1}. \qquad (7.249)$$

With the help of (7.249) we can prove the following.

Theorem 7.8 [34]

(i) Both $\mathcal{R}_0$ and its antisymmetric part $\tilde{\mathcal{R}}_0 = \frac{1}{2}(\mathcal{R}_0 - \mathcal{R}_0^*)$ satisfy YB(1).
(ii) Both $\mathcal{R}_1$ and its antisymmetric part $\tilde{\mathcal{R}}_1 = \frac{1}{2}(\mathcal{R}_1 - \mathcal{R}_1^*)$ satisfy YB(1).

The proof of the theorem consist of a long, but straightforward calculation. Observe that $\tilde{\mathcal{R}}_1 = \tilde{\mathcal{R}}_0$, so part (ii) of the theorem follows easily from part (i). Hence, from the general considerations of Sect. 7.2 we immediately find that Θ_i is Poisson for all $i = 1, 2, 3$ and for both $\mathcal{R}$-matrices $\mathcal{R}_0$ and $\mathcal{R}_1$. Since we are interested in reduction (7.232), the next step is to answer the question: are the reductions (7.232) Poisson, i.e. do they preserve the Hamiltonian structure of the hierarchies given by formulas (7.66)–(7.68) and (7.77)? The answer in general is no, as we shall see. Let us then investigate all three structures separately.

According to (7.66) and (7.249) the linear tensor Θ_1 on $g \cong g^*$ can be written as

$$\Theta_1(L) : \nabla H \to -ad_L(\nabla H)_{\geq k} + (ad_L \nabla H)_{>-k}$$

$$= ad_L(\nabla H)_{<-k} - (ad_L \nabla H)_{\leq k}. \tag{7.250}$$

If L has the reduced form (7.232) (case $k = 0$)

$$L = \mathcal{E}^{\alpha+n} + u_{\alpha+n-1}\mathcal{E}^{\alpha+n-1} + \ldots + u_\alpha \mathcal{E}^\alpha \tag{7.251}$$

then the expression on the right hand side of (7.250) is of the form field $\cdot$ $\mathcal{E}^\alpha + \ldots +$ field $\cdot \mathcal{E}^{\alpha+n-1}$, i.e. it represents a vector tangent to the (affine) subspace generated by all the L's of the form (7.251). This means that in the case $k = 0$ the image of $\Theta_1(L)$ for each L belonging to the above subspace is itself a subspace (not necessarily proper) of the space tangent to this affine subspace. Similarly, in the case of $k = 1$ reduction (7.232)

$$\bar{L} = \bar{u}_{\alpha+n}\mathcal{E}^{\alpha+n} + \bar{u}_{\alpha+n-1}\mathcal{E}^{\alpha+n-1} + \ldots + \bar{u}_{\alpha+1}\mathcal{E}^{\alpha+1} + \mathcal{E}^\alpha, \tag{7.252}$$

expression (7.250) attains the form: field $\cdot \mathcal{E}^{\alpha+1} + \ldots +$ field $\cdot \mathcal{E}^{\alpha+n}$, which is the vector tangent to the affine subspace given by (7.252), so again the image of Θ_1 is at each point a subspace of the space tangent to the affine subspace (7.252). It follows that all the reductions (7.232) are Poisson with respect to Θ_1. It is now very easy to find the Hamiltonian formulation for the systems presented in this section. Actually, the linear Lax Poisson structure (7.250) translates into the following field representatives.

Two-Field Systems

1. $k = 0$, $\alpha = -1$. Then

$$L = v\mathcal{E}^{-1} + p + \mathcal{E}$$

and

$$\theta_1 = \begin{pmatrix} 0 & v(1 - E^{-1}) \\ (E - 1)v & 0 \end{pmatrix}. \tag{7.253}$$

The Hamiltonian generating the first flow ($q = 1$) is given by formula (7.76)

$$H_2 = \frac{1}{2}\text{tr}(L^2) = \frac{1}{2}\sum_{n=-\infty}^{+\infty}[2v(n) + p^2(n)]. \tag{7.254}$$

2. $k = 1$, $\alpha = -1$

$$\bar{L} = \mathcal{E}^{-1} + \bar{p} + \bar{v}\mathcal{E}$$

$$\bar{\theta}_1 = \begin{pmatrix} 0 & (1 - E^{-1})\bar{v} \\ \bar{v}(E - 1) & 0 \end{pmatrix}. \tag{7.255}$$

The Hamiltonian generating the first flow is of the form

$$\bar{H}_2 = \frac{1}{2}\text{tr}(\bar{L}^2) = \frac{1}{2}\sum_{n=-\infty}^{+\infty}[2\bar{v}(n) + \bar{p}^2(n)]. \tag{7.256}$$

Three-Field Systems

1. $k = 0$, $\alpha = -1$

$$L = u\mathcal{E}^{-1} + v + p\mathcal{E} + \mathcal{E}^2$$

$$\theta_1 = \begin{pmatrix} 0 & u(1 - E^{-1}) & 0 \\ (E - 1)u & 0 & 0 \\ 0 & 0 & E - E^{-1} \end{pmatrix} \tag{7.257}$$

and

$$H_2 = \frac{1}{2} \sum_{n=-\infty}^{+\infty} [v^2(n) + 2p(n)u(n + 1)]. \tag{7.258}$$

2. $k = 0$, $\alpha = -2$

$$L = w\mathcal{E}^{-2} + v\mathcal{E}^{-1} + u + \mathcal{E}$$

$$\theta_1 = \begin{pmatrix} 0 & 0 & w(1 - E^{-2}) \\ 0 & Ew - wE^{-1} & v(1 - E^{-1}) \\ (E^2 - 1)w & (E - 1)v & 0 \end{pmatrix} \tag{7.259}$$

and

$$H_2 = \frac{1}{2} \sum_{n=-\infty}^{+\infty} [u^2(n) + 2v(n)]. \tag{7.260}$$

3. $k = 1$, $\alpha = -1$

$$\bar{L} = \mathcal{E}^{-1} + \bar{u} + \bar{v}\mathcal{E} + \bar{w}\mathcal{E}^2$$

$$\bar{\theta}_1 = 2 \begin{pmatrix} 0 & (1 - E^{-1})\bar{v} & (1 - E^{-2})\bar{w} \\ \bar{v}(E - 1) & \bar{w}E - E^{-1}\bar{w} & 0 \\ \bar{w}(E^2 - 1) & 0 & 0 \end{pmatrix} \tag{7.261}$$

and

$$\overline{H}_2 = \frac{1}{2} \sum_{n=-\infty}^{+\infty} [2\bar{v}(n) + \bar{u}^2(n)]. \tag{7.262}$$

4. $k = 1$, $\alpha = -2$

$$\bar{L} = \mathcal{E}^{-2} + \bar{p}\mathcal{E}^{-1} + \bar{v} + \bar{u}\mathcal{E}$$

$$\bar{\theta}_1 = \begin{pmatrix} E - E^{-1} & 0 & 0 \\ 0 & 0 & (1 - E^{-1})\bar{u} \\ 0 & \bar{u}(E - 1) & 0 \end{pmatrix} \tag{7.263}$$

and

$$\overline{H}_2 = \frac{1}{2} \sum_{n=-\infty}^{+\infty} [\bar{v}^2(n) + 2\bar{u}(n)\bar{p}(n + 1)]. \tag{7.264}$$

With the help of formulas (7.249) we can write the quadratic tensor Θ_2 as

$$\Theta_2(L) : \nabla H \to -ad_L(ad_L^+\nabla H)_{\geq k} + ad_L^+(ad_L\nabla H)_{>-k}$$

$$= ad_L(ad_L^+\nabla H)_{<k} - ad_L^+(ad_L\nabla H)_{\leq -k}. \qquad (7.265)$$

Consider, for example, the case $k = 1$. Then, taking the reduction (7.232) we get that the right hand side of (7.265) has the form: field$\cdot\mathcal{E}^{\alpha+n}+\ldots+$field$\cdot\mathcal{E}^\alpha$. An identical result is true for $k = 0$. The reduction (7.232) is thus not Poisson for Θ_2, but the one given by (7.231) :

$$L = u_{\alpha+n}\mathcal{E}^{\alpha+n} + u_{\alpha+n-1}\mathcal{E}^{\alpha+n-1} + \ldots + u_\alpha\mathcal{E}^\alpha \ , \quad -n < \alpha \leq -1. \quad (7.266)$$

is surely the proper reduction.

Taking into account that for the L (7.266) the appropriate form of ∇H reads: $\nabla H = \mathcal{E}^{-\alpha}\gamma_\alpha+\ldots+\mathcal{E}^{-n-\alpha}\gamma_{n+\alpha}$, where $\gamma_k = \frac{\delta H}{\delta u_k}$, the quadratic tensor Θ_2 can be simplified to the form

$$\tfrac{1}{2}\Theta_2(L) : \nabla H \quad \to \quad (L\nabla H)_{>0}L - L(\nabla HL)_{>0} + \tfrac{1}{2}(L\nabla H)_0 L$$

$$-\tfrac{1}{2}L(\nabla HL)_0 + (\tfrac{1}{2} - k)(\nabla HL)_0 L - (\tfrac{1}{2} - k)L(L\nabla H)_0$$

$$= -(L\nabla H)_{<0}L + L(\nabla HL)_{<0} - \tfrac{1}{2}(L\nabla H)_0 L$$

$$+\tfrac{1}{2}L(\nabla HL)_0 + (\tfrac{1}{2} - k)(\nabla HL)_0 L - (\tfrac{1}{2} - k)L(L\nabla H)_0$$

$$(7.267)$$

Now, applying the procedure of Dirac reduction to the lowest ($k = 1$) or highest ($k = 0$) field in L, we finally reduce the original Poisson structure Θ_2 to affine subspaces (7.232). Below we list the second Hamiltonian structures for two- and three-field systems of our hierarchies. As the first flow ($q = 1$) is generated by $\Theta_2\nabla H_1$, and ∇H_1 is the trivial one, we list only the Poisson tensors corresponding to the previously listed hierarchies.

Two-field systems

1. $k = 0, \alpha = -1$

   Here, we start with the form (7.266)

$$L = v\mathcal{E}^{-1} + p + q\mathcal{E}. \qquad (7.268)$$

The standard procedure of reduction leads to the following 'intermediate' Poisson tensor Θ on the subspace (7.268)

$$\theta = \frac{1}{2}\begin{pmatrix} v(E - E^{-1})v & 2v(1 - E^{-1})p & v(1 - E^{-2})q \\ 2p(E - 1)v & 2(qEv - vE^{-1}q) & 0 \\ q(E^2 - 1)v & 0 & q(E^{-1} - E)q \end{pmatrix}. \qquad (7.269)$$

Now we reduce θ to the affine subspace of the space (7.268) by introducing the constraint $q = 1$. According to the formula (7.82)

$$\theta_2 = \frac{1}{2} \begin{pmatrix} v(E - E^{-1})v & 2v(1 - E^{-1})p \\ 2p(E - 1)v & 2(Ev - vE^{-1}) \end{pmatrix}$$

$$- \frac{1}{2} \begin{pmatrix} v(1 - E^{-2}) \\ 0 \end{pmatrix} (E^{-1} - E)^{-1} (\ (E^2 - 1)v \quad 0 \ )$$

which yields

$$\theta_2 = \begin{pmatrix} v(E - E^{-1})v & v(1 - E^{-1})p \\ p(E - 1)v & Ev - vE^{-1} \end{pmatrix} \tag{7.270}$$

which is the well known second Hamiltonian structure of the Toda lattice. The formal inverse $(E - E^{-1})^{-1}$ occurring during the calculation makes sense, since $(E - E^{-1})^{-1}$ acts from the left on the expression $E - E^{-1}$, i.e. it always acts on something lying in its domain of definition. It turns out that this is a typical behaviour of nonlocalities in such calculations.

2. $k = 1$, $\alpha = -1$.

This corresponds to the subspace

$$\bar{L} = \mathcal{E}^{-1} + \bar{p} + \bar{v}\mathcal{E}$$

and the reduction of θ_2 to this subspace can be done exactly as in the above example. The result is

$$\bar{\theta}_2 = 2 \begin{pmatrix} \bar{v}E - E^{-1}\bar{v} & \bar{p}(1 - E^{-1})\bar{v} \\ \bar{v}(E - 1)\bar{p} & \bar{v}(E - E^{-1})\bar{v} \end{pmatrix}. \tag{7.271}$$

Three-Field Systems

In case of three-field systems a new feature enters the stage: strong nonlocalities of structures. The structures Θ_i are thus defined in the whole space $C^\infty(g^*)$, but differentials of Casimirs (7.76) belong to their domain of definition. In all the examples below the procedure of calculations was exactly as the one described above.

1. $k = 0$, $\alpha = -1$

$$L = u\mathcal{E}^{-1} + v + p\mathcal{E} + \mathcal{E}^2$$

$$\theta_2 = \begin{pmatrix} \theta^{11} & \theta^{12} & \theta^{13} \\ -(\theta^{12})^\dagger & \theta^{22} & \theta^{23} \\ -(\theta^{13})^\dagger & -(\theta^{23})^\dagger & \theta^{33} \end{pmatrix},$$

$$\theta^{11} = u[E - E^{-1} + (E - 1)(E + E^{-1})(1 + E)^{-1}]u, \quad \theta^{12} = 2u(1 - E^{-1})v,$$

$$\theta^{13} = u[1 - E^{-2} - (E + E^{-1})(1 - E^{-1})(1 + E)^{-1}]p,$$

$$\theta^{22} = 2pEu - 2uE^{-1}p, \quad \theta^{23} = 2Eu - 2uE^{-2},$$

$$\theta^{33} = p[E^{-1} - E + (E + E^{-1})(E - 1)(1 + E)^{-1}]p + 2Ev - 2vE^{-1}. \quad (7.272)$$

2. $k = 0, \ \alpha = -2$

$$L = wE^{-2} + vE^{-1} + u + E$$

$$\theta_2 = \begin{pmatrix} \theta^{11} & \theta^{12} & \theta^{13} \\ -(\theta^{12})^\dagger & \theta^{22} & \theta^{23} \\ -(\theta^{13})^\dagger & -(\theta^{23})^\dagger & \theta^{33} \end{pmatrix},$$

$$\theta^{11} = w(E^2 - E^{-2} + E - E^{-1})w, \quad \theta^{12} = w(1 + E - E^{-1} - E^{-2})v,$$

$$\theta^{13} = w(1 - E^{-2})u, \quad \theta^{22} = uEw - wE^{-1}u + v(E - E^{-1})v,$$

$$\theta^{23} = Ew - wE^{-2} + v(1 - E^{-1})u, \quad \theta^{33} = Ev - vE^{-1}. \quad (7.273)$$

3. $k = 1, \ \alpha = -1$

$$L = E^{-1} + \bar{u} + \bar{v}E + \bar{w}E^2$$

$$\bar{\theta}_2 = \begin{pmatrix} \theta^{11} & \theta^{12} & \theta^{13} \\ -(\theta^{12})^\dagger & \theta^{22} & \theta^{23} \\ -(\theta^{13})^\dagger & -(\theta^{23})^\dagger & \theta^{33} \end{pmatrix},$$

$$\theta^{11} = \bar{v}E - E^{-1}\bar{v}, \quad \theta^{12} = \bar{u}(1 - E^{-1})\bar{v} + \bar{w}E - E^{-2}\bar{w},$$

$$\theta^{13} = \bar{u}(1 - E^{-2})\bar{w}, \quad \theta^{22} = \bar{w}E\bar{u} - \bar{u}E^{-1}\bar{w} + \bar{v}(E - E^{-1})\bar{v},$$

$$\theta^{23} = \bar{v}(E - E^{-1} - E^{-2} + 1)\bar{w}, \quad \theta^{33} = \bar{w}(E^2 - E^{-2} + E - E^{-1})\bar{w}. \quad (7.274)$$

4. $k = 1, \ \alpha = -2$

$$L = E^{-2} + \bar{p}E^{-1} + \bar{v} + \bar{u}E$$

$$\theta_2 = \begin{pmatrix} \theta^{11} & \theta^{12} & \theta^{13} \\ -(\theta^{12})^\dagger & \theta^{22} & \theta^{23} \\ -(\theta^{13})^\dagger & -(\theta^{23})^\dagger & \theta^{33} \end{pmatrix},$$

$$\theta^{11} = \bar{p}[E^{-1} - E + (E + E^{-1})(E - 1)(1 + E)^{-1}]\bar{p}, \quad \theta^{12} = 2\bar{u}E - 2E^{-2}\bar{u},$$

$$\theta^{13} = \bar{p}[1 - E^{-2} + (E + E^{-1})(E^{-1} - 1)(1 + E)^{-1}]\bar{u},$$

$$\theta^{22} = 2\bar{u}E\bar{p} - 2\bar{p}E^{-1}\bar{u}, \quad \theta^{23} = 2\bar{v}(1 - E^{-1})\bar{u},$$

$$\theta^{33} = 2\bar{u}[E - E^{-1} + (E - E^{-1})(E + E^{-1})(1 + E)^{-1}]\bar{w}. \tag{7.275}$$

Observe that θ_2 is local for $k = 0$, $\alpha = -2$ and $k = 1$, $\alpha = -1$, and non-local for $k = 0$, $\alpha = -1$ and for $k = 1$, $\alpha = -2$ due to the presence of the term $(1 + E)^{-1}$.

Using formulas (7.249) and (7.68) the cubic Poisson structure can be presented in the following form

$$\Theta_3(L) : \nabla H \to \quad -ad_L(L\nabla HL)_{\geq k} + L(ad_L\nabla H)_{> -k}L$$

$$= ad_L(L\nabla HL)_{< k} - L(ad_L\nabla H)_{\leq -k}L. \tag{7.276}$$

It can be shown immediately that L reduced to the subspace (7.232) produces on the right hand side of (7.276) an expression field$\cdot \mathcal{E}^{2n+2\alpha} + \ldots + $field$\cdot \mathcal{E}^{2\alpha+1}$ (for $k = 0$) or field $\cdot \mathcal{E}^{2n+2\alpha-1} + \ldots + $ field $\cdot \mathcal{E}^{2\alpha}$ (for $k = 1$). Evidently, to produce the third Hamiltonian structure on the reduced subspaces one has to use the Dirac reduction.

We are now ready to establish some gauge transformations, which transform the above hierarchies into one another. We begin with a theorem whose validity allows us to apply gauge transformations.

Theorem 7.9 *Let $L \in g$ satisfy the hierarchy of Lax equations (7.229) with $k = 0$*

$$L_{t_q} = [(L^q)_{\geq 0}, L].$$

Let $\Phi = \Phi(u, t_q)$, $\Phi \neq 0$ satisfies the condition

$$\Phi_{t_q} = (L^q)_0\Phi$$

(where the subscript 0 denotes the coefficient field of $\mathcal{E}^0$). Then $\bar{L} = \Phi^{-1}L\Phi$ satisfies the hierarchy

$$\bar{L}_{t_q} = [(\bar{L}^q)_{\geq 1}, \bar{L}] \quad q = 1, 2, \ldots \tag{7.277}$$

Proof.

$$\bar{L}_{t_q} - [(\bar{L}^q)_{\geq 1}, \bar{L}]$$

$$= \Phi^{-1}(L_{t_q} - [(L^q)_{\geq 0}, L])\Phi + [\Phi^{-1}((L^q)_0\Phi - \Phi_{t_q}), \bar{L}]$$

$$= 0$$

because of the assumptions. $\square$

This Theorem gives us a tool for generating new hierarchies from old ones. However the function Φ remains undetermined. The best choice to determine it is to put an additional, but perfectly reasonable assumption: we demand that after the above gauge transformations L attains the form of the lower expression from (7.232) with the same α; otherwise we do not obtain a nontrivial hierarchy after the gauge transformation. This demand leads to the following set of equations

$$
\begin{aligned}
\bar{u}_{\alpha+n} &= \Phi^{-1}(E^{\alpha+n}\Phi), \\
\bar{u}_{\alpha+n-1} &= \Phi^{-1}u_{\alpha+n-1}(E^{\alpha+n-1}\Phi), \\
&\vdots \\
\bar{u}_{\alpha+1} &= \Phi^{-1}u_{\alpha+1}(E^{\alpha+1}\Phi), \\
1 &= \Phi^{-1}u_{\alpha}(E^{\alpha}\Phi).
\end{aligned}
\tag{7.278}
$$

It is possible to eliminate the Φ's from the above equations. This leads to a certain map between the old and new variables which can be considered as the lattice analogue of the Miura map for integrable soliton systems. However, as we shall see, this analogy is not complete. For $\alpha = -1$ (7.278) can be solved to give the map (for arbitrary n)

$$
\begin{aligned}
\bar{u}_0 &= u_0, \\
\bar{u}_1 &= u_1 E(u_{-1}), \\
&\vdots \\
\bar{u}_r &= u_r(E^r u_{-1})(E^{r-1}u_{-1})\ldots(Eu_{-1}), \\
&\vdots \\
\bar{u}_{n-2} &= u_{n-2}(E^{n-2}u_{-1})(E^{n-3}u_{-1})\ldots(Eu_{-1}), \\
\bar{u}_{n-1} &= (E^{n-1}u_{-1})(E^{n-2}u_{-1})\ldots(Eu_{-1}),
\end{aligned}
\tag{7.279}
$$

going from $k = 0$ to $k = 1$, and for $\alpha = -n + 1$ we obtain

$$
\begin{aligned}
u_0 &= \bar{u}_0, \\
u_{-1} &= \bar{u}_{-1}(E^{-1}\bar{u}_1), \\
&\vdots \\
u_{-r} &= \bar{u}_{-r}(E^{-r}\bar{u}_1)(E^{-r+1}\bar{u}_1)\ldots(E^{-2}\bar{u}_1)(E^{-1}\bar{u}_1), \\
&\vdots \\
u_{-n+2} &= \bar{u}_{-n+2}(E^{-n+2}\bar{u}_1)(E^{-n+3}\bar{u}_1)\ldots(E^{-2}\bar{u}_1)(E^{-1}\bar{u}_1), \\
u_{-n+1} &= (E^{-n+1}\bar{u}_1)(E^{-n+2}\bar{u}_1)\ldots(E^{-2}\bar{u}_1)(E^{-1}\bar{u}_1),
\end{aligned}
\tag{7.280}
$$

going from $k = 1$ to $k = 0$ (the same α). In the case of other α's the formulas obtained from (7.278) are not so regular; nevertheless they can still be considered as Miura-like transformations. Often they are reduced to some linear transformations.

Let us illustrate the above concepts by some examples.

Example 7.12 First let us consider the case of two-field systems (7.233) and (7.235) and gauge $k = 0 \to k = 1$. In this case $\bar{L}$ attains the form:

$$
\bar{L} = \Phi^{-1}v(E^{-1}\Phi)\mathcal{E}^{-1} + p + \Phi^{-1}(E\Phi)\mathcal{E}
$$

and (7.279) has the form

$$\begin{aligned} \bar{p} &= p, \\ \bar{v} &= (Ev), \end{aligned}$$

which is linear.

Example 7.13 In a similar way one can obtain from (7.280) the Miura map between four-field systems (7.240) and (7.244)

$$\begin{aligned} w &= (E^{-1}\bar{u})(E^{-2}\bar{u}), \\ v &= \bar{p}(E^{-1}\bar{u}), \\ u &= \bar{v}, \end{aligned}$$

which is no longer linear.

Example 7.14 The map between systems (7.246) and (7.248) must be calculated directly from (7.278), and it gives again a linear transformation:

$$\begin{aligned} \bar{u} &= (E^{-1}w), \\ \bar{v} &= v, \\ \bar{w} &= (Eu), \\ \bar{q} &= (E^2 p). \end{aligned}$$

Moreover from Theorem 7.9 we obtain immediately:

Lemma 7.12 *The map (7.278) preserves the dynamics of systems, i.e. it maps Hamiltonian flows onto Hamiltonian flows.*

More delicate in its nature is the question of whether (7.278) preserves the Poisson structure of the corresponding equations. Here we have

Theorem 7.10 *The map (7.278) maps the pth Poisson structure of the case $k = 0$ to the pth Poisson structure of the case $k = 1$ ($p = 1, 2, 3$).*

Proof. One can show this by direct calculations. For the sake of simplicity we perform calculations for the specific case of systems (7.238) and (7.242), and only for the first Poisson structures. The corresponding map F has the form

$$\begin{aligned} \bar{u} &= v, \\ \bar{v} &= p(Eu), \\ \bar{w} &= (Eu)(E^2 u), \end{aligned}$$

and its derivative F' is given by

$$F' = \begin{pmatrix} 0 & 1 & 0 \\ pE & 0 & (Eu) \\ (Eu)E^2 + (E^2 u)E & 0 & 0 \end{pmatrix},$$

so the induced Poisson tensor has the form

$$F' \circ \theta_1 \circ F'^\dagger = \begin{pmatrix} 0 & (1 - E^{-1})\bar{v} & (1 - E^{-2})\bar{w} \\ \bar{v}(E - 1) & \bar{w}E - E^{-1}\bar{w} & 0 \\ \bar{w}(E^2 - 1) & 0 & 0 \end{pmatrix}$$

where F'^* denotes the pull-back of F. In the above formula we also simultaneously substitute new variables in place of the old ones. The resulting tensor is just the corresponding $\bar{\theta}_1$.

As is clear from the above, the transformation (7.278) does not behave exactly like ordinary Miura maps. Firstly, in some cases it degenerates to a linear transformation; secondly, it maps the first Hamiltonian structure into the first and the second into the second so we cannot obtain any new Hamiltonian structures. Nevertheless, it has the most remarkable feature of the Miura transformation: in spite of its nonlinearity each of its derivatives can be expressed in terms of new variables, the old ones being combined in the final expressions exactly in the desired combinations.

Finally, we establish certain linear relations between the hierarchies generated in the above approach. First, let us introduce the following notation:

$$\text{if } L = \sum_i u_i \mathcal{E}^i, \quad \text{then } L^\dagger \overset{\mathrm{df}}{=} \sum_i \mathcal{E}^{-i} u_i. \tag{7.281}$$

By direct calculations, one can prove

Lemma 7.13

$$((L^q)_{\geq k})^\dagger = ((L^q)^\dagger)_{\leq k}$$

for any integers q and k.

Now we are ready to formulate the following Lemma.

Lemma 7.14

$$\text{If } L_{t_q} = [(L^q)_{\geq 1} \, , \, L] \text{ then } L^\dagger_{t_q} = [((L^q)^\dagger)_{\geq 0}, \, L^\dagger] \tag{7.282}$$

The proof of the above Lemma is a straightforward consequence of the following lemma.

Lemma 7.15 *Consider the family of the n-field Lax operators. Denote by L_γ the Lax operator corresponding to $k = 0$ and $\alpha = \gamma$, and by $\bar{L}_\gamma$ the Lax operator corresponding to $k = 1$ and $\alpha = \gamma$. Then*

$$\bar{L}_{-n-\alpha} = L^\dagger_\alpha \tag{7.283}$$

if and only if

$$\bar{u}_i = (E^i u_{-i}). \tag{7.284}$$

This can be proved by direct calculations performed over (7.232). A moment's consideration leads us now to the following theorem.

Theorem 7.11 *The n-field hierarchies generated by the Lax operators L_α and $\bar{L}_{-n-\alpha}$ are equivalent up to a linear transformation (7.284).*

This theorem allows us to establish many linear transformations between the existing hierarchies with the same number of fields generated by Lax operators connected by (7.283).

Example 7.15 We shall illustrate the theorem in the case of $n = 4$ and $\alpha = -2$. Then $-n - \alpha = -2$. Moreover

$$L_{-2} = p\mathcal{E}^{-2} + u\mathcal{E}^{-1} + v + w\mathcal{E} + \mathcal{E}^2,$$

$$\bar{L}_{-2} = \mathcal{E}^{-2} + \bar{u}\mathcal{E}^{-1} + \bar{v} + \bar{w}\mathcal{E} + \bar{q}\mathcal{E}^2$$

and the formula (7.284) reads

$$\begin{aligned}
\bar{q} &= (E^2 p), \\
\bar{w} &= (Eu), \\
\bar{v} &= v, \\
\bar{u} &= (E^{-1}w),
\end{aligned}$$

which is identical to the corresponding map between the hierarchies from Example 7.14.

Theorem 7.11 asures us that in fact the only independent case of an $\mathcal{R}$-matrix is $\mathcal{R}_0$. Before we pass onto searching for a proper $\mathcal{R}_1$-matrix, we construct the Dirac constraint quadratic Poisson structure in the explicit form applying the Suris scheme.

Theorem 7.12 *Let $\Pi : g_0 \to g_0$ be an arbitrary skew-symmetric linear map on the algebra g_0 of zero order terms of g. Then*

$$A_1 : \xi \to \xi_{\geq 1} - \xi_{<0} - \Pi(\xi_0), \quad S : \xi \to (1 - 2k)\xi_0 + \Pi(\xi_0),$$

$$A_2 : \xi \to \xi_{\geq 1} - \xi_{<0} + \Pi(\xi_0), \quad S^* : \xi \to (1-2k)\xi_0 - \Pi(\xi_0), \quad k = 0, 1, \tag{7.285}$$

satisfy all the conditions of Theorem 7.2. The resulting Poisson structure reads

$$\overline{\Theta}_2(L) : \nabla H \to \Theta_2(L) + [\Pi([\nabla H, L]_0), L], \tag{7.286}$$

where $\Theta_2(L)$ is the quadratic structure (7.267) compatible with the linear structure (7.250) and admits a proper restriction to Lax operators of the form (7.231).

Conditions (7.71) are easily verified. With

$$A_1 + S = A_2 + S^* = \mathcal{R}_k : \xi \to \xi_{\geq 1} + (1 - 2k)\xi_0 - \xi_{<0}$$

then guarantee the Poisson property of $\overline{\Theta}_2(L)$ according to Theorem 7.3. We note that an additional term generated by Π drops out for Casimir functions, so that $\overline{\Theta}_2$ generates the same integrable hierarchy as Θ_2. Since the associated integrable hierarchies $L_{t_q} = [(L^q)_{\geq 0}, L]$ (i.e. those for $k = 0$, as these for $k - 1$ are, according to Theorem 7.11, equivalent to the case $k = 0$) can be restricted to (7.232), one would like $\overline{\Theta}_2$ to admit a proper restriction to such Lax operators. For $\overline{\Theta}_2$ the constraint $u_{n+\alpha} = 1$ can be achieved by an appropriate choice of Π.

Theorem 7.13 *With*

$$\Pi(\xi_0) = (E^{n+\alpha} + 1)(E^{n+\alpha} - 1)^{-1}\xi_0 \qquad (7.287)$$

the Poisson structure $\overline{\Theta}_2(L)$ (7.286) can be restricted properly to Lax operators of the form (7.232, $k = 0$), where it coincides with $\Theta_2^{\mathrm{red}}(L)$, i.e. the Dirac reduction of quadratic structure (7.267, $k = 0$).

The reader can find the proof in [155].

Remark 7.3 For the case $k = 0$, besides the considered restriction (7.232) of a spectral operator, there exists another important restriction of L to the form [155]

$$L = \mathcal{E}^N + u_{N-1}\mathcal{E}^{N-1} + \dots + u_M\mathcal{E}^M + \psi\mathcal{E}^M\mathcal{D}^{-1}\varphi, \quad N \geq 1 \geq M, \quad (7.288)$$

where $\mathcal{D} = \mathcal{E} - 1$ is a difference operator and $\mathcal{D}^{-1} = \mathcal{E}^{-1} + \mathcal{E}^{-2} + \dots$ its formal inverse, and where the functions ψ and φ inherit the dynamics

$$\psi_{t_q} = ((L^q)_{\geq 0}\psi)_0, \qquad \varphi_{t_q} = -(((L^q)_{\geq 0})^\dagger\varphi)_0. \qquad (7.289)$$

(Compare with Remark 7.1). Moreover, the linear structure $\Theta_1(L)$ (7.250) and the quadratic one $\Theta_2^{\mathrm{red}}(L)$ admit the proper restriction to Lax operators of the form (7.288).

Example 7.16 The simplest example of the Lax operator (7.288) with $N = M = 1$ reads

$$L = \mathcal{E} + \psi\mathcal{E}\mathcal{D}^{-1}\varphi, \qquad L_{\geq 0} = \mathcal{E} + \psi\varphi. \qquad (7.290)$$

The associated hierarchy of Lax equations $L_{t_q} = [(L^q)_{\geq 0}, L]$ is equivalent to (7.289) with the basic system

$$\psi(n)_{t_1} = \psi(n+1) + \varphi(n)\psi^2(n),$$

$$\varphi(n)_{t_1} = -\varphi(n-1) - \psi(n)\varphi^2(n), \qquad (7.291)$$

related to the relativistic Toda lattice (see Example 7.17). Two Poisson structures of (7.291) are the following

$$\theta_1 = \begin{pmatrix} 0 & -1 \\ 1 & 0 \end{pmatrix}, \quad \theta_2 = \begin{pmatrix} -\psi \Pi \psi & E + \psi(1 + \Pi)\varphi \\ -E^{-1} + \varphi(\Pi - 1)\psi & -\varphi \Pi \varphi \end{pmatrix},$$

$$(7.292)$$

with

$$\Pi = (E + 1)\Delta^{-1} : u(n) \rightarrow \sum_{i<n} u(i) - \sum_{i>n} u(i). \qquad (7.293)$$

The Hamiltonians of (7.291) are

$$H_1 = \text{tr}(L) = \sum_{n=-\infty}^{n=+\infty} \psi(n)\varphi(n),$$

$$H_2 = \frac{1}{2}\text{tr}(L^2) = \sum_{n=-\infty}^{n=+\infty} \left[\psi(n)\varphi(n+1) + \frac{1}{2}\psi^2(n)\varphi^2(n) \right].$$

The nontrivial $\overline{\mathcal{R}}_1$-matrix for modified hierarchies comes from the following consideration [155]. One can rewrite the operators from the g algebra (7.223) in an alternative form, that is, being ordered w.r.t. the powers of $\mathcal{D} = \mathcal{E} - 1$ instead of the powers of $\mathcal{E}$:

$$\xi = u_N \mathcal{E}^N + \dots + u_1 \mathcal{E} + u_0 + u_{-1}\mathcal{E}^{-1} + \dots$$

$$= v_N \mathcal{D}^N + \dots + v_1 \mathcal{D} + v_0 + v_{-1}\mathcal{D}^{-1} + \dots \qquad (7.294)$$

with

$$v_N = u_N, \quad v_{N-1} = u_{N-1} + N u_N, \dots, v_0 = u_0 + u_1 + \dots + u_N, \dots .$$

There are two decompositions of g into sub-(Lie) algebras, leading to two $\mathcal{R}$-structures. The first consists of the powers ≥ 0 of $\mathcal{D}$ versus the powers < 0. This is the same as splitting the operators into the powers ≥ 0 and < 0 of E, so that the corresponding integrable systems are covered by our previous considerations. The second decomposition $g = g_{\geq 1} \oplus g_{<1}$ into the powers ≥ 1 of $\mathcal{D}$ versus the powers < 1 is new. The $\overline{\mathcal{R}}_1$-matrix and its adjoint $\overline{\mathcal{R}}_1^*$ w.r.t. the duality map (7.225) are the following

$$\overline{\mathcal{R}}_1(\xi) = \frac{1}{2}(\xi_{\geq 1} - \xi_{<1}) - (\xi \mathcal{D}^{-1})_0,$$

$$\overline{\mathcal{R}}_1^*(\xi) = \frac{1}{2}(\xi_{<0} - \xi_{\geq 0}) - \mathcal{D}^{-1}\xi_0, \qquad (7.295)$$

where $\xi_{\geq 1}, \xi_0, \xi_{<1}$, etc. denote the projection to various powers of the shift operator $\mathcal{E}$. Note that the $\overline{\mathcal{R}}_1$-matrix (7.295) is the identity on $g_{\geq 1}$ and minus the identity on $g_{<1}$, since $(\xi \mathcal{D}^{-1})_0 = u_N + \dots + u_1 = v_0 - u_0$.

The linear Poisson structure (7.66) with the $\overline{\mathcal{R}}_1$-matrix (7.295)

$$\overline{\Theta}_1(L) : \nabla H \to -ad_L \left(\nabla H_{\geq 1} - (\nabla H \mathcal{D}^{-1})_0\right) + (ad_L \nabla H)_{\geq 0} + \mathcal{D}^{-1}(ad_L \nabla H)_0$$

$$= ad_L \left(\nabla H_{<1} + (\nabla H \mathcal{D}^{-1})_0\right) - (ad_L \nabla H)_{<0} - \mathcal{D}^{-1}(ad_L \nabla H)_0 \quad (7.296)$$

produces the integrable hierarchy

$$L_{t_q} = [(L^q)_{\geq 1} - (L^q \mathcal{D}^{-1})_0, L] = -[(L^q)_{<1} + (L^q \mathcal{D}^{-1})_0, L] \quad (7.297)$$

with the commuting Hamiltonians $H = \operatorname{tr}(L^{q+1})/(q+1)$. Equations (7.297) are Miura related to the hierarchy (7.229, $k = 0$). For the Lax operators the link is given by a gauge transformation.

Theorem 7.14 *If $L_{t_q} = [(L^q)_{\geq 0}, L]$ and $\phi_{t_q} = \left((L^q)_{\geq 0}\phi \mathcal{D}^{-1}\right)_0$, then $\overline{L} = \phi^{-1} L \phi$ satisfies (7.297).*

Proof. One finds

$$\overline{L}_{t_q} = [\phi^{-1}(L^q)_{\geq 0}\phi - \phi^{-1}\phi_{t_q}, \overline{L}]$$

$$= [(\overline{L}^q)_{\geq 1} - \phi^{-1}\left((L^q)_{\geq 1}\phi \mathcal{D}^{-1}\right)_0, \overline{L}],$$

where

$$\phi^{-1}\left((L^q)_{\geq 1}\phi \mathcal{D}^{-1}\right)_0 = (\phi^{-1} L^q \phi \mathcal{D}^{-1})_0 = (\overline{L}^q \mathcal{D}^{-1})_0. \square$$

One can check that the dynamics (7.297) can be restricted to Lax operators of the form

$$L = v_N \mathcal{E}^N + v_{N-1}\mathcal{E}^{N-1} + \ldots + v_M \mathcal{E}^M + \mathcal{E}^M \mathcal{D}^{-1} w. \quad (7.298)$$

Lemma 7.16 *The linear Poisson structure (7.296) admits a proper restriction to Lax operators of the form (7.298) with $N \geq 1 \geq M$.*

This is verified by straightforward calculations using the two representations (7.296) to check the highest and lowest orders of $\overline{\Theta}_1(L)\nabla H$.

Now we turn to a discussion of the quadratic Poisson structure associated with (7.297). As the skew-symmetric part of the $\overline{\mathcal{R}}_1$-matrix

$$(\overline{\mathcal{R}}_1 - \overline{\mathcal{R}}_1^*) : \xi \to \xi_{\geq 1} - \xi_{<0} - (\xi \mathcal{D}^{-1})_0 + \mathcal{D}^{-1}\xi_0$$

does not satisfy the modified YB(1) equation, so the quadratic structure (7.67) does not enjoy any Poisson properties. However, the Suris construction of Theorem 7.2 is capable of producing a suitable quadratic Poisson bracket [155].

Theorem 7.15 *Let $\Pi = (E+1)(E-1)^{-1} : u(n) \to \sum_{j<n} u(j) - \sum_{j>n} u(j)$. Then*

$$A_1 : \xi \to \xi_{\geq 1} - \xi_{<0} - 2(\xi \mathcal{D}^{-1})_0 - \Pi(\xi_0) + 2\mathcal{D}^{-1}\xi_0,$$

$$A_2 : \xi \to \xi_{\geq 1} - \xi_{<0} + \Pi(\xi_0),$$

$$S : \xi \to \Pi(\xi_0) - \xi_0 - 2\mathcal{D}^{-1}\xi_0,$$

$$S^* : \xi \to \Pi(\xi_0) - \xi_0 - 2(\xi \mathcal{D}^{-1})_0$$

satisfy all the conditions of Theorem 7.2. The resulting Poisson structure

$$\overline{\Theta}_2(L) : \nabla H \to -ad_L(L\nabla H)_{\geq 1} + \frac{1}{2}ad_L^+(ad_L\nabla H)_0$$

$$+ \mathcal{D}^{-1}[L, \nabla H]_0 L + ad_L(L\nabla H \mathcal{D}^{-1})_0 - \frac{1}{2}ad_L \Pi([\nabla H, L]_0)$$

$$= ad_L(L\nabla H)_{<0} + \frac{1}{2}ad_L(ad_L^+\nabla H)_0$$

$$+ \mathcal{D}^{-1}[L, \nabla H]_0 L + ad_L(L\nabla H \mathcal{D}^{-1})_0 - \frac{1}{2}ad_L \Pi([\nabla H, L]_0) \qquad (7.299)$$

is compatible with the linear Poisson structure (7.296) and admits a proper restriction to Lax operators of the form (7.298) $N \geq 1 \geq M$.

Proof. With $A_1 + S = A_2 + S^* = \overline{\mathcal{R}}_1$ one has to verify (7.71) in Theorem 7.3. The reduction properties of $\overline{\Theta}_2(L)$ can be checked using the two representations (7.98) and the identities

$$(A\varphi \mathcal{E}^M \mathcal{D}^{-1})_{<M} = (A\varphi)\mathcal{E}^M \mathcal{D}^{-1}, \quad (\mathcal{E}^M \mathcal{D}^{-1}\psi A)_{<M} = \mathcal{E}^M \mathcal{D}^{-1}(A^\dagger \psi),$$
$$(7.300)$$

which hold for arbitrary functions φ, ψ and symbols $A \in g_{\geq 0}$.

Example 7.17 The simplest Lax operator of the form (7.298) reads

$$\overline{L} = v\mathcal{E} + \mathcal{E}\mathcal{D}^{-1}w. \qquad (7.301)$$

The basic equation of the integrable hierarchy (7.297) takes the form

$$v(n)_{t_1} = v(n)[v(n+1) - v(n) + w(n+1) - w(n)],$$

$$w(n)_{t_1} = v(n)w(n) - v(n-1)w(n-1) \qquad (7.302)$$

and is known as the relativistic Toda lattice. The first two Hamiltonian structures can be reconstructed from (7.296) and (7.299) as

$$\theta_1 = \begin{pmatrix} 0 & v(E-1) \\ (1-E^{-1})v & 0 \end{pmatrix},$$

$$\theta_2 = \begin{pmatrix} v(E-E^{-1})v & v(E-1)(vE+w) \\ (E^{-1}v+w)(1-E^{-1})v & vwE - E^{-1}vw \end{pmatrix}. \qquad (7.303)$$

The Hamiltonians of (7.302) are

$$H_1 = \mathrm{tr}(L) = \sum_{n=-\infty}^{n=+\infty} w(n), \quad H_2 = \frac{1}{2}\mathrm{tr}(L^2) = \sum_{n=-\infty}^{n=+\infty} \left[v(n)w(n) + \frac{1}{2}w^2(n) \right].$$

$$(7.304)$$

The gauge transformation $L \to \overline{L} = \phi^{-1}L\phi$ of Theorem 7.14 applied to the Lax operator of Example 7.16 yields

$$L = \mathcal{E} + \psi\mathcal{E}\mathcal{D}^{-1}\varphi \to \overline{L} = v\mathcal{E} + \mathcal{E}\mathcal{D}^{-1}w$$

which induces the Miura map in the form

$$v = \psi^{-1}(E\psi), \quad w = \phi\psi. \tag{7.305}$$

8. Towards a Multi-Hamiltonian Theory of (2+1)-Dimensional Field Systems

The last chapter of this book is devoted to the algebraic theory of $(2 + 1)$-dimensional field systems. The reason why the algebraic theory of integrable field systems in multidimensions is remarkably different from that in one space dimension was an important observation made in 1985 by Konopelchenko and Zakharov [195]. Actually, they proved that for $(n + 1)$-dimensional field theory for $n > 1$ there are no recursion operators expressible by pseudo-differential matrix operators as for the $(1 + 1)$ case. As a consequence, there are no second Poisson structures in the same form. Hence, one can try to find them only in the form of bilocal objects, i.e. in a kernel representation, for example. It makes the algebraic theory of integrable field (lattice) systems in multidimensions more complex when compared to these in one space dimension.

We present two different approaches to two dimensional field systems. The first one is the so called Sato approach [56],[99],[156],[169],[170] and allows the construction of infinite hierarchies of commuting vector fields in two space dimensions, together with respective Lax operators. This method was proposed by Sato [169],[170] in 1981. The second approach is a direct extension of the $\mathcal{R}$−matrix formalism from one to more space dimensions. This is possible if one passes from the dynamics on a space of field variables to the dynamics on a space of operator variables. Such a lifting allows the construction of bi-Hamiltonian operator dynamics. Unfortunately it is 'nonphysical' dynamics, so the last step consists of constraining the dynamics to 'physical' invariant subspace of field variables. Consequently, some invariant objects do not survive such a restriction. The reader can notice some analogy to the reduction process from a field level to an invariant finite dimensional manifold considered in Chap. 6. The idea of operator field dynamical systems was introduced by Dorfman and Fokas [59],[76] in 1992, although the kernel representation of a second Poisson structure and a recursion operator was derived for the first time by Fokas and Santini [74],[75],[168] in 1988.

8.1 The Sato Theory

In this section we demonstrate the construction of three classes of integrable equations in 2+1 dimensions within the so called Sato approach

[56],[99],[156],[169],[170]. The idea essentially is that one considers Lax operators spanned over infinitely many fields. Hence, for a fixed k and varying q,
each of the equations (7.91) represents a (1+1)-dimensional integrable system with infinitely many fields. As all equations commute for different values
of q, they may be considered simultaneously. Then one tries to extract a
closed equation for a single field by eliminating the other fields using the
evolution equations. We consider operators of the form (7.106) and without
losing generality, we may assume $N = 1$, as the Lax equation (7.91) for any
$L = u_N \partial^N +$ lower for a given power q leads to the same equation as the
Lax operator $\overline{L} = L^{\frac{1}{N}} = (u_N)^{\frac{1}{N}} \partial +$ lower, with the power Nq. Thus, the
construction depends only on the choice of the nondynamical leading order
coefficients in L. We again choose $c_N = 1$, $c_{N-1} = 0$ in the case $k = 0$ and
$c_N = 1$ in the case $k = 1$. Further, we will fix $N = 1$, and the powers q
in (7.91) will be restricted to the natural numbers. As we will consider the
equations simultaneously, we introduce a different time variable t_n for each
choice of $q = n \in N$. Thus, we will consider the hierarchy of equations

$$\frac{\mathrm{d}}{\mathrm{d}t_n} L = [(L^n)_{\geq k}, L], \qquad n = 1, 2, 3, \dots \qquad (8.1)$$

where we choose

$$k = 0: \qquad L_{KP} = \partial + \frac{1}{2} u \partial^{-1} + u_2 \partial^{-2} + u_3 \partial^{-3} + \dots, \qquad (8.2)$$

as in the standard Sato theory, together with

$$k = 1: \qquad L_{mKP} = \partial + v + v_1 \partial^{-1} + v_2^{-2} + v_3 \partial^{-3} + \dots, \qquad (8.3)$$

$$k = 2: \qquad L_{HD2} = w\partial + w_0 + w_1 \partial^{-1} + w_2 \partial^{-2} + \dots \ . \qquad (8.4)$$

Here, the first fields u, v and w in the L operators carry no indices, since
they are distinguished fields which satisfy the Kadomtsev–Petviashvili (KP)
hierarchy for $k = 0$, the modified KP hierarchy for $k = 1$ and a (2+1)-
dimensional Harry–Dym (HD2) hierarchy for $k = 0$. The remaining fields
$u_2, u_3, \dots, v_1, v_2, \dots, w_0, w_1, \dots$, may be regarded as auxiliary fields to be eliminated in the following procedure.

The Kadomtsev–Petviashvili Hierarchy: $k = 0$

We review the construction of the KP hierarchy via the standard Sato theory.
Thus with $L = L_{KP}$ one readily calculates the differential operators

$$L_{\geq 0} = \partial, \quad (L^2)_{\geq 0} = \partial^2 + u, \quad (L^3)_{\geq 0} = \partial^3 + \frac{3}{2} u\partial + \frac{3}{2} u_x + 3u_2. \qquad (8.5)$$

Obviously the operator $(L^n)_{\geq 0}$ is a differential operator of order n involving
fields $u, u_2, u_3, \dots, u_{n-1}$. Thus, the first of the evolution equations (8.1) reads

$$u_{t_1} = u_x,$$

$$\frac{\mathrm{d}}{\mathrm{d}t_1} L = [(L^1)_{\geq 0}, L] \Leftrightarrow \begin{array}{l} (u_2)_{t_1} = (u_2)_x, \\ (u_3)_{t_1} = (u_3)_x, \\ \vdots \end{array} \qquad (8.6)$$

$$\frac{\mathrm{d}}{\mathrm{d}t_2} L = [(L^2)_{\geq 0}, L] \Leftrightarrow \begin{array}{l} u_y = u_{xx} + 4(u_2)_{xx}, \\ (u_2)_y = (u_2)_{xx} + 2(u_3)_x + \frac{1}{2}uu_x, \\ (u_3)_y = (u_3)_{xx} + 2(u_4)_x - \frac{1}{2}uu_{xx} + 2(u_2)u_x, \\ \vdots \end{array} \qquad (8.7)$$

$$\frac{\mathrm{d}}{\mathrm{d}t_3} L = [(L^3)_{\geq 0}, L]$$

$$\Updownarrow$$

$$\begin{array}{l} u_t = u_{xxx} + 3uu_x + 6(u_2)_{xx} + 6(u_3)_x, \\ (u_2)_t = (u_2)_{xxx} + 3(uu_2)_x + 3(u_3)_{xx} + 3(u_4)_x, \\ \vdots \end{array} \qquad (8.8)$$

with the more convenient notation $y = t_2, t = t_3$. These are the first equations of a hierarchy of commuting coupled equations, involving the infinitely many fields $u, u_2, u_3, \ldots$. For each n, the corresponding evolution equations represent a $(1+1)$-dimensional system in two independent variables t_n and x and infinitely many dependent fields. To eliminate all auxiliary fields u_i from each system one makes use of the particular structure of the equations associated with $t_2 = y$. It is not difficult to note that by a simple integration with respect to x the fields $u_2, u_3, \ldots$, can be recursively expressed in terms of the field u and its derivatives with respect to y and hence eliminated from the higher equations of the hierarchy (8.1). For example, eliminating u_2 and u_3 via (8.7), one can rewrite the first component of (8.8) in terms of u, and its x- and y-derivatives. The resulting equation is the KP equation in the form

$$u_t = \frac{1}{4}(u_{3x} + 6uu_x) + \frac{3}{4}D_x^{-1}u_{yy}. \qquad (8.9)$$

Elimination of the auxiliary fields from the other components of (8.8) simply gives differential consequences of (8.9). One may also consider the higher equations for u arising from (8.1). They represent the members of the KP hierarchy. Using the zero-curvature equation (7.80) the n-th member of the KP hierarchy is encoded in the operator equation

$$\frac{\mathrm{d}}{\mathrm{d}t_n}(L^2)_{\geq 0} - \frac{\mathrm{d}}{\mathrm{d}y}(L^n)_{\geq 0} + [(L^2)_{\geq 0}, (L^n)_{\geq 0}] = 0. \qquad (8.10)$$

The lowest differential order defines the time evolution u_{t_n} in terms of the auxiliary fields entering $(L^n)_{\geq 0}$. The coefficients of the higher differential orders define the constraints, from which the auxiliary fields can be eliminated in terms of u and its y-derivatives.

Now, considering the L_{KP} operator in the form (8.2) as a basic operator for the (2+1)-dimensional KP dynamical system, one can reconstruct the Gelfand-Dikii spectral operator L_{GD} (7.184) for (1+1)-dimensional systems imposing the constraint

$$(L_{KP})^N = L_{GD} = \text{purely differential}, \qquad (8.11)$$

from which all field variables of L_{KP} can be expressed as differential expressions of the $N-1$ field variables of L_{GD}. Moreover, the restriction (7.167) from Remark 7.1 can be obtained by imposing another constraint on L_{KP}, i.e.

$$(L_{KP})^N = L_{GD} + \psi \partial^{-1} \varphi. \qquad (8.12)$$

We interpret this constraint in terms of the prime field u of (8.2) satisfying the KP equation and its hierarchy. Calculating the N-th flow for L_{KP} one obtains

$$\frac{d}{dt_N} L_{KP} = [(L_{KP}^N)_{\geq 0}, L_{KP}] = -[(L_{KP}^N)_{<0}, L_{KP}] = [\psi \partial^{-1} \varphi, L_{KP}]. \quad (8.13)$$

Inserting $L_{KP} = \partial + \frac{1}{2} u \partial^{-1} + \dots$ one extracts the first negative differential order from (8.13) leading to

$$\frac{du}{dt_N} = 2(\psi \varphi)_x. \qquad (8.14)$$

Hence, for Lax operators of the form (7.167), the related evolution equations may be regarded as the KP hierarchy constrained by (8.14), where ψ and φ are eigenfunctions and adjoint eigenfunctions of the KP hierarchy. The reader can find the details of this type of reduction in the series of papers [49],[50],[105],[106],[177],[189],[190].

The Modified Kadomtsev–Petviashvili Hierarchy: $k = 1$

A similar analysis, now with $k = 1$ and $L = L_{mKP}$ (8.3) leads to the modified KP hierarchy for the field v. In this case

$$L_{\geq 1} = \partial, \quad (L^2)_{\geq 1} = \partial^2 + 2v\partial, \quad (L^3)_{\geq 1} = \partial^3 + 3v\partial^2 + 3(v_x + v^2 + v_1)\partial. \qquad (8.15)$$

In general, the operator $(L^n)_{\geq 1}$ is a differential operator of order n which involves the fields $v, v_1, v_2, \dots, v_{n-2}$. The first of the evolution equations (8.1) is

$$\frac{d}{dt_1} L = [L_{\geq 1}, L]$$

$$\updownarrow$$

$$\begin{aligned} v_{t_1} &= v_x, \\ (v_1)_{t_1} &= (v_1)_x, \\ (v_2)_{t_1} &= (v_2)_x, \end{aligned} \qquad (8.16)$$

$$\vdots$$

$$\frac{d}{dt_2}L = [(L^2)_{\geq 1}, L]$$

$$\Updownarrow$$

$$\begin{aligned}
v_y &= v_{xx} + 2vv_x + 2(v_1)_x, \\
(v_1)_y &= (v_1)_{xx} + 2(vv_1)_x + 2(v_2)_x, \\
(v_2)_y &= (v_2)_{xx} - 2(v_1)v_{xx} + 2v(v_2)_x + 4v_x(v_2) + 2(v_3)_x, \\
&\vdots
\end{aligned} \tag{8.17}$$

$$\frac{d}{dt_3}L = [(L^3)_{\geq 1}, L]$$

$$\Updownarrow$$

$$v_t = v_{xxx} + 3(v_1)_{xx} + 3(v_2)_x + 3(vv_x)_x + 3v^2 v_x + 6(vv_1)_x, \tag{8.18}$$
$$\vdots$$

where again we have set $y = t_2$, $t = t_3$. Now, the auxiliary fields $v_1, v_2, ...,$ can be eliminated by expressing them in terms of the field v and its y-derivatives. For example, eliminating v_1 and v_2 via (8.17), one can rewrite the first component of (8.18) in terms of v. The obtained equation is the modified Kadomtsev–Petviashvili (mKP) equation

$$4v_{tx} = (v_{xxx} - 6v^2 v_x)_x + 3v_{yy} + 6v_x v_y + 6v_{xx} D_x^{-1} v_y. \tag{8.19}$$

Considering the higher equations for v arising from $L_{t_n} = [(L^n)_{\geq 1}, L]$ with $n = 4, 5, ...,$ after elimination of the auxiliary fields via (8.17) one gets consecutive members of the mKP hierarchy. Again, the necessary calculations are simplified by considering directly the corresponding zero-curvature equations (7.80) (with $P_{\geq 0}$ replaced by $P_{\geq 1}$).

Now, in analogy to the case $k = 0$, one can reconstruct the spectral operator (7.108) of (1+1)-dimensional systems for the case $k = 1$ imposing the appropriate constraint on the L_{mKP} operator. Actually the constraint takes the form

$$(L_{mKP})^N = \partial^N + u_{N-1}\partial^{N-1} + ... + u_0 + \partial^{-1}u_{-1}. \tag{8.20}$$

We interpret this constraint in terms of the prime field v of (8.3) satisfying the modified KP equation (8.19) and its hierarchy. Calculating the N-th flow for L_{mKP} one obtains

$$\frac{d}{dt_N}L_{mKP} = [(L_{mKP}^N)_{\geq 1}, L_{mKP}] = -[(L_{mKP}^N)_{<1}, L_{mKP}]$$

$$= -[u_0 + \partial^{-1}u_{-1}, L_{mKP}] . \tag{8.21}$$

Inserting L_{mKP} in the form (8.3) one extracts the zero differential order from (8.21) leading to $dv/dt_N = (u_0)_x$. According to the result of Remark 7.2, using $u_0 = \psi - D_x^{-1}\varphi$, one concludes that for spectral Lax operators of the form (7.108) the (1+1)-dimensional bi-Hamiltonian evolution equations (7.112) (for $k = 1$) may be regarded as the mKP hierarchy constrained by

$$\frac{dv}{dt_N} = \psi_x - \varphi. \tag{8.22}$$

Here ψ and φ are eigenfunctions and adjoint eigenfunctions of the modified KP hierarchy satisfying

$$\frac{d}{dt_n}\psi = (B_n\psi)_0, \quad \frac{d}{dt_n}\varphi = -(B_n^\dagger\varphi)_0, \quad B_n = (L_{mKP}^n)_{\geq 1}, \quad n = 1, 2, \dots . \tag{8.23}$$

The (2+1)-Dimensional Harry–Dym Hierarchy: k = 2

A similar analysis, but now with $k = 2$ and $L = L_{HD2}$ (8.4), leads to a hierarchy of (2+1)-dimensional Harry–Dym equations for the field w. Now

$$L_{\geq 2} = 0, \quad (L^2)_{\geq 2} = w^2\partial^2, \quad (L^3)_{\geq 2} = w^3\partial^3 + 3w^2(w_x + w_0)\partial^2. \tag{8.24}$$

The first of the evolution equations (8.1) are

$$\frac{d}{dt_1}L = [L_{\geq 2}, L]$$

$$\Updownarrow$$

$$w_{t_1} = 0$$
$$(w_0)_{t_1} = 0 \tag{8.25}$$
$$\vdots$$

$$\frac{d}{dt_2}L = [(L^2)_{\geq 2}, L]$$

$$\Updownarrow$$

$$w_y = w^2 w_{xx} + 2w^2(w_0)_x,$$
$$(w_0)_y = w^2(w_0)_{xx} + 2w(ww_1)_x,$$
$$(w_1)_y = w^2(w_1)_{xx} + 2(w^2w_2)_x - 2w_1(ww_x)_x, \tag{8.26}$$
$$\vdots$$

$$\frac{d}{dt_3}L = [(L^3)_{\geq 2}, L]$$

$$\Updownarrow$$

$$w_t = w^2 \left(\tfrac{1}{2}(w^2)_{xx} + 3(ww_0)_x + 3w_0^2 + 3ww_1 \right)_x$$
$$\vdots$$
(8.27)

where again $y = t_2$ and $t = t_3$. The auxiliary fields $w_0, w_1, ...$, can be eliminated and expressed in terms of the field w and its y-derivatives. For instance, eliminating w_0 and w_1 via (8.26), one can rewrite the first equation of (8.27) in terms of the field w. It is (2+1)-dimensional version of the Harry–Dym (HD2) equation

$$4w_t = w^3 w_{xxx} - 3w^{-1}(w^2 D_x^{-1}(w^{-1})_y)_y, \tag{8.28}$$

constructed for the first time in [104]. As before, considering the higher equations for w arising from $L_{t_n} = [(L^n)_{\geq 2}, L]$ with $n = 4, 5, ...$, after elimination of the auxiliary fields by (8.26) one obtains the whole hierarchy of the HD2.

In order to derive the (2+1)-dimensional version of the AKNS system, the Davey–Stewardson (DS) equation or the Ishimori equation (the Heisenberg ferromagnet in (2+1)-dimensions) together with their Lax representations, we briefly sketch the matrix Sato theory [107],[154]. The basic objects to construct the integrable equations of the matrix (modified) KP hierarchy are the following *dressing operators*

$$k = 0 : \quad W = 1 + w_1 \partial^{-1} + w_2 \partial^{-2} + ...,$$

$$k = 1 : \quad W = w_0 + w_1 \partial^{-1} + w_2 \partial^{-2} + ..., \tag{8.29}$$

where the matrix fields $w_0, w_1, ...$, which are functions of a variable x, taking values in some finite dimensional (matrix) algebra g_0, are the dynamical fields of the (modified) KP hierarchy satisfying a certain set of nonlinear differential equations. The label $k = 0$ is attributed to the matrix KP hierarchy, whereas $k = 1$ yields the modified matrix KP hierarchy.

For an invertible leading order coefficient these operators can be inverted by means of the ansatz $W^{-1} = 1 + v_1 \partial^{-1} + v_2 \partial^{-2} + ...$ and $W^{-1} = w_0^{-1} + v_1 \partial^{-1} + v_2 \partial^{-2} + ...$ respectively. The requirement $WW^{-1} = 1$ leads to a straightforward recursive scheme for the coefficients v_j, from which they can be expressed as differential expressions of the coefficients w_j. In particular, the first terms of the expressions W^{-1} are given as

$$k = 0 : \quad W^{-1} = 1 - w_1 \partial^{-1} + (w_1^2 - w_2)\partial^{-2} + ...,$$

$$k = 1 : \quad W^{-1} = w_0^{-1} - w_0^{-1} w_1 w_0^{-1} \partial^{-1} + (..)\partial^{-2} + ... \tag{8.30}$$

The coefficients w_j of W are to be endowed with a certain dynamics. For this we choose a collection of commuting elements of g_0, and define a collection of *undressed operators*

$$\mathcal{A} = \{A_n = a_n \partial^n, \ n \in N, \ a_n \in g_0, \ [a_n, a_m] = 0\}. \tag{8.31}$$

With each element $A \in \mathcal{A}$ we associate the dynamics of the dressing operator W by requiring that

$$W_{t_A} = -(WAW^{-1})_{<k}W = -WA + (WAW^{-1})_{\geq k}W, \qquad (8.32)$$

where t_A denotes the evolution parameter ('time') of this flow. Equation (8.31) is known as the *Sato equation*. Note that the Sato equation defines a consistent dynamics for all the coefficients w_j in W, as the right-hand side of (8.32) is a pseudo-differential operator with highest differential order $k-1$. For $k=0$ the set of nonlinear equations for the coefficients w_j in W obtained from (8.32) is called the multi-component KP hierarchy (associated with the collection $\mathcal{A}$ of undressed operators), whereas we refer to the case $k=1$ as the modified multicomponent KP hierarchy. The justification for these names is the fact that for the scalar case $g_0 = \mathbb{R}$ and $k=0$ the field w_1 satisfies the potential KP equation, whereas for $k=1$ the field $u = \ln w_0$ satisfies the potential version of the modified KP equation.

To rephrase the system (8.32) as a hierarchy of zero-curvature formulations one defines the differential operators

$$M(A) = (WAW^{-1})_{\geq k} \qquad (8.33)$$

as the dressed versions of the undressed operators $A \in \mathcal{A}$. A straightforward calculations yields the identities

$$W_{t_A t_B} - W_{t_B t_A} = -\left(W(A_{t_B} - B_{t_A} - [B,A])W^{-1}\right)_{<k}W,$$

$$M(A)_{t_B} - M(B)_{t_A} - [M(B), M(A)] = \left(W(A_{t_B} - B_{t_A} - [B,A])W^{-1}\right)_{\geq k},$$
$$(8.34)$$

for $k = 0, 1$ and any undressed operators A, B, which in principle may depend on the time variables. The simple proof of this identity only makes use of the fact that in both cases $k=0$ and $k=1$ the subspaces $g_{\geq k} = \left\{\sum_{i \geq k} u_i \partial^i\right\}$ and $g_{<k} = \left\{\sum_{i<k} u_i \partial^i\right\}$ form proper Lie subalgebras in the space of all pseudo-differential symbols. As the undressed operators in $\mathcal{A}$ are assumed to be constant and commuting, the flows (8.32) associated with two different undressed operators $A, B \in \mathcal{A}$ commute: $W_{t_A t_B} = W_{t_B t_A}$. The *dressed operators* (8.33) satisfy the zero-curvature equation

$$M(A)_{t_B} - M(B)_{t_A} = [M(B), M(A)] \qquad (8.35)$$

for any choice of the pair A, B.

We remark that the Sato equation (8.32) immediately leads to the Lax equation. To show this one fixes an arbitrary element $C \in \mathcal{A}$ and defines a pseudo-differential Lax operator

$$L(C) = WCW^{-1}. \qquad (8.36)$$

With the identity

$$L(C)_{t_A} = -(WAW^{-1})_{<k}WCW^{-1} + WCW^{-1}(WAW^{-1})_{<k} + WC_{t_A}W^{-1}$$
$$= [M(A) - WAW^{-1}, L(C)] + WC_{t_A}W^{-1}$$
$$= [M(A), L(C)] + W(C_{t_A} - [A, C])W^{-1} \qquad (8.37)$$

one finds that $L(C)$ satisfies the hierarchy of Lax equations $L(C)_{t_A} = [M(A), L(C)]$.

Example 8.1 The KP hierarchy: the scalar case.
With the choice $A = \partial^2$, $B = \partial^3$ and the notation $y = t_A, t = t_B$ the Sato equation (8.32) with $k = 0$ implies

$$(w_1)_y = (w_1)_{xx} + 2(w_2)_x - 2(w_1)_x w_1,$$

$$(w_2)_y = (w_2)_{xx} + 2(w_3)_x - 2(w_1)_x w_2, \qquad (8.38)$$

and

$$(w_1)_t = (w_1)_{xxx} + 3(w_2)_{xx} + 3(w_3)_x - 3(w_1)_x^2 - 3(w_1)_x(w_2)$$
$$- 3[(w_1)_{xx} + (w_2)_x - (w_1)(w_1)_x]w_1. \qquad (8.39)$$

The set of equations (8.38) can be used to express the fields w_2, w_3 in terms of w_1, and its x- and y-derivatives. Insertion into (8.39) yields the potential KP equation

$$4(w_1)_t = (w_1)_{4x} - 12(w_1)_x(w_1)_{xx} + 3(w_1)_{yy}. \qquad (8.40)$$

Alternatively, one may consider the dressed operators

$$M(A) = (W\partial^2 W^{-1})_{\geq 0} = \partial^2 - 2(w_1)_x,$$

$$M(B) = (W\partial^3 W^{-1})_{\geq 0} = \partial^3 - 3(w_1)_x\partial - \frac{3}{2}[(w_1)_{xx} + (w_1)_y]. \qquad (8.41)$$

and derive (8.40) directly from the zero-curvature condition (8.35). Application of the undressed operators $B_n = \partial^n$, $n \geq 4$, generates the hierarchy of higher KP flows.

Example 8.2 The modified KP hierarchy: the scalar case.
With the choice $A = \partial^2$, $B = \partial^3$ and the notation $y = t_A, t = t_B$ the Sato equation (8.32) with $k = 1$ implies

$$(w_0)_y = (w_0)_{xx} - (w_0)_x^2 w_0^{-1} + 2(w_1 w_0^{-1})_x w_0,$$

$$(w_1)_y = (w_1)_{xx} - 2(w_1)_x(w_0)_x w_0^{-1} + 2(w_2 w_0^{-1})_x w_0, \qquad (8.42)$$

and

$$(w_0)_t = (w_0)_{xxx} + 6(w_0)_x^3 w_0^{-2} - 6(w_0)_x(w_0)_{xx}w_0^{-1} + 3(w_1 w_0^{-1})_{xx}$$

$$- 3(w_0)_x w_0^{-1}(w_1 w_0^{-1})_x - 3w_1 w_0^{-1}(w_1 w_0^{-1})_x + 3(w_2 w_0^{-1})_x. \qquad (8.43)$$

Using (8.42) one can eliminate the fields w_1 and w_2 from (8.43). The analysis is simplified by introducing the new fields u, v_1, v_2 via

$$w_0 = e^u, \quad w_1 = v_1 e^u, \quad w_2 = v_2 e^u. \tag{8.44}$$

Elimination of v_1 and v_2 as indicated above results in the potential form of the modified KP equation

$$4u_{tx} = u_{4x} - 6u_x^2 u_{xx} - 6u_{xx} u_y + 3u_{yy}. \tag{8.45}$$

The associated dressed operators are calculated as

$$M(A) = (W\partial^2 W^{-1})_{\geq 1} = \partial^2 - 2u_x\partial,$$

$$M(B) = (W\partial^3 W^{-1})_{\geq 1} = \partial^3 - 3u_x\partial^2 - \frac{1}{2}(u_{xx} - u_x^2 + u_y)\partial. \tag{8.46}$$

Application of the undressed operators $B_n = \partial^n$, $n \geq 4$, generates the hierarchy of higher modified KP flows for $u = \ln w_0$.

Example 8.3 The KP hierarchy: the 2×2 case.
For the simplest multi-component case the coefficients $w_1, w_2, \ldots$ of the dressing operator are 2×2 matrices. Let us consider the collection

$$\mathcal{A} = \{a_n\partial^n, \ a_n \in \{1, \sigma_3\}, \ n \in N\} \tag{8.47}$$

of commuting undressed operators, where the matrix coefficients a_n are either the 2×2 identity matrix or the third Pauli matrix $\sigma_3 = \text{diag}(1, -1)$. With the choice $A = \sigma_3\partial$ and $B = \sigma_3\partial^2$ we shall use the zero-curvature formulation in order to derive the nonlinear equations in question. The coefficient w_2 may be found directly in terms of w_1 by evaluating $W_y = -(W\sigma_3 W^{-1})_{<0}W$. Its highest order reads

$$(w_1)_y = [\sigma_3, w_2] + \sigma_3(w_1)_x + [w_1, \sigma_3]w_1. \tag{8.48}$$

Hence, we find the following parametrization of the dressed operators

$$M(A) = (W\sigma_3\partial W^{-1})_{\geq 0} = \sigma_3\partial + [w_1, \sigma_3],$$

$$M(B) = (W\sigma_3\partial^2 W^{-1})_{\geq 0} = \sigma_3\partial^2 + [w_1, \sigma_3]\partial - (w_1)_y - \sigma_3(w_1)_x \tag{8.49}$$

in terms of the field w_1. Let us introduce

$$U = \begin{pmatrix} 0 & q \\ r & 0 \end{pmatrix} = [w_1, \sigma_3], \quad D = \begin{pmatrix} a & 0 \\ 0 & b \end{pmatrix} \tag{8.50}$$

in order to decompose w_1 into its off-diagonal and diagonal part according to $w_1 = -\frac{1}{2}(\sigma_3 U + D)$. The zero-curvature condition (8.35) then leads to the evolution equation

$$2U_t = \sigma_3(U_{xx} + U_{yy} + UD_x + D_xU) + [D_y, U]$$

$$+ D_{yy} - D_{xx} + (U^2)_x + \sigma_3(U^2)_y. \tag{8.51}$$

From the diagonal part of (8.48) one has

$$\sigma_3 D_y - D_x + U^2 = 0, \tag{8.52}$$

hence

$$D_{yy} - D_{xx} + (U^2)_x + \sigma_3(U^2)_y = 0 \tag{8.53}$$

and the diagonal part of (8.51) is satisfied identically. The combined system (8.51)/(8.52) may be rewritten as

$$U_t = \frac{1}{2}\sigma_3(U_{xx} + U_{yy}) + \sigma_3 U^3 + [D_y, U], \quad \sigma_3 D_y - D_x + U^2 = 0, \tag{8.54}$$

i.e.

$$q_t = \frac{1}{2}(q_{xx} + r_{yy}) + q^2 r + (a_y - b_y)q, \quad a_y - a_x + qr = 0,$$

$$r_t = \frac{1}{2}(r_{xx} + q_{yy}) - r^2 q - (a_y - b_y)r, \quad b_y + b_x - qr = 0, \tag{8.55}$$

which represents a (2+1)-dimensional version of the first member of the AKNS hierarchy. As we have $(a+b)_y = (a-b)_x$ one may introduce the field χ by $a + b = \chi_x$, $a - b = \chi_y$, which has to satisfy $\chi_{xx} - \chi_{yy} = 2qr$. If we replace t by $-it$ and impose the constraints $r = q^*$, $\chi = \chi^*$, then this system reduces to the 'real' version of the Davey–Stewardson equation

$$iq_t = \frac{1}{2}(q_{xx} + q_{yy}) + \frac{1}{2}q(\chi_{xx} + \chi_{yy}), \quad \chi_{xx} - \chi_{yy} = 2|q|^2. \tag{8.56}$$

Commuting flows of higher order may be generated by using undressed operators of the form ∂^n and $\sigma_3\partial^m$.

Example 8.4 The modified KP hierarchy: the 2×2 case.
As in the KP case, we consider the collection (8.47) of undressed operators. With the choice $A = \sigma_3\partial$, $B = \sigma_3\partial^2$ the coefficient w_1 may be found directly in terms of w_0 by evaluating $W_y = -(W\sigma_3\partial W^{-1})_{<1}W$, i.e.

$$(w_0)_y = [S, w_1 w_0^{-1}]w_0 + S(w_0)_x, \quad S = w_0\sigma_3 w_0^{-1}. \tag{8.57}$$

The dressed operators $M(A) = (W\sigma_3\partial W^{-1})_{\geq 1}$ and $M(B) = (W\sigma_3\partial^2 W^{-1})_{\geq 1}$ are calculated as

$$M(A) = S\partial, \quad M(B) = S\partial^2 + T\partial, \tag{8.58}$$

where

$$T = [w_1 w_0^{-1}, S] - 2S(w_0)_x w_0^{-1} \overset{(8.57)}{=} -(w_0)_y w_0^{-1} - S(w_0)_x w_0^{-1}. \tag{8.59}$$

The zero-curvature equation (8.35) leads to

$$S_t - T_y = SS_{xx} + TS_x - ST_x, \quad S_y + SS_x = [S, T]. \tag{8.60}$$

A more compact formulation of these equations is obtained by the representation

$$S = \mathbf{S} \cdot \boldsymbol{\sigma} = S_1 \sigma_1 + S_2 \sigma_2 + S_3 \sigma_3, \quad T = \mathbf{S} \cdot \boldsymbol{\sigma} + \beta I, \qquad (8.61)$$

where

$$\sigma_1 = \begin{pmatrix} 0 & 1 \\ 1 & 0 \end{pmatrix}, \quad \sigma_2 = \begin{pmatrix} 0 & -i \\ i & 0 \end{pmatrix}, \quad \sigma_3 = \begin{pmatrix} 1 & 0 \\ 0 & -1 \end{pmatrix}$$

are the three Pauli matrices. We note that $S = w_0 \sigma_3 w_0^{-1}$ is traceless and satisfies $S^2 = I$. From (8.57) one finds

$$\mathrm{Tr}[w_0^{-1}(w_0)_y] = \mathrm{Tr}[w_0^{-1}(w_0)_x \sigma_3], \quad \mathrm{Tr}[w_0^{-1}(w_0)_x] = \mathrm{Tr}[w_0^{-1}(w_0)_y \sigma_3]. \quad (8.62)$$

With $\mathbf{S} = (S_1, S_2, S_3)$ and $\mathbf{T} = (T_1, T_2, T_3)$ one concludes

$$\beta = \frac{1}{2}\mathrm{Tr}(T) = \mathrm{Tr}[w_0^{-1}(w_0)_y] = \tau_y,$$

$$\mathbf{S} \cdot \mathbf{T} = \frac{1}{2}\mathrm{Tr}(ST) = -\mathrm{Tr}[w_0^{-1}(w_0)_x] = -\tau_x, \qquad (8.63)$$

where $\tau = \ln \det w_0$. Decomposition of the matrix equations (8.60) yields

$$\mathbf{S}_y + i\mathbf{S} \times \mathbf{S}_x = 2i\mathbf{S} \times \mathbf{T}, \qquad (8.64)$$

and

$$\mathbf{S}_t - \mathbf{T}_y = i\mathbf{S} \times \mathbf{S}_{xx} + i(\mathbf{T} \times \mathbf{S})_x + \beta \mathbf{S}_x - \beta_x \mathbf{S},$$

$$-\beta_y = \mathbf{S} \cdot \mathbf{S}_{xx} + \mathbf{T} \cdot \mathbf{S}_x - \mathbf{T}_x \cdot \mathbf{S}, \qquad (8.65)$$

Since $\mathbf{T}$ is completely determined by (8.63) and (8.64), one can solve it for $\mathbf{T}$ and obtain

$$\mathbf{T} = \frac{1}{2} i\mathbf{S} \times \mathbf{S}_y + \frac{1}{2}\mathbf{S}_x - \tau_x \mathbf{S}, \qquad (8.66)$$

having used $\mathbf{S}^2 = 1$. Insertion into (8.65) yields the Ishimori equation ((2+1)-dimensional Heisenberg ferromagnet)

$$\mathbf{S}_t = \frac{1}{2} i\mathbf{S} \times (\mathbf{S}_{xx} + \mathbf{S}_{yy}) - \tau_x \mathbf{S}_y - \tau_y \mathbf{S}_x,$$

$$\tau_{xx} - \tau_{yy} = i\mathbf{S} \cdot (\mathbf{S}_x \times \mathbf{S}_y). \qquad (8.67)$$

8.2 Multi-Hamiltonian Lax Dynamics
for Noncommutative Operator Valued Field Variables

In the previous section we have derived a few best known $(2+1)$-dimensional integrable field systems together with their zero-curvature representation. In order to extend the $\mathcal{R}$−matrix formalism to such systems we first have to find their Lax representations. This can be done in the following way. The zero-curvature representation for each hierarchy of the previous section takes the form

$$U_{t_n} - V_{n_y} + [U, V_n] = 0, \tag{8.68}$$

where U and V represent appropriate differential operators. Noticing that

$$V_y = ad_{\partial_y} V = [\partial_y, V], \tag{8.69}$$

the zero-curvature representation can be put into the Lax form

$$L_{t_n} + [L, V_n] = 0, \qquad L = U - \partial_y. \tag{8.70}$$

Example 8.5 The KP hierarchy.

$$L = \partial_x^2 - \partial_y + u, \qquad V_n = (L_{KP}^n)_{\geq 0}. \tag{8.71}$$

Example 8.6 The modified KP hierarchy.

$$L = \partial_x^2 + 2v\partial_x - \partial_y, \qquad V_n = (L_{mKP}^n)_{\geq 1}. \tag{8.72}$$

Example 8.7 The $(2+1)$-dimensional Harry–Dym hierarchy.

$$L = w^2\partial_x^2 - \partial_y, \qquad V_n = (L_{HD2}^n)_{\geq 2}. \tag{8.73}$$

Remark 8.1 Note that for systems like KP (8.9) and HD2 (8.28) for example, the change $y \to -y$ does not affect the form of the dynamics. Thus, they have alternative Lax pairs $(\overline{L}, \overline{V})$ where $\overline{L} = L(y \to -y)$ and $\overline{V} = V(y \to -y)$. For instance in the KP case, the dynamics (8.9) is generated by both pairs

$$L = \partial_x^2 - \partial_y + u, \quad V = \partial_x^3 + \frac{3}{2}u\partial_x + \frac{5}{4}u_x + \frac{1}{4}D_x^{-1}u_y,$$

and

$$\overline{L} = \partial_x^2 + \partial_y + u, \quad \overline{V} = \partial_x^3 + \frac{3}{2}u\partial_x + \frac{5}{4}u_x - \frac{1}{4}D_x^{-1}u_y.$$

Example 8.8 The $(2+1)$-dimensional AKNS hierarchy.

$$L = \sigma_3\partial_x - I_y + U, \quad U = \begin{pmatrix} 0 & q \\ r & 0 \end{pmatrix},$$

$$V_n = (W\sigma_3\partial_x^n W^{-1})_{\geq 0}. \tag{8.74}$$

Example 8.9 The Ishimori hierarchy.

$$L = S\partial_x - \partial_y, \quad S = \begin{pmatrix} S_3 & S_1 - iS_2 \\ S_1 + iS_2 & -S_3 \end{pmatrix},$$

$$V_n = (W\sigma_3\partial_x^n W^{-1})_{\geq 1}. \tag{8.75}$$

The form of the known Lax operators for (2+1)-dimensional systems suggests that one can try to extend the $\mathcal{R}$-matrix formalism to the algebra of pseudo-differential operators

$$g = \left\{ L = \sum_{-\infty < i < N} \sum_{-\infty < j < M} u_{i,j}(x,y)\partial_x^i\partial_y^j \right\}, \tag{8.76}$$

where $u_{i,j}$ are basic dynamical fields. One can check that the appropriate trace form

$$\mathrm{tr}(L) = \int_R \int_R u_{-1,-1}\mathrm{d}x\,\mathrm{d}y \tag{8.77}$$

yields a symmetric and nondegenerate pairing on g.

The $\mathcal{R}$-matrix formalism developed in the previous chapter comes from the admissible decomposition of the algebra of x-pseudo-differential operators into a direct sum of two Lie subalgebras. Unfortunately, for the algebra (8.76) such a decomposition does not exist. Thus, what one can do is the following [59],[76].

Let us consider a ring R consisting of smooth (matrix-valued in general) functions of the basic dynamical variables $u_i(x,y)$ $i = 0,...,m$, of its x and y derivatives and of primitives (integrals). Typical elements of R are $u, u_y, D_x^{-1}u_y, u_x y_{yy}$, e.c.t. With the ring R we relate the ring $R\{\partial_y\}$ of formal pseudo-differential operators in ∂_y with coefficients in the ring R

$$v = \sum_{-\infty < k}^{M} a_k\partial_y^k, \quad a_k \in R. \tag{8.78}$$

Consider the derivatives D_x and D_y of the ring $R\{\partial_y\}$ defined by the formulas

$$D_x v = \sum_{-\infty < k}^{M} (a_k)_x\partial_y^k, \quad D_y v = \sum_{-\infty < k}^{M} (a_k)_y\partial_y^k. \tag{8.79}$$

Any derivation $\partial : R\{\partial_y\} \to R\{\partial_y\}$ that commutes with D_x and D_y forms a Lie algebra. Further, each such derivation is completely defined by its action on the basic dynamical variables u_i. Thus, the Lie algebra of derivations can be identified with the linear space of vectors $v = (v^0,...v^m)^{\mathrm{T}}$, $v^i \in R\{\partial_y\}$, endowed with the Lie bracket

$$[v_1, v_2] = v_2'[v_1] - v_1'[v_2], \tag{8.80}$$

where the prime denotes the Frechet derivative. We denote this algebra by $\mathcal{V}$.

To construct the space of zero-forms ('operator functionals'), we introduce in $R\{\partial_y\}$ a relation of equivalence: $F_1 \sim F_2$ iff F_1 can be obtained from F_2 by cyclic permutation of basic variables in any term of it, or by overthrowing the x and y derivatives in a conventional way. For example $u_{xx}u\partial_y^{-1} \sim \partial_y^{-1}u_{xx}u \sim u\partial_y^{-1}u_{xx} \sim -u_x\partial_y^{-1}u_x$. We denote the class of equivalences of F by $\uparrow F\downarrow$ and $\uparrow F\downarrow \in \Omega_0$. The basic properties of elements of Ω_0 are

$$\uparrow F_1 F_2 \downarrow = \uparrow F_2 F_1 \downarrow, \qquad \uparrow D_x F\downarrow = 0, \qquad \uparrow D_y F\downarrow = 0. \tag{8.81}$$

Vector fields v act on the space Ω_0 of these equivalence classes by the formula

$$v\uparrow F\downarrow = \uparrow F'[v]\downarrow = \uparrow\left(\frac{\delta F}{\delta u}\right)v\downarrow, \qquad v, F, \delta F/\delta u \in R\{\partial_y\}. \tag{8.82}$$

For example, when $m = 0$, $u_0 = u$ and $\uparrow F\downarrow = \uparrow u\partial_y^{-1}u_{xx}\downarrow$, then $\uparrow F'[v]\downarrow = \uparrow v\partial_y^{-1}u_{xx} + u\partial_y^{-1}v_{xx}\downarrow = \uparrow\left(\partial_y^{-1}u_{xx} + u_{xx}\partial_y^{-1}\right)v\downarrow$. $\delta F/\delta u$ is the pseudo-differential operator $\partial_y^{-1}u_{xx} + u_{xx}\partial_y^{-1} = 2u_{xx}\partial_y^{-1} - u_{3x}\partial_y^{-2} + \cdots$.

Similarly we define Ω^1 as the space of one-forms whose elements, denoted by γ, are linear mappings from $\mathcal{V}$ to Ω^0 represented by the formula

$$\gamma(v) = \uparrow\gamma v\downarrow, \qquad \gamma, v \in R\{\partial_y\}. \tag{8.83}$$

Note that Ω^1 contains the differentials of all the elements Ω^0. Indeed

$$(d\uparrow F\downarrow)(v) = v\uparrow F\downarrow = \uparrow\left(\frac{\delta F}{\delta u}\right)v\downarrow. \tag{8.84}$$

Now we can consider Hamiltonian operators $\theta : \Omega^1 \to \mathcal{V}$. Evidently, any operator of this kind can be represented by a matrix $\theta^{ij} : R\{\partial_y\} \to R\{\partial_y\}$, $i, j = 0, ..., m$, so that a Hamiltonian vector field K takes the usual form

$$K_i = \sum_j \theta^{ij}\frac{\delta F}{\delta u_j}. \tag{8.85}$$

The ring $R\{\partial_y\}$ itself consists of operators, so θ^{ij} acts in the space of operators. To emphasize this fact, the operators acting in $R\{\partial_y\}$ are called *operands* (i.e. operators acting in the space of operators).

Now we shall come back to the $\mathcal{R}$-matrix formalism. Let us consider the following operator algebra

$$g = \left\{L = \sum_{-\infty < i}^{N} V_i\partial_x^i\right\}, \tag{8.86}$$

where $V_i \in R\{\partial_y\}$ are elements such that

$$V_i'[v] = v_i, \qquad v \in \mathcal{V}. \tag{8.87}$$

For example one may put $V_i = u_i + \{\partial_y\}_i$, where $\{\partial_y\}_i$ denotes some constant pseudo-differential operator with respect to ∂_y. At this stage of the considerations $R\{\partial_y\}$ represents a noncommutative ring of y-pseudo-differential operators (8.78) built up on an infinite number of field variables $u_i = u_i(x, y)$. The trace form

$$\mathrm{tr}(L) = \uparrow V_{-1}\downarrow \tag{8.88}$$

yields a symmetric pairing on $g : (L_1, L_2) = \mathrm{tr}(L_1 L_2)$.

Applying the $\mathcal{R}$-matrix formalism to the algebra (8.86), together with the results of the noncommutative case from the end of Sect. 7.3, the following conclusions can be drawn. There are two possible decompositions of g

$$g_{\geq k} = \left\{ \sum_{i \geq k} V_i \partial_x^i \right\}, \qquad g_{< k} = \left\{ \sum_{i < k} V_i \partial_x^i \right\}, \quad k = 0, 1 \tag{8.89}$$

with the following forms of the L operators

$$k = 0: \quad L = C_N \partial_x^N + V_{N-1} \partial_x^{N-1} + \dots + V_0 + V_{-1} \partial^{-1} + \dots,$$

$$L = I \partial_x^N + C_{N-1} \partial_x^{N-1} + \dots + V_0 + V_{-1} \partial^{-1} + \dots,$$

$$k = 1: \quad L = V_N \partial_x^N + V_{N-1} \partial_x^{N-1} + \dots + V_0 + V_{-1} \partial^{-1} + \dots,$$

$$L = I \partial_x^N + V_{N-1} \partial_x^{N-1} + \dots + V_0 + V_{-1} \partial^{-1} + \dots, \tag{8.90}$$

where I is a unit operator and C_N, C_{N-1} are constant operators different from I. The basic admissible restrictions to a finite number of field operators V_i are the following

$$k = 0: \quad L = C_N \partial_x^N + V_{N-1} \partial_x^{N-1} + \dots + V_0, \tag{8.91}$$

$$L = I \partial_x^N + C_{N-1} \partial_x^{N-1} + \dots + V_0, \tag{8.92}$$

$$k = 1: \quad L = V_N \partial_x^N + V_{N-1} \partial_x^{N-1} + \dots + V_0 + \partial^{-1} V_{-1}, \tag{8.93}$$

$$L = I \partial_x^N + V_{N-1} \partial_x^{N-1} + \dots + V_0 + \partial^{-1} V_{-1}. \tag{8.94}$$

Constructing Hamiltonian structures via the $\mathcal{R}$-matrix formalism we get the bi-Hamiltonian formulation for $(2 + 1)$-dimensional systems in the form

$$V_t = K(V) = \theta_1(V) \frac{\delta H_2(V)}{\delta u} = \theta_2(V) \frac{\delta H_1(V)}{\delta u}, \tag{8.95}$$

where

$$V = (V_1, \dots, V_m)^{\mathrm{T}}, \quad V_i = u_i + \{\partial_y\}_i.$$

The price one pays for the possibility of finding a bi-Hamiltonian formulation in $(2 + 1)$-dimension is its 'nonphysical' representation. Actually, the representation (8.95) is given for operator field variables V_i instead of function fields u_i.

Example 8.10 The Kadomtsev–Petviashvili (KP) Lax operator: $\partial_x^2 + (u + \partial_y)$. We consider the case (8.92) with $N = 2, C_1 = 0, V_0 = V = u + \partial_y$. Then $\nabla H(L) = \partial_x^{-1}\gamma$, where now $\gamma = \delta H(V)/\delta u$ is an operator which does not commute with the field function $u(x, y)$. Two Hamiltonian operands calculated from formulas (7.185) and (7.188) are found in the form

$$\theta_1 = D_x,$$

$$\theta_2 = D_x^3 + ad_V^+ D_x + D_x ad_V^+ + ad_V D_x^{-1} ad_V$$
$$= D_x^3 + ad_u^+ D_x + D_x ad_u^+ + ad_u D_x^{-1} ad_u + D_y D_x^{-1} ad_u + ad_u D_x^{-1} D_y$$
$$+ D_x^{-1} D_y^2 + 2D_x D_y + 4\rho_{\partial_y} D_x. \tag{8.96}$$

The recursion operand $\Psi = \theta_1^{-1} \circ \theta_2$ (for cosymmetries), when acting on $\gamma_0 \in \ker \theta_1$, generates the hierarchy γ_n of closed one-forms if γ_0 is a closed one form. We choose as γ_0 an arbitrary constant operator

$$\gamma_0 = \sum_{i=-M}^{N} c_i \partial_y^i = \frac{\delta}{\delta u} \uparrow \gamma_0 u \downarrow \in \ker \theta_1. \tag{8.97}$$

Hence, each Casimir of θ_1 generates an infinite hierarchy of bi-Hamiltonian operator vector fields

$$K_{n+1} = \theta_2 \circ \Psi^n \circ \gamma_0 = \Phi^n \circ \theta_2 \circ \gamma_0 = \Phi^n \circ K_1, \tag{8.98}$$

where $\Phi = \theta_2 \circ \theta_1^{-1}$ is an appropriate recursion operand generating symmetries.

Example 8.11 The nonsymmetric Novikov–Veselov Lax operator: $\partial_y \partial_x + u$. We consider the case (8.91) with $N = 1, C_1 = \partial_y$ and $V_0 = u$. Two Hamiltonian operands generated by (7.185) and (7.188) are found in the form

$$\theta_1 = D_y,$$

$$\theta_2 = \lambda_{\partial_y} \rho_u - \lambda_u \rho_{\partial_y} - \lambda_{\partial_y} \rho_{\partial_y} D_x. \tag{8.99}$$

Again each Casimir of θ_1 of the form (8.97) generates an infinite hierarchy of bi-Hamiltonian operator vector fields $K_{n+1} = \Phi^n \circ \theta_2 \circ \gamma_0$.

Example 8.11 is a special case of a general form of the first order Lax operator (8.91)

$$L = C\partial_x + V, \tag{8.100}$$

where C is a constant and V is a nonconstant element of $R\{\partial_y\}$. The second Hamiltonian operand according to (7.188) takes the form

$$\theta_2 = \lambda_C \rho_V - \lambda_V \rho_C - \lambda_C \rho_C D_x. \tag{8.101}$$

In order to produce a compatible pair of operands we consider a deformation of V, $V \to V + \mu B$, $\mu \in \mathbb{R}$, B a constant element of $R\{\partial_y\}$. This gives us a Hamiltonian operand

$$\theta_1 = \lambda_C \rho_B - \lambda_B \rho_C, \tag{8.102}$$

which together with θ_2 constitutes a compatible Hamiltonian pair. Note that for the special case when $B = I$ the Hamiltonian operand (8.102) is equal to that produced by the formula (7.185). Furthermore, for $C = I$ we still have a bi-Hamiltonian representation, this gives a generalization of the case (8.91) with $N = 1$ to all constant C including $C = I$.

As shown in the examples, the bi-Hamiltonian recursion scheme generates the hierarchy of vector fields being elements of the ring $R\{\partial_y\}$. Therefore in order to construct flows which would remain in the (matrix) function part of $R\{\partial_y\}$, i.e. 'physical' dynamical systems, we must arrange for cancellations of the operator parts in the recursion scheme. Actually, the appropriate combination of operator flows must be found which would produce ordinary ones. How to do it for an arbitrary Lax operator is at present an open problem. Nevertheless, there exists a special class of Lax operators for which it can be done [59],[16].

Theorem 8.1 *Suppose we are given a compatible pair of Hamiltonian (Poisson) operands θ_1 and θ_2 and $\Phi = \theta_2 \circ \theta_1^{-1}$ constitutes the recursion operand. Suppose further that*

(i) θ_1 is of the ff-type, i.e. takes functions to functions,
(ii) $[\theta_1, \rho_{\partial_y}] = 0$,
(iii) $\theta_2 = \theta - \epsilon \theta_1 \rho_{\partial_y}$, where $\epsilon \in \mathbb{R}$ and θ is ff-type,

then for an arbitrary function $A \in \ker \theta_1$ the operators

$$K_n = \sum_{k=0}^{n} \epsilon^k \binom{n}{k} \Phi^{n-k} \circ \theta_2 \circ (A \partial_y^k) \tag{8.103}$$

are functions.

Proof. Let

$$S_n = \sum_{k=0}^{n} \epsilon^k \binom{n}{k} \Psi^{n+1-k} \circ \rho_{\partial_y}^k, \quad \Psi = \theta_1^{-1} \circ \theta_2. \tag{8.104}$$

Then the formula

$$S_{n+1} = \Psi_1 \circ S_n + \epsilon [S_n, \rho_{\partial_y}], \tag{8.105}$$

where $\Psi_1 = \theta \circ \theta_1^{-1}$ is *ff*-type, can be checked directly using the properties of the binomial coefficients. Applying θ_1 to (8.105) using condition (ii) we obtain

$$\theta_1 \circ S_{n+1} = \Psi_1 \circ \theta_1 \circ S_n + [\theta_1, \Psi_1] \circ S_n + \epsilon [\theta_1 \circ S_n, \rho_{\partial_y}]. \tag{8.106}$$

One may use this to show that, for arbitrary n,

$$\theta_1 \circ S_n = P_n + Q_n \circ \rho_{\partial_y} \circ \theta_1, \tag{8.107}$$

where P_n and Q_n are generalized operators of $f\!f$-type. This can be done by induction. For $n = 0$ we have

$$\theta_1 \circ S_0 = \theta_1 \circ \Psi = \theta_1 \circ \Psi_1 - \epsilon \rho_{\partial_y} \circ \theta_1$$

and $P_0 = \theta_1 \circ \Psi_1$ is $f\!f$-type by (i) and (ii), and $Q_0 = -\epsilon$. Now assume that (8.107) is valid for some n. Using (8.106) and (ii), we have

$$\theta_1 \circ S_{n+1} = P_{n+1} + Q_{n+1} \circ \rho_{\partial_y} \circ \theta_1, \tag{8.108}$$

where

$$P_{n+1} = \Psi_1 \circ P_n + [\theta_1, \Psi_1] \circ \theta_1^{-1} \circ P_n + \epsilon[P_n, \rho_{\partial_y}],$$

$$Q_{n+1} = \Psi_1 \circ Q_n + [\theta_1, \Psi_1] \circ \theta_1^{-1} \circ Q_n + \epsilon[Q_n, \rho_{\partial_y}]. \tag{8.109}$$

It is easy to see by hypothesis and conditions (i) and (iii) that P_{n+1} and Q_{n+1} are $f\!f$-type and (8.106) is thus proved by induction. Finally, for an element $F \in \ker \theta_1$, $\theta_1 \circ S_n(A) = P_n(A)$ is a function. But

$$\theta_1 \circ S_n = \sum_{k=0}^{n} \epsilon^k \binom{n}{k} \theta_1 \circ \Psi^{n+1-k} \circ \rho_{\partial_y}^k \tag{8.110}$$

and so

$$K_n = \theta_1 \circ S_n(A) = \sum_{k=0}^{n} \epsilon^k \binom{n}{k} \Phi^{n-k} \circ \theta_2 \circ (A \partial_y^k), \tag{8.111}$$

the expression given in the statement of the theorem. Consequently this expression is a function. $\square$

Lemma 8.1 *The following residue representation of the hierarchy (8.103) can be given*

$$K_n = \text{res} \, (\Phi^n \circ \theta_2 \circ A \partial_y^{-1}), \tag{8.112}$$

where $\text{res} \equiv \text{res}_{\partial_y}$.

Proof. From the previous Theorem we find that

$$\Psi_1 = \bar{\Psi} + \epsilon \rho_{\partial_y}, \qquad \Psi = \theta_1^{-1} \circ \theta_2. \tag{8.113}$$

Then from relation (8.109) for the $f\!f$-type function P_{n+1} we get the following representation

$$P_{n+1} = \Phi \circ P_n + \epsilon P_n \circ \rho_{\partial_y}. \tag{8.114}$$

Hence we obtain the following

$$K_n = \text{res}[\rho_{\partial_y^{-1}} \circ \theta_1 \circ S_n(A)] = \text{res}[\rho_{\partial_y^{-1}} \circ (P_n + Q_n \circ \rho_{\partial_y} \circ \theta_1)(A)]$$

$$= \text{res}(P_n \circ A\partial_y^{-1}) = \text{res}(\Phi^n \circ P_n \circ A\partial_y^{-1}) = \text{res}(\Phi^n \circ \theta_1 \circ \Psi_1 \circ A\partial_y^{-1})$$

$$= \text{res}(\Phi^n \circ \theta_1 \circ \Psi \circ A\partial_y^{-1}) = \text{res}\ (\Phi^n \circ \theta_2 \circ A\partial_y^{-1}). \ \square$$

Remark 8.2 Note that the schema proposed in Theorem 8.1 on the basis of R-matrix approach with operator value field variables, was proposed for the first time by Magri, Morosi and Tondo in a different language of the so called Nijenhuis G-manifolds and Lenard bicomplexes [126],[127], [137],[138].

Example 8.12 The KP hierarchy.
It is readily verifiable that the pair of Hamiltonian operands (8.96) satisfies the conditions of Theorem 8.1 with $\epsilon = -4$. Hence, choosing $A = \frac{1}{2}$ one gets

$$K_1 = \frac{1}{2}\Phi \circ \theta_2 1 - 2\theta_2\partial_y = u_{xxx} + 6uu_x + 3D_x^{-1}u_{yy}, \tag{8.115}$$

i.e. the KP vector field. For $n > 1$ one finds the hierarchy of KP symmetries.

Example 8.13 The nonsymmetric Novikov–Veselov hierarchy [60],[37].
We note that the pair of Hamiltonian operands (8.99) does not satisfy the requirements of Theorem 8.1, so we cannot expect to obtain a hierarchy of functions described by the formula (8.103). What one can do is the following. Let us take the generic Lax operator in the form [60]

$$L = \begin{pmatrix} 0 & 0 \\ 0 & 1 \end{pmatrix} \partial_x + \begin{pmatrix} \partial_y & u \\ v & 0 \end{pmatrix}. \tag{8.116}$$

Notice that the reduction $v = 1$ is equivalent to the NV Lax operator $L = \partial_y\partial_x + u$. As L of (8.116) contains only two from four possible dependent variables, the Hamiltonian formalism must be subjected to a reduction.

According to formula (7.214) with the dropped factor $-\frac{1}{2}$ in θ_1, in the matrix representation for the dynamics

$$\begin{pmatrix} 0 & u \\ v & 0 \end{pmatrix}_t = \bar{\theta}_1 \begin{pmatrix} \gamma_{11} & \gamma_{12} \\ \gamma_{21} & \gamma_{22} \end{pmatrix}, \tag{8.117}$$

we have

$$\bar{\theta}_1 \begin{pmatrix} \gamma_{11} & \gamma_{12} \\ \gamma_{21} & \gamma_{22} \end{pmatrix} = \begin{pmatrix} 0 & -\gamma_{12} \\ \gamma_{21} & 0 \end{pmatrix},$$

$$\bar{\theta}_2 \begin{pmatrix} \gamma_{11} & \gamma_{12} \\ \gamma_{21} & \gamma_{22} \end{pmatrix} = \begin{pmatrix} 0 & -\partial_y.\gamma_{12} - u \cdot \gamma_{22} \\ \gamma_{21}.\partial_y + \gamma_{22} \cdot v & \gamma_{21} \cdot u - v \cdot \gamma_{12} - (\gamma_{22})_x \end{pmatrix}, \tag{8.118}$$

with the restriction

$$\gamma_{22} = D_x^{-1}(\gamma_{21} \cdot u - v \cdot \gamma_{12}). \tag{8.119}$$

It is easy to verify that the sufficient conditions from Theorem 8.1 are indeed satisfied with $\epsilon = -1$ and for

$$A = \begin{pmatrix} 0 & 0 \\ 0 & 1 \end{pmatrix} \in \ker \bar{\theta}_1. \tag{8.120}$$

For the vector representation of the dynamics

$$\begin{pmatrix} u \\ v \end{pmatrix}_t = \theta \begin{pmatrix} \gamma_u \\ \gamma_v \end{pmatrix}, \tag{8.121}$$

where $\gamma_{12} = \gamma_v$ and $\gamma_{21} = \gamma_u$, one finds

$$\theta_1 = \begin{pmatrix} 0 & -1 \\ 1 & 0 \end{pmatrix},$$

$$\theta_2 = \begin{pmatrix} -\lambda_u D_x^{-1}\rho_u & -\lambda \partial_y + \lambda_u D_x^{-1}\lambda_v \\ \rho \partial_y + \rho_v D_x^{-1}\rho_u & -\rho_v D_x^{-1}\lambda_v \end{pmatrix} \tag{8.122}$$

and hence

$$\Phi = \theta_2 \theta_1^{-1} = \begin{pmatrix} \lambda \partial_y - \lambda_u D_x^{-1}\lambda_v & -\lambda_u D_x^{-1}\rho_u \\ \rho_v D_x^{-1}\lambda_v & \rho \partial_y + \rho_v D_x^{-1}\rho_u \end{pmatrix}. \tag{8.123}$$

Then the hierarchy of mutually commuting vector fields given by formula (8.103) reads

$$K_n = \sum_{k=0}^{n} (-1)^k \begin{pmatrix} n \\ k \end{pmatrix} \Phi^{n-k} \sigma_k, \quad \sigma_k = \begin{pmatrix} -u\partial_y^k \\ \partial_y^k v \end{pmatrix}, \tag{8.124}$$

where σ_k is the vector representation of the element $\bar{\theta}_2(A\partial_y^k)$. The first two nontrivial members K_2 and K_3 correspond to the following systems of evolution equations

$$\begin{pmatrix} u \\ v \end{pmatrix}_t = \begin{pmatrix} -u_{yy} + 2u D_x^{-1}(uv)_y \\ v_{yy} - 2v D_x^{-1}(vu)_y \end{pmatrix} \tag{8.125}$$

and

$$\begin{pmatrix} u \\ v \end{pmatrix}_t = \begin{pmatrix} -u_{yyy} + 3[u D_x^{-1}(u_y v)]_y + 3u_y D_x^{-1}(uv_y) \\ -v_{yyy} + 3[v D_x^{-1}(v_y u)]_y + 3v_y D_x^{-1}(u_y v) \end{pmatrix}. \tag{8.126}$$

Under the reduction $v = 1$, the system (8.126) yields

$$u_t = -u_{yyy} + 3(u D_x^{-1} u_y)_y, \tag{8.127}$$

which is the nonsymmetric version of the Novikov–Veselov equation.

Let us come back to the first order Lax operators (8.100) and assume them to be in the form

$$L = \partial_x + A\partial_y + Q, \tag{8.128}$$

where A is a constant operator and $V = A\partial_y + Q$ belongs to $R\{\partial_y\}$. Then θ_2 is given by

$$\theta_2 = \rho_{Q+A\partial_y} - \lambda_{Q+A\partial_y} - D_x = -D_x - ad_Q - AD_y - ad_A\rho_{\partial_y}. \quad (8.129)$$

In order to apply Theorem 8.1 we must carefully choose the deformation of V. Deforming it in the A direction gives

$$\theta_1 = -ad_A \quad (8.130)$$

and consequently the requirements of the Theorem are satisfied for $\epsilon = -1$.

All considerations of this section are unchanged under the assumption of the interchanged role of differential operators ∂_x and ∂_y . Actually one can consider the ring $R\{\partial_x\}$ instead of $R\{\partial_y\}$ and the algebra of y-pseudo-differential operators

$$g = \left\{ L = \sum_{-\infty < i}^{N} V_i \partial_y^i \right\}, \quad (8.131)$$

instead of the algebra (8.86). Let us illustrate the situation on the example of the KP hierarchy.

Example 8.14 The KP Lax operator $L = \partial_y + (\partial_x^2 + u)$.
Consider L as the first order Lax operator (8.128) with respect to ∂_y, with $A = \partial_x$ and $Q = u$. From (8.129) and (8.130) we get a new compatible pair of Hamiltonian operands

$$\theta_1 = D_x, \qquad \theta_2 = ad_u + 2\rho_{\partial_x} D_x + D_x^2 + D_y, \quad (8.132)$$

which is much simpler then (8.96). In fact nonlocal terms in (8.96) are due to the reduction process caused by the absence of the first order ∂_x term in the Lax operator, whereas no reductions are made in deriving the pair (8.132) and so there are no nonlocalities. The pair (8.132) obviously satisfies the requirements of Theorem 8.1 with $\epsilon = -2$, so the hierarchy of vector fields K_n can be obtained by the formula (8.103) with A taken to be $\frac{1}{2}$. The K_n is again the KP hierarchy with K_3 equal to (8.115).

In the same way one can find the second pair of Hamiltonian operands for the hierarchy of Example 8.14 taking now

$$L = \begin{pmatrix} 1 & 0 \\ 0 & 0 \end{pmatrix} \partial_y + \begin{pmatrix} 0 & 0 \\ 0 & 1 \end{pmatrix} \partial_x + \begin{pmatrix} 0 & u \\ v & 0 \end{pmatrix}. \quad (8.133)$$

Moreover, with respect to

$$L = \begin{pmatrix} 1 & 0 \\ 0 & 1 \end{pmatrix} \partial_y + \begin{pmatrix} -1 & 0 \\ 0 & 1 \end{pmatrix} \partial_x + \begin{pmatrix} 0 & q \\ r & 0 \end{pmatrix} \quad (8.134)$$

one can get the Hamiltonian pair and commuting hierarchy for the $(2 + 1)$-AKNS system and the Davey-Stewardson equation as its special case.

Let us note that the same results were obtaine in the papers [161],[162], where the $\mathcal{R}$-matrix formalism was applied to the zero-curvature representation with operator valued field variables.

We complete our considerations comparing the results of this section to the alternative approach to multi-Hamiltonian dynamics in (2+1)-dimensions given by the kernel (bilocal) representation [168],[74],[75]. In this approach operators are represented by kernels, i.e. to any $K \in R\{\partial_y\}$ there corresponds an $\widehat{K}(x, y, y')$ such that

$$(Kf)(x, y) = \int_{-\infty}^{+\infty} \widehat{K}(x, y, y')f(x, y')dy'. \tag{8.135}$$

The multiplicative law in terms of kernels is represented by the formula

$$KG(x, y, y') = \int_R \widehat{K}(x, y, y'')\widehat{G}(x, y'', y')dy'' := \widehat{K} \circ \widehat{G}. \tag{8.136}$$

We have the following correspondence between integro-differential and bilocal representation:

$$a(x, y)\partial_y^k \to a(x, y)\delta^{(k)}(y - y'), \quad k \geq 0,$$

$$b(x, y)\partial_y^{-1} \to b(x, y)\theta(y - y'),$$

$$c(x, y)\partial_y^{-k} \to \frac{1}{(k-1)!}c(x, y)(y - y')^{k-1},$$

$$V = u(x, y) + \partial_y \to \widehat{V} = u(x, y)\delta(y - y') + \delta'(y - y'), \tag{8.137}$$

where δ is the Dirac delta function, $\frac{\partial}{\partial y}\theta(y - y') = \delta(y - y')$ and $\frac{\partial^k \delta(y-y')}{\partial y^k} = \delta^{(k)}(y - y')$. Operands are also represented in the space of kernels. To any operand O there corresponds $\widehat{O}$ such that $\widehat{O} \circ \widehat{K} = \widehat{OK}$. As an illustration we compute $\widehat{(ad_V^\pm)}$. We have

$$\widehat{(ad_V^\pm)} \circ \widehat{K} = \widehat{V} \circ \widehat{K} \pm \widehat{K} \circ \widehat{V}. \tag{8.138}$$

Substituting the expression for $\widehat{V}$ (8.137) and using the multiplication law (8.136) we get

$$\widehat{(ad_V^\pm)} = u(x, y) \pm u(x, y') + D_y \mp D_{y'}. \tag{8.139}$$

Hence, for example, the operand θ_2 given by (8.96) is converted into

$$\widehat{\theta_2} = D_x - \widehat{ad_V^+}D_x - D_x\widehat{ad_V^+} + \widehat{ad_V}D_x^{-1}\widehat{ad_V}, \tag{8.140}$$

where expression (8.139) must be substituted. This is the form in which this operator appeared in [74], [75].

Lemma 8.2 [59] *Let G be an operator whose kernel $\widehat{G}(x, y, y') = \theta(y - y')g(x, y, y')$. Then*

$$\operatorname{res} G = g(x, y, y) = \int_R \delta(y - y')g(x, y, y')dy'. \qquad (8.141)$$

Proof.

$$(Gf)(x,y) = \int_{-\infty}^{+\infty} \theta(y - y')g(x,y,y')f(x,y')dy'$$

$$= \int_{-\infty}^{y} g(x,y,y')f(x,y')dy'$$

$$= \int_{-\infty}^{y} \sum_{k=0}^{\infty} \frac{\partial^k g}{\partial y'^k}(x,y,y) \frac{(y - y')^k}{k!} f(x,y')dy'$$

$$= \sum_{k=0}^{\infty} \frac{\partial^k g}{\partial y'^k}(x,y,y)\partial_y^{-k-1} f,$$

so we have $\operatorname{res} G = g(x, y, y)$. $\square$

From Lemma 8.1 it follows that the bilocal representation of the hierarchy (8.112) is given by the formula

$$u_{t_n}(x,y) = \int_R \delta(y - y')(\widehat{\Phi}^n \circ \widehat{\theta}_2 A)(x,y,y')dy', \qquad (8.142)$$

which indeed is the form which particular cases were derived in [74], [75],[168].

References

1. Ablowitz M.J., Kaup D.J., Newell A.C. and Segur H. (1974): *The inverse scattering transforme-Fourier analysis for nonlinear problems*, Stud.Appl.Math. **53** 249
2. Ablowitz M.J. and Segur H. (1981): *Solitons and the inverse scattering transform*, Philadelphia: SIAM
3. Abraham R. and Marsden J.E. (1978): *Fundations of Mechanics*, Benjamin/Cummings, New York, Second Edition
4. Adler M. (1979): *On a trace functional for pseudo differential operators and the symplectic structure of the Korteweg-de Vries equation*, Invent. Math. **50** 219
5. Antonowicz M., Fordy A.P. and Wojciechowski S. (1987): *Integrable stationary flows: Miura maps and bi-Hamiltonian structures*, Phys. Lett. **A 124** 143
6. Antonowicz M. and Fordy A.P. (1987): *Coupled KdV equations with multi-Hamiltonian structures*, Physica **D 28** 345
7. Antonowicz M. and Fordy A.P. (1988): *Coupled Harry–Dym equations with multi-Hamiltonian structures*, J. Phys. A: Math. Gen. **21** 2269
8. Antonowicz M. and Fordy A.P. (1989): *Factorization of energy dependent Schrödinger operators: Miura maps and modified systems*, Comm. Math. Phys. **124** 465
9. Antonowicz M. and Fordy A.P. (1989): in *Soliton theory: a survey of results*, ed. A. Fordy, (Manchester University Press) p. 273
10. Antonowicz M. and Rauch-Wojciechowski S. (1990): *Constrained flows of integrable PDE's and bi-Hamiltonian structure of Garnier system*, Phys. Lett. **A 147** 455
11. Antonowicz M. and Błaszak M. (1990): *On a non-standard Hamiltonian description of NLEE*, in: Nonlinear Evolution Equations and Dynamical Systems, eds. S.Carillo and O.Ragnisco, Springer-Verlage p. 152
12. Antonowicz M. and Rauch-Wojciechowski S. (1991): *Restricted flows of soliton hierarchies: coupled KdV and Harry–Dym case*, J. Phys. A: Math. Gen. **24** 5043
13. Antonowicz M. and Rauch-Wojciechowski S. (1992): *How to construct finite-dimensional bi-Hamiltonian systems from soliton equations: Jacobi integrable potentials*, J. Math. Phys. **33** 2115
14. Antonowicz M. and Rauch-Wojciechowski S. (1992): *Bi-Hamiltonian formulation of the Henon-Heiles system and its multidimensional extension*, Phys. Lett. **A 163** 167
15. Arnold V.I. (1978): *Mathematical methods of classical mechanics*, Springer-Verlag, New York
16. Athorne C. and Dorfman I.Ya. (1993): *The Hamiltonian structure of the (2+1)-dimensional AKNS hierarchy*, J. Math. Phys. **34** 3507

17. Baker S., Enolskii V.Z. and Fordy A.P. (1995): *Integrable quartic potentials and coupled KdV equations*, Phys. Lett. A **201** 167
18. Błaszak M. (1986): *The method of finding Lax pairs for higher order equations of soliton hierarchies*, Acta Phys. Polon. A **70** 487
19. Błaszak M. (1988): *On interacting solitons*, Acta Phys. Polon. A **74** 439
20. Błaszak M. (1989): *Benjamin–Ono interacting solitons as field representatives of Galilean point particles*, J. Phys. A: Math. Gen. **22** 451
21. Błaszak M. (1989): *Theory of classical soliton particles*, A. Mickiewicz University Press **60** (University of Poznań)
22. Błaszak M. and Wojciechowski S. (1989): *Bi-Hamiltonian dynamical systems related to low-dimensional Lie algebras*, Physica **155A** 545
23. Błaszak M and Oevel G. (1991): *Action-angle representation of complex multisolitons*, Prog. Theor. Phys. **86** 29
24. Błaszak M. (1991): *Symmetries on a soliton manifold* in Nonlinear fields: Classical, Random, Semiclassical, World Scientific Publishing, Singapore p. 215
25. Błaszak M. (1991): *Symmetries, conservation laws and multisoliton perturbation theory*, J. Phys. A: Math. Gen. **24** 4459
26. Błaszak M. (1991): *Multisoliton adiabatic perturbation theory. Algebraic approach.* in Nonlinear evolution equations and dynamical systems, Springer-Verlag
27. Błaszak M. and Basarab-Horwath P. (1992): *Bi-Hamiltonian formulation of a finite dimensional integrable system reduced from a Lax hierarchy of the KdV*, Phys. Lett. A **171** 45
28. Błaszak M. (1993): *On a non-standard algebraic description of integrable nonlinear systems*, Physica **198A** 637
29. Błaszak M. (1993): *Lagrangian–Hamiltonian formulation for stationary flows of some classes of nonlinear dynamical systems*, J. Phys. A: Math. Gen. **26** L263
30. Błaszak M. and Rauch-Wojciechowski S. (1993): *Newton representation of nonlinear ordinary differential equations*, Physica **197A** 191
31. Błaszak M. (1993): *Miura map and bi-Hamiltonian formulation for restricted flows of the KdV hierarchy.*, J. Phys. A: Math. Gen. **26** 5985
32. Błaszak M. (1993): *Bi-Hamiltonian field Garnier system*, Phys. Lett. A **174** 85
33. Błaszak M. and Rauch-Wojciechowski S. (1994): *A generalized Henon–Heiles system and related integrable Newton equations*, J. Math. Phys. **35** 1693
34. Błaszak M. and Marciniak K. (1994): *R-matrix approach to lattice integrable systems*, J. Math. Phys. **35** 4661
35. Błaszak M. (1995): *Newton representation for stationary flows of some class of nonlinear dynamical systems*, Physica **215A** 201
36. Błaszak M. (1995): *Bi-Hamiltonian formulation for the Kortweg-de Vries hierarchy with sources*, J. Math. Phys. **36** 4826
37. Błaszak M. (1997): *R-matrix approach to multi-Hamiltonian Lax dynamics*, Rep. Math. Phys. **40** 395
38. Błaszak M. (1998): *On separability of bi-Hamiltonian chain with degenerated Poisson structures*, J. Math. Phys. **39** 3213
39. Brouzet R. (1993): *About the existence of recursion operators for completely integrable Hamiltonian systems near the Liouville torus*, J. Math. Phys. **34** 1309
40. Brouzet R., Caboz R., Rabenivo J. and Ravoson V. (1996): *Two degrees of freedom quasi-bi-Hamiltonian systems*, J. Phys. A: Math. Gen. **29** 2069
41. Bullough R.K. and Caudrey eds. (1980): *Solitons (Topics in Current Physics)* Springer-Verlag, Berlin

42. Calogero F. (1971): *Solution of the one-dimensional n-body problem with quadratic and/or inversely quadratic pair potentials*, J. Math. Phys. **12** 419

43. Calogero F. and Degasperis A. (1982): *Spectral transforms and solitons*, North-Holland Publishing Company, Amsterdam

44. Calogero F. and Nucci M.C. (1991): *Lax pairs galore*, J. Math. Phys. **32** 72

45. Cao C. and Geng X. (1990): *Classical integrable systems generated through nonlinearization of eigenvalue problems*, in: Research Reports in Physics. Nonlinear Physics, editors: Gu Chaohao, Li Yishen and Tu Guizhang, Springer-Verlag

46. Cao C. (1990): *Nonlinearization of the Lax system for the AKNS hierarchy*, Science in China **A** 528

47. Cao C. and Geng X. (1990): *Neumann and Bargmann systems associated with the coupled KdV soliton hierarchy*, J. Phys. A: Math. Gen. **23** 4117

48. Cao C. and Geng X. (1991): *A nonconfocal generator of involutive systems and three associated soliton hierarchies*, J. Math. Phys. **32** 2323

49. Cheng Yi and Li Yishen (1992): *Constraints of the 2+1 dimensional integrable soliton systems*, J. Phys. A: Math. Gen. **25** 419

50. Cheng Yi. (1992): *Constraints of the Kadomtsev–Petviashvili hierarchy*, J. Math. Phys. **33** 3774

51. Choquet-Bruchat Y., de Vith-Morette C. and Dillard-Bleick (1982): *Analysis, Manifolds and Physics*, North-Holland Publishing Company, Amsterdam

52. Crampin M. and Pirani F.A.E. (1988): *Applicable Differential Geometry*, Cambridge University Press, Lecture Note Series 59

53. Damianou P.A. (1993): *Symmetries of Toda equations*, J. Phys. A: Math. Gen. **26** 3791

54. Damianou P.A. (1994): *Multiple Hamiltonian structures for Toda-type systems*, J. Math. Phys. **35** 5511

55. Das A. and Okubo S. (1989): *A systematic study of Toda lattice* , Ann. Phys. **190** 215

56. Date E.,Jimbo M., Kashiwara M. and Miwa T. (1983): *Transformation groups for soliton equations*, in: Nonlinear Integrable Systems – Classical Theory and Quantum Theory, eds. M.Jimbo and T.Miwa, World Scientific, Singapore, p.39

57. Davydov A.S. (1984): *Solitons in molecular systems*, Naukova Dumka, Kiev (in Russian)

58. Dickey L.A. (1991): *Soliton equations and Hamiltonian systems*,World Scientific Publishing, Singapore

59. Dorfman I.Ya. and Fokas A.S. (1992): *Hamiltonian theory over noncommutative rings and integrability in multidimensions*, J. Math. Phys. **33** 2504

60. Dorfman I.Ya. and Athorne C. (1993): *On the nonsymmetric Novikov-Veselov hierarchy*, Phys. Lett. **A 182** 369

61. Dorfman I.Ya. (1993): *Dirac structures and integrability of nonlinear equations*, Wiley

62. Drinfeld V.G. (1983): *Hamiltonian structures on Lie groups, Lie bialgebras and the geometric meaning of classical Yang–Baxter equations*, Soviet Math. Dokl. **27** 68

63. Eisenhart L.P. (1926): *Riemannian geometry*, Princeton University Press, Princeton

64. Faddeev L.D. and Takhatajan L.A. (1987): *Hamiltonian methods in the theory of solitons*, Springer-Verlag, New York-Berlin-Heidelberg

65. Ferapontov E.V. and Fordy A.P. (1997): *Separable Hamiltonians and integrable systems of hydrodynamic type*, J. Geom. and Phys. **21** 169

66. Fernandes R.L. (1993): *On the master symmetries and bi-Hamiltonian structure of the Toda lattice*, J. Phys. A: Math. Gen. **26** 3797

67. Fernandes R.L. (1994): *Completely integrable bi-Hamiltonian systems*, J. Dynam. Differential Equations **6** 53
68. Flaschka H. (1974): *On the Toda lattice* , Phys. Rev. **9** 1924
69. Flaschka H., Newell A.C. and Ratiu T. (1983): *Kac-Moody Lie algebras and soliton systems*, Physica **9D** 300
70. Fokas A.S. and Fuchssteiner B. (1980): *On the structure of symplectic operators and hereditary symmetries*, Lett. Al Nuovo Cimento **28** 299
71. Fokas A.S. and Fuchssteiner B. (1981): *The hierarchy of the Benjamin-Ono equation*, Phys. Lett. **A 86** 341
72. Fokas A.S. and Anderson R.L. (1982): *On the use of isospectral eigenvalue problems for obtaining hereditary symmetries for Hamiltonian systems*, J. Math. Phys. **23** 1066
73. Fokas A.S. (1987): *Symmetries and integrability*, Stud. Appl. Math. **77** 253
74. Fokas A.S. and Santini P.M. (1988): *Recursion operators and bi-Hamiltonian structures in multisolitons II*, Commun. Math. Phys. **115** 449
75. Fokas A.S. and Santini P.M. (1988): *Bi-Hamiltonian formulation of the Kadomtsev-Petviashvili and Benjamin-Ono equations*, J. Math. Phys. **29** 604
76. Fokas A.S. and Gelfand I.M. (1992): *Bi-Hamiltonian structures and integrability* in: Recent developments in soliton theory, edited by A.S.Fokas and V.E.Zakharov, Springer-Verlage, New York
77. Fokas A.S. and Gelfand I.M. Eds. (1997): *Algebraic aspects of integrable systems*, Birkhauser
78. Fordy A.P.F. and Gibbons J. (1980): *Some remarkable nonlinear transformations*, Phys. Lett. **A 75** 325
79. Fordy A.P.F. (1991): *The Henon-Heiles system revisited*, Physica **52D** 204
80. Fordy A.P. (1995): *Stationary flows: Hamiltonian structures and canonical transformations*, Physica D **87** 20
81. Fordy A.P. and Harris S.D. (1997): *Hamiltonian flows on stationary manifolds*, Methods and Applications of Analysis **2** 212
82. Fuchssteiner B. (1979): *Application of hereditary symmetries to nonlinear evolution equations*, Nonl. Anal. TMA **3** 849
83. Fuchssteiner B. and Fokas A.S. (1981): *Symplectic structures, their Bäcklund transformations and hereditary symmetries*, Physica **4D** 47
84. Fuchssteiner B. (1982): *The Lie algebra structure of degenerate Hamiltonian and bi-Hamiltonian systems*, Prog. Theor. Phys. **68** 1082
85. Fuchssteiner B. and Oevel W. (1982): *The bi-Hamiltonian structure of some nonlinear fifth and seventh order differential equations and recursion formulas for their symmetries and conserved covariants*, J. Math. Phys. **23** 358
86. Fuchssteiner B. (1983): *Mastersymmetries, higher order time-dependent symmetries and conserved densities of nonlinear evolution rquations*, Prog. Theor. Phys. **70** 1508
87. Fuchssteiner B. (1987): *Solitons in interaction*, Prog. Theor. Phys. **78** 1022
88. Fuchssteiner B. and Aiyer R.N. (1987): *Multisolitons, or the discrete eigenfunctions of the recursion operator of non-linear evolution equations: II. Background*, J. Phys. A: Math. Gen. **20** 375
89. Fuchssteiner B. and Oevel W. (1987): *New hierarchies of nonlinear completely integrable systems related to a change of variables for the evolution parameter*, Physica **A 68** 67
90. Gardner S.C., Green J.M., Kruskal M.D. and Miura R.M. (1974): *Korteweg-de Vries equation and generalization. VI Methods for exact solution*, Comm. Pure Appl. Math. **27** 97
91. Gelfand I.M. and Dikii L.A. (1976): *Fractional powers of operators and Hamiltonian systems*, Funct. Anal. Appl. **10** 259

92. Gelfand M. and Dorfman I.Ya. (1979): *Hamiltonian operators and algebraic structure related to them*, Funct. Anal. Appl. **13** 248
93. Gelfand I.M. and Dorfman I.Ya. (1980): *The Schouten bracket and Hamiltonian operators*, Funct. Anal. Appl. **14** 223
94. Gelfand I.M. and Dorfman I.Y. (1982): *Hamiltonian operators and the classical Yang–Baxter equation*, Funct. Anal. Appl. **17** 241
95. Gu Z. and Baocai Z. (1991): *The coupled HD hierarchy and a classical integrable system of the complex form*, J. Phys. A: Math. Gen. **24** 963
96. Gu Z. (1991): *Complex confocal involutive systems associated with the solutions of the AKNS evolution equations*, J. Math. Phys. **32** 1498
97. Gu Z. (1991): *Two finite-dimensional completely integrable Hamiltonian systems associated with the solutions of the MKdV hierarchy*, J. Math. Phys. **32** 1531
98. Hongwei Zhang, Gui-zhang Tu, Oevel W. and Fuchssteiner B.(1991): *Symmetries, conserved quantities and hierarchies for some lattice systems with soliton structure*, J. Math. Phys. **32** 1908
99. Jimbo M. and Miwa T. (1983): *Solitons and infinite-dimensional Lie algebras*, Publ. RIMS, Kyoto Univ. **19** 943
100. Kadomtsev B.B. and Petviashvili V.I. (1970): *On the stability of waves in a medium with waek dispersion*, Sov. Phys. Dokl. **192** 753
101. Karpman I. and Maslow (1977): *Perturbation theory for solitons*, Sov. Phys. JETP **48** 281
102. Karpman I. and Solovev V.V. (1981): *A perturbation theory for soliton systems*, Physica **3D** 142
103. Kazhadan D., Kostant B. and Sternberg S. (1978): *Hamiltonian group actions and dynamical systems of Calogero type*, Comm. Pure Appl. Math. **31** 481
104. Konopelchenko B. and Dubrovsky V. (1984): *Some new integrable nonlinear evolution equations in 2+1 dimensions*, Phys. Lett. A **102** 15
105. Konopelchenko B. and Strampp W. (1991): *The AKNS hierarchy as symmetry constraint of the KP hierarchy*, Inverse Problems **7** L17
106. Konopelchenko B. and Strampp W. (1992): *New reductions of the Kadomtsev–Petviashvili and two dimensional Toda lattice hierarchies via symmetry constraints*, J. Math. Phys. **33** 3676
107. Konopelchenko B. and Oevel V. (1992): *Matrix Sato theory and integrable equations in 2+1 dimensions*, in: Nonlinear evolution equations and dynamical systems, eds. M.Boiti, L.Martina and F.Pempinelli, World Scientific, Singapore, p. 87
108. Konopelchenko B.G. and Oevel W. (1993): *An r-matrix approach to nonstandard classes of integrable equations*, Publ. RIMS., Kyoto Univ. **29** 581
109. Kosmann-Schwarzbach Y. and Magri F. (1996): *Lax-Nijenhuis operators for integrable systems*, J. Math. Phys. **37** 6173
110. Kruskal M.D. (1975): *Lecture notes in physics*, V. 38 Springer-Verlage, Berlin p. 310
111. Kupershmidt B.A. and Wilson G. (1981): *Modified Lax equations and the second Hamiltonian structure*, Invent. Math. **62** 403
112. Kupershmidt B.A. (1985): *Discrete Lax equations and differential-difference calculus*, Asterisque
113. Kupershmidt B.A. (1988): *Mathematics of dispersive water waves*, Commu. Math. Phys. **99** 51
114. Lamb C.L. (1980): *Elements of soliton theory*, Wiley, New York
115. Lamb C.L. (1971): *Analytical descriptions of ultrashort optical pulse propagation in a resonant medium*, Rev. Mod. Phys. **43** 99

342   References

116. Landi G., Marmo G. and Vilasi G. (1994): *Recursion operator: meaning and existence for completely integrable systems*, J. Math. Phys. **35** 808
117. Li L.C. and Parmentier S. (1989): *Nonlinear Poisson structures and R-matrices*, Comm. Math. Phys. **125** 545
118. Libermann P. and Marle C.M. (1987): *Symplectic geometry and analytical mechanics*, D.Reidel
119. Lichnerowicz A. (1977): *Les variétés de Poisson et leurs algèbres associées*, J. Diff. Geom. **12** 253
120. Ma W.X. (1992): *Lax representation and Lax operator algebras of isospectral and nonisospectral hierarchies of evolution equations*, J. Math. Phys. **33** 2464
121. Ma W.X. (1992): *The algebraic structures of isospectral Lax operators and applications to integrable equations*, J. Phys. A: Math. Gen. **25** 5329
122. Ma W.X. (1993): *A simple scheme for generating nonisospectral flows from the zero curvature representation*, Phys. Lett. **A 179** 179
123. Magri F. (1978): *A simple model of the integrable Hamiltonian equation*, J. Math. Phys. **19** 1156
124. Magri F. and Morosi C. (1984): *A geometric characterization of integrable Hamiltonian systems through the theory of Poisson–Nijenhuis manifolds*, Preprint, University Milano
125. Magri F., Morosi C. and Ragnisco O. (1985): *Reduction techniques for infinite dimensional Hamiltonian systems: Some ideas and applications*, Comm. Math. Phys. **99** 115
126. Magri F., Morosi C. and Tondo G. (1988): *Nijenhuis G-manifolds and Lenard bicomplexes: a new approach to KP system*, Commun. Math. Phys. **116** 457
127. Magri F., Morosi C. and Tondo G. (1988): *On the relation between the bicomplex and bilocal formalism for KP system*, in: Nonlinear Evolutions ed. Leon J. Singapore: World Scientific p. 231
128. Magri F. and Marsico T. (1995): *Some developments of the concept of Poisson manifold in the sense of A.Lichnerowicz* in: Gravitation, electromagnetism and geometrical structures, ed. G.Ferrarese, Pitagora Editrice Bologna p. 207
129. Marciniak K. (1996): *Newton decomposition of stationary flows of the coupled Korteweg-de Vries hierarchy*, Linköping Studies in Science and Technology. Theses No. 533
130. Marciniak K. (1997): *Coupled KdV with sources and its Newton decomposition*, J. Math. Phys. **38** 5739
131. Marshall I. and Wojciechowski S. (1988): *When is a Hamiltonian system separable*, J.Math. Phys. **29** 1338
132. Matsuno Y. (1984): *The bilinear transformation method*, Academic Press, New York
133. Maxon S. and Viecelli J. (1974): *Cylindrical solitons*, Phys. Fluids **17** 1614
134. McKean H.P. (1993): *Compatible brackets in Hamiltonian mechanics*, in: Important Developments in Soliton Theory, edited by A.S.Fokas and V.E.Zakharov, Springer-Verlag, New York p. 344
135. Melnikov V.K. (1990): *Integration of the Korteveg-de Vries equation with source*, Inverse Problems **6** 233
136. Miura R.M. (1968): *Korteweg-de Vries equation and generalizations I. A remarkeble explicit nonlinear transformation*, J. Math. Phys. **9** 1202
137. Morosi C. and Tondo G. (1989): *Yang–Baxter equations and intermediate long wave hierarchies*, Commun. Math. Phys. **122** 91
138. Morosi C. and Tondo G. (1991): *Integrability structures and master symmetries*, Inverse Problems **7** 127
139. Morosi C. and Pizzocchero L. (1995): *On the bi-Hamiltonian interpretation of the Lax formalism*, Rev. Math. Phys. **7** 389

140. Morosi C. and Tondo G. (1997): *Quasi-bi-Hamiltonian systems and separability*, J. Phys. A: Math. Gen. **30** 2799
141. Moser J. (1975): *Three integrable Hamiltonian systems connected with isospectral deformations*, Adv. Math. **16** 1
142. Moser J. (1978): *Various aspects of integrable Hamiltonian systems*, Proc. CIME Conf. Bressanone, Italy
143. Newell A.C. (1985): *Solitons in mathematics and physics*, Lectures in Apply Math. 15, Philadelphia
144. Nijenhuis A. (1951): $X_{n-1}-$*forming sets of eigenvectors*, Proc. Kon. Ned. Akad. Amsterdam **54** 200
145. Oevel G., Fuchssteiner B. and Błaszak M. (1990): *Action/angle representation of multisolitons by potentials of master symmetries*, Prog. Theor. Phys. **83** 395
146. Oevel W. and Fuchssteiner B. (1982): *Explicit formulas for symmetries and conservation laws of the Kadomtsev–Petviashvili equation*, Phys. Lett. **A 88** 323
147. Oevel W. and Fokas A.S (1984): *Infinitely many commuting symmetries and constants of motion in involution for explicitly time-dependent evolution equations*, J. Math. Phys. **25** 918
148. Oevel W. (1988): *Master symmetries: weak action/angle structure for Hamiltonian and non-Hamiltonian systems*, preprint, University of Paderborn
149. Oevel W., Zhang H. and Fuchssteiner B. (1989): *Mastersymmetries and multi-Hamiltonian formulation for some integrable lattice systems*, Prog. Theor. Phys. **81** 294
150. Oevel W. and Ragnisco O. (1990): *R-matrices and higher Poisson brackets for integrable systems*, Physica **161A** 181
151. Oevel W. and Strampp W. (1993): *Constrained KP hierarchy and bi-Hamiltonian structures*, Commun. Math. Phys. **157** 51
152. Oevel W. and Rogers C. (1993): *Gauge transformations and reciprocal links in 2+1 dimensions*, Rev. Math. Phys. **5** 299
153. Oevel W. (1994): *A note on the Poisson brackets associated with Lax operators*, Phys. Lett. **A 186** 79
154. Oevel W. and Schief W. (1995): *Squered eigenfunctions of the (modified) KP hierarchy and scattering problems of Loewner type*, Rev. Math. Phys.
155. Oevel W. (1996): *Poisson brackets for integrable lattice systems*, in: Algebraic Aspects of Integrable Systems:In Memory of Irene Dorfman, eds. A.S.Fokas and I.M.Gelfand, Progress in Nonlinear Differential Equations, Vol. 26, Birkhäuser, Boston p. 261
156. Ohta Y., Satsuma J., Takahashi D. and Tokihiro T. (1988): *An elementary introduction to Sato theory*, Progr. Theor. Phys. Suppl. **94** 210
157. Okubo S. (1989): *Integrable dynamical systems with hierarchy. I. Formulation*, J. Math. Phys. **30** 834
158. Olver P.J. (1977): *Evolution equations possessing infinitely many symmetries*, J. Math. Phys. **18** 1212
159. Olver P.J. (1986): *Application of Lie groups to differential equations*, Springer-Vrlage, Berlin
160. Ono H. (1975): *Algebraic solitary waves in stratified fluids*, J. Phys. Soc. Japan **39** 1082
161. Prykarpatsky A.K., Samoilenko V.Hr., Andrushkiw R.I., Mitropolski Yu.O. and Prytula M.M. (1994): *Algebraic structure of the gradient-holonomic algorithm for Lax integrable nonlinear dynamical systems. I.*, J. Math. Phys. **35** 1763

162. Prykarpatsky A.K., Samoilenko V.Hr. and Andrushkiw R.I. (1994): *Algebraic structure of the gradient-holonomic algorithm for Lax integrable nonlinear dynamical systems. II. The reduction via Dirac and canonical quantization procedure.*, J. Math. Phys. **35** 4088

163. Ragnisco O. and Rauch-Wojciechowski S. (1992): *Restricted flows of the AKNS hierarchy*, Inverse Problems **8** 245

164. Ragnisco O. and Rauch-Wojciechowski S. (1994): *Integrable mechanical systems related to the Harry Dym hierarchy*, J. Math. Phys. **35** 834

165. Rauch-Wojciechowski S. (1991): *A bi-Hamiltonian formulation for separable potentials and its application to the Kepler problem and the Euler problem of thwo centers of gravitation*, Phys. Lett. **A 160** 149

166. Rauch-Wojciechowski S. (1992): *Newton representation for stationary flows of the KdV hierarchy*, Phys. Lett. **A 170** 91

167. Rauch-Wojciechowski S. Marciniak K. and Błaszak M. (1996): *Two Newton decompositions of stationary flows of KdV and Harry–Dym hierarchies*, Physica **233A** 307

168. Santini P.M. and Fokas A.S. (1988): *Recursion operators and bi-Hamiltonian structures in multisolitons I*, Commun. Math. Phys. **115**

169. Sato M. (1981): *Soliton equations as dynamical systems on infinite Grassmann manifold*, RIMS Kokyuroku, Kyoto Univ. **439** 30

170. Sato M. and Sato Y. (1983): *Soliton equations as dynamical systems on infinite Grassmann manifold*, in: Nonlinear Partial Differential Equations in Applied Science, eds. H.Fujita, P.D.Lax and G.Strang, Kinokuniya/North Holland, Tokyo, p. 259

171. Sawada K. and Kotera T. (1974): *A method for finding N-soliton solutions of th KdV equation and KdV-like equations*, Prog. Theor. Phys. **51** 1355

172. Schouten J.A. (1953): *On the differential operators of first order in tensor calculus*, Conv. Int. Geom. Diff. ed.Cremones, Rome p. 1

173. Scott A.C. and Jonson W.J. (1969): *"Internal flux motion in large Josephson junction"*, Apply Phys. Lett. **14** 316

174. Scott A.C. (1984): *Launching a Davydov soliton: I. Soliton analysis*, Physica Scripta **29** 279

175. Semenov-Tian-Shansky M.A. (1983): *What is a classical r-matrix?*, Funct. Anal. Appl. **17** 259

176. Semenov-Tian-Shansky M.A. (1985): *Dressing transformations and Poisson group action*, Publ. RIMS, Kyoto Univ. **21** 1237

177. Sidorenko J. and Strampp W. (1991): *Symmetry constraints of the KP hierarchy*, Inverse Problems **7** L37

178. Smirnov R.G. (1996): *On the master symmetries related to certain classes of integrable Hamiltonian systems*, J. Phys. A: Math. Gen. **29** 8133

179. Smirnov R.G. (1997): *Magri–Morosi–Gelfand–Dorfman bi-Hamiltonian constructions in the action-angle variables*, J. Math. Phys **38** 6444

180. Suris Y.B. (1993): *On the bi-Hamiltonian structure of Toda and relativistic Toda lattices*, Phys. Lett. **A 180** 419

181. Ten Eikelder H.M.M. (1985): *Symmetries for dynamical and Hamiltonian systems*, CWI trace 17, CWI, Amsterdam

182. Ten Eikelder H.M.M. (1986): *On the local structure of recursion operators*, Proc. Kon. Ned. Akad. A **89** 386

183. Toda M. (1981): *Theory of nonlinear lattice*, Springer-Verlag, New York

184. Tulczyjew W.M. (1974): *Poisson brackets for canonical manifolds*, Bull. Acad. Pol. Sci. **160** 931

185. Weiss J. (1983): *The Painleve property for partial differential equations. II. Bäcklund transformation, Lax pairs, and the Schwarzian derivative*, J.Math. Phys. **24** 1405

186. von Westenholz C. (1981): *Differential forms in mathematical physics*, Studies in mathematics and its applications, Vol.3, North-Holland Publishing Company

187. Whittaker E.T. (1944): *A treatise on the analytical dynamics of particles and rigid bodies*, Dover, New York ch. 1

188. Wojciechowski S. (1986): in: Proc. SMS on Systemes dynamiques nonlineaires, ed. P.Winternitz Press Universite Montreal, Montreal p.294

189. Xu Bing (1992): *A unified approach to recursion operators of the reduced (1+1)-dimensional Hamiltonian systems*, Inverse Problems **8** L13

190. Xu Bing and Li Yishen (1992): *(1+1)-dimensional Hamiltonian systems as symmetry constraints of the Kadomtsev–Petviashvili equation*, J. Phys. A: Math. Gen. **25** 2957

191. Yamamuro S. (1974): *Differential calculus in topological linear spaces*, Springer-Verlag, Berlin-Heidelberg-New York

192. Ymosa S. (1984): *Solitary excitation in DNA double helice*, Phys. Rev. **A 30** 474

193. Zabusky N.J. (1967): in Nonlinear Waves, ed. W.Ames Academic Press, New York

194. Zakharov V.E., Manakov S.V., Novikov S.P. and Pitajevski L.P. (1980): *Theory of solitons*, Nauka (in Russian)

195. Zakharov V.E. and Konopelchenko B.G. (1984): *On the theory of recursion operator*, Commun. Math. Phys. **94** 483

196. Zakharov V.E. (1967): in Nonlinear Waves, ed. W.Ames Academic Press, New York

197. Zeng Y. and Li Y. (1989): *The constraints of potentials and finite-dimensional integrable systems*, J. Math. Phys. **30** 1679

198. Zeng Y. and Li Y. (1990): *Two kinds of finite-dimensional systems related to the generalized Schrödinger equation*, in: Research Reports in Physics. Nonlinear Physics, editors: Gu Chaohao, Li Yishen and Tu Guizhang, Springer-Verlag

199. Zeng Y. and Li Y. (1990): *Integrable Hamiltonian systems related to the polynomial eigenvalue problem*, J. Math. Phys. **31** 2835

200. Zeng Y. (1991): *Gauge transformation and bi-Hamiltonian structure of a finite-dimensional integrable system reduced from a soliton equations*, J. Phys. A: Math. Gen. **24** L11

201. Zeng Y. (1994): *New factorization of the Kaup–Newell hierarchy*, Physica **73D** 171

Index